ASHRAE POCKET GUIDE

for Air Conditioning, Heating, Ventilation, Refrigeration

SI

AF831227

10th Edition

ASHRAE · 180 Technology Parkway, Peachtree Corners, GA 30092 · www.ashrae.org

ISBN 978-1-947192-79-9 (softcover)
ISBN 978-1-947192-80-5 (PDF)

Product code: 90072   12/22

Library of Congress Cataloging-in-Publication Data

Names: ASHRAE (Firm), author.
Title: ASHRAE pocket guide for air conditioning, heating, ventilation, refrigeration.
Other titles: Pocket guide for air conditioning, heating, ventilation, refrigeration
Description: 10th edition, SI. | Peachtree Corners : ASHRAE, 2022. | Includes index. | Summary: "Comprehensive yet consolidated reference for HVAC engineers with data from the ASHRAE Handbooks and ASHRAE Standards 62.1, 62.2, 15, and 55; revised from the 2017 edition"-- Provided by publisher.
Identifiers: LCCN 2021018079 | ISBN 9781947192799 (paperback) | ISBN 9781947192805 (adobe pdf)
Subjects: LCSH: Heating--Equipment and supplies--Handbooks, manuals, etc. | Ventilation--Handbooks, manuals, etc. | Air conditioning--Handbooks, manuals, etc. | Refrigeration and refrigerating machinery--Handbooks, manuals, etc.
Classification: LCC TH7011 .A38 2022 | DDC 697.9--dc23
LC record available at https://lccn.loc.gov/2021018079

Updates and errata for this publication will be posted on the ASHRAE website at www.ashrae.org/publicationupdates.

CONTENTS

PREFACE

The ASHRAE Pocket Guide was developed to serve as a ready, offline reference for engineers without easy access to complete ASHRAE Handbook volumes.

This tenth edition has been revised in 2022 to include updates from the current editions of the ASHRAE Handbook series as well as from various ASHRAE standards. This edition continues to emphasize basic design information.

The ASHRAE Pocket Guide was first published in 1987. This edition was compiled by Walter Grondzik. Major contributors to previous editions include Carl W. MacPhee; Griffith C. Burr, Jr.; Harry E. Rountree; and Frederick H. Kohloss.

Throughout this Pocket Guide, original sources of figures and tables are indicated where applicable. Due to space constraints, a shorthand for ASHRAE publications has been adopted. ASHRAE sources are noted after figure captions or table titles in brackets using the following abbreviations:

Fig	Figure
Tbl	Table
Ch	Chapter
Std	ASHRAE Standard
Gdl	ASHRAE Guideline
2021F, 2017F, etc	*ASHRAE Handbook—Fundamentals*
2020S, 2016S, etc.	*ASHRAE Handbook—HVAC Systems and Equipment*
2019A, 2015A, etc.	*ASHRAE Handbook—HVAC Applications*
2018R, 2014R, etc.	*ASHRAE Handbook—Refrigeration*

Complete entries for all references cited in tables and figures are available in the original source publications.

1. AIR HANDLING AND PSYCHROMETRICS

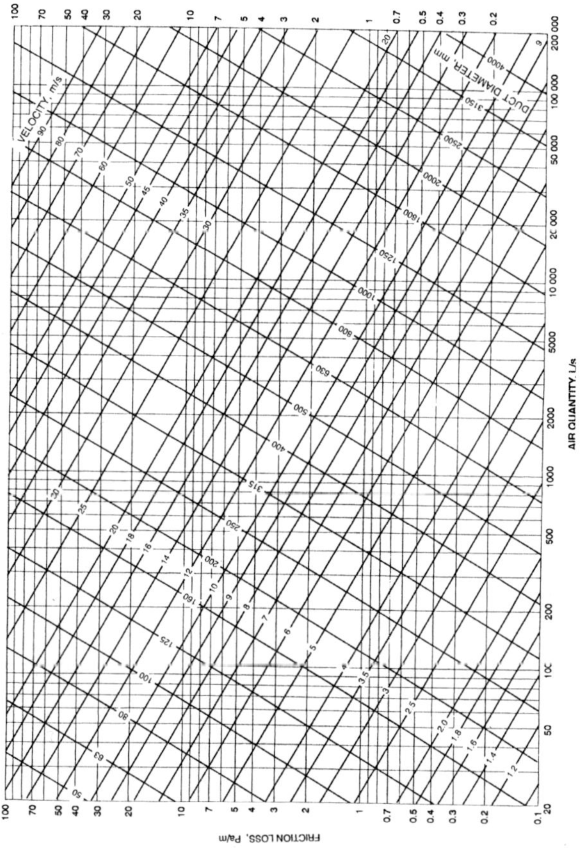

Figure 1.1 Friction Chart for Round Duct (ρ = 1.20 kg/m^3 and ε = 0.09 mm)
[2021F, Ch 21, Fig 9]

Table 1.1 Velocities vs. Velocity Pressures

Velocity V, m/s	Velocity Pressure P_v, Pa
1.0	0.6
2.0	2.4
3.0	5.4
4.0	9.6
5.0	15.1
5.5	18.3
6.0	21.7
6.5	25.5
7.0	29.5
7.5	33.9
8.0	38.5
8.5	43.5
9.0	48.8
9.5	54.3
10.0	60.2
11.0	72.9
12.0	86.7
13.0	101.8
14.0	118.0
15.0	135
17.5	184
20.0	241
22.5	305
25.0	376

$P_v = 0.602\ V^2$

Noncircular Ducts

Hydraulic diameter $D_h = 4A/P$, where A = duct area (mm) and P = perimeter (mm). Ducts having the same hydraulic diameter will have approximately the same fluid resistance at equal velocities.

Fittings

Resistance to flow through fittings can be expressed by fitting loss coefficients C. The friction loss in a fitting in inches of water is CP_v. The more radically the airflow is changed in direction or velocity, the greater the fitting loss coefficient. See *ASHRAE Duct Fitting Database* for a complete list. 90° mitered elbows with vanes will usually have C between 0.11 and 0.33.

Round Flexible Ducts

Nonmetallic flexible ducts fully extended have friction losses approximately three times that of galvanized steel ducts. The loss rises rapidly for unextended ducts; with a correction factor of 2 if 90% extended, 3 if 80% extended, and 4 if 70% extended. For centerline bend radius ratio to diameter of 1 to 4, the approximate loss coefficient is between 0.82 and 0.87.

Table 1.2 Duct Leakage Classification[a]

Duct Type	Sealed[b,c]		Unsealed[c]	
	Predicted Leakage Class C_L	Leakage Rate, $L/(s \cdot m^2)$ at 250 Pa	Predicted Leakage Class C_L	Leakage Rate, $L/(s \cdot m^2)$ at 250 Pa
Metal (flexible excluded)				
Round and flat oval	4	0.14	42	1.5
			(8 to 99)	(0.3 to 3.6)
Rectangular				
≤500 Pa	17	0.62	68	2.5
(both positive and negative pressures)			(17 to 155)	(0.6 to 5.6)
>500 and ≤2500 Pa	8	0.29	68	2.5
(both positive and negative pressures)			(17 to 155)	(0.6 to 5.6)
Flexible				
Metal, aluminum	11	0.40	42	1.5
			(17 to 76)	(0.6 to 2.8)
Nonmetal	17	0.62	30	1.5
			(6 to 76)	(0.2 to 2.8)
Fibrous glass				
Round	4	0.14	na	na
Rectangular	8	0.29	na	na

[a] The leakage classes listed in this table are averages based on tests conducted by AISI/ SMACNA (1972), ASHRAE/SMACNA/TIMA (1985), and Swim and Griggs (1995).

[b] The leakage classes listed in the sealed category are based on the assumptions that for metal ducts, all transverse joints, seams, and openings in the duct wall are sealed at pressures over 750 Pa, that transverse joints and longitudinal seams are sealed at 500 and 750 Pa, and that transverse joints are sealed below 500 Pa. Lower leakage classes are obtained by careful selection of joints and sealing methods.

[c] Leakage classes assigned anticipate about 0.82 joints per metre of duct. For systems with a high fitting to straight duct ratio, greater leakage occurs in both the sealed and unsealed conditions.

Table 1.3 Recommended Ductwork Leakage Class by Duct Type

Duct Type	Leakage Class C_L	Leakage Rate, $L/(s \cdot m^2)$ at 250 Pa
Metal		
Round	4	0.14
Flat oval	4	0.14
Rectangular	8	0.29
Flexible	8	0.29
Fibrous glass		
Round	4	0.14
Rectangular	8	0.29

$$\text{Leakage Class } C_L = Q/\Delta P_S^{0.65} \quad (1.1)$$

where

Q = leakage rate, L/s/100 m² surface area

ΔP_s = static pressure difference, Pa between inside and outside of duct

Table 1.4 Duct Sealing Requirement Levels

Duct Seal Level	Sealing Requirements[a]
A	All transverse joints, longitudinal seams, and duct wall penetrations
B	All transverse joints and longitudinal seams
C	Transverse joints only

[a] Transverse joints are connections of two duct or fitting elements oriented perpendicular to flow. Longitudinal seams are joints oriented in the direction of airflow. Duct wall penetrations are openings made by screws, non-self-sealing fasteners, pipe, tubing, rods, and wire. Round and flat oval spiral lock seams need not be sealed prior to assembly, but may be coated after assembly to reduce leakage. All other connections are considered transverse joints, including but not limited to spin-ins, taps and other branch connections, access door frames, and duct connections to equipment.

Table 1.5 Duct Leakage per Unit Length

Unsealed Longitudinal Seam Leakage, Metal Ducts		Leakage, L per metre Seam Length at 250 Pa Static Pressure	
Type of Duct/Seam		Range	Average
Rectangular	Pittsburgh lock		
	26 gage	0.015 to 0.03	0.025
	22 gage	0.0015 to 0.003	0.0025
	Button punch snaplock		
	26 gage	0.05 to 0.23	0.12
	22 gage	NA (1 test)	0.005
Round	Spiral (26 gage)	NA (1 test)	0.023
	Snaplock	0.06 to 0.22	0.17
	Grooved	0.17 to 0.28	0.19

Per Section 6.4.4.2.1 of ANSI/ANSI/ASHRAE/IES Standard 90.1, ductwork and all plenums with pressure class ratings must be constructed to Seal Class A. Openings for rotating shafts must be sealed with bushings or other devices that seal off air leakage. Pressure-sensitive tape must not be used as the primary sealant unless it has been certified to comply with UL-181A or UL-181B by an independent testing laboratory, and the tape is used in accordance with that certification. All connections must be sealed, including but not limited to spin-ins, taps, other branch connections, access doors, access panels, and duct connections to equipment. Sealing that would void product listings is not required. Spiral lock seams need not be sealed.

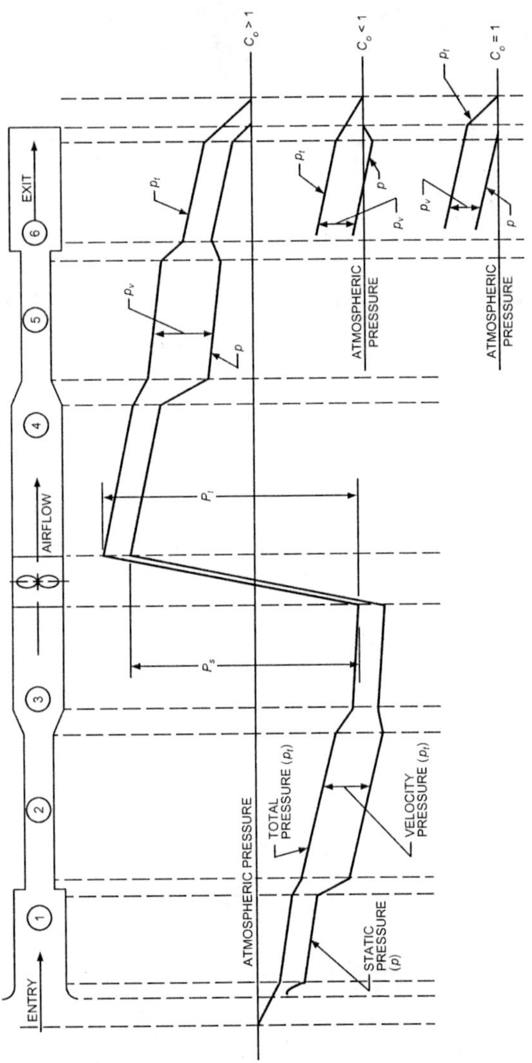

Figure 1.2 Pressure Changes During Flow in Ducts [2021F, Ch 21, Fig 6]

Table 1.6 Circular Ecuivalents of Rectangular Duct for Equal Friction and Airflow* [2021F, Ch21, Tbl 3]

Side Lengths, mm	Length of One Side of Rectangular Duct *a*, mm										
	100	150	200	250	300	400	500	600	800	1 000	1 200
	Circular Duct Diameter, mm										
200	152	189	219								
250	169	210	244	273							
300	183	229	265	299	328						
400	207	260	305	343	378	437					
500		287	337	381	420	488	547				
600		310	365	414	457	533	598	656			
800			414	470	520	609	687	755	875		
1 000				517	574	674	762	840	976	1 093	
1 200					620	731	827	914	1 066	1 196	1 312
1 400						781	886	980	1 146	1 289	1 416
1 600						827	939	1 041	1 219	1 373	1 511
1 800							988	1 096	1 286	1 451	1 598
2 000							1 034	1 147	1 348	1 523	1 680

*Table based on Equation (25) from Chapter 21 of the 2021 *ASHRAE Handbook—Fundamentals*.

Table 1.7 Equivalent Flat Oval Duct Dimensions* [2021F, Ch 21, Tbl 4]

Circular Duct Diameter, mm	Minor Axis *a*, mm																
	70	100	125	150	175	200	250	275	300	325	350	375	400	450	500	550	600
	Major Axis *A*, mm																
125	205																
140	265	180															
160	360	235	190														
180	475	300	235	200													
200		380	290	245	215												
224		490	375	305	—	240											
250			475	385	325	290											
280				485	410	360	—	285									
315				635	525	—	—	345	325								
355				840	—	580	460	425	395	375							
400				1 115	—	760	—	530	490	460	435						
450				1 490	—	995	—	675	—	570	535	505					
500						1 275	—	845	—	700	655	615	580				
560						1 680	—	1 085	—	890	820	765	720				
630								1 425	—	1 150	1 050	970	905	810			

Table 1.7 Equivalent Flat Oval Duct Dimensions* [2021F, Ch 21, Tbl 4] ***(Continued)***

Circular Duct Diameter, mm	Minor Axis *a*, mm																
	70	100	125	150	175	200	250	275	300	325	350	375	400	450	500	550	600
	Major Axis *A*, mm																
710										1 505	1 370	1 260	1 165	1 025			
800											1 800	1 645	1 515	1 315	1 170	1 065	
900												2 165	1 985	1 705	1 500	1 350	
1 000														2 170	1 895	1 690	
1 120															2 455	2 170	1 950
1 250																2 795	2 495

*Table based on Equation (26) of Chapter 21 of 2021 *ASHRAE Handbook—Fundamentals*.

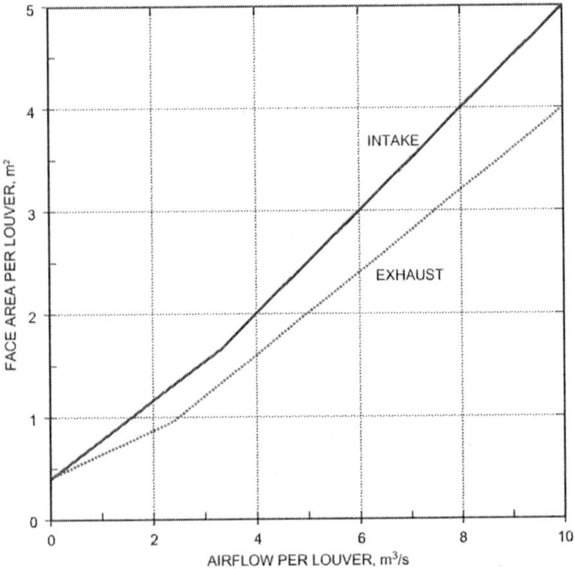

Parameters Used to Establish Figure	Intake Louver	Exhaust Louver
Minimum free area (1220 mm square test section), %	45	45
Water penetration, mL/($m^2 \cdot$0.25 h)	Negligible (less than 0.3)	N/A
Maximum static pressure drop, Pa	35	60

Figure 1.3 Criteria for Louver Sizing [2021F, Ch 21, Fig 18]

Table 1.8 Typical Design Velocities for HVAC Components

Duct Element	Face Velocity, m/s
Louvers	
Intake	
7000 cfm3300 L/s and greater	2
Less than 7000 cfm3300 L/s	See Figure 1.3 for criteria
Exhaust	
5000 cfm2400 L/s and greater	2.5
Less than 5000 cfm2400 L/s	See Figure 1.3 for criteria
Filters	
Panel filters	
Viscous impingement	1 to 4
Dry-type, extended-surface	
Flat (low efficiency)	Duct velocity
Pleated media (intermediate efficiency)	Up to 3.8
HEPA	1.3
Renewable media filters	
Moving-curtain viscous impingement	2.5
Moving-curtain dry media	1
Electronic air cleaners	
Ionizing type	0.8 to 1.8
Heating Coils	
Steam and hot water	2.5 to 5 1 min., 8 max.
Electric	
Open wire	Refer to mfg. data
Finned tubular	Refer to mfg. data
Dehumidifying Coils	2 to 3
Air Washers	
Spray type	Refer to mfg. data
Cell type	Refer to mfg. data
High-velocity spray type	6 to 9

Table 1.9 Fan Laws [a,b] [Adapted from 2020S, Ch 21, Tbl 2]
For All Fan Laws: $\eta_{t1} = \eta_{t2}$ and $(\text{point of rating})_1 = (\text{point of rating})_2$

No.	Dependent Variables	Independent Variables	
1a	$Q_1 = Q_2$	$\times\left(\frac{D_1}{D_2}\right)^3 \times \frac{N_1}{N_2}$	$\times 1$
1b	$\text{Pressure}_1 = \text{Pressure}_2{}^{c}$	$\times\left(\frac{D_1}{D_2}\right)^2 \times \left(\frac{N_1}{N_2}\right)^2$	$\times \frac{\rho_1}{\rho_2}$
1c	$W_1 = W_2$	$\times\left(\frac{D_1}{D_2}\right)^5 \times \left(\frac{N_1}{N_2}\right)^3$	$\times \frac{\rho_1}{\rho_2}$
2a	$Q_1 = Q_2$	$\times\left(\frac{D_1}{D_2}\right)^2 \times \left(\frac{\text{Press.}_1}{\text{Press.}_2}\right)^{1/2}$	$\times\left(\frac{\rho_2}{\rho_1}\right)^{1/2}$
2b	$N_1 = N_2$	$\times\left(\frac{D_2}{D_1}\right) \times \left(\frac{\text{Press.}_1}{\text{Press.}_2}\right)^{1/2}$	$\times\left(\frac{\rho_2}{\rho_1}\right)^{1/2}$
2c	$W_1 = W_2$	$\times\left(\frac{D_1}{D_2}\right)^2 \times \left(\frac{\text{Press.}_1}{\text{Press.}_2}\right)^{3/2}$	$\times\left(\frac{\rho_2}{\rho_1}\right)^{1/2}$
3a	$N_1 = N_2$	$\times\left(\frac{D_2}{D_1}\right)^3 \times \frac{Q_1}{Q_2}$	$\times 1$
3b	$\text{Pressure}_1 = \text{Pressure}_2$	$\times\left(\frac{D_2}{D_1}\right)^4 \times \left(\frac{Q_1}{Q_2}\right)^2$	$\times \frac{\rho_1}{\rho_2}$
3c	$W_1 = W_2$	$\times\left(\frac{D_2}{D_1}\right)^4 \times \left(\frac{Q_1}{Q_2}\right)^3$	$\times \frac{\rho_1}{\rho_2}$

where
D = fan size
N = rotational speed
ρ = gas density
Q = volumetric airflow rate
P = pressure
η = fan efficiency
W = power

a. The subscript 1 denotes that the variable is for the fan under consideration.
b. The subscript 2 denotes that the variable is for the tested fan.
c. Fan total pressure P_{tf}, fan velocity pressure P_{vf}, or fan static pressure P_{sf}.

Unless otherwise identified, fan performance data are based on dry air at standard conditions 101.325 kPa and 20°C (1.204 kg/m^3). In actual applications, the fan may be required to handle air or gas at some other density. The change in density may be because of temperature, composition of the gas, or altitude. As indicated by the Fan Laws, the fan performance is affected by gas density. With constant size and speed, the horsepower and pressure varies directly as the ratio of gas density to the standard air density.

The application of the Fan Laws for a change in fan speed N for a specific size fan is shown in Figure 1.4. The computed P_t curve is derived from the base curve. For example, point E(N_1 = 650) is computed from point D(N_2= 600) as follows:

At D,

$$Q_2 = 3 \text{ m}^3\text{/s and } P_{tf_2} = 228 \text{ Pa} \tag{1.2}$$

Using Fan Law 1a at Point E

$$Q_1 = 3 \times 650/600 = 3.25 \text{ m}^3\text{/s} \tag{1.3}$$

Using Fan Law 1b

$$P_{tf_1} = 228 \times (650/600)^2 = 268 \text{ Pa} \tag{1.4}$$

The completed P_{tf_1}, N = 650 curve thus may be generated by computing additional points from data on the base curve, such as point G from point F.

$$\text{fan power, kW} = \frac{\text{L/s} \times \text{pressure difference, kPa}}{40350 \times \text{fan efficiency} \times \text{motor efficiency}} \tag{1.5}$$

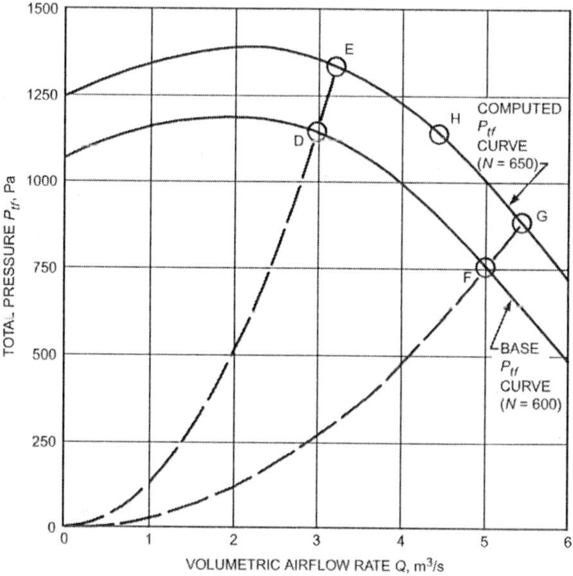

Figure 1.4 Example Application of Fan Laws [2020S, Ch 21, Fig 4]

Table 1.10 Types of Fans [2020S, Ch 21, Tbl 1]

	Type	Impeller Design	Housing Design	Performance Characteristics	Applications
Centrifugal Fans	Backward-Inclined (includes AF, BC, and BI)	Blades inclined away from direction of rotation and can be single-thickness flat (BI), single-thickness curved (BC), or airfoil (AF) contour. Deep blades allow efficient expansion in blade passages. Air leaves impeller at velocity less than tip speed. For given duty, has highest speed of centrifugal fan designs. 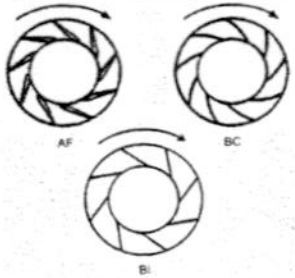	Single- or double-inlet scroll design for efficient conversion of tangential velocity pressure to static pressure. Maximum efficiency requires close clearance and alignment between impeller and inlet. 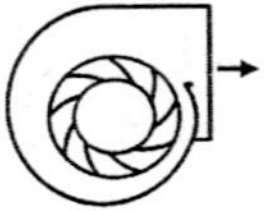	Highest efficiency of all centrifugal fan designs, with peak efficiencies occurring at 50 to 60% of wide-open volume. Fan has a non-overloading characteristic, which means power reaches maximum near peak efficiency and becomes lower, or self-limiting, toward free delivery. Airfoil blades are most efficient, followed by curved and then flat blades.	General HVAC. Used in ducted applications covering a large range of pressures. Airfoil blades can be used for clean air industrial operations. For industrial applications where the environment may corrode or erode airfoil blades, consider using single thickness blades instead.
	Radial and Radial Tipped	Blades are either fully radial (R) or backward inclined with a radial curve at outer edge (RT). Fully radial blades can have back plate and shroud, back plate only, or neither (open). Radial tipped blades normally have back plate and shroud. 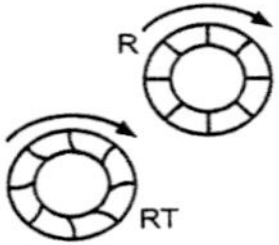	Single or double inlet scroll similar to other centrifugal fan designs. Fit between impeller and inlet not as critical as for backward inclined fans. 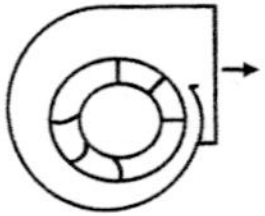	Higher pressure characteristics than airfoil and backward curved fans. Power rises continually to free delivery, which is an overloading characteristic. Radial tipped blades are slightly more efficient than straight blades. Shrouded impellers are more efficient than nonshrouded.	Primarily for material handling in industrial plants where high duct velocities are required to keep materials airborne. Also for some high-pressure industrial requirements. Rugged impeller is simple to repair in the field. Choice of impeller type generally depends on materials being transported. Impeller sometimes coated with special material.

Table 1.10 Types of Fans [2020S, Ch 21, Tbl 1] ***(Continued)***

	Type	Impeller Design	Housing Design	Performance Characteristics	Applications
Centrifugal Fans *(Continued)*	Forward-Curved	Large number of thin curved blades with outer edge tipped toward direction of rotation. Air leaves impeller with a high tangential velocity. Relies on scroll housing to convert this velocity pressure to static pressure. For given duty, has lowest speed of centrifugal fan designs. 	Single or double inlet scroll shaped design necessary for conversion of tangential velocity pressure to static pressure. Fit between impeller and inlet not as critical as for backward inclined fans. 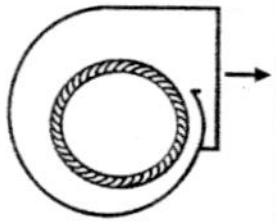	Pressure curve less steep than that of backward-inclined fans. Curve dips to left of peak pressure. Highest efficiency occurs at 40 to 50% of wide open volume. Operate fan to right of peak pressure. Power rises continually to free delivery which is an overloading characteristic.	Primarily for low- to medium-pressure HVAC applications, such as residential furnaces, central station units, and packaged air conditioners. Slower speed and lower tonal noise characteristics allow them to be used closer to occupied spaces. A possible exception is central station air handling units because of low frequency noise that is more difficult to attenuate.
	Plenum/Plug	Single inlet centrifugal impellers can be airfoil, backward curved, or backward inclined. Mixed flow or radial impellers can also be used for specific applications. 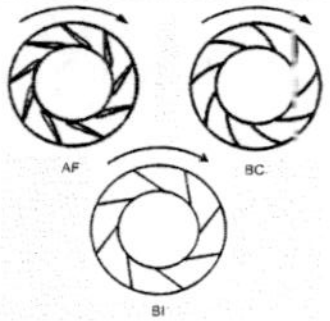	No integral housing. The equivalent of a housing, or plenum chamber (dashed line), depends on the application. Due to the lack of a housing, all pressure development occurs in the impeller. The components of the drive system for the plug fan are located outside the airstream. 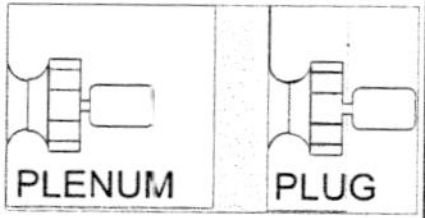	Similar to housed airfoil/backward curved/backward inclined fans but slightly less efficient due to the lack of conversion of kinetic energy in the discharge airstream. Because plenum and plug fans do not have usable outlet velocity pressure, they should always be selected based on fan static pressure. Both fans are susceptible to performance degradation caused by poor inlet installation and insufficient clearance to surrounding walls.	Plenum fans are used for HVAC equipment such as air handlers. Advantages include flexibility of equipment discharge and potential for smaller equipment footprint. Multiple plenum fans in parallel (fan arrays) can be used to further reduce the axial length of air handling equipment. Plug fans are commonly used for high-temperature process applications.

Table 1.10 Types of Fans [2020S, Ch 21, Tbl 1] ***(Continued)***

	Type	Impeller Design	Housing Design	Performance Characteristics	Applications
Axial Fans	Propeller	Small number of blades (two to six), either airfoil or single thickness curved, attached to a relatively small hub. 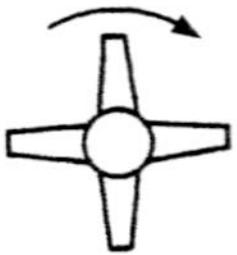	Simple circular ring, orifice plate, or venturi. Optimum design is close to blade tips and forms smooth airflow into impeller. Improved static efficiency can be obtained with an extension of the venturi into an expanding conical discharge. 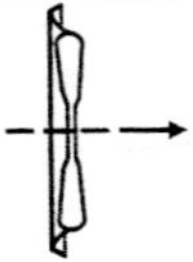	High flow rate, but very low pressure capabilities. Discharge pattern circular or conical and airstream swirls.	For low-pressure, high-volume air-moving applications, such as ventilation through a wall without ductwork or air-cooled condenser fans.
	Tubeaxial	Usually has four to eight blades with airfoil or single thickness cross section. Blades may have fixed or adjustable pitch. Hub is usually less than half the fan tip diameter. 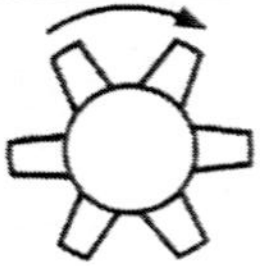	Cylindrical tube with close clearance to blade tips. Free inlet applications should use a bell mouth inlet to minimize entrance losses. 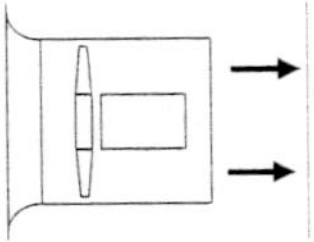	High flow rate, medium pressure capabilities. Pressure curve can dip to left of peak pressure at higher blade pitches. Avoid operating fan in this stall region. Discharge pattern circular and airstream rotates or swirls downstream of fan.	Low- and medium-pressure ducted HVAC applications where straight through flow and compact installation are required. Used in some industrial applications, such as drying ovens, paint spray booths, and fume exhausts.
	Vaneaxial	Typically has more blades than tubeaxial fans. Blades may have fixed, adjustable, or controllable pitch. Hub is usually greater than half fan tip diameter. Most efficient designs have airfoil blades. 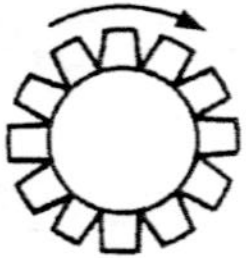	Cylindrical tube with close clearance to blade tips. Guide vanes upstream or downstream from impeller increase pressure capability and efficiency. 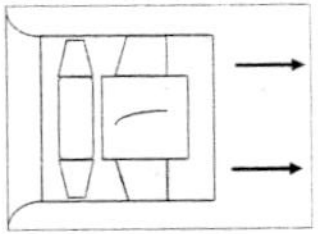	High pressure characteristics with medium volume flow capabilities. Pressure curve can dip to left of peak pressure. Avoid operating fan in this stall region. Guide vanes correct circular motion imparted by impeller, which improves pressure characteristics and fan efficiency.	General HVAC systems in medium- and high-pressure applications where straight-through flow and compact installation are required. Has swirl-free downstream airflow. Used in industrial applications in place of tubeaxial fans. More compact than centrifugal fans for same duty.

Table 1.10 Types of Fans [2020S, Ch 21, Tbl 1] *(Continued)*

	Type	Impeller Design	Housing Design	Performance Characteristics	Applications
Mixed-Flow	Mixed-Flow	Combination of axial and centrifugal characteristics. Contoured back or hub; airfoil, curved, or straight blades, either shrouded or not. Airflow through impeller has both radial and axial components. 	The majority of mixed-flow fans have a tubular housing for use in ducted applications and include outlet turning vanes. 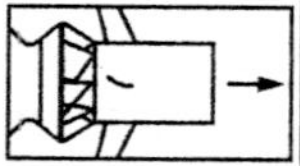	Characteristic pressure curve between axial fans and centrifugal fans. Higher pressure than axial fans and higher-volume flow than centrifugal fans.	Similar HVAC applications to centrifugal fans or in applications where an axial fan cannot generate sufficient pressure rise.
Cross-flow	Cross-flow (Tangential)	Impeller with forward-curved blades. Typically impeller width is much greater than diameter. 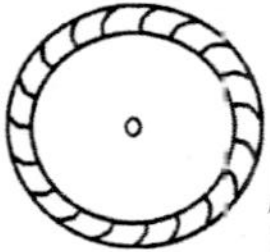	Specially designed housing for 90° or straight-through airflow. Housing causes air to enter the impeller at its periphery, flow through the impeller, and discharge at a different point of its periphery. 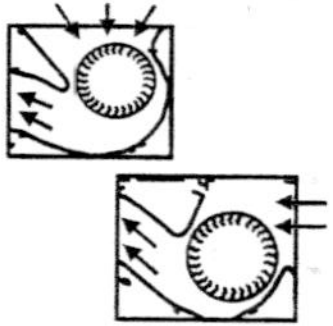	Similar to forward-curved fans. Power rises continually to free delivery, which is an overloading characteristic. Lowest efficiency of any centrifugal fan type.	Low-pressure HVAC components (e.g., fan heaters, fireplace inserts, electronic cooling, air curtains) where a long, narrow discharge is needed.

Table 1.10 Types of Fans [2020S, Ch 21, Tbl 1] ***(Continued)***

	Type	Impeller Design	Housing Design	Performance Characteristics	Applications
Other Designs	Inline Centrifugal	Single-inlet centrifugal impellers can be airfoil, backward curved, or backward inclined. 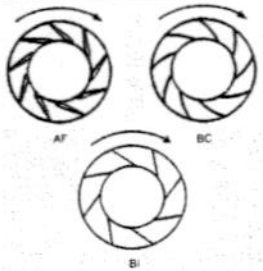	Cylindrical tube similar to vaneaxial fan, except clearance to impeller is not as close. Air discharges radially from impeller and turns 90° to flow through guide vanes. Variations include cylindrical and square housings with or without guide vanes. Square housings can include side discharge. 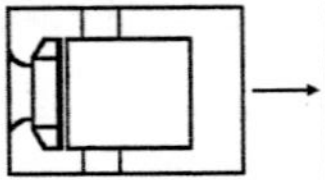	Performance similar to backward-inclined fan, except capacity and pressure are lower. Lower efficiency than backward-inclined fan because air turns 90°.	Ducted HVAC applications with air discharging in axial direction (e.g., low- to medium-pressure return air systems in HVAC applications).
	Power Roof Ventilators — Centrifugal	Single-inlet centrifugal impellers can be airfoil, backward curved, or backward inclined. 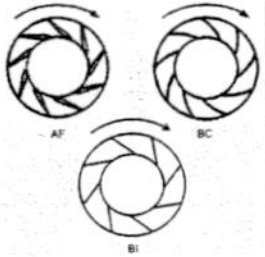	Weather-protected housing with means of mounting to a building opening. Usually does not include configuration to recover velocity pressure component. Radial discharge from impeller can be directed either toward building or away from building. 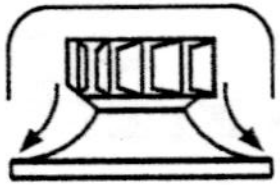	Less efficient than scroll-type housed fan. Centrifugal units are slightly quieter than axial units.	Low-pressure ducted exhaust systems (e.g., general factory, kitchen, warehouse, some commercial installations).

Table 1.10 Types of Fans [2020S, Ch 21, Tbl 1] *(Continued)*

	Type		Impeller Design	Housing Design	Performance Characteristics	Applications
Other Designs (*Continued*)	Power Roof Ventilators (*Continued*)	Axial	Small number of blades (two to six), similar to an axial propeller fan. 	Similar housing to an axial propeller fan with added weather protection and means of mounting to a building opening. Can supply or exhaust air from a building, and housing can direct air either toward building or away from building. 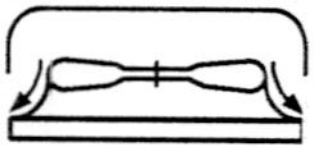	Usually installed without ductwork; therefore, operates at very low pressure and high volume.	Low-pressure exhaust or supply systems (e.g., general factory, kitchen, warehouse, some commercial installations).
		Induced-Flow	Centrifugal backward inclined or mixed flow. 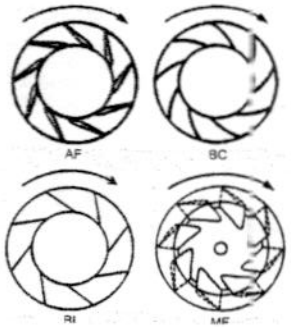	Can be scroll or tubular inline with converging cone or nozzle to increase nozzle velocity, which induces additional airflow through outlet. 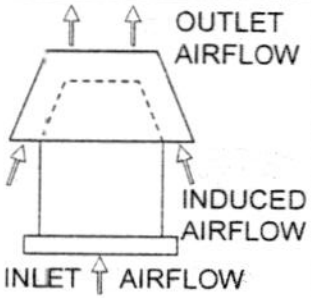	Performance similar to BI and MF but with reduced efficiency due to high velocity needed for induction. Outlet airflow greater than inlet airflow.	Typically used for laboratory or hazardous chemical exhaust where dilution is required and tall stacks are not desired.

Table 1.10 Types of Fans [2020S, Ch 21, Tbl 1] ***(Continued)***

	Type		Impeller Design	Housing Design	Performance Characteristics	Applications
Other Designs *(Continued)*	Jet Fan	Centrifugal	Single-inlet centrifugal impellers can be airfoil, backward curved, or backward inclined. 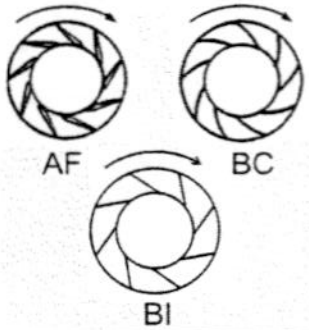	Rectangular, low profile. Outlet tapers to increase discharge velocity. May contain outlet straighteners and inlet protection guards. 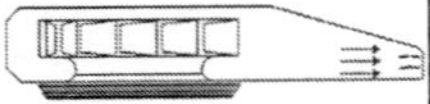	Performance measured and rated in thrust F_t. Designed for high-velocity discharge to add momentum to larger volume of air. Note that performance curves shown are not at a single speed, but at variable speeds N.	Unducted for space-constrained parking garages and emergency fire and smoke evacuation. Comparable initial cost with ducted solutions but more flexibility for exhaust and inlet air locations.
		Axial	Could be bidirectional or reversible in operation; similar flow rates in either direction is common. 	Tubular housing may or may not have guide vanes. Typically fitted with inlet and outlet silencers and protection screens. 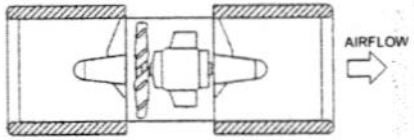	Performance measured and rated in thrust F_t. Designed for high-velocity discharge to add momentum to larger volume of air. Typically operates at high speed, which necessitates inlet and outlet silencers. Note that performance curves shown are not at a single speed, but at variable speeds N.	Road or rail tunnels, parking garages, and emergency fire and smoke evacuation. Comparable initial cost with ducted solutions but more flexibility for the exhaust and inlet air locations.
		Circulating	Includes small propeller fans up to large-diameter ceiling mounted fans. Typically two to eight blades, either airfoil or single-thickness curved, attached to a comparatively small-diameter hub. 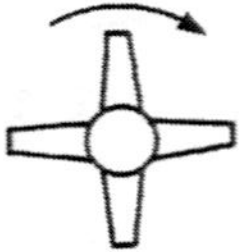	Often unhoused, but can have housing. 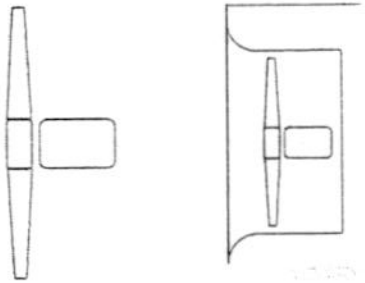	Performance typically measured and rated in airflow rate versus input power and impeller rotational speed N. Usually operated at free air with high-volume flow rate. Note that performance curves shown are not at a single speed, but at variable speeds.	Used for generating elevated air speeds to provide cooling of people or animals and prevent stratification of temperature or humidity (e.g., inside warehouses, airports, gymnasiums, barns). Can also be used for drying processes.

Fan System Effect

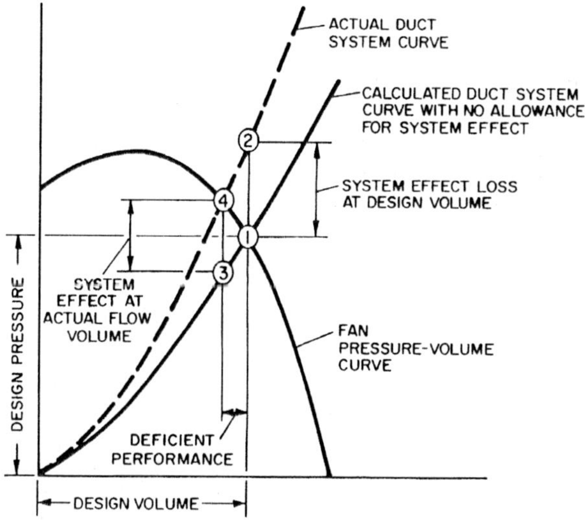

Figure 1.5 Deficient Fan/System Performance

Figure 1.5 illustrates deficient fan/system performance. System pressure losses have been determined accurately, and a fan has been selected for operation at point 1. However, no allowance has been made for the effect on fan performance of system connections. To compensate, a fan system effect must be added to the calculated system pressure losses to determine the actual system curve. The point of intersection between the fan performance curve and the actual system curve is point 4. The actual flow volume is, therefore, deficient by the difference from 1 to 4. To achieve design flow volume, a fan system effect pressure loss equal to the pressure difference between points 1 and 2 should be added to the calculated system pressure losses, and the fan should be selected to operate at point 2.

For rated performance, air must enter a fan uniformly over the inlet area in an axial direction without prerotation.

Fans within plenums and cabinets or next to walls should be located so that air may flow unobstructed into the inlets.

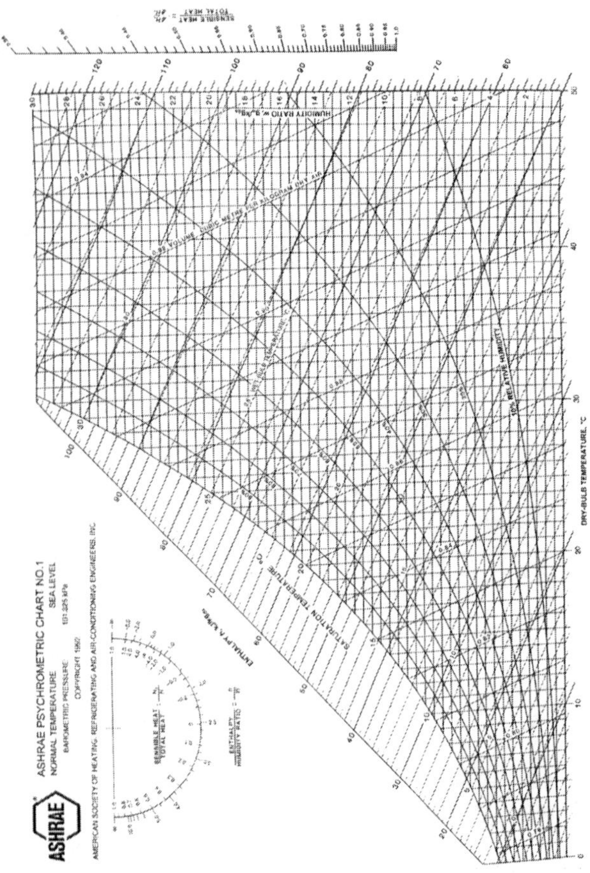

Figure 1.6 Psychrometric Chart for Normal Temperature, Sea Level
[2021F, Ch 1, Fig 2]

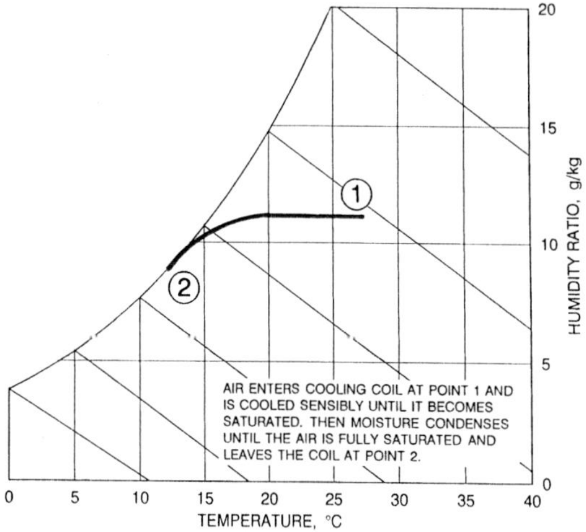

Figure 1.7 Direct Expansion or Chilled Water Cooling and Dehumidification

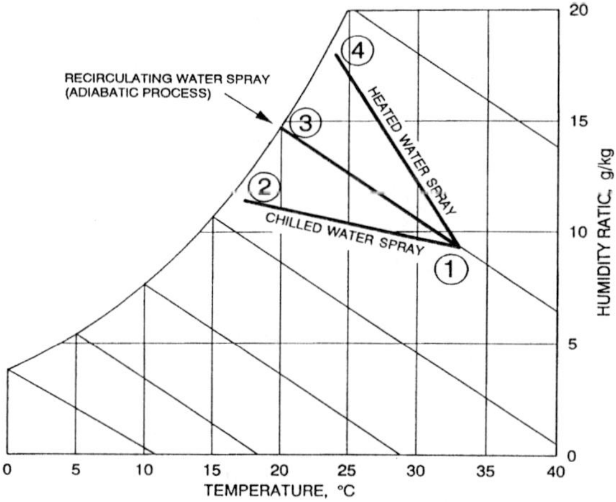

Figure 1.8 Water Spray Cooling

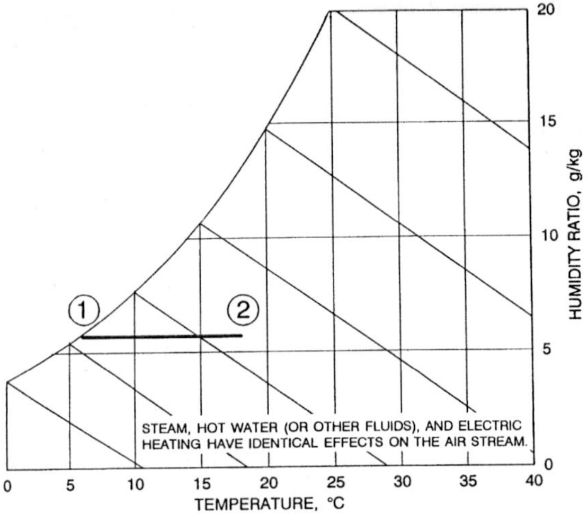

Figure 1.9 Steam, Hot-Water, or Electric Heating

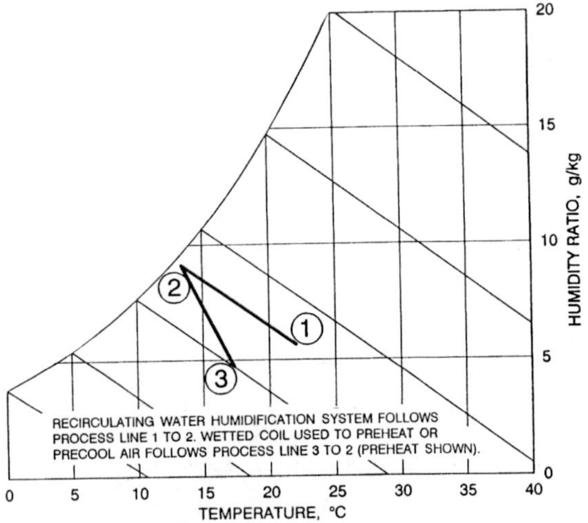

Figure 1.10 Water Humidification

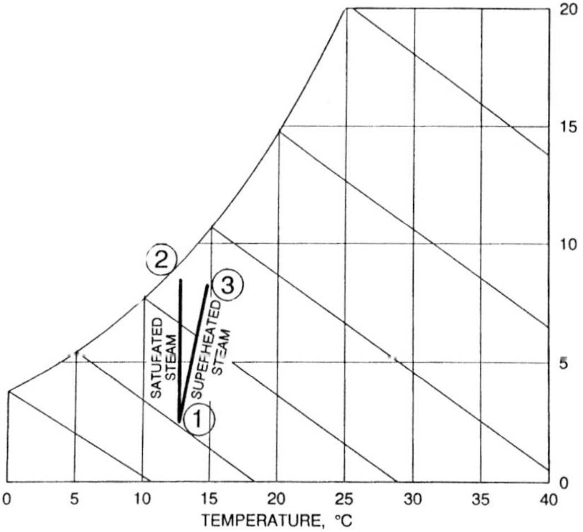

Figure 1.11 Steam Humidification

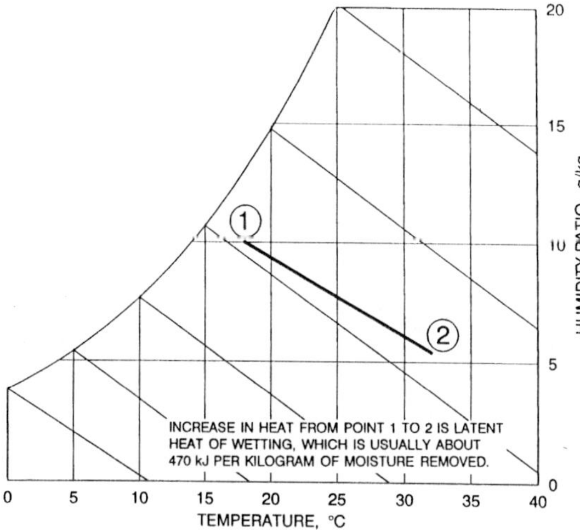

Figure 1.12 Chemical Dehumidification

Table 1.11 Specific Enthalpy of Moist Air at Standard Atmospheric Pressure, 101.325 kPa
[Adapted from 2021F, Ch 1, Tbl 2]

Temp., °C	Specific Enthalpy, kJ/kg_{da}	Temp., °C	Specific Enthalpy, kJ/kg_{da}
–60	–60.325	26	80.801
–55	–55.280	27	85.289
–50	–50.222	28	89.979
–45	–45.144	29	94.882
–40	–40.031	30	100.009
–35	–34.859	31	105.372
–30	–29.593	32	110.985
–25	–24.181	33	116.860
–20	–18.542	34	123.013
–10	–6.070	35	129.458
–8	–3.282	36	136.213
–6	–0.356	37	143.294
–4	2.728	38	150.720
–2	5.995	39	158.510
0	9.475	40	166.685
2	12.981	45	214.169
4	16.696	50	275.349
6	20.644	55	355.144
8	24.853	60	460.880
10	29.354	70	803.464
12	34.181	80	1541.765
14	39.371	90	3867.556
16	44.966		
18	51.011		
20	57.558		
21	61.037		
22	64.663		
23	68.444		
24	72.388		
25	76.503		

Table 1.12 Standard Atmospheric Data for Altitudes to 10 000 m
[2021F, Ch 1, Tbl 1]

Altitude, m	Temperature, °C	Pressure, kPa
–500	18.2	107.478
0	15.0	101.325
500	11.8	95.461
1000	8.5	89.875
1500	5.2	84.556
2000	2.0	79.495
2500	–1.2	74.682
3000	–4.5	70.108
4000	–11.0	61.640
5000	–17.5	54.020
6000	–24.0	47.181
7000	–30.5	41.061
8000	–37.0	35.600
9000	–43.5	30.742
10 000	–50	26.436

Source: Adapted from NASA (1976).

Space Air Diffusion

Room air diffusion methods can be classified as one of the following:

- **Fully mixed systems** produce little or no thermal stratification of air within the space. Overhead air distribution is an example of this type of system.
- **Fully (thermally) stratified systems** produce little or no mixing of air within the occupied space. Thermal displacement ventilation is an example of this type of system.
- **Partially mixed systems** provide some mixing within the occupied and/or process space while creating stratified conditions in the volume above. Most underfloor air distribution and task/ambient conditioning designs are examples of this type of system.

Local temperature and carbon dioxide (CO_2) concentration have similar stratification profiles.

Air distribution systems, such as thermal displacement ventilation (TDV) and underfloor air distribution (UFAD), that deliver air in cooling mode at or near floor level and return air at or near ceiling level produce varying amounts of room air stratification. For floor-level supply, thermal plumes that develop over heat sources in the room play a major role in driving overall floor-to-ceiling air motion. The amount of stratification in the room is primarily determined by the balance between total room airflow and heat load. In practice, the actual temperature and concentration profile depends on the combined effects of various factors, but is largely driven by the characteristics of the room supply airflow and heat load configuration.

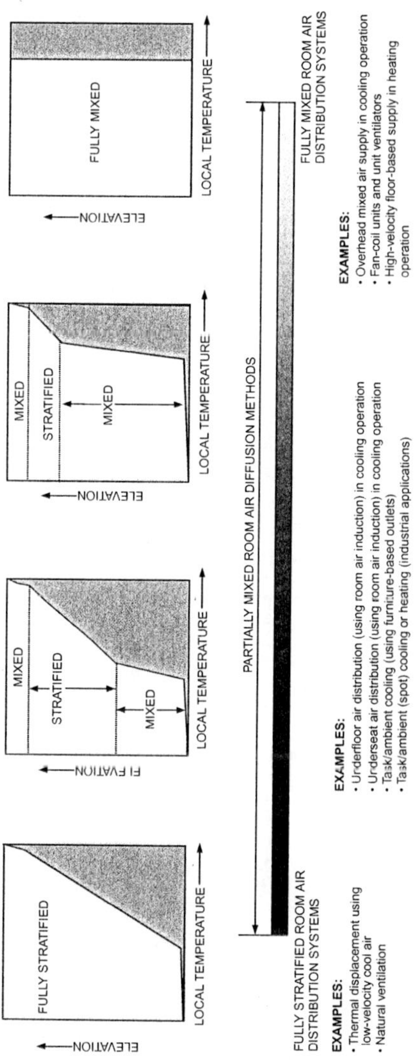

Figure 1.13 Classification of Air Diffusion Methods [2021F, Ch 20, Fig 1]

Principles of Jet Behavior

Air Jet Fundamentals

Air supplied to rooms through various types of outlets can be distributed by turbulent air jets (mixed and partially mixed systems) or in a low-velocity, unidirectional manner (stratified systems).

If an air jet is not obstructed or affected by walls, ceiling, or other surfaces, it is considered a **free jet**. When outlet area is small compared to the dimensions of the space normal to the jet, the jet may be considered free as long as

$$X \le 1.5\sqrt{A_R} \tag{1.6}$$

where

X = distance from face of outlet, m

A_R = cross-sectional area of confined space normal to jet, m^2

Characteristics of the air jet in a room might be influenced by reverse flows created by the same jet entraining ambient air. If the supply air temperature is equal to the ambient room air temperature, the air jet is called an **isothermal jet**. A jet with an initial temperature different from the ambient air temperature is called a **nonisothermal jet**. The air temperature differential between supplied and ambient room air generates thermal forces (buoyancy) in jets, affecting the jet's (1) trajectory, (2) location at which it attaches to and separates from the ceiling/floor, and (3) throw. The significance of these effects depends on the ratio between the thermal buoyancy of the air and jet momentum.

Jet Expansion Zones. The full length of an air jet, in terms of the maximum or centerline velocity and temperature differential at the cross section, can be divided into four zones:

- **Zone 1** is a short core zone extending from the outlet face, in which the maximum velocity and temperature of the airstream remains practically unchanged.
- **Zone 2** is a transition zone, with its length determined by the type of outlet, aspect ratio of the outlet, initial airflow turbulence, etc.
- **Zone 3** is a zone of jet degradation, where centerline air velocity and temperature decrease rapidly. Turbulent flow is fully established and may be 25 to 100 equivalent air outlet diameters (i.e., widths of slot air diffusers) long.
- **Zone 4** is of major engineering importance because, in most cases, the jet enters the occupied area in this zone. Distance to this zone and its length depend on the velocities and turbulence characteristics of ambient air. In a few diameters or widths, air velocity becomes less than 0.25 m/s.

Centerline Velocities in Zones 1 and 2. In zone 1, the ratio V_x/V_o is constant and ranges between 1.0 and 1.2, equal to the ratio of the center velocity of the jet at the start of expansion to the average velocity. The ratio V_x/V_o varies from approximately 1.0 for rounded entrance nozzles to about 1.2 for straight pipe discharges; it has much higher values for diverging discharge outlets.

Experimental evidence indicates that, in zone 2,

$$\frac{V_x}{V_o} = \sqrt{\frac{K_{c2}H_o}{X}} \tag{1.7}$$

where

V_x = centerline velocity at distance X from outlet, m/s

V_o = $V_c/C_d R_{fa}$ = average initial velocity at discharge, m/s

V_c = nominal velocity of discharge based on core area, m/s

C_d = discharge coefficient (usually between 0.65 and 0.90)

R_{fa} = ratio of free area to gross (core) area

H_o = width of jet at outlet or at vena contracta, m

K_{c2} = centerline velocity constant, depending on outlet type and discharge pattern

X ≥ $(1/K_{c2}H_o)^{1/2}$ = distance from outlet to measurement of centerline velocity V_x, m

Centerline Velocity in Zone 3. In zone 3, centerline velocities of radial and axial isothermal jets can be determined accurately from Equation 1.8:

$$V_x = \frac{K_{c3} V_o \sqrt{A_o}}{X} = \frac{K_{c3} Q_o}{X \sqrt{A_o}} \tag{1.8}$$

where

K_{c3}	=	centerline velocity constant (see Table 1.13 for generic values)
V_o	=	$V_c/C_d R_{fa}$ = average initial velocity at discharge, m/s
A_o	=	free area, core area, or neck area as shown in Table 1.13 (obtained from outlet manufacturer), m^2
A_c	=	measured gross (core) area of outlet, m^2
Q_o	=	volumetric flow rate of supply air, m^3/s
X	=	distance from face of outlet, m

The effective area, according to ANSI/ASHRAE Standard 70, can be used in place of A_o in Equation 1.8 with the appropriate value of K_{c3}.

Throw. Equation 1.8 can be transposed to determine the throw X of an outlet if the discharge volume and the centerline velocity are known:

$$X = \frac{K_{c3} Q_o}{V_x \sqrt{A_o}} \tag{1.9}$$

Comparison of Free Jet to Attached Jet

Most manufacturers' throw data obtained in accordance with ANSI/ASHRAE Standard 70 assume the discharge is attached to a surface. An attached jet induces air along the exposed side of the jet, whereas a free jet can induce air on all its surfaces. Because a free jet's induction rate is larger compared to that of an attached jet, a free jet's throw distance will be shorter. To calculate the throw distance X for a noncircular free jet from catalog data for an attached jet, the following estimate can be used.

$$X_{free} = X_{attached} \times 0.707 \tag{1.10}$$

Circular free jets generally have longer throws compared to noncircular jets.

Jets from ceiling diffusers initially tend to attach to the ceiling surface, because of the force exerted by the Coanda effect. However, cold air jets will detach from the ceiling if the airstream's buoyancy forces are greater than the inertia of the moving air stream.

Table 1.13 Generic Values for Centerline Velocity Constant K_{c3}[a] for Commercial Supply Outlets for Fully and Partially Mixed Systems, Except UFAD [2021F, Ch 20, Tbl 1]

Outlet Type	Discharge Pattern	A_0	K_{c3}[a]
High sidewall grilles (2021F, Ch 20, Fig 4)	0° deflection[b]	Free	5.7
	Wide deflection	Free	4.2
High sidewall linear	Core less than 100 mm high[c]	Free	4.4
	Core more than 100 mm high	Free	5.0
Low sidewall (2021F, Ch 20, Fig 7)	Up and on wall, no deflection	Free	4.5
	Wide deflection	Free	3.0
Baseboard	Up and on wall, no deflection	Core	4.0
	Wide deflection	Core	2.0
Floor grille (2021F, Ch 20, Fig 5)	No deflection	Free	4.7
	Wide deflection	Free	1.6
Ceiling (2021F, Ch 20, Fig 2)	360° horizontal[d]	Neck	1.1
	Four-way; little deflection	Neck	3.8
Ceiling linear slot (2021F, Ch 20, Fig 3)	Horizontal/vertical along surface[c]	Free	5.5
	Horizontal/vertical free jet[c]	Free	3.9
	Free jet (air curtain units)	Free	6.0

[a]Generic values shown for example purposes only. See manufacturer's data for specific K_{c3} values.
[b]Free area is about 80% of core area.
[c]Free area is about 50% of core area.
[d]Cone free area is greater than duct area.

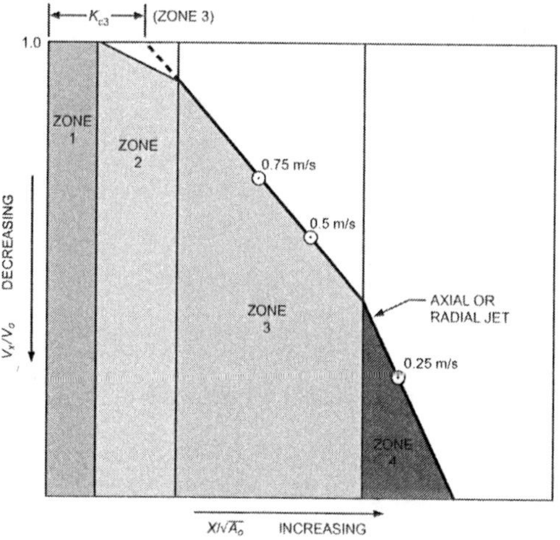

Figure 1.14 Zones of Expansion for Axial or Radial Air Jets
[2021F, Ch 20, Fig 11]

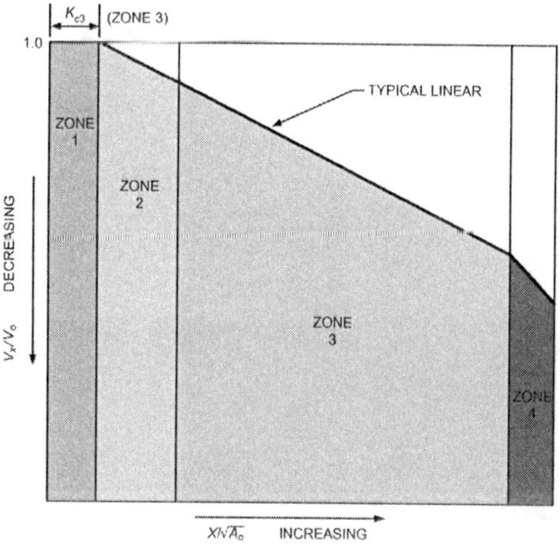

Figure 1.15 Zones of Expansion for Linear Air Jets
[2021F, Ch 20, Fig 12]

Outlet Types and Characteristics

Straub and Chen (1957) and Straub et al. (1956) classified outlets into five major groups (the subgrouping was added in 2017 and was not part of the original research):

- **Group A1.** Outlets mounted in or near the ceiling that discharge air horizontally (Figures 1.16 and 1.17).
- **Group A2.** Outlets discharging horizontally that are not influenced by an adjacent surface (free jet; Figure 1.18).
- **Group B.** Outlets mounted in or near the floor that discharge air vertically in a linear jet (Figure 1.19).
- **Group C.** Outlets mounted in or near the floor that discharge air vertically in a spreading jet (Figure 1.20).
- **Group D.** Outlets mounted in or near the floor that discharge air horizontally (Figures 1.21 and 1.22).
- **Group E.** Outlets that project supply air vertically downward (Figures 1.23 and 1.24). When used in partially stratified systems (e.g., laminar flow outlets, TDV), these outlets use low discharge velocities; in mixed systems (e.g., air curtain units, other downward directed ceiling devices, etc.), they use higher discharge velocities.

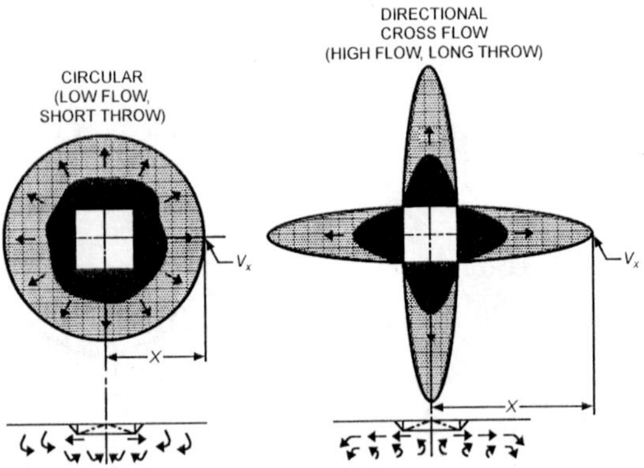

Figure 1.16 Example Airflow Patterns of Outlet Group A1 [2021F, Ch 20, Fig 2]

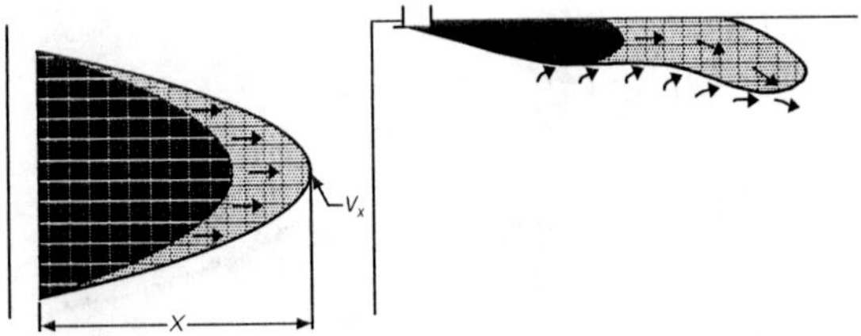

Figure 1.17 Example Airflow Patterns (Nonisothermal) of Outlet Group A1 [2021F, Ch 20, Fig 3]

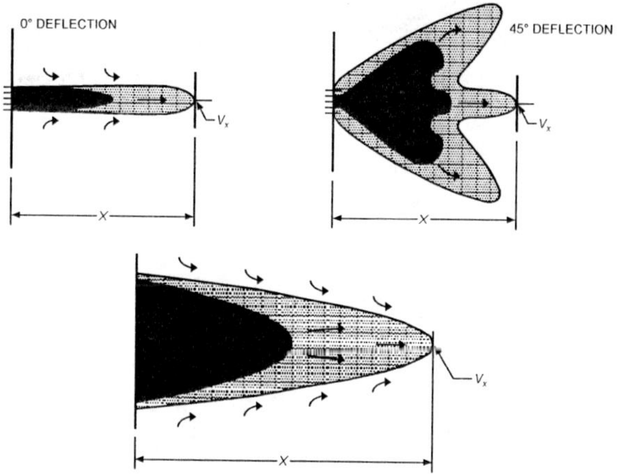

Figure 1.18 Example Airflow Patterns (Isothermal) of Outlet Group A2
[2021F, Ch 20, Fig 4]

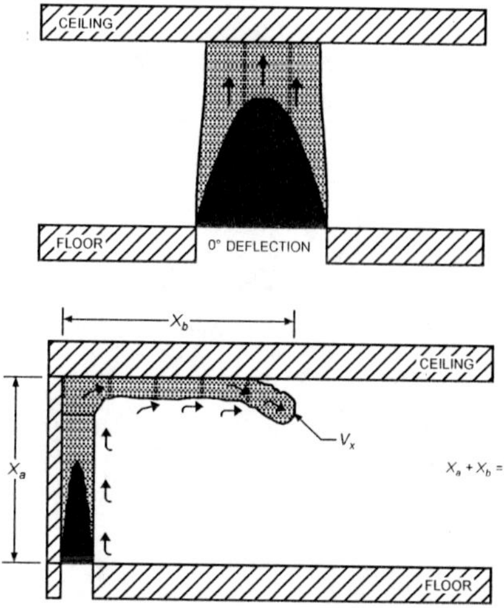

Figure 1.19 Example Airflow Patterns (Nonisothermal) of Outlet Group B
[2021F, Ch 20, Fig 5]

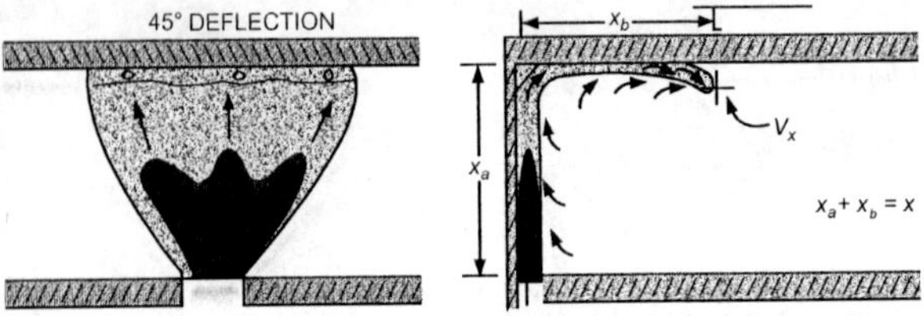

Figure 1.20 Example Airflow Patterns (Nonisothermal) of Outlet Group C [2021F, Ch 20, Fig 6]

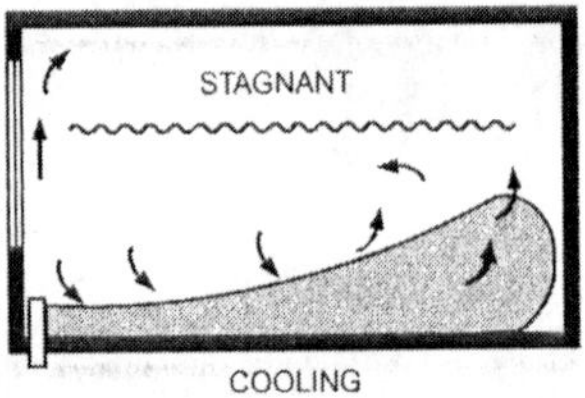

Figure 1.21 Example Airflow Patterns (Nonisothermal) of Outlet Group D (High Velocity) [2021F, Ch 20, Fig 7]

Figure 1.22 Example Airflow Patterns (Nonisothermal) of Outlet Group D (Low Velocity) [2021F, Ch 20, Fig 8]

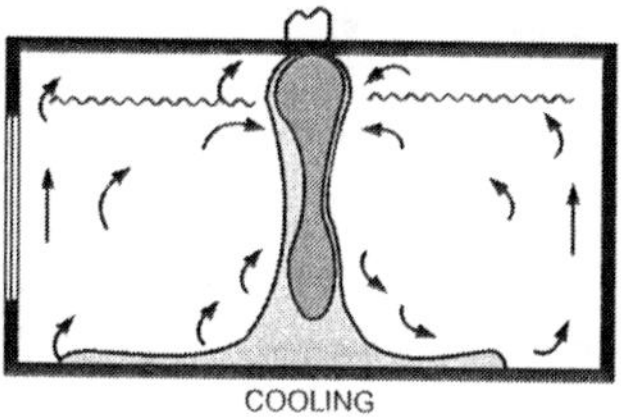

Figure 1.23 Example Airflow Patterns (Nonisothermal) of Outlet Group E (High Velocity) [2021F, Ch 20, Fig 9]

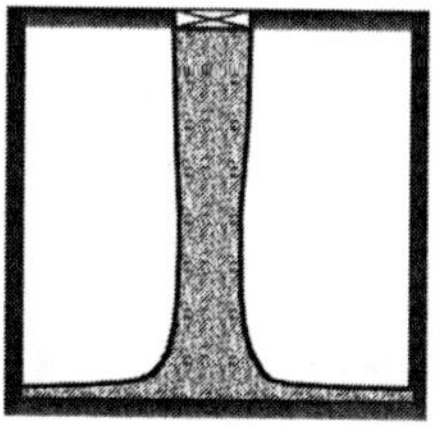

Figure 1.24 Example Airflow Patterns (Nonisothermal) of Outlet Group E (Low Velocity) [2021F, Ch 20, Fig 10]

Room Air Distribution

Air distribution systems affect not only indoor air quality (IAQ) and thermal comfort but also energy consumption over the entire life of a project. ANSI/ASHRAE/IES Standard 90.1 provides energy efficiency requirements that affect supply air characteristics.

Ventilation effectiveness is affected directly by the room air distribution system's design, construction, and operation, but it is very difficult to predict. ANSI/ASHRAE Standard 62.1 provides a table of typical values. As an example, well-designed ceiling-based air diffusion systems produce near-perfect air mixing in cooling mode and yield an air change effectiveness of almost 1.0.

Fully Mixed Air Distribution

In mixed air systems, high-velocity supply jets from air outlets maintain comfort by mixing room air with supply air. This air mixing, heat transfer, and resultant velocity reduction should occur outside the occupied zone. Occupant comfort is maintained not directly by motion of air from outlets, but from secondary air motion from mixing in the unoccupied zone. Comfort is maximized when uniform temperature distribution and room air velocities of less than 0.25m/s are maintained in the occupied zone.

Maintaining velocities less than 0.25m/s in the occupied zone is often overlooked by designers, but is critical to maintaining comfort. The outlet's selection, location, supply air volume, discharge velocity, and air temperature differential determine the resulting air motion in the occupied zone.

Principles of Operation

Mixed systems generally provide comfort by entraining room air into discharge jets located outside occupied zones, mixing supply and room air. Ideally, these systems generate low-velocity air motion (less than 0.25m/s) throughout the occupied zone to provide uniform temperature gradients and velocities. Proper selection of an air outlet is critical for proper air distribution; improper selection can result in room air stagnation, unacceptable temperature gradients, and unacceptable velocities in the occupied zone, possibly leading to occupant discomfort or poor air quality.

The location of a discharge jet relative to surrounding surfaces is important. Discharge jets attach to parallel surfaces, given sufficient velocity and proximity. When a jet is attached, the throw increases by about 30% over a jet discharged in an open area. This difference is important when selecting an air outlet. For detailed discussion of the surface effect on discharge jets, see Chapter 20 of the 2021 *ASHRAE Handbook—Fundamentals*.

Mixed air systems typically use either ceiling or sidewall outlets discharging air horizontally, or floor- or sill-mounted outlets discharging air vertically. They are the most common method of air distribution in North America.

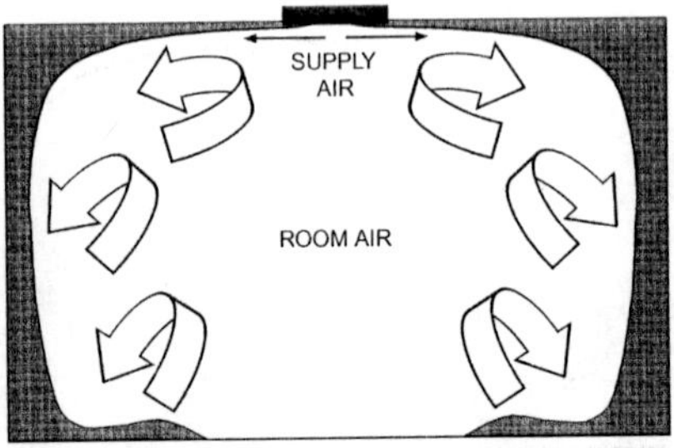

Figure 1.25 Air Supplied at Ceiling Induces Room Air into Supply Jet
[2019A, Ch 58, Fig 5]

Horizontal-Discharge Cooling with Ceiling-Mounted Outlets

Ceiling-mounted outlets typically use the surface effect to transport supply air in the unoccupied zone. The supply air projects across the ceiling and, with sufficient velocity, can continue down wall surfaces and across floors. In this application, supply air should remain outside the occupied zone until it is adequately mixed and tempered with room air.

Overhead outlets may also be installed on exposed ducts, in which case the surface effect does not apply. Typically, if the outlet is mounted 300 mm or more below a ceiling surface, discharge air will not attach to the surface. The unattached supply air has a shorter throw and can project downward, resulting in high air velocities in the occupied zone. Some outlets are designed for use in exposed duct applications. Typical outlet performance data presented by manufacturers are for outlets with surface effect; consult manufacturers for information on exposed duct applications.

Vertical-Discharge Cooling or Heating with Ceiling-Mounted Outlets

Vertically projected outlets are typically selected for high-ceiling applications that require forcing supply air down to the occupied zone. It is important to keep cooling supply air velocity below 0.25 m/s in the occupied zone. For heating, supply air should reach the floor.

There are outlets specifically designed for vertical projection and it is important to review the manufacturer's performance data notes to understand how to apply catalog data. Throws for heating and cooling differ and also vary depending on the difference between supply and room air temperatures.

Cooling with Sidewall Outlets

Sidewall outlets are usually selected when access to the ceiling plenum is restricted. Sidewall outlets within 300 mm of a ceiling and set for horizontal or a slightly upward projection the sidewall outlet provide a discharge pattern that attaches to the ceiling and travels in the unoccupied zone. This pattern entrains air from the occupied zone to provide mixing.

In some applications, the outlet must be located 0.5 to 1.25 m below the ceiling. When set for horizontal projection, the discharge at some distance from the outlet may drop into the occupied zone. Most devices used for sidewall application can be adjusted to project the air pattern upwards toward the ceiling. This allows the discharge air to attach to the ceiling, increasing throw distance and minimizing drop. This application provides occupant comfort by inducing air from the occupied zone into the supply air.

Some outlets may be more than 1.25 m below the ceiling (e.g., in high-ceiling applications, the outlet may be located closer to the occupied zone to minimize the volume of the conditioned space). Most devices used for sidewall applications can be adjusted to project the air pattern upward or downward, which allows the device's throw distance to be adjusted to maximize performance.

When selecting sidewall outlets, it is important to understand the manufacturer's data. Most manufacturers offer data for outlets tested with surface effect, so they only apply if the device is set to direct supply air toward the ceiling. When the device is 1.25 m or more below a ceiling, or supply air is directed horizontally or downward, the actual throw distance of the device is typically shorter. Many sidewall outlets can be adjusted to change the spread of supply air, which can significantly change throw distance. Manufacturers usually publish throw distances based on specific spread angles.

Cooling with Floor-Mounted Air Outlets

Although not typically selected for nonresidential buildings, floor-mounted outlets can be used for mixed system cooling applications. In this configuration, room air from the occupied zone is induced into the supply air, providing mixing. When cooling, the device should be selected to discharge vertically along windows, walls, or other vertical surfaces. Typical nonresidential applications include lobbies, long corridors, and houses of worship.

It is important to select a device that is specially designed for floor applications. It must be able to withstand both the required dynamic and static structural loads (e.g., people walking on them, loaded carts rolling across them). Also, many manufacturers offer devices designed to reduce the possibility of objects falling into the device. It is strongly recommended that obstructions are not located above these in-floor air terminals, to avoid restricting their air jets.

Long floor-mounted grilles generally have both functioning and nonfunctioning segments. When selecting air outlets for floor mounting, it is important to note that the throw distance and sound generated depend on the length of the active section. Most manufacturers' catalog data include correction factors for length's effects on both throw and sound. These corrections can be significant and should be evaluated. Understanding manufacturers' performance data and corresponding notes is imperative.

Cooling with Sill-Mounted Air Outlets

Sill-mounted air outlets are commonly used in applications that include unit ventilators and fan coil units. The outlet should be selected to discharge vertically along windows, walls, or other vertical surfaces, and project supply air above the occupied zone.

As with floor-mounted grilles, when selecting and locating sill grilles, consider selecting devices designed to reduce the nuisance of objects falling inside them. It is also recommended that sills be designed to prevent them from being used as shelves.

Heating and Cooling with Perimeter Ceiling-Mounted Outlets

When air outlets are used at the perimeter with vertical projection for heating and/or cooling, they should be located near the perimeter surface, and selected so that the published 0.75 m/s isothermal throw extends at least halfway down the surface or 1.4 m above the floor, whichever is lower. In this manner, during heating, warm air mixes with the cool downdraft on the perimeter surface, to reduce or even eliminate drafts in the occupied space.

If a ceiling-mounted air outlet is located away from the perimeter wall, in cooling mode, the high-velocity cool air reduces or overcomes the thermal updrafts on the perimeter surface. To accomplish this, the outlet should be selected for horizontal discharge toward the wall. Outlet selection should be such that isothermal throw to the terminal velocity of 0.75 m/s should include the distance from the outlet to the perimeter surface. For heating, the supply air temperature should not exceed 8.5°C above the room air temperature.

Space Temperature Gradients and Airflow Rates

A fully mixed system creates homogeneous thermal conditions throughout the space. As such, thermal gradients should not be expected to exist in the occupied zone. Improper selection, sizing, or placement may prevent full mixing and can result in stagnant areas, or having high-velocity air entering the occupied zone.

Supply airflow requirements to satisfy space sensible heat gains or losses are inversely proportional to the temperature difference between supply and return air. Equation 1.11 can be used to calculate space airflow requirements (at standard conditions):

$$Q = \frac{q_s}{1.2(t_r - t_s)} \tag{1.11}$$

where

Q	=	required supply airflow rate to meet sensible load, L/s
q_s	=	net sensible heat gain in the space, W
t_r	=	return or exhaust air temperature, °C
t_s	=	supply air temperature, °C

For fully mixed systems with conventional ceiling heights, the return (or exhaust) and room air temperatures are the same; for example, a room with a set-point temperature of 24°C has, on average, a 24°C return or exhaust air temperature.

The objective of air diffusion is to create the proper combination of room air temperature, humidity, and air motion to provide thermal comfort and acceptable indoor environmental quality in the occupied zone. There are three recommended methods of selecting outlets for mixed-air systems using manufacturers' data:

- by appearance, flow rate, and sound data
- by isovels (lines of constant velocity) and mapping
- by comfort criteria

Table 1.14 Characteristic Room Length for Several Diffusers (Measured from Center of Air Outlet) [2019A, Ch 58, Tbl 5]

Diffuser Type	Characteristic Length L
High sidewall grille	Distance to wall perpendicular to jet
Adjustable blade	
Fixed blade	
Linear bar	
Nozzle	
Horizontal-throw ceiling diffuser	Distance to closest wall, midplane between outlets or intersecting air jet
Round	
Square	
Perforated	
Louvered	
Plaque	
Swirl	
Sill grille	Length of room in direction of jet flow
Ceiling slot diffuser	Distance to wall perpendicular to jet or midplane between outlets
Light troffer diffusers	Distance to midplane between outlets plus distance from ceiling to top of occupied zone

Air Diffusion Performance Index (ADPI)

ADPI was developed as a way to quantify the comfort level in heating and cooling for a space conditioned by a mixed-air system. High ADPI values generally correlate to high space thermal comfort levels, with the maximum obtainable value of 100. Selecting outlets to provide a minimum ADPI value of 80 generally results in a well-mixed space.

Table 1.15 ADPI Selection Guide for Typical Cooling Loads [2019A, Ch 58, Tbl 6A]

Terminal Device in Cooling Mode	Installation	Load, W/m²	Max. ADPI $T_{0.25}/L$	Max. ADPI	*T/L* Low Limit for ADPI > 80%	*T/L* High Limit for ADPI > 80%
Adjustable-blade grilles	45° upward blades, High sidewall	25	0.8	98	0.4	1.3
		50	0.9	96	0.5	1.2
	0° horizontal blades, High sidewall	25	1.7	94	1.2	2.2
		50	1.8	88	1.4	2.2
	45° downward blades, High sidewall	25	0.9	76	NA	NA
		50	1	70	NA	NA
Fixed-blade grilles (high sidewall installation)	15° upward blades, High sidewall	25	1.4	96	0.5	2.4
		50	2.1	94	1.2	2.9
	15° downward blades, High sidewall	25	1.9	85	1.5	2.2
		50	2	82	1.8	2.2
Linear-bar grilles (high sidewall installation)	High sidewall	25	1.3	92	0.7	1.8
		50	1.3	88	1.0	1.6
	Sill	25	1.3	94	0.9	1.7
		50	1.3	90	1.0	1.6
Nozzles (high sidewall installation)	High sidewall	25	0.7	96	0.4	2.0
		50	1	89	0.4	1.9
Round ceiling diffuser	Ceiling	25	1.6	99	0.4	3.2
		50	1.9	98	0.5	3.2
Square ceiling diffuser	Ceiling	25	1.8	100	0.8	2.8
		50	1.8	100	0.6	3.1
Perforated diffusers, round pattern	Ceiling	25	1.9	95	0.5	3.3
		50	2.1	95	0.9	3.4
Perforated diffusers, directional pattern (4-way)	Ceiling	25	2.1	100	1.2	3.1
		50	2	95	1.0	2.9
Louvered face diffusers, with lip on deflector blade	Ceiling	25	2.5	100	0.5	4.4
		50	2.6	100	0.6	4.5
Louvered face diffusers, without lip on deflector blade	Ceiling	25	2	100	0.5	3.6
		50	1.8	100	0.4	3.4
Plaque face diffusers	Ceiling	25	1.6	100	0.3	3.0
		50	1.6	100	0.4	3.2
Linear-slot diffusers	Ceiling	25	1.8	100	0.5	3.0
		50	1.8	100	0.5	3.1
T-bar slot diffusers	Ceiling, periphery of a wall	25	1.3	96	0.7	1.9
		50	1.5	90	1.1	1.9
Swirl diffusers	Ceiling	25	1.3	100	0.4	2.4
		50	1.3	98	0.4	2.4
N-slot diffusers	Ceiling	25	1.8	100	1.3	2.4
		50	1.8	95	1.3	2.3

Source: Data developed by Liu et al. (2016) for this chapter from ASHRAE research project RP-1546 (Liu 2016), and air speed limit (0.36 m/s) extrapolated from data. Additional data point used to create new regressions for ADPI curves to better represent current diffusers/grilles. Table applies to spaces with maximum 3.7 m ceiling.

Table 1.16 ADPI Selection Guide for Typical Heating Loads [2019A, Ch 58, Tbl 6B]

Terminal Device in Heating Mode	Installation	Load, W/m²	Max. ADPI $T_{0.25}/L$	Max. ADPI	*T/L* Low Limit for ADPI > 80%	*T/L* High Limit for ADPI > 80%
Adjustable-blade grilles	45° upward blades, High sidewall	30 to 40	1.1	95	0.6	1.9
	0° horizontal blades, High sidewall	30 to 40	1.6	94	1.1	2.4
	45° downward blades, High sidewall	30 to 40	0.7	84	0.6	0.8
Fixed-blade grilles	15° upward blades, High sidewall	30 to 40	1.8	96	1.2	2.8
	15° downward blades, High sidewall	30 to 40	1.4	88	0.6	2.2
Linear-bar grilles	High sidewall	30 to 40	1.2	94	0.6	1.7
	Sill	30 to 40	1.2	100	0.7	1.8
Nozzles (high sidewall installation)	High sidewall	30 to 40	1.5	92	1.0	2.0
Round ceiling diffuser	Ceiling	30 to 40	1.4	93	1.0	2.3
Square ceiling diffuser	Ceiling	30 to 40	1.7	91	2.5	3.4
Perforated diffusers, round pattern	Ceiling	30 to 40	2.1	90	2.0	2.8
Perforated diffusers, directional pattern (4-way)	Ceiling	30 to 40	2.5	87	2.5	3.4
Louvered face diffusers, with lip on deflector blade	Ceiling	30 to 40	2.6	88	2.5	4.4
Louvered face diffusers, without lip on deflector blade	Ceiling	30 to 40	2.1	88	2.1	3.2
Plaque face diffusers	Ceiling	30 to 40	2.1	93	2.1	3.0
Linear-slot diffusers	Ceiling	30 to 40	1.7	90	1.7	3.1
T-bar slot diffusers	Ceiling, periphery of a wall	30 to 40	1.6	91	1.3	2.0
Swirl diffusers	Ceiling	30 to 40	1.4	100	1.4	2.1
N-slot diffusers	Ceiling	30 to 40	1.9	100	1.5	2.4

Source: Data developed by Liu and Novoselac (2015) for this chapter from ASHRAE research project RP-1546 (Liu 2016), and air speed limit (0.36 m/s) extrapolated from data. Additional data point used to create new regressions for ADPI curves to better represent current diffusers/grilles. Table applies to spaces with maximum 3.7 m ceiling.

Fully Stratified Air Distribution

Fully stratified air distribution systems are characterized by a vertical temperature gradient throughout the space, where the coolest temperature is at the floor level, and the warmest temperature is at the ceiling height. Displacement ventilation (DV) systems are the most common example of a fully stratified air distribution system. DV systems typically use floor or low sidewall (sometimes ceiling-mounted) diffusers delivering low-velocity, cool air across the floor. The low-velocity air, in conjunction with room loads and buoyancy effects, creates the characteristic vertical thermal stratification.

Principles of Operation

DV systems (see Figure 1.26) use very low discharge velocities, with diffusers typically sized to provide outlet velocities between 0.20 and 0.35 m/s. In addition to the low velocity discharge, the temperature of the supply air is also different from a fully mixed system, with temperatures generally above 16°C; lower temperatures may be used in industrial applications, exercise or sports facilities, and transient areas where comfort concerns are minimal. This cool supply air is more dense than the ambient air and drops to the floor after discharge, whether from floor, low sidewall, or ceiling mounted locations, spreading across the lower level of the space (typically less than 0.2 m in height).

As convective heat sources (Figure 1.26) in the space transfer heat to the cooler air around them, natural convection currents form and rise along the heat transfer boundary. Without significant room air movement, these currents rise to form a convective heat plume (thermal plume) around and above the heat source; as the plume rises, it expands by entraining surrounding air. Its growth and velocity are proportional to the heat source's size and sensible load, as well as the temperature of the ambient air above it. As the plume rises, ambient air from below and around the heat source fills the void. An occupant in a DV system entrains the cool, conditioned air directly into their breathing zone. As the occupant exhales, the spent air, being warmer and more humid than the ambient air, is pulled out of the breathing zone by the rising plume. Convective heat from sources located above the occupied zone has little effect on occupied-zone air temperature.

At a certain height, where plume temperature equals ambient temperature, the plume dissipates and spills horizontally. Two distinct zones are thus formed in the room: a lower occupied zone with little or no recirculation flow (close to displacement flow), and an upper zone with recirculation flow. The boundary between these two zones is often called the **shift zone** (or **stratification height**). The height of this boundary layer between the upper and lower

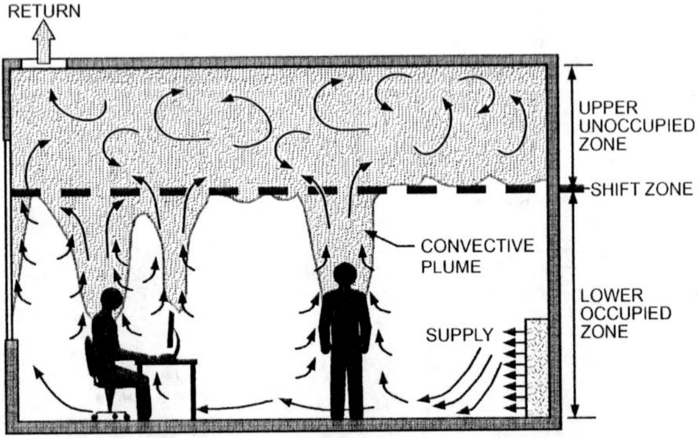

Figure 1.26 Displacement Ventilation System Characteristics
[2019A, Ch 58, Fig 6]

zones is determined based on the convection flow rates of the thermal plumes in relation to the airflow rate supplied by the diffusers. In a DV system, increasing the airflow rate limits stratification and lowers the boundary layer height, with a decrease in airflow providing the opposite effect Actual and simplified representations of the temperature gradient in the space are shown in Figure 1.27.

Outlet Characteristics

Displacement outlets are designed for average face velocities between 0.25 to 0.35 m/s, and are typically in a low sidewall or floor location. Return or exhaust air intakes should always be located above the occupied zone for human thermal comfort applications.

Displacement outlets are available in a number of configurations and sizes. Some models are designed to fit in corners or along sidewalls, or stand freely as columns. It is important to consider the degree of flow equalization the outlet achieves, because use of the entire outlet surface for air discharge is paramount to minimizing clear zones and maintaining acceptable temperatures at the lower levels of the space.

Stationary occupants should not be subjected to discharge velocities exceeding about 0.25 m/s because air at the ankle level within this velocity envelope tends to be quite cool. As such, most outlet manufacturers define an adjacent zone (also called a *clear zone*) in which locating stationary, low-activity occupants is strongly discouraged, but transient occupancy, such as in corridors or aisles, is possible. Occupants with high activity levels may also find the clear zone acceptable.

Unlike mixed systems, outlets in thermal displacement systems discharge air at very low velocities, resulting in very little mixing. As such, design of these systems primarily involves determining a supply airflow rate to manage the thermal gradients in the space in accordance

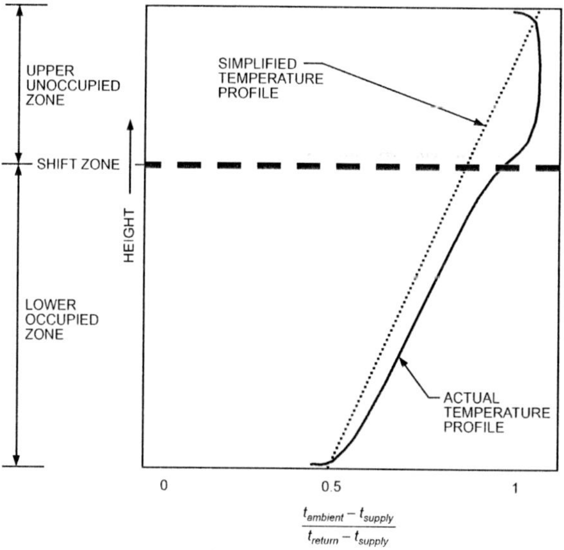

Figure 1.27 Temperature Profile of Displacement Ventilation System
[2019A, Ch 58, Fig 7]

with ASHRAE comfort guidelines. ANSI/ASHRAE Standard 55 recommends that the vertical temperature difference between the ankle and head levels of space occupants be limited to no more than 3 K to maintain a high degree (>95%) of occupant satisfaction.

Application Considerations

For heating, either a displacement diffuser with integrated heating or a separate system is generally recommended. Displacement ventilation can be used successfully in combination with perimeter fan-coils, hydronic systems, or radiators and convectors installed at the exterior walls to offset space heat losses. Radiant heating panels and heated floors also can also be used with displacement ventilation. To maintain displacement ventilation, outlets should supply ventilation air about 2 K lower than the desired room temperature.

Thermal displacement ventilation systems can be either constant or variable air volume. A thermostat in a representative location in the space or return plenum should determine the delivered air volume or temperature. If the time-averaged requirements of ANSI/ASHRAE Standard 62.1 are met, intermittent on/off airflow control can be used.

Avoid using displacement ventilation and mixed air systems in the same space, because mixing destroys the natural stratification that drives the thermal displacement ventilation system. Displacement ventilation systems can be complemented by hydronic systems such as chilled floors. Use caution when combining chilled ceilings, beams, or panels with fully stratified systems, because cold surfaces in the upper zone of the space may recirculate contaminants stratified in the upper zone back into the occupied zone.

Partially Mixed Air Distribution

A partially mixed system's characteristics fall between a fully mixed system and a fully stratified system. It includes both a high-velocity mixed air zone and a low-velocity stratified zone where room air motion is caused by thermal forces. For example, floor-based outlets, when operating in a cooling mode with relatively high discharge velocities (>0.75 m/s), create mixing, thus affecting the amount of stratification in the lower portions of the room. In the upper portions of the room, away from the influence of floor outlets, room air often remains thermally stratified in much the same way as displacement ventilation systems.

Principles of Operation

Supply air is discharged, usually vertically, at relatively high velocities and entrains room air in a similar fashion to outlets used in mixed air systems. This entrainment, as shown in Figure 1.28 reduces the temperature and velocity differentials between supply and ambient room air. This discharge results in a vertical plume that rises until its velocity is reduced to about 0.25 m/s. At this point, its kinetic energy is insufficient to entrain much more room air, so mixing stops. Because air in the plume is still cooler than the surrounding air, the supply air spreads horizontally across the space, where it is entrained by rising thermal plumes generated by nearby heat sources.

Research and experience have shown that the amount of room air stratification varies depending on design, commissioning, and operation. Control of stratification includes the following considerations:

- By reducing airflow and mixing in the occupied zone, fan energy can be reduced and stratification can be increased, approaching a reasonable target at 1.5 to 2.5 K temperature difference from head to ankle height, which satisfies ANSI/ASHRAE Standard 55.
- By increasing airflow and mixing in the occupied zone, excessive stratification can be avoided, thereby improving thermal comfort.

Figure 1.28 shows one example of the resulting room air distribution in which the room air is mixed in the **lower mixed zone**, which is bounded by the floor and the elevation (**throw height**) at which the 0.25 m/s terminal velocity occurs. At this elevation, stratification begins to occur and a linear temperature gradient, similar to that found in thermal displacement systems, forms and extends through the **stratified zone**. As with thermal displacement ventilation, convective heat plumes from space heat sources draw conditioned air from the lower (mixed) level through the stratified zone and to the overhead return location. A third zone,

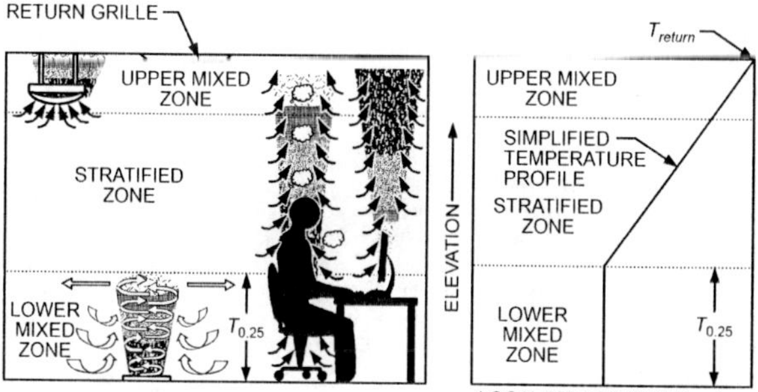

Figure 1.28 UFAD System in Partially Stratified Application
[2019A, Ch 58, Fig 9]

referred to as the **upper mixed zone**, may exist where the volume of rising heat plumes terminate. Although velocities in this area are quite low, the air tends to be mixed.

Outlet Characteristics

One outlet type is a **swirl diffuser** with a high-induction core, which induces large amounts of room air to quickly reduce supply to ambient air velocity and temperature differentials. Supply air is injected into the room as a swirling vertical plume close to the outlet. Properly selected, these outlets produce a limited vertical projection of the supply air plume, restricting mixing to the lower portions of the space. Most of these outlets allow occupants to adjust the outlet airflow rate easily. Other versions incorporate automatically controlled dampers that are repositioned by a signal from the space thermostat and/or central control system.

Another category includes more conventional **floor grilles** designed for directional discharge of supplied airflow. These grilles may be either linear or modular in design, and may allow occupants to adjust the discharge air pattern by repositioning the core of the outlet. Most floor grilles include an integral actuated damper, or other means, that automatically throttles the volume of air in response to the zone conditioning requirements.

Room air induction allows UFAD diffusers to comfortably deliver supply air a few degrees cooler than possible with outlets used for thermal displacement ventilation outlets. The observance of clear, or adjacent, zones above and around the diffusers, where stationary occupants should not reside, is recommended. Outlet manufacturers typically identify such restrictive areas in their product literature.

As with thermal displacement systems, design involves determining a supply airflow rate that limits thermal gradients in the occupied zone in accordance with ANSI/ASHRAE Standard 55 guidelines. ASHRAE Standard 55 recommends that the vertical temperature difference between the ankle and head level of space occupants be limited to no more than 3 K if a high degree (>95%) of occupant comfort is to be maintained.

Application Considerations

Some considerations include the following:

- Supply temperatures in the access floor cavity should be kept at 16°C or above, to minimize the risk of condensation and subsequent mold growth.
- Most UFAD outlets can be adjusted automatically by a space thermostat or other control system, or manually by the occupant. In the latter case, outlets should be located within the workstation they serve.
- Use of manually adjusted outlets should be restricted to open office areas where cooling loads do not tend to vary considerably or frequently. Perimeter areas and conference rooms require automatic control of supply air temperatures and/or flow rates because their thermal loads are highly transient.
- Heat transfer to and from the floor slab affects discharge air temperature and should be considered when calculating space airflow requirements. Floor plenums should be well sealed to minimize air leakage, and exterior walls should be well insulated and have good vapor retarders. Night and holiday temperature setbacks should likely be avoided, or at least reduced, to minimize plenum condensation and thermal mass effect problems. With air-side economizers, using enthalpy control rather than dry-bulb control can help reduce hours of admitting high moisture-content air, thus also reducing the potential for condensation in the floor plenums.
- Avoid using stratified and mixed air systems in the same space, because mixing destroys the natural stratification that drives the stratified system.
- Return static pressure drop should be relatively equal throughout the spaces being served by a common UFAD plenum. This reduces the chance of unequal pressurization in the UFAD plenum.

See *UFAD Guide: Design, Construction and Operation of Underfloor Air Distribution Systems* for a more thorough discussion of issues involved with the design and application of UFAD systems.

Air Dispersion Systems

Textile air dispersion systems are low-pressure extended plenum systems with pressurized tubing and air distributed along the path of least resistance. Systems constructed with metal follow the same principles.

Various supply air outlet styles are available: porous fabric weave used as an outlet, microperforations, linear vents, orifices, and nozzles. Many of these air outlet styles can be used together to achieve specific results.

Return Air Inlets

The success of a mixed air distribution system depends primarily on supply diffuser location. Return grille location is far less critical than the location of air outlets. In fact, the return air intake affects room air motion only in the area immediately around the grille. Measurements of velocity near a return air grille show a rapid decrease in magnitude as the measuring device is moved away from the grille face. Table 1.17 shows recommended maximum (to avoid excessive noise) return inlet face velocities as a function of grille location. Every enclosed space should have return/transfer inlets of adequate size per this table.

For stratified and partially mixed air distribution systems, it can be advantageous to place the return air inlet in the ceiling. ANSI/ASHRAE Standard 62.1 allows ventilation effectiveness greater than 1.0 for some stratified and partially mixed air distribution systems in cooling mode if the return air inlet is located in the ceiling.

Supply air short circuiting is normally not a problem if the outlet is selected to provide adequate throw and directed away from returns or exhausts. The success of this practice is confirmed by the availability and use of combination supply and return diffusers.

Table 1.17 Recommended Return Inlet Face Velocities
[2019A, Ch 58, Tbl 1]

Inlet Location	Velocity Across Gross Area, m/s
Above occupied zone	>4
In occupied zone, not near seats	3 to 4
In occupied zone, near seats	2 to 3
Door or wall louvers	1 to 1.5
Through undercut area of doors	1 to 1.5

2. AIR CONTAMINANTS AND CONTROL

Contaminants

Bioaerosols

Bioaerosols are a diverse class of particles of biological origin. They include bacteria, fungal spores, fungal fragments, pollen grains, subpollen particles, viruses, pet- and pest-associated allergens, and plant debris. They are of particular concern in indoor air because of their association with allergies and asthma and their ability to cause disease. Chapters 10 and 11 of the 2021 *ASHRAE Handbook—Fundamentals* contain more detailed descriptions of these contaminants.

Bioaerosols range in size from 0.01 to 100 μm. Single bacterial cells are approximately 1 μm or less in aerodynamic diameter; however, they are often transported as larger bacterial cell agglomerates (~2 to 5 μm) or attached to other biological and abiotic particles. Unicellular fungal spores are generally 2 to 5 μm in aerodynamic diameter and multicellular fungal spores are larger than 10 μm. Fungal fragments are typically less than 1 μm in size. Pollen grains range in size from 10 to 100 μm, whereas subpollen particles span ~0.01 μm to several micrometers in size. Individual viruses can be as small as 0.02 to 0.03 μm, but are typically carried on larger particles. Pet- and pest-associated allergens are approximately 1 to over 20 μm.

Air Filters and Cleaners

ASHRAE Air Filtration Standards

ANSI/ASHRAE Standard 52.2 contains minimum efficiency reporting values (MERVs) for air cleaner particle size efficiency. In 2008, the standard incorporated arrestance testing from the discontinued ASHRAE Standard 52.1. Table 2.1 provides an approximate cross-reference for air cleaners tested under ASHRAE Standards 52.1 and 52.2. Currently, there is no ASHRAE standard for testing electronic air cleaners.

Filter Installation

In-service efficiency of an air filter is sharply reduced if air leaks through the bypass dampers or poorly designed filter-holding frames. The higher the filter efficiency, the more careful attention that must be paid to the rigidity and sealing effectiveness of the frame. In addition, high-efficiency filters must be handled and installed with care.

Published performance data for all air filters are based on straight-through unrestricted airflow. Filters should be installed so that the face area is at right angles to the airflow whenever possible. Provide sufficient space in front of (upstream) or behind (downstream of) the filter, or both, depending on its type, to make the filter accessible for inspection and service. A distance of 500 to 1000 mm is required, depending on the filter chosen

Filter Types

Major factors influencing filter design and selection include (1) degree of air cleanliness required, (2) specific particle size range or aerosols that require filtration, (3) aerosol concentration, (4) resistance to airflow through the filter, (5) design face velocity to achieve published performance, (6) change-out cycle requirements, (7) energy consumption requirements, (8) special disposal mandates, and (9) resistance to certain conditions (physical, chemical, or biological).

Common air cleaners are broadly grouped as follows:

In **fibrous media unit filters**, the accumulating dust load causes pressure drop to increase up to some maximum recommended or predetermined value before changing filters. During this period, efficiency normally increases. However, at high dust loads, dust may adhere poorly to filter fibers and efficiency drops because of offloading. Filters in this condition should be replaced or reconditioned, as should filters that have reached their final (maximum recommended) pressure drop. This category includes viscous impingement and dry air filters, available in low-efficiency to ultrahigh-efficiency construction.

Renewable media filters are fibrous media filters where fresh media is introduced into the airstream as needed to maintain essentially constant resistance and, consequently, constant average efficiency.

Electronic air cleaners require a power source and, if maintained properly by regular cleaning, have relatively constant pressure drop and efficiency. If they are not cleaned regularly, the accumulated dust can build up to the point that arcing can occur in the collection section. This may reduce collection efficiency as well as produce unwanted extraneous noise. Dust buildup in the ionizer section can also reduce efficiency.

Combination air cleaners combine the other types. For example, an electronic air cleaner may be used as an agglomerator with a fibrous media downstream to catch the agglomerated particles blown off the plates. Low-efficiency pads, throwaway panels and automatically renewable media roll filters, or low- to medium-efficiency pleated prefilters may be used upstream of a high-efficiency filter to extend the life of the better and more costly final filter. Charged media filters are also available that increase particle deposition on media fibers by an induced electrostatic field. With these filters, pressure loss increases as it does on a noncharged fibrous media filter. The benefits of combining different air cleaning processes vary.

Contaminant Control Applications

See Tables 12.2 and 12.3 for residential exhaust airflow requirements.

For information on laboratory applications, see Chapter 17 of the 2019 *ASHRAE Handbook—HVAC Applications*; also see *ASHRAE Laboratory Design Guide: Planning and Operation of Laboratory HVAC Systems*, Second Edition.

For cleanroom applications, see *ASHRAE Design Guide for Cleanrooms: Fundamentals, Systems, and Performance*; also see Chapter 19 of the 2019 *ASHRAE Handbook—HVAC Applications*.

Tables 2.6 to 2.9 provide data for hoods used in commercial kitchens.

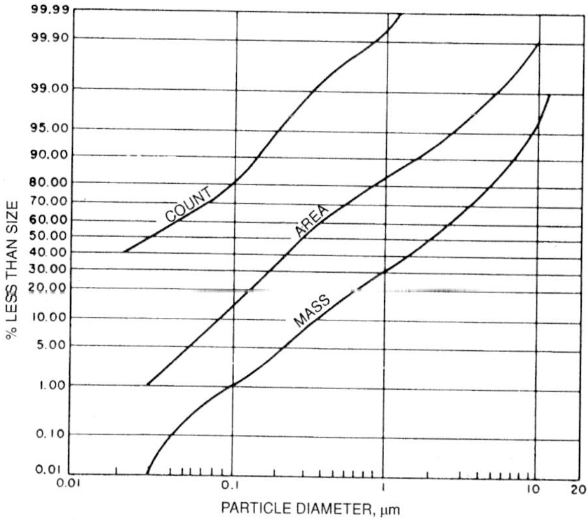

Count curve: Based on measurements by electron microscope.
Area curve: Calculated.
Mass curve: Solid section based on measurements by sedimentation.

Figure 2.1 Particle Size Distribution of Atmospheric Dust

Table 2.1 Air Filter Cross-Reference and Application Guidelines [2020S, Ch 29, Tbl 3]

Standard 52.2 MERV	Arrestance Value	Example Range of Contaminants Controlled	Example Applications	Sample Air Cleaner Type(s)
E-1 Range				
MERV 16	N/A	0.3 to 1.0 μm size range: bacteria, smoke (ETS), paint pigments, face powder, some virus, droplet nuclei, insecticide dusts, soldering fumes	Day surgery, general surgery, hospital general ventilation, turbo equipment, compressors, welding/soldering air cleaners, prefilters to HEPAs, LEED for existing (EB) and new (NC) commercial buildings, smoking lounges	Box-style wet-laid or lofted fiberglass, box-style synthetic media, minipleated synthetic or fiberglass paper, depths from 50 to 300 mm. Pocket filters of fiberglass or synthetic media 300 to 900 mm.
MERV 15	N/A			
MERV 14	N/A			
MERV 13	N/A			
E-2 Range				
MERV 12	N/A	1.0 to 3.0 μm size range: milled flour, lead dust, combustion soot, *Legionella*, coal dust, some bacteria, process grinding dust	Food processing facilities, air separation plants, commercial buildings, better residential, industrial air cleaning, prefiltration to higher-efficiency filters, schools, gymnasiums	Box-style wet-laid or lofted fiberglass, box-style synthetic media, minipleated synthetic or fiberglass paper, depths from 50 to 300 mm. Pocket filters either rigid or flexible in synthetic or fiberglass, depths from 300 to 900 mm.
MERV 11	N/A			
MERV 10	N/A			
MERV 9	N/A			
E-3 Range				
MERV 8	N/A	3.0 to 10 μm size range: pollens, earth-origin dust, mold spores, cement dust, powdered milk, snuff, hair spray mist	General HVAC filtration, industrial equipment filtration, commercial property, schools, prefilter to high-efficiency filters, paint booth intakes, electrical/phone equipment protection	Wide range of pleated media, ring panels, cubes, pockets in synthetic or fiberglass, disposable panels, depths from 25 to 600 mm.
MERV 7	N/A			
MERV 6	N/A			
MERV 5	N/A			
MERV 4	>70%	Arrestance method	Protection from blowing large particle dirt and debris, industrial environment ventilation air	Inertial separators
MERV 3	>70%			
MERV 2	>65%			
MERV 1	<65%			

Note: MERV for non-HEPA/ULPA filters also includes test airflow rate, but it is not shown here because it is of no significance for the purposes of this table.
N/A = not applicable.

Table 2.2 NAAQS for the United States [2021F, Ch 11, Tbl 12]

Contaminant	Primary or Secondary Standard	Averaging Time	Level	Details
Carbon monoxide	Primary	1 h	35 ppm	Not to be exceeded more than once per year
		8 h	9 ppm	
Nitrogen dioxide	Primary	1 h	100 ppb	98th percentile, averaged over 3 years
	Primary/ secondary	1 yr	53 ppb	Annual mean
Ozone	Primary/ secondary	8 h	75 ppb	Annual fourth-highest daily maximum 8 h concentration, averaged over 3 years
Sulfur dioxide	Primary	1 h	75 ppb	99th percentile of 1 h daily maximum concentrations, averaged over 3 years
	Secondary	3 h	500 ppb	Not to be exceeded more than once per year
Particulate, $PM_{2.5}$[a]	Primary/ secondary	24 h	35 $\mu g/m^3$	98th percentile, averaged over 3 years
		1 yr	15 $\mu g/m^3$	Annual mean, averaged over 3 years
Particulate, PM_{10}[b]	Primary/ secondary	24 h	150 $\mu g/m^3$	Not to be exceeded more than once per year on average over 3 years
Lead (Pb) in particles	Primary/ secondary	3 mo	0.15 $\mu g/m^3$	Not to be exceeded

[a] $PM_{2.5}$ = particulates below 2.5 μm diameter.
[b] PM_{10} = particulates below 10 μm diameter.
ppb = parts per 10^9
Source: *National Ambient Air Quality Standards* (NAAQS), U.S. Environmental Protection Agency, Washington, DC, 2015.

Table 2.3 Sources and Indoor and Outdoor Concentrations of Selected Indoor Contaminants [2021F, Ch 11, Tbl 13]

Contaminant	Sources of Indoor Contaminants	Typical Indoor Concentration	Typical Outdoor Concentration	Locations
Carbon monoxide	Combustion equipment, engines, faulty heating systems	0.5 to 5 ppm[a] (without gas stoves) 5 to 15 ppm[a] (with gas stoves)	2 ppm[a]	Indoor ice rinks, homes, cars, vehicle repair shops, parking garages
$PM_{2.5}$	Stoves, fireplaces, cigarettes, condensation of volatiles, aerosol sprays, cooking	7 to 10 $\mu g/m^3$[a]	<10 $\mu g/m^3$[a]	Homes, offices, cars, public facilities, bars, restaurants
PM_{10}	Combustion, heating system, cooking	40 to 60 $\mu g/m^3$[a]	60 $\mu g/m^3$[a]	Homes, offices, transportation, restaurants
Organic vapors	Combustion, solvents, resin products, pesticides, aerosol sprays, cleaning products, building materials, paints	Different for each VOC[c] (2 to 5 times outdoor levels)	See 2021F, Ch 11, Tbl 11	Homes, restaurants, public facilities, offices, hospitals
Nitrogen dioxide	Combustion, gas stoves, water heaters, gas-fired dryers, cigarettes, engines	<8 ppb[a] (without combustion appliances) >15 ppb with combustion appliances	15 ppb[a]	Homes, indoor ice rinks
Nitric oxide	Combustion, gas stoves, water heaters, gas-fired dryers, cigarettes, engines	—	Various	Homes, any building with combustion source
Sulfur dioxide	Heating system	20 $\mu g/m^3$[b]	<20 $\mu g/m^3$[b] 3 ppb[a]	Mechanical/furnace rooms
Formaldehyde	Insulation, product binders, pressed wood products, carpets	0.1 to 0.3 ppm[a]	NA	Homes, schools, offices
Radon and progeny	Building materials, groundwater, soil	1.3 pCi/L[a]	4 pCi/L[a]	Homes, schools
Carbon dioxide	Combustion appliances, humans, pets	600 to 1000 ppm[c]	300 to 500 ppm[c]	
Biological contaminants	Humans, pets, rodents, insects, plants, fungi, humidifiers, air conditioners	NA	NA (lower than indoor levels)	Homes, hospitals, schools, offices, public facilities
Ozone	Electric arcing, electronic air cleaners, copiers, printers	42 ppb[d]	70 ppb[a]	Airplanes, offices, homes

NA = not applicable
ppb = parts per 10^9

Sources:
[a]EPA (2011)
[b]NRC (1981)
[c]Seppänen et al. (1999) and ANSI/ASHRAE Standard 62.1, Appendix C
[d]Weschler (2000)

Table 2.4 Media Selection by Contaminant [2019A, Ch 47, Tbl 9]

Gaseous Contaminant	PIA	AC	AIC	BIC
Acetaldehyde	1	2		
Acetic acid (!)	1	2		2,1
Acetic anhydride (!)	1,2	1		2
Acetone (!)	1	2		
Acetylene	1			
Acrolein	1	2		
Acrylic acid (!)	1	1		2
Allyl sulfide	1	2		
Ammonia (NH_3)			1	
Aniline	2	1		
Arsine	1			
Benzene		1		
Borane (!)	1	2,2		
Bromine		1		
1,3 Butadiene	1	2		
Butane		1		
2-Butanone	1	2		
2-Butoxyethanol	2	1		
Butyl acetate (!)	1,2	2,1		
Butyl alcohol	2	1		
Butyl mercaptan	2	1		
Butylene	2	1		

Gaseous Contaminant	PIA	AC	AIC	BIC
Dichlorofloromethane		1		
R-114 (see note)		1		
Diethylamine	2	1		
Dimethylamine		1	2	
Dioctyl phthalate		1		
Dioxane	1	2		
Ethanol	1	2		
Ethyl acetate	2	1		
Ethyl chloride (!)	1,2	2,1		
Ethylene (C_2H_4)	1			
Ethylene oxide	1	2		
Ethyl ether	2	1		
Ethyl mercaptan (!)	1,1	2		2
Formaldehyde	1			
Gasoline	1			
General halocarbons		1		
General hydrocarbons	2	1		
General VOC	2	1		
Heptane		1		
Hydrogen bromide		2		1
Hydrogen chloride		2		1
Hydrogen cyanide	1			

Gaseous Contaminant	PIA	AC	AIC	BIC
Methyl formate	2	1		
Methyl isobutyl ketone	2	1		
Methyl sulfide	1			
Methyl vinyl ketone	2	1		
Naphtha		1		
Naphthalene		1		
Nicotine	1	2		
Nitric acid				1
Nitric oxide (NO)	1			2
Nitrobenzene		1		
Nitrogen dioxide	1			2
Nitromethane	1			
Nitrous oxide				1
Octane (!)	2	1,1		
Ozone (O_3) (!)	2	1,1		
Perchloroethylene	2	1		
Peroxy acetyl nitrate (PAN)		1		
Phenol	2	1		
Phosgene	2	1		
Phosphine	1			
Putrescine	1	2		
Pyridine (!)	1	1		

Table 2.4 Media Selection by Contaminant [2019A, Ch 47, Tbl 9] ***(Continued)***

Gaseous Contaminant	PIA	AC	AIC	BIC	Gaseous Contaminant	PIA	AC	AIC	BIC	Gaseous Contaminant	PIA	AC	AIC	BIC
Butyne	2	1			Hydrogen fluoride	1			1	Skatole	2	1		
Butyraldehyde	2	1			Hydrogen iodide	2				Silane	1			
Butyric acid		1		2	Hydrogen selenide					Stoddard solvent	1			
Cadaverine	2	1			Hydrogen sulfide	1			1	Stibine	1			
Camphor		1			Iodine		1			Styrene (!)	2	1,1		
Carbon dioxide (CO_2)	Carbon w/catalyst				Iodoform	2	1			Sulfur dioxide	1			1
Carbon disulfide	2	1			Isopropanol	2	1			Sulfur trioxide	1			1
Carbon monoxide (CO)	Carbon w/catalyst				Kerosene		1			Sulfuric acid		2		1
Carbon tetrachloride		1			Lactic acid		1			Toluene		1		
Chlorine (Cl_2)				1	Menthol	2	1			Triethylamine		2	1	
Chloroform		1			Mercury vapor		Impreg. AC			Trichlorethylene		1		
Creosote (!)	1,2	2,1			Methanol	2	1			1,1,1, trichloroethane (!)	1	2,1		
Cyclohexane		1			Methyl acrylate	2	1			R-11 (see below)		1		
Cyclohexanol	2	1			Methyl bromide (!)	2,1	1			Turpentine	2	1		
Cyclohexanone	2	1			Methyl butyl ketone (!)	1,2	2,1			Urea (!)	2	1,1		
Cyclohexene		1			Methyl cellosolve acetate	2	1			Uric acid (!)	1	1		2,2
Decane		1			Methylchloroform		1			Vinyl chloride		1		
Diborane	1				Methylcyclohexane		1			Xylene		1		
Dichlorobenzene		1			Methylene chloride		1							

1 = primary media selection for contaminant; 2 = secondary media selection.
PIA = permanganate-impregnated alumina; AC = activated carbon; AIC = acid-impregnated carbon; BIC = base-impregnated carbon.
R-114 is dichlorotetrafluoroethane; R-11 is trichlorofluoromethane.

Comments: Some contaminant molecules have isomers that, because they have different physical properties (boiling point, vapor pressures), require different treatment methods. For some contaminants, preferred treatment is ion exchange or another (nonlisted) impregnated carbon. For some contaminants, manufacturer recommendations differ. "!" is used to identify these cases.

Table 2.5 Example Generation of Gaseous Contaminants by Building Materials [2019A, Ch 47, Tbl 2]

Contaminant	CAS Number	Emission Factor Averages (ranges), $\mu g/(h \cdot m^2)$					
		Acoustic Ceiling Panels	Carpets	Fiberboards	Gypsum Boards	Paints on Gypsum Board	Particle Boards
4-Phenylcyclo-hexene (PCH)	4994-16-5		8.4 (n.d.-85)				
Acetaldehyde	75-07-0		2.8 (n.d.-37)	9.0 (n.d.-32)			28 (n.d.-55)
Acetic acid	64-19-7			8.4 (n.d.-26)			
Acetone	67-64-1	12 (n.d.-33)		35 (n.d.-67)	37 (n.d.-110)	35 (n.d.-120)	
Ethylene glycol	107-21-1			140 (n.d.-290)		19 (n.d.-190)	160 (140-200)
Formaldehyde	50-00-0	5.8 (n.d.-25)	3.6 (n.d.-41)	220 (n.d.-570)	6.8 (n.d.-19)		49 (n.d.-97)
Naphthalene	91-20-3		11 (n.d.-59)	3.0 (n.d.-8.2)			
n-Heptane	142-82-5			21 (n.d.-53)			
Nonanal	124-19-6	4.9 (1.7-11)	11 (n.d.-68)		10 (n.d.-28)	3.7 (n.d.-24)	
Toluene	108-88-3			19 (n.d.-46)			
TotalVOC (TVOC)*	N/A	32 (3.2-150)	1900 (270-9100)	400 (52-850)	15 (n.d.-61)	2500 (170-6200)	420 (240-510)

Table 2.5 Example Generation of Gaseous Contaminants by Building Materials [2019A, Ch 47, Tbl 2] ***(Continued)***

Contaminant	CAS Number	Emission Factor Averages (ranges), $\mu g/(h \cdot m^2)$					
		Plastic Laminates and Assemblies	Non-Rubber-Based Resilient Flooring	Rubber-Based Resilient Flooring	Tackable Wall Panels	Thermal Insulations	Wall Bases (Rubber-Based)
1,2,4-Trimethylbenzene	95-63-6			210 (n.d.-590)			
2-Butoxy-ethanol	111-76-2		2.7 (n.d.-24)	1.6 (n.d.-24)			
Acetaldehyde	75-07-0		11 (n.d.-49)				
Acetone	67-64-1	75 (4.8-150)	120 (n.d.-830)			12 (1.8-21)	220 (30-400)
Butyric acid	107-92-6		0.51 (n.d.-5.1)				
Dodecane	112-40-3			1.3 (n.d.-20)			
Ethylene glycol	107-21-1		38 (n.d.-210)				
Formaldehyde	50-00-0	13 (n.d.-29)	6.8 (n.d.-79)			5.9 (0.35-14)	32 (3.6-61)
Naphthalene	91-20-3		3.4 (n.d.-14)	5.6 (n.d.-28)	6.6 (6.6)		
n-Butanol	71-36-3						100 (n.d.-200)
Nonanal	124-19-6		5.7 (n.d.-19)	1.4 (n.d.-11)		1.8 (0.57-4)	
Octane	111-65-9						150 (n.d.-300)
Phenol	108-95-2	9.4 (4.4-19)	35 (n.d.-310)				340 (n.d.-680)
Toluene	108-88-3		5.1 (n.d.-12)				
Undecane	1120-21-4						140 (13-270)
TVOC*	N/A	160 (6.3-310)	680 (100-2100)	15000 (1500-100000)	270 (100-430)	7.5 (0.57-26)	7100 (1200-13000)

Source: CIWMB (2003).
n.d. = nondetectable N/A = not applicable
*TVOC concentrations calculated from total ion current (TIC) from GC/MS analysis by adding areas of integrated peaks with retention times greater than 5 min, subtracting from sum of area of internal standard chlorobenzene-d5, and using response factor of chlorobenzene-d5 as calibration.

Table 2.6 Type I Hood Requirements by Appliance Type [Std 154-2016, Tbl 1]

Appliance Description	Size	Type I Hoods[a]			
		Light Duty	Medium Duty	Heavy Duty	Extra-Heavy Duty
Braising pan/tilting skillet, electric	All	•			
Oven, baking, electric and gas	All	•			
Oven, rotisserie, electric and gas	All	•			
Oven, combination, electric and gas	All	•			
Oven, convection, full-size, electric and gas	All	•			
Oven, convection, half-size, electric and gas (protein cooking)	All	•			
Oven, conveyor, electric	All	•			
Oven, deck, electric and gas	All	•			
Oven, duck, electric and gas	All	•			
Oven, revolving rack, electric and gas	All	•			
Oven, rapid cook, electric	All	•			
Oven, roasting, electric and gas	All	•			
Oven, rotisserie, electric and gas	All	•			
Oven, stone hearth, gas	All	•			
Range, cook-top, induction	All	•			
Range, discrete element, electric (with or without oven)	All	•			
Salamander, electric and gas	All	•			
Braising pan/tilting skillet, gas	All		•		
Broiler, chain conveyor, electric	All		•		
Broiler, electric, underfired	All		•		
Fryer, doughnut, electric and gas	All		•		
Fryer, kettle, electric and gas	All		•		
Fryer, open deep-fat, electric and gas	All		•		
Fryer, pressure, electric and gas	All		•		
Griddle, double-sided, electric and gas	All		•		
Griddle, flat, electric and gas	All		•		
Oven, conveyor, gas	All		•		
Range, open burner, gas (with or without oven)	All		•		
Range, hot top, electric and gas	All		•		
Smoker, electric and gas	All		•		
Broiler, chain conveyor, gas	All			•	
Broiler, electric and gas, over-fired (upright)	All			•	
Broiler, gas, underfired	All			•	
Grill, plancha, electric and gas	All			•	
Oven, tandoor, gas	All			•	
Range, wok, gas and electric	All			•	
Oven, stone hearth, wood-fired or wood for flavoring	All				•
Solid fuel cooking appliances combusting a solid fuel (such as wood, charcoal, or coal) to provide all or part of the heat for the cooking process[b]	All				•

[a] Where recirculating systems or recirculating hoods are used, the additional heat and moisture loads generated by such appliances shall be accounted for in the sensible and latent loads for the HVAC system.

[b] Solid-fuel flavoring cooking appliances shall comply with this table as if they do not combust solid fuel.

Table 2.7 Type II Hood Requirements by Appliance Description
[Std 154-2016, Tbl 2]

Appliance Description	Size	Hood Not Required[a]	Type II Hoods[b]	
			Light Duty	Medium Duty
Cabinet, holding, electric	All	•		
Cabinet, proofing, electric	All	•		
Cheese-melter, electric	All	•		
Coffee maker, electric	All	•		
Cooktop, induction, electric	All	•		
Dishwasher, door-type rack, hot water sanitizing, heat recovery and vapor reduction, electric	All	•		
Dishwasher, door-type rack, chemical sanitizing, heat recovery and vapor reduction, electric	All	•		
Dishwasher, door-type dump and fill, hot water sanitizing, electric	All	•		
Dishwasher, door-type dump and fill, chemical sanitizing, electric	All	•		
Dishwasher, pot and pan, hot water sanitizing, heat recovery and vapor reduction, electric	All	•		
Dishwasher, powered sink, electric	All	•		
Dishwasher, under-counter, chemical sanitizing, electric	All	•		
Dishwasher, under-counter, electric	All	•		
Dishwasher, undercounter, hot water sanitizing, heat recovery and vapor reduction, electric	All	•		
Drawer warmer, 2 drawer, electric	All	•		
Egg cooker, electric	All	•		
Espresso machine, electric	All	•		
Grill, panini, electric	All	•		
Hot dog cooker, electric	All	•		
Hot plate, countertop, electric	All	•		
Ovens, microwave, electric	All	•		
Popcorn machine, electric	All	•		
Rethermalizer, electric	All	•		
Rice cooker, electric	All	•		
Steam table, electric	All	•		
Steamers, bun, electric	All	•		
Steamer, compartment atmospheric, countertop, electric	All	•		
Steamer, compartment pressurized, countertop, electric	All	•		
Table, hot food, electric	All	•		
Toaster, electric	All	•		
Waffle iron, electric	All	•		
Kettle, steam jacketed, tabletop, electric, gas and direct steam	<75 L		•	
Oven, convection, half-size, electric and gas (non-protein cooking)	All		•	

Table 2.7 Type II Hood Requirements by Appliance Description [Std 154-2016, Tbl 2] ***(Continued)***

Appliance Description	Size	Hood Not Required[a]	Type II Hoods[b]	
			Light Duty	Medium Duty
Pasta cooker, electric	All		•	
Rethermalizer, gas	All		•	
Rice cooker, gas	All		•	
Steamer, atmospheric, gas	All		•	
Steamer, pressurized, gas	All		•	
Steamer, atmospheric, floor-mounted, electric	All		•	
Steamer, pressurized, floor-mounted, electric	All		•	
Kettle, steam-jacketed floor mounted, electric, gas, and direct steam	<75 L		•	
Dishwasher, conveyor rack, chemical sanitizing	All			•
Dishwasher, conveyor rack, hot water sanitizing	All			•
Dishwasher, door-type rack, chemical sanitizing	All			•
Dishwasher, door-type rack, hot water sanitizing	All			•
Dishwasher, pot and pan, hot water sanitizing	All			•
Pasta cooker, gas	All			•
Steam-jacketed kettle, floor mounted, electric and gas	≥75 L			•

[a] Where hoods are not required, the additional heat and moisture loads generated by such appliances shall be accounted for in the sensible and latent loads for the HVAC system.

[b] Where recirculating systems or recirculating hoods are used, the additional heat and moisture loads generated by such appliances shall be accounted for in the sensible and latent loads for the HVAC system.

Table 2.8 Minimum Overhang Requirements for Type II Hoods [Std 154-2016, Tbl 3]

Type of Hood	End Overhang, mm	Front Overhang, mm	Rear Overhang, mm
Wall-mounted canopy	152	305	N/A
Single-island canopy	305	305	305
Double-island canopy	305	305	N/A
Eyebrow	N/A	305	N/A
Backshelf/proximity/pass-over	152	254 (setback)	N/A

N/A = Not Applicable

Table 2.9 Type II Hood Minimum Net Exhaust Airflow Rates [Std 154-2016, Tbl 4]

Type of Hood	Minimum Net Exhaust Flow Rate per Linear Hood Length, L/s/m	
	Light Duty Equipment	Medium Duty Equipment
Wall-mounted canopy	310	465
Single island	620	775
Double island	388	388
Eyebrow	388	388
Backshelf (pass-over)	310	465

3. WATER

Table 3.1 Common Pump Terms, Symbols, and Formulas

Term	Symbol	Units	Formula
Velocity	v	m/s	
Volume	V	m^3	
Flow rate	Q_v	m^3/s or L/s	
Pressure	p	kPa	
Density	ρ	kg/m^3	
Acceleration of gravity	g	$9.807\ m/s^2$	
Speed	n	rpm or rad	
Specific gravity	SG	—	$= \frac{\text{Mass of liquid}}{\text{Mass of water}}$
Head	H	m	$2.31\ p/SG$
Net positive suction head (NPSH)	H	m	
Efficiency (percent)			
Pump	η_p		
Electric motor	η_m		
Variable speed drive	η_v		
Equipment (constant-speed pumps)	η_e		$\eta_e = \eta_p \eta_m / 100$
Equipment (variable-speed pumps)	η_e		$\eta_e = 10^{-4} \eta_p \eta_m \eta_v$
Utilization	η_u		
Q_D = design flow Q_A = actual flow H_D = design head H_A = actual head			$\eta_u = 100 \frac{Q_D H_D}{Q_A H_A}$
System Efficiency Index (decimal)			$SEI = 10^{-4} \eta_\varepsilon \eta_u$
Output power (pump)	P_o	kW	$Q_v \rho m / 101$ (Q_v in L/s)
Shaft power	P_s	kW	$100 P_o / \eta_p$
Input power	P_i	kW	$100 P_s / \eta_m$

Water

The theoretical power to circulate water in a hydronic system is the **water power P_w** and is calculated as follows:

$$P_w = \dot{m}\ \Delta p/\rho \tag{3.1}$$

where

$\dot{m}$ = mass flow of fluid, kg/s
Δp = pressure increase, Pa
ρ = density of fluid, kg/m^3

Pump efficiency is determined by comparing the output power to the input power:

$$\text{Efficiency} = \frac{\text{Output}}{\text{Input}} = \frac{P_w}{P_t} \times 100\% \tag{3.2}$$

Table 3.2 Pump Affinity Laws [2020S, Ch 44, Tbl 1]

Function	Speed Change	Impeller Diameter Change
Flow	$Q_2 = Q_1 \left(\frac{N_2}{N_1}\right)$	$Q_2 = Q_1 \left(\frac{D_2}{D_1}\right)$
Pressure	$p_2 = p_1 \left(\frac{N_2}{N_1}\right)^2$	$p_2 = p_1 \left(\frac{D_2}{D_1}\right)^2$
Power	$P_2 = P_1 \left(\frac{N_2}{N_1}\right)^3$	$P_2 = P_1 \left(\frac{D_2}{D_1}\right)^3$

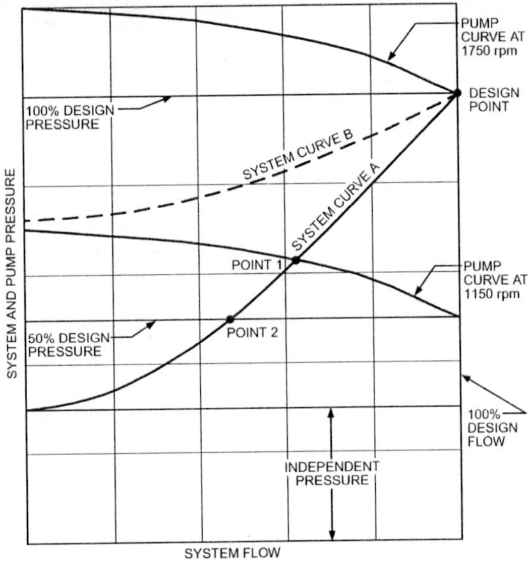

Figure 3.1 Example Application of Affinity Law [2020S, Ch 44, Fig 28]

If the hydronic system has a system head curve as shown in curve A, the pump at 1150 rpm will operate at point 1, not at point 2, as would be predicted by the affinity laws alone. If the hydronic system has a system head curve like curve B of Figure 3.1, the pump at 1150 rpm will run at shutoff head and deliver no water. This demonstrates that the affinity laws should be used to develop new pump head/capacity curves, but not to predict performance with a particular hydronic system unless its system head curve is known.

Net Positive Suction Characteristics

Particular attention must be given to the pressure and temperature of the water as it enters the pump, especially in condenser towers, steam condensate returns, and steam boiler feeds.

If the absolute pressure at the suction nozzle approaches the vapor pressure of the liquid, vapor pockets form in the impeller passages. The collapse of the vapor pockets (cavitation) is noisy and can be destructive to the pump impeller.

The pressure in excess of that required to prevent vapor pockets from forming is the net positive suction head required (NPSHR). NPSHR is a characteristic of a given pump and varies with pump speed and flow. It is determined by the manufacturer and is included on the pump performance curve.

NPSHR is particularly important when a pump is operating with hot liquids or is applied to a circuit having a suction lift. The vapor pressure increases with water temperature and reduces the net positive suction head available (NPSHA). Each pump has its NPSHR, and the installation has its NPSHA, which is the total useful energy above the vapor pressure at the pump inlet (see Figure 3.2).

$$\text{NPSHA} = p_p + p_z - p_{vpa} - p_f \tag{3.3}$$

where

p_p = absolute pressure on surface of liquid that enters pump, Pa
p_z = static elevation of liquid above center line of pump, Pa
p_{vpa} = absolute vapor pressure at pumping temperature, Pa
p_f = friction and head losses in suction piping, Pa

To determine the NPSHA in an existing installation, Equation 3.4 may be used (see Figure 3.2):

$$\text{NPSHA} = p_a + p_s + \frac{V^2\rho}{2} - p_{vpa} \tag{3.4}$$

where

p_a = atmospheric head for elevation of installation, Pa
p_s = head at inlet flange corrected to center line of pump (h_s is negative if below atmospheric pressure), Pa
$V^2\rho/2$ = velocity head at point of measurement of h_s, Pa
ρ = density of fluid, kg/m^3

If the NPSHA is less than the pump's NPSHR, cavitation, noise, inadequate pumping, and mechanical problems will result. **For trouble-free design, the NPSHA must always be greater than the pump's NPSHR.** In closed hot- and chilled-water systems where sufficient system fill pressure is exerted on the pump suction, NPSHR is normally not a factor.

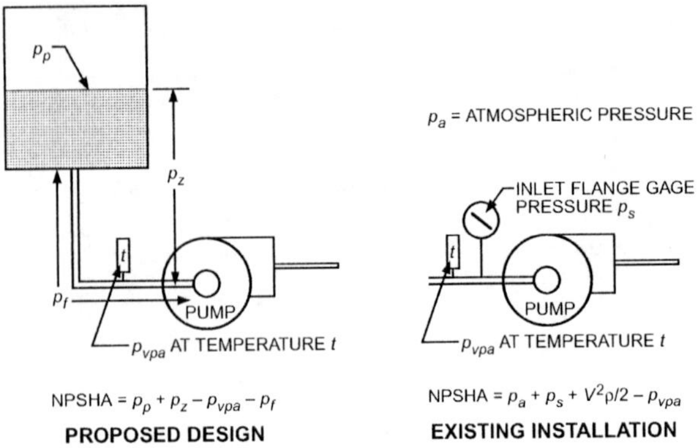

Figure 3.2 Net Positive Suction Head Available [2020S, Ch 44, Fig 33]

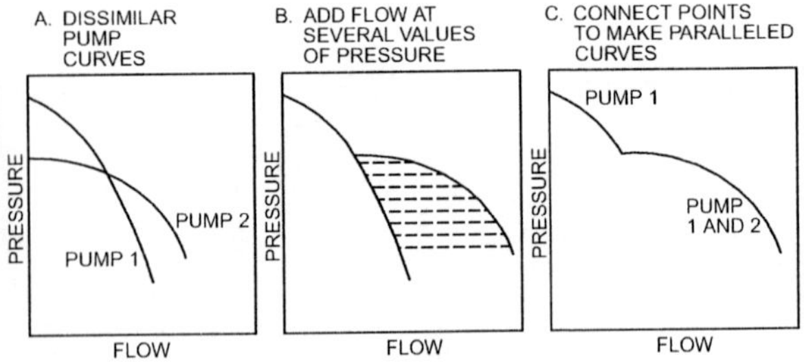

Figure 3.3 Construction of Curve for Dissimilar Parallel Pumps [2020S, Ch 44, Fig 38]

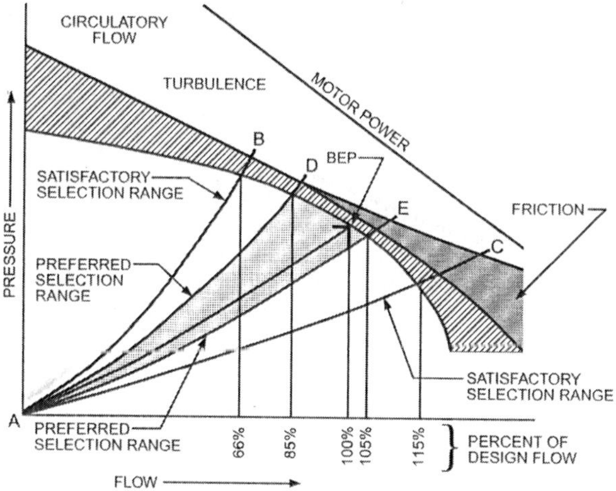

Figure 3.4 Pump Selection Regions [2020S, Ch 44, Fig 35]

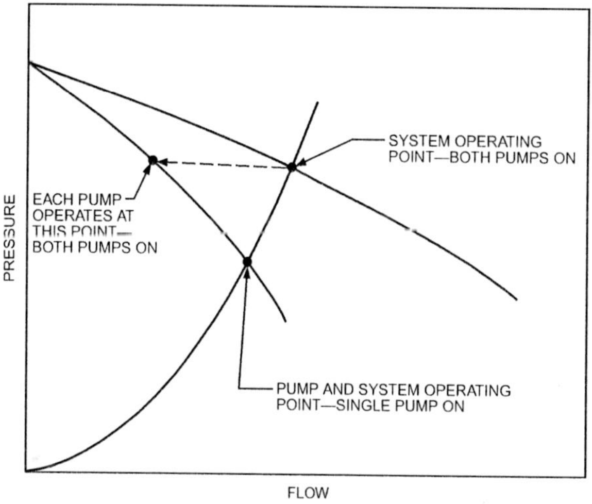

Figure 3.5 Operating Conditions for Parallel Operation [2020S, Ch 44, Fig 37]

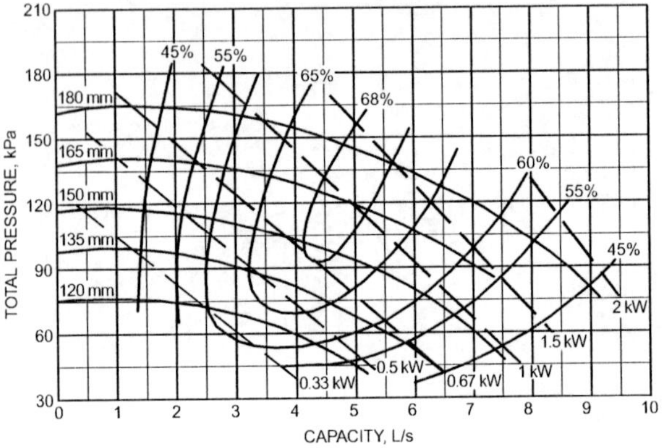

Figure 3.6 Typical Pump Performance Curves [2020S, Ch 44, Fig 13]

Table 3.3 Properties of Water, 100°C to 210°C
[Adapted from 2020S, Ch 15, Tbl 1]

Temperature, °C	Absolute Pressure, kPa*	Density, kg/m³	Specific Heat, kJ/(kg·K)
100	101.3	958.1	4.22
110	143.3	950.7	4.23
120	198.5	942.8	4.25
130	270.1	934.6	4.27
140	361.4	925.9	4.29
150	476.0	916.9	4.32
160	618.1	907.0	4.35
170	792.0	897.4	4.38
180	1002.7	887.0	4.42
190	1255.1	875.8	4.46
200	1554.9	864.5	4.51
210	1907.7	853.0	4.56

**Source*: *Thermodynamic Properties of Steam*, J.H. Keenan and F.G. Keyes, John Wiley & Sons, 1936.

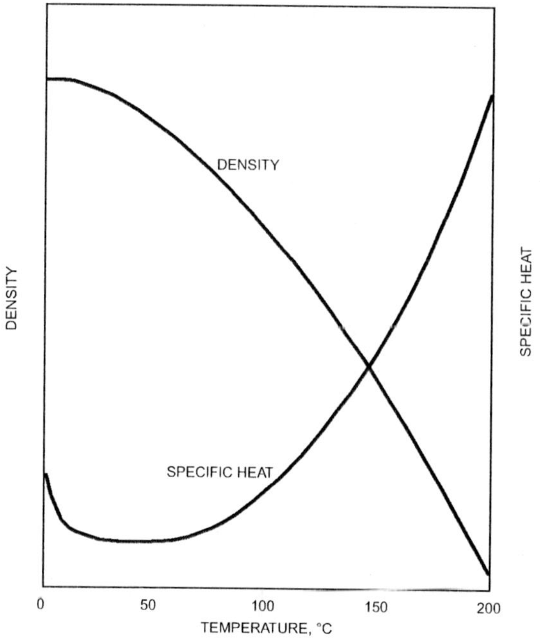

Figure 3.7 Density and Specific Heat of Water [2020S, Ch 15, Fig 3]

Table 3.4 Freezing Points for Solutions of Ethylene Glycol and Propylene Glycol

Glycol, % by mass	Ethylene Glycol °C	Propylene Glycol °C
10	–3.2	–1.3
15	–5.4	–5.1
20	–7.8	–7.1
25	–10.7	–9.6
30	–14.1	–12.7
40	–22.3	–21.1
50	–33.8	–33.5
60	–48.3	–51.1

Ethylene glycol solutions are less viscous than propylene glycol solutions at the same concentration. Less toxic propylene glycol is preferred for applications involving possible human contact.

Table 3.5 Volume of Cylindrical Tank per Metre of Depth

Diameter, m	Volume, m^3	Diameter, m	Volume, m^3
0.1	0.0079	3.1	7.548
0.3	0.0707	303	8.553
0.5	0.1963	3.5	9.621
0.7	0.3848	3.7	10.75
0.9	0.6362	3.9	11.95
1.1	0.9503	4.1	13.20
1.3	1.327	4.3	14.52
1.5	1.767	4.5	15.90
1.7	2.270	4.7	17.35
1.9	2.835	4.9	18.86
2.1	3.464	5.1	20.43
2.3	4.155	5.3	22.06
2.5	4.909	5.5	23.76
2.7	5.726	5.7	25.52
2.9	6.605	5.9	27.34

Volume per metre of depth = $\pi\,(\text{Diameter}/2)^2$

Table 3.6 Quantities for Various Depths of Vertical Cylindrical Tanks in Horizontal Position

% Depth Filled	% of Capacity	% Depth Filled	% of Capacity	% Depth Filled	% of Capacity	% Depth Filled	% of Capacity
1	.20	26	20.73	51	51.27	76	81.50
3	.90	28	23.00	53	53.81	78	83.68
5	1.87	30	25.31	55	56.34	80	85.77
7	3.07	32	27.66	57	58.86	82	87.76
9	4.45	34	30.03	59	61.36	84	89.68
11	5.98	36	32.44	61	63.86	86	91.50
13	7.64	38	34.90	63	66.34	88	93.20
15	9.40	40	37.36	65	68.81	90	94.80
17	11.27	42	39.89	67	71.16	92	96.26
19	13.23	44	42.40	69	73.52	94	97.55
21	15.26	46	44.92	71	75.93	96	98.66
23	17.40	48	47.45	73	78.14	98	99.50
25	19.61	50	50.00	75	80.39	100	100.0

Table 3.7 Volume of Water in Standard Pipe and Tube

Nominal Pipe Size, mm	Schedule No.	Standard Steel Pipe Inside Diameter, mm	Standard Steel Pipe Volume, L/m	Type *L* Copper Tube Inside Diameter, mm	Type *L* Copper Tube Volume, L/m
(10)	—	—	—	(10.9)	(0.09)
(15)	40	(15.8)	(0.19)	(13.8)	(0.15)
(16)	—	—	—	(16.9)	(0.22)
(20)	40	(20.9)	(0.34)	(19.9)	(0.31)
(25)	40	(26.6)	(0.56)	(26.0)	(0.53)
(32)	40	(35.0)	(0.97)	(32.1)	(0.81)
(40)	40	(40.9)	(1.32)	(38.2)	(1.15)
(50)	40	(52.5)	(2.16)	(50.4)	(2.00)
(65)	40	(62.7)	(3.09)	(62.6)	(3.08)
(80)	40	(77.9)	(4.77)	(74.8)	(4.40)
(90)	40	(90.1)	(6.38)	(87.0)	(5.95)
(100)	40	(102.3)	(8.21)	(99.2)	(7.73)
(125)	40	(128.2)	(12.92)	(123.8)	(12.05)
(150)	40	(154.1)	(18.63)	(148.5)	(17.26)
(200)	30	(205.0)	(33.03)	(196.2)	(30.18)
(250)	30	(257.5)	(52.04)	(244.5)	(46.95)
(300)	30	(307.1)	(74.02)	(293.8)	(67.81)

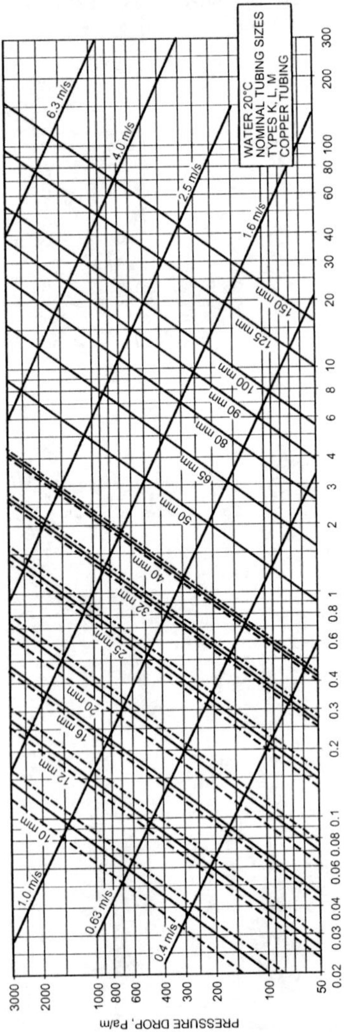

Figure 3.8 Friction Loss for Water in Copper Tubing (Types K, L, M)
[2021F, Ch 22, Fig 15]

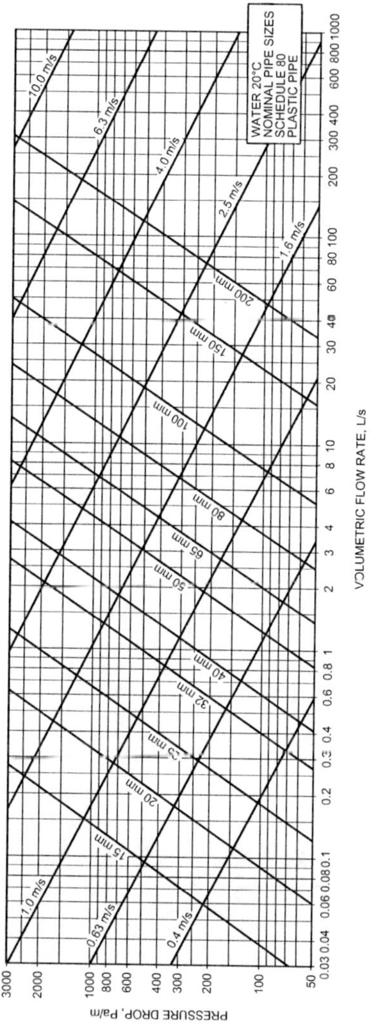

Figure 3.9 Friction Loss for Water in Plastic Pipe (Schedule 80)
[2021F, Ch 22, Fig 16]

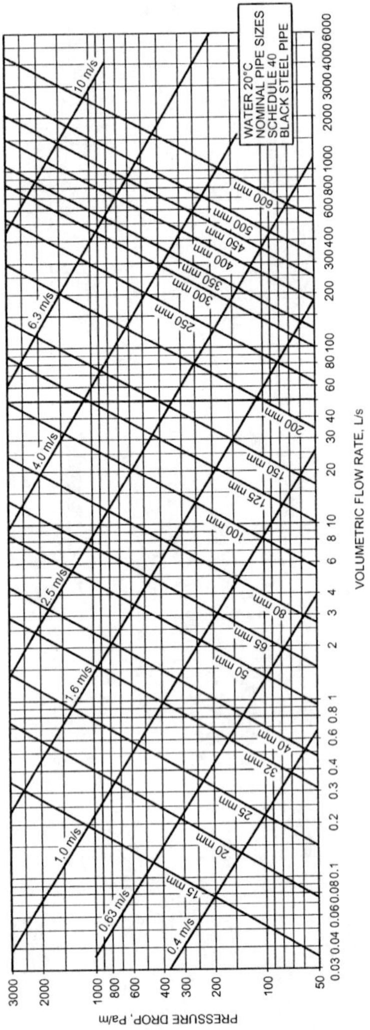

Figure 3.10 Friction Loss for Water in Commercial Steel Pipe (Schedule 40)
[2021F, Ch 22, Fig 14]

Valve and Fitting Losses

Valves and fittings cause pressure losses greater than those caused by the pipe alone. One formulation expresses losses as

$$\Delta p = K\rho\left(\frac{V^2}{2}\right) \quad \text{or} \quad \Delta h = K\left(\frac{V^2}{2g}\right) \tag{3.5}$$

where K = geometry- and size-dependent loss coefficient (see Tables 3.8 to 3.13).

ASHRAE research project RP-1193 found that, in general, PVC fitting geometry varied much more from one manufacturer to another than steel fittings did.

Calculating Pressure Losses

The most common engineering design flow loss calculation selects a pipe size for the desired total flow rate and available or allowable pressure drop.

Because formulation of fitting losses requires a known diameter, pipe size must be selected before calculating the detailed influence of fittings. A frequently used rule of thumb assumes that the design length of pipe is 50% to 100% longer than actual to account for fitting losses. After a pipe diameter has been selected on this basis, the influence of each fitting can be evaluated.

Table 3.8 ***K*** **Factors: Threaded Pipe Fittings** [2021F, Ch 22, Tbl 3]

Nominal Pipe Dia., mm	90° Ell Reg.	90° Ell Long	45° Ell	Return Bend	Tee-Line	Tee-Branch	Globe Valve	Gate Valve	Angle Valve	Swing Check Valve	Bell Mouth Inlet	Square Inlet	Projected Inlet
10	2.5	—	0.38	2.5	0.90	2.7	20	0.40	—	8.0	0.05	0.5	1.0
15	2.1	—	0.37	2.1	0.90	2.4	14	0.33	—	5.5	0.05	0.5	1.0
20	1.7	0.92	0.35	1.7	0.90	2.1	10	0.28	6.1	3.7	0.05	0.5	1.0
25	1.5	0.78	0.34	1.5	0.90	1.8	9	0.24	4.6	3.0	0.05	0.5	1.0
32	1.3	0.65	0.33	1.3	0.90	1.7	8.5	0.22	3.6	2.7	0.05	0.5	1.0
40	1.2	0.54	0.32	1.2	0.90	1.6	8	0.19	2.9	2.5	0.05	0.5	1.0
50	1.0	0.42	0.31	1.0	0.90	1.4	7	0.17	2.1	2.3	0.05	0.5	1.0
65	0.85	0.35	0.30	0.85	0.90	1.3	6.5	0.16	1.6	2.2	0.05	0.5	1.0
80	0.80	0.31	0.29	0.80	0.90	1.2	6	0.14	1.3	2.1	0.05	0.5	1.0
100	0.70	0.24	0.28	0.70	0.90	1.1	5.7	0.12	1.0	2.0	0.05	0.5	1.0

Source: *Engineering Data Book* (Hydraulic Institute 1990).

Table 3.9 *K* Factors: Flanged Welded Pipe Fittings [2021F, Ch 22, Tbl 4]

Nominal Pipe Dia., mm	90° Ell Reg.	90° Ell Long	45° Ell Long	Return Bend Standard	Return Bend Long-Radius	Tee-Line	Tee-Branch	Globe Valve	Gate Valve	Angle Valve	Swing Check Valve
25	0.43	0.41	0.22	0.43	0.43	0.26	1.0	13	—	4.8	2.0
32	0.41	0.37	0.22	0.41	0.38	0.25	0.95	12	—	3.7	2.0
40	0.40	0.35	0.21	0.40	0.35	0.23	0.90	10	—	3.0	2.0
50	0.38	0.30	0.20	0.38	0.30	0.20	0.84	9	0.34	2.5	2.0
65	0.35	0.28	0.19	0.35	0.27	0.18	0.79	8	0.27	2.3	2.0
80	0.34	0.25	0.18	0.34	0.25	0.17	0.76	7	0.22	2.2	2.0
100	0.31	0.22	0.18	0.31	0.22	0.15	0.70	6.5	0.16	2.1	2.0
150	0.29	0.18	0.17	0.29	0.18	0.12	0.62	6	0.10	2.1	2.0
200	0.27	0.16	0.17	0.27	0.15	0.10	0.58	5.7	0.08	2.1	2.0
250	0.25	0.14	0.16	0.25	0.14	0.09	0.53	5.7	0.06	2.1	2.0
300	0.24	0.13	0.16	0.24	0.13	0.08	0.50	5.7	0.05	2.1	2.0

Source: *Engineering Data Book* (Hydraulic Institute 1990).

Table 3.10 **Summary of *K* Values for Steel Ells, Reducers, and Expansions**
[Adapted from 2021F, Ch 22, Tbl 6]

	ASHRAE Research[a,b]		
	1.2 m/s	**2.4 m/s**	**3.6 m/s**
50 mm S.R.[c] ell (*R*/*D* = 1) thread	0.60	0.68	0.736
100 mm S.R. ell (*R*/*D* = 1) weld	0.37	0.34	0.33
25 mm L.R. ell (*R*/*D* = 1.5) weld	—	—	—
50 mm L.R. ell (*R*/*D* = 1.5) weld	—	—	—
100 mm L.R. ell (*R*/*D* = 1.5) weld	0.26	0.24	0.23
150 mm L.R. ell (*R*/*D* = 1.5) weld	0.26	0.24	0.24
200 mm L.R. ell (*R*/*D* = 1.5) weld	0.22	0.20	0.19
250 mm L.R. ell (*R*/*D* = 1.5) weld	0.21	0.17	0.16
300 mm L.R. ell (*R*/*D* = 1.5) weld	0.17	0.17	0.17
400 mm L.R. ell (*R*/*D* = 1.5) weld	0.12	0.12	0.11
500 mm L.R. ell (*R*/*D* = 1.5) weld	0.12	0.10	0.10
600 mm L.R. ell (*R*/*D* = 1.5) weld	0.098	0.089	0.089
Reducer (50 by 40 mm) thread	0.53	0.28	0.20
(100 by 80 mm) weld	0.23	0.14	0.10
(150 by 100 mm) weld	0.62	0.54	0.53
(200 by 150 mm) weld	0.31	0.28	0.26
(250 by 200 mm) weld	0.16	0.14	0.14
(300 by 250 mm) weld	0.14	0.14	0.14
(400 by 300 mm) weld	0.17	0.16	0.17
(500 by 400 mm) weld	0.16	0.13	0.13
(600 by 500 mm) weld	0.053	0.053	0.055
Expansion (40 by 50 mm) thread	0.16	0.13	0.02
(80 by 100 mm) weld	0.11	0.11	0.11
(100 by 150 mm) weld	0.28	0.28	0.29
(150 by 200 mm) weld	0.15	0.12	0.11
(200 by 250 mm) weld	0.11	0.09	0.08
(250 by 300 mm) weld	0.11	0.11	0.11
(300 by 400 mm) weld	0.073	0.076	0.073
(400 by 500 mm) weld	0.024	0.021	0.022
(500 by 600 mm) weld	0.020	0.023	0.020

Source: Rahmeyer (2003a).
[b]Rahmeyer (1999a, 2002a).
[c] S.R.—short radius or regular ell; L.R.—long-radius ell.

Table 3.11 Summary of *K* Factors for Pipe Tees [2021F, Ch 22, Tbl 7]

		ASHRAE Research[a,b]		
		1.2 m/s	2.4 m/s	3.6 m/s
50 mm thread tee,	100% branch	0.93	—	—
	100% line (flow-through)	0.19	—	—
	100% mix	1.19	—	
100 mm weld tee,	100% branch	—	0.57	—
	100% line (flow-through)	—	0.06	—
	100% mix	—	0.49	—
150 mm weld tee,	100% branch	—	0.56	—
	100% line (flow-through)	—	0.12	—
	100% mix	—	0.88	—
200 mm weld tee,	100% branch	—	0.53	—
	100% line (flow-through)	—	0.08	—
	100% mix	—	0.70	—
250 mm weld tee,	100% branch	—	0.52	—
	100% line (flow-through)	—	0.06	—
	100% mix	—	0.77	—
300 mm weld tee,	100% branch	0.70	0.63	0.62
	100% line (flow-through)	0.062	0.091	0.096
	100% mix	0.88	0.72	0.72
400 mm weld tee,	100% branch	0.54	0.55	0.54
	100% line (flow-through)	0.032	0.028	0.028
	100% mix	0.74	0.74	0.76

[a]Rahmeyer (1999b, 2002b).
[b]Ding et al. (2005).

Table 3.12 Test Summary for Loss Coefficients *K* and Equivalent Loss Lengths [2021F, Ch 22, Tbl 8]

Schedule 80 PVC Fitting		*K*	*L*, m
Injected molded elbow,	50 mm	0.91 to 1.00	2.6 to 2.8
	100 mm	0.86 to 0.91	5.6 to 5.9
	150 mm	0.76 to 0.91	8.0 to 9.5
	200 mm	0.68 to 0.87	10.0 to 12.8
200 mm fabricated elbow, Type I, components		0.40 to 0.42	5.9 to 6.2
Type II, mitered		0.073 to 0.76	10.8 to 11.2
150 by 100 mm injected molded reducer		0.12 to 0.59	1.2 to 6.2
Bushing type		0.49 to 0.59	5.2 to 6.2
200 by 150 mm injected molded reducer		0.13 to 0.63	1.9 to 9.3
Bushing type		0.48 to 0.68	7.1 to 10.0
Gradual reducer type		0.21	3.1
100 by 150 mm injected molded expansion		0.069 to 1.19	0.46 to 7.7
Bushing type		0.069 to 1.14	0.46 to 7.4
150 by 200 mm injected molded expansion		0.95 to 0.96	10.0 to 10.1
Bushing type		0.94 to 0.95	9.9 to 10.0
Gradual reducer type		0.99	10.4

Table 3.13 Test Summary for Loss Coefficients *K* of PVC Tees
[2021F, Ch 22, Tbl 9]

Branching		
Schedule 80 PVC Fitting	K_{1-2}	K_{1-3}
50 mm injection molded branching tee, 100% line flow	0.13 to 0.26	—
50/50 flow	0 to 0.12	0.74 to 1.02
100% branch flow	—	0.98 to 1.39
100 mm injection molded branching tee, 100% line flow	0.07 to 0.22	—
50/50 flow	0.03 to 0.13	0.74 to 0.82
100% branch flow	—	0.97 to 1.12
150 mm injection molded branching tee, 100% line flow	0.01 to 0.14	—
50/50 flow	0.06 to 0.11	0.70 to 0.84
100% branch flow	—	0.95 to 1.15
150 mm fabricated branching tee, 100% line flow	0.21 to 0.22	—
50/50 flow	0.04 to 0.09	1.29 to 1.40
100% branch flow	—	1.74 to 1.88
200 mm injection molded branching tee, 100% line flow	0.04 to 0.09	—
50/50 flow	0.04 to 0.07	0.64 to 0.75
100% branch flow	—	0.85 to 0.96
200 mm fabricated branching tee, 100% line flow	0.09 to 0.16	—
50/50 flow	0.08 to 0.13	1.07 to 1.16
100% branch flow	—	1.40 to 1.62
Mixing		
PVC Fitting	K_{1-2}	K_{3-2}
50 mm injection molded mixing tee, 100% line flow	0.12 to 0.25	—
50/50 flow	1.22 to 1.19	0.89 to 1.88
100% mix flow	—	0.89 to 1.54
100 mm injection molded mixing tee, 100% line flow	0.07 to 0.18	—
50/50 flow	1.19 to 1.88	0.98 to 1.88
100% mix flow	—	0.88 to 1.02
150 mm injection molded mixing tee, 100% line flow	0.06 to 0.14	—
50/50 flow	1.26 to 1.80	1.02 to 1.60
100% mix flow	—	0.90 to 1.07
150 mm fabricated mixing tee, 100% line flow	0.19 to 0.21	—
50/50 flow	2.94 to 3.32	2.57 to 3.17
100% mix flow	—	1.72 to 1.98
200 mm injection molded mixing tee, 100% line flow	0.04 to 0.09	—
50/50 flow	1.10 to 1.60	0.96 to 1.32
100% mix flow	—	0.81 to 0.93
200 mm fabricated mixing tee, 100% line flow	0.13 to 0.70	—
50/50 flow	2.36 to 10.62	2.02 to 2.67
100% mix flow	—	1.34 to 1.53

Coefficients based on average velocity of 2.43 m/s. Range of values varies with fitting manufacturers. Line or straight flow is Q_2/Q_1 = 100%. Branch flow is Q_2/Q_1 = 0%.

4. STEAM

Table 4.1 Properties of Saturated Steam[a]

Absolute Pressure p, kPa	Temperature t_s, °C	Specific Enthalpy, kJ/kg — In Saturated Liquid h_f	Specific Enthalpy, kJ/kg — Latent Heat of Evaporation h_{fg}	Specific Enthalpy, kJ/kg — In Saturated Vapor h_g	Spec. Vol. v, m^3/kg
1	6.98	29.3	2484.3	2513.6	129.205
2	17.51	73.5	2459.5	2533.0	67.010
4	28.98	121.4	2432.4	2553.9	34.805
6	36.18	151.5	2415.3	2566.8	23.742
8	41.53	173.9	2404.5	2576.4	18.104
10	45.83	191.8	2392.2	2584.1	14.673
20	60.09	251.5	2357.7	2609.1	7.648
30	69.13	289.3	2335.4	2624.8	5.228
40	75.89	317.7	2318.6	2636.3	3.992
50	81.35	340.6	2304.9	2645.4	3.239
60	85.95	359.9	2293.2	2653.1	2.731
70	89.96	376.8	2282.9	2659.7	2.364
80	93.51	391.7	2273.7	2665.4	2.087
90	96.71	405.2	2265.4	2670.6	1.869
100	99.63	417.5	2257.7	2675.2	1.694
110	102.32	428.8	2250.6	2679.5	1.549
120	104.81	439.3	2244.0	2683.3	1.428
130	107.13	449.2	2237.8	2686.9	1.325
140	109.32	458.4	2231.9	2690.3	1.237
150	11.37	467.5	2226.3	2693.4	1.159
160	113.32	475.4	2221.0	2696.4	1.091
170	115.17	483.2	2215.9	2699.1	1.031
180	116.93	490.7	2211.1	2701.8	0.977
190	118.62	497.9	2206.4	2704.2	0.929
200	120.23	504.7	2201.9	2706.6	0.886
210	121.78	511.3	2197.6	2708.9	0.846
220	123.27	517.6	2193.4	2711.0	0.810
230	124.71	523.7	2189.3	2713.1	0.777
240	126.09	529.6	2185.4	2715.0	0.747
250	127.43	535.4	2181.6	2716.9	0.719
260	128.73	540.9	2177.8	2718.7	0.693
270	129.99	546.2	2174.2	2720.5	0.669
280	131.21	551.5	2170.7	2722.2	0.646
290	132.39	556.5	2167.3	2723.8	0.625
300	133.54	561.4	2163.9	2735.4	0.606
310	134.66	566.2	2160.6	2726.9	0.587
320	135.76	570.9	2157.4	2728.4	0.570
330	136.82	575.5	2154.3	2729.8	0.554
340	137.86	579.9	2151.2	2731.1	0.539
350	138.88	584.3	2148.2	2732.5	0.524
360	139.87	588.5	2145.2	2733.8	0.510
370	140.84	592.7	2142.3	2735.0	0.497
380	141.79	596.8	2139.5	2736.3	0.485
390	142.72	600.8	2136.7	2737.5	0.473
400	143.63	604.7	2133.9	2738.6	0.462

Table 4.1 Properties of Saturated Steam[a] *(Continued)*

Absolute Pressure p, kPa	Temperature t_s, °C	Specific Enthalpy, kJ/kg In Saturated Liquid h_f	Latent Heat of Evaporation h_{fg}	In Saturated Vapor h_g	Spec. Vol. v, m^3/kg
410	144.52	608.5	2131.2	2739.8	0.452
420	145.39	612.3	2138.6	2740.9	0.442
430	146.25	616.0	2125.9	2741.9	0.432
440	147.09	619.6	2123.4	2743.0	0.423
450	147.92	623.2	2120.8	2744.0	0.414
460	148.73	626.7	2118.2	2745.0	0.405
470	149.53	630.1	2115.8	2746.0	0.397
480	150.31	633.5	2113.4	2746.9	0.389
490	151.09	636.8	2111.0	2747.8	0.382
500	151.85	640.1	2108.6	2748.7	0.375
520	153.33	646.5	2104.0	2750.5	0.361
540	154.77	652.8	2099.4	2752.2	0.349
560	156.16	658.8	2095.0	2753.8	0.337
580	157.52	664.7	2090.7	2755.4	0.326
600	158.84	670.4	2086.4	2756.8	0.316
620	160.12	676.0	2082.3	2758.3	0.306
640	161.38	681.5	2078.2	2759.9	0.297
660	162.60	686.8	2074.2	2761.0	0.288
680	163.79	692.0	2070.3	2762.3	0.280
700	164.96	697.1	2066.4	2763.5	0.273
720	166.10	702.0	2062.7	2764.7	0.266
740	167.21	706.9	2058.9	2765.8	0.259
760	168.30	711.7	2055.3	2767.0	0.252
780	169.37	716.4	2051.7	2768.0	0.246
800	170.41	720.9	2048.2	2769.1	0.240
820	171.44	725.4	2044.7	2770.1	0.235
840	172.45	729.8	2041.2	2771.1	0.229
860	173.43	734.2	2037.8	2772.0	0.224
880	174.40	738.4	2034.5	2772.9	0.220
900	175.36	742.6	2031.2	2773.8	0.215
920	176.29	746.8	2028.0	2774.7	0.210
940	177.21	750.8	2024.7	2775.6	0.206
960	178.12	754.8	2021.6	2776.4	0.202
980	179.01	758.7	2018.4	2777.2	0.198
1000	179.88	762.6	2015.3	2777.9	0.194
1100	184.06	781.1	2000.4	2781.5	0.177
1200	187.96	798.4	1986.2	2784.6	0.163
1300	191.60	814.7	1972.6	2787.3	0.151
1400	195.04	830.0	1959.6	2789.7	0.141
1500	198.28	844.6	1947.1	2791.8	0.132
1600	201.37	858.5	1935.1	2793.6	0.124
1700	204.30	871.8	1923.4	2795.2	0.117
1800	207.10	884.5	1912.1	2796.6	0.110
1900	209.79	896.8	1901.1	2797.8	0.105
2000	212.37	908.6	1890.4	2798.9	0.100

a. Reprinted by permission of CIBSE.

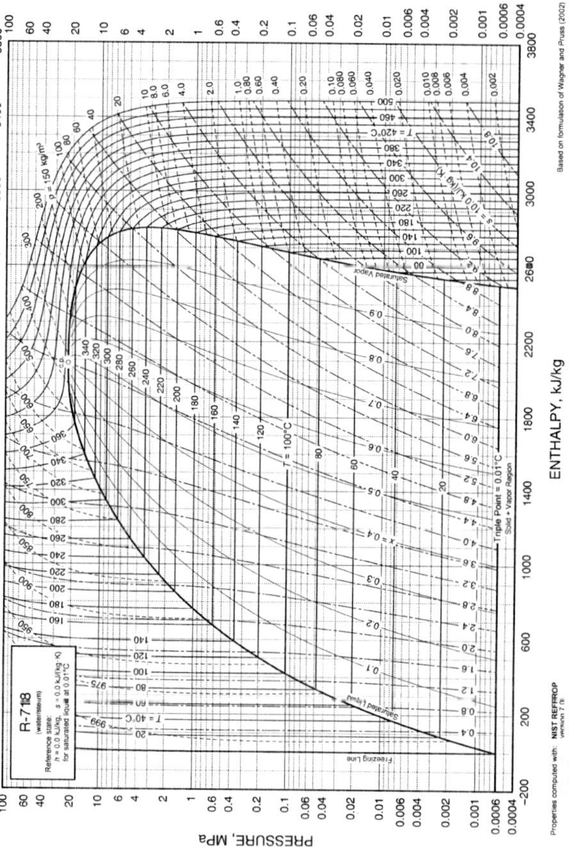

Figure 4.1 Pressure-Enthalpy Diagram for Refrigerant 718 (Water/Steam)
[2021F, Ch 30, Fig 20]

Table 4.2 Flow Rate of Steam in Schedule 40 Pipe [2021F, Ch 22, Tbl 32]

Nominal Pipe Size, mm	Pressure Drop, Pa/m													
	14 Pa/m		28 Pa/m		58 Pa/m		113 Pa/m		170 Pa/m		225 Pa/m		450 Pa/m	
	Sat. Press., kPa		Sat. Press., kPa		Sat. Press., kPa		Sat. Press., kPa		Sat. Press., kPa		Sat. Press., kPa		Sat. Press., kPa	
	25	85	25	85	25	85	25	85	25	85	25	85	25	85
20	1.1	1.4	1.8	2.0	2.5	3.0	3.7	4.4	4.5	5.4	5.3	6.3	7.6	9.2
25	2.1	2.6	3.3	3.9	4.7	5.8	6.8	8.3	8.6	10	10	12	14	17
32	4.5	5.7	6.7	8.3	9.8	12	14	17	18	21	20	25	29	35
40	7.1	8.8	11	13	15	19	22	26	27	33	31	38	45	54
50	14	17	20	24	29	36	42	52	53	64	60	74	89	107
65	22	27	33	39	48	58	68	83	86	103	98	120	145	173
80	40	48	59	69	83	102	121	146	150	180	174	210	246	302
90	58	69	84	101	125	153	178	214	219	265	252	305	372	435
100	81	101	120	146	178	213	249	302	309	378	363	436	529	617
125	151	180	212	265	307	378	450	536	552	662	643	769	945	1 080
150	242	290	355	422	499	611	718	857	882	1 080	1 060	1 260	1 500	1 790
200	491	605	702	882	1 020	1 260	1 440	1 800	1 830	2 230	2 080	2 580	3 020	3 720
250	907	1 110	1 290	1 590	1 890	2 290	2 650	3 280	3 300	4 030	3 780	4 660	5 380	6 550
300	1 440	1 730	2 080	2 460	2 950	3 580	4 160	5 040	5 170	6 240	6 050	7 250	8 540	10 200

Notes:
1. Flow rate is in g/s at initial saturation pressures of 25 and 85 kPa (gage). Flow is based on Moody friction factor, where the flow of condensate does not inhibit the flow of steam.
2. The flow rates at 25 kPa cover saturated pressure from 7 to 41 kPa, and the rates at 85 kPa cover saturated pressure from 55 to 110 kPa with an error not exceeding 8%.
3. The steam velocities corresponding to the flow rates given in this table can be found from Figures 18 and 19 of Chapter 22 of the 2021 *ASHRAE Handbook—Fundamentals*.

Table 4.3 Steam Pipe Capacities for Low-Pressure Systems
[2021F, Ch 22, Tbl 33]

Nominal Pipe Size, mm	Capacity, g/s				
	Two-Pipe System		One-Pipe Systems		
	Condensate Flowing Against Steam		Supply Risers Upfeed	Radiator Valves and Vertical Connections	Radiator and Riser Runouts
	Vertical	Horizontal			
A	B[a]	C[b]	D[c]	E	F[b]
20	1.0	0.9	0.8	—	0.9
25	1.8	1.8	1.4	0.9	0.9
32	3.9	3.4	2.5	2.0	2.0
40	6.0	5.3	4.8	2.9	2.0
50	12	11	9.1	5.3	2.9
65	20	17	14	—	5.3
80	36	25	25	—	8.2
90	49	36	36	—	15
100	64	54	48	—	23
125	132	99	—	—	35
150	227	176	—	—	69
200	472	378	—	—	—
250	882	718	—	—	—
300	1450	1200	—	—	—
400	2770	2390	—	—	—

Notes:
1. For one- or two-pipe systems in which condensate flows against steam flow.
2. Steam at average pressure of 7 kPa (gage) used as basis of calculating capacities.

[a] Do not use column B for pressure drops of less than 13 Pa per metre of equivalent run. Use Figure 18 or Table 33 of Chapter 22 of 2021 *ASHRAE Handbook—Fundamentals* instead.

[b] Pitch of horizontal runouts to risers and radiators should be not less than 40 mm/m. Where this pitch cannot be obtained, runouts over 2.5 m in length should be one pipe size larger than that called for in this table.

[c] Do not use column D for pressure drops of less than 9 Pa per metre of equivalent run, except on sizes 80 mm and over. Use Figure 18 or Table 33 of Chapter 22 of 2021 *ASHRAE Handbook—Fundamentals* instead.

Table 4.4 Return Main and Riser Capacities for Low-Pressure Systems, kg/h [2021F, Ch 22, Tbl 34]

	Pipe Size, mm	Pressure Drop																	
		7 Pa/m			9 Pa/m			14 Pa/m			28 Pa/m			57 Pa/m			113 Pa/m		
		Wet	Dry	Vac.	Wet	Dry	Vac.	Wet	Dry	Vac.	Wet	Dry	Vac.	Wet	Dry	Vac.	Wet	Dry	Vac.
	G	H	I	J	K	L	M	N	O	P	Q	R	S	T	U	V	W	X	Y
Return Main	20						19			45			64			91			128
	25	57	28		66	32	65	79	36	79	113	47	113	159	52	159			224
	32	97	59		112	68	111	136	76	136	193	98	193	272	109	272			385
	40	153	93		178	107	176	215	120	215	306	154	306	431	171	431			608
	50	318	213		367	243	370	454	261	454	635	336	644	907	374	907			1280
	65	535	345		717	394	616	762	431	762	1070	558	1080	1520	617	1520			2150
	80	853	662		966	708	989	1220	794	1220	1700	1020	1720	2430	1130	2430			3430
	90	1250	894		1500	998	1400	1810	1130	1810	2490	1470	2580	3630	1620	3630			5130
	100	1760	1330		2080	1520	2040	2490	1700	2490	3520	2190	3540	4990	2440	4990			7030
	125						3570			4390			6210			8800			12400
	150						5720			7030			9980			14100			19900
Riser	20		22			22	65		22	79		22	113		22	159			224
	25		51			51	111		51	136		51	193		51	272			385
	32		112			112	176		112	215		112	306		112	431			608
	40		170			170	370		170	454		170	644		170	907			1280
	50		340			340	616		340	762		340	1080		340	1520			2150
	65						989			1220			1720			2430			3430
	80						1470			1810			2580			3630			5130
	70						2030			2490			3540			4990			7030
	100						3570			4390			6210			8800			12400
	125						5720			7030			9980			14100			19900

5. PIPING

Table 5.1 Common Applications of Pipe, Fittings, and Valves for Heating and Air Conditioning [2021F, Ch 22, Tbl 1]

Application	Size, mm	Material	Type	Joint Type	Fitting Material	Class (When Applicable)	System[g] Temp., °C	Maximum Pressure at Temperature,[a,b] kPa
Chilled water	≤51	Steel Type F (CW)	Schedule 40	Thread	Cast iron	125	121	862
	62.5 to 305	Steel A or B, Type E (ERW)	Schedule 40	Weld	Wrought steel	Standard	121	2758
				Flange	Wrought steel	150	121	1724
					Cast iron	125	121	1207
					Cast iron	250	121	2758
		Copper, hard or soft	Type K or L	Solder Flared (soft) Rolled groove (51 to 203) Press-connect (13 to 102) Push connect (13 to 51) Mechanical formed	Wrought or cast Cu		38	2586 Type K soft 4378 Type K hard 1724 Type L soft 2999 Type L hard
				Braze Weld	Wrought or cast Cu		38	1724 Type L soft 2586 Type K soft
		Copper, hard	Type M	Solder Rolled groove (51 to 203) Press-connect (13 to 102) Push connect (13 to 51) Mechanical formed	Wrought or cast Cu		38	2724 Type M hard
				Braze Weld	Wrought or cast Cu		38	1586 Type M soft
	10 to 25	PEX (barrier)	SDR-9	Crimp Clamp Expansion Compression Push fit Proprietary	Bronze Brass Copper Engineered plastic		23	1000

Table 5.1 Common Applications of Pipe, Fittings, and Valves for Heating and Air Conditioning [2021F, Ch 22, Tbl 1] ***(Continued)***

Application	Size, mm	Material	Type	Joint Type	Fitting Material	Class (When Applicable)	System[g] Temp., °C	Maximum Pressure at Temperature,[a,b] kPa
Chilled water (cont'd)	13 to 152	PE	Schedule 40,[f] 80, SDR	Thermal fusion, compression	PE		49 (60) limit for some applications)	Varies with pipe wall thickness, grade, schedule, size. Check manufacturer's documentation for design ratings 207 to 758 at 54°C
Heating and recirculating	51 and smaller	Steel Type F (CW)	Schedule 40	Thread	Cast iron	125	121	862
	6 to 305	Steel B Type E (ERW)	Schedule 40	Weld	Wrought steel	Standard	121	2758
				Flange	Wrought steel	150	121	1724
					Cast iron	125	121	862
					Cast iron	250	121	2758
		Copper, hard or soft	Type K or L	Solder Braze Flared (soft) Rolled groove (51 to 203) Press-connect (13 to 102) Push connect (13 to 51) Mechanical formed	Wrought or cast Cu		93	2069 Type K soft 4378 Type K hard 1413 Type L soft 2999 Type L hard
				Braze Weld	Wrought or cast Cu		93	2069 Type K soft 1413 Type L soft
	6 to 305	Copper, hard	Type M	Solder Rolled groove (51 to 203) Press-connect (13 to 102) Push connect (13 to 51) Mechanical formed	Wrought or cast Cu		93	2724 Type M hard
				Braze Weld	Wrought or cast Cu		93	1379 Type M soft

Table 5.1 Common Applications of Pipe, Fittings, and Valves for Heating and Air Conditioning [2021F, Ch 22, Tbl 1] ***(Continued)***

Application	Size, mm	Material	Type	Joint Type	Fitting Material	Class (When Applicable)	System[g] Temp., °C	Maximum Pressure at Temperature,[a,b] kPa
Heating and recirculating (cont'd)	10 to 25	PEX (barrier)	SDR-9	Crimp Clamp Expansion Compression Push fit Proprietary	Bronze Brass Copper Engineered plastic		93	545
Steam and condensate	51 and smaller	Steel Type F (CW) or S	Schedule 40[d]	Thread	Cast iron	125		621
				Thread	Malleable iron	150		621
				Socket	Forged steel	3000		621
		Steel B Type E (ERW) or S	Schedule 40[d]	Thread	Cast iron	125		690
				Thread	Malleable iron	150		862
				Socket	Forged steel	3000		2758
		Steel B Type E (ERW) or S	Schedule 80	Thread Socket	Cast iron	250		1379
				Thread	Malleable iron	300		1724
				Socket	Forged steel	3000		2758
	51 to 305	Steel B Type E (ERW) or S	Schedule 40	Weld	Wrought steel	Standard		1724
				Flange	Wrought steel	150		1379
					Cast iron	125		690
		Steel B Type E (ERW) or S	Schedule 80	Weld	Wrought steel	XS		4826
				Flange	Wrought steel	300		3448
					Cast iron	250		1379

Table 5.1 Common Applications of Pipe, Fittings, and Valves for Heating and Air Conditioning [2021F, Ch 22, Tbl 1] ***(Continued)***

Application	Size, mm	Material	Type	Joint Type	Fitting Material	Class (When Applicable)	System[g] Temp., °C	Maximum Pressure at Temperature,[a,b] kPa
Ground-source heat pump	6 to 51	Copper, hard or soft	Type L or ACR	Flared or brazed	Wrought or cast Cu		93	1413 Type L soft, 2999 Type L hard, 4655 ACR soft, 3448 ACR hard
	10 to 25	PEX (barrier)	SDR-9	Crimp Clamp Expansion Compression Push fit Proprietary	Bronze Brass Copper Engineered plastic		82	690
Refrigerant	10 to 105	Steel B Type E (ERW)	Schedule 40	Weld			Wrought steel	
		Copper, hard	Type L or ACR	Braze	Wrought or forged Cu		93	2999 Type L hard, 4655 ACR soft
Natural gas and LP	6 to 305	Copper, hard or soft	Type K or L	Solder Rolled groove (51 to 203) Press-connect (13 to 102) Push connect (13 to 51) Mechanically formed	Wrought or cast Cu		38	2551 Type K soft 4378 Type K hard 1724 Type L soft 2999 Type L hard
				Braze Weld	Wrought or cast Cu		38	2551 Type K soft 1724 Type L soft
	10 to 105	Copper, hard	ACR	Solder	Wrought or cast Cu		38	3448 Type ACR hard
				Braze	Wrought or cast Cu		38	2000 Type ACR Soft

Table 5.1 Common Applications of Pipe, Fittings, and Valves for Heating and Air Conditioning [2021F, Ch 22, Tbl 1] ***(Continued)***

Application	Size, mm	Material	Type	Joint Type	Fitting Material	Class (When Applicable)	System[g] Temp., °C	Maximum Pressure at Temperature,[a,b] kPa
Natural gas and LP (cont'd)	10 to 25	PEX	SDR-9	Crimp Clamp Expansion Compression Push fit Proprietary	Bronze Brass Copper Engineered plastic		23	1000
	13 to 152	PE	Schedule 40, 80, SDR	Thermal fusion, compression	PE		49 (60) limit for some applications)	Depends on pipe, grade, schedule, size. Generally 207 to 758 at 54°C
	13 to 152	HDPE	SDR	Thermal fusion, compression	HDPE		49	Depends on pipe, grade, schedule, size. Generally 441 for SDR 11 at 49°C
Fuel oil, above-ground	51 to 305	Black Steel, B Type E (ERW) or S (seamless)	Schedule 40	Thread or weld	Black malleable iron Wrought steel weld Forged steel flanges	150 150		
	6 to 305	Copper, hard or soft	Type K or L	Solder Flared (soft) Rolled groove (51 to 203) Press-connect (13 to 102) Push connect (13 to 51) Mechanical formed	Wrought or cast Cu		38	2069 Type K soft 4378 Type K hard 1724 Type L soft 2999 Type L hard
				Braze or weld	Wrought or cast Cu		38	2069 Type K soft, 1724 Type L soft

Table 5.1 Common Applications of Pipe, Fittings, and Valves for Heating and Air Conditioning [2021F, Ch 22, Tbl 1] ***(Continued)***

Application	Size, mm	Material	Type	Joint Type	Fitting Material	Class (When Applicable)	System[g] Temp., °C	Maximum Pressure at Temperature,[a,b] kPa
Fuel oil, above-ground (cont'd)	6 to 305 (cont'd)	Copper, hard	Type M	Solder Braze Rolled groove (51 to 203) Press-connect (13 to 102) Push connect (13 to 51) Mechanical formed	Wrought or cast Cu		38	2724 Type M hard
	6 to 305	ABS	Schedule 40,[f] 80, SDR	Solvent weld, thread, flange	ABS		71 limit	Depends on pipe class: approximately 345 at 71°C
	13 to 152	HDPE	SDR-9	Thermal fusion, compression	HDPE		49	Depends on pipe, grade, schedule, size. Generally 441 for SDR 11 at 49°C
Compressed air	≤62.5 and smaller	Black steel	Schedule 40	Thread	Black malleable iron	150	177	
	>62.5	Black steel	Schedule 40	Flange or weld	Black malleable iron	150	177	
	10 to 105	Copper, hard	ACR	Solder Flared (soft) Mechanical formed Braze	Wrought or cast Cu		93 93	4655 ACR soft 3448 ACR hard 4655 ACR hard
	13 to 102	ABS HDPE	Schedule 40 Schedule 40, 80, SDR	Solvent weld	ABS HDPE		23	1276
	10 to 25	PEX	SDR-9					

Table 5.1 Common Applications of Pipe, Fittings, and Valves for Heating and Air Conditioning [2021F, Ch 22, Tbl 1] ***(Continued)***

Application	Size, mm	Material	Type	Joint Type	Fitting Material	Class (When Applicable)	System[g] Temp., °C	Maximum Pressure at Temperature,[a,b] kPa
Potable water, inside building	6 to 305	Steel, galvanized	Schedule 40	Thread	Galv. cast iron Galv. cast iron	150 150	38 38	862 1034
		Copper, hard or soft	Type K or L	Solder[c] Flared (soft) Rolled groove (51 to 203) Press-connect (13 to 102) Push connect (13 to 51) Mechanical formed	Wrought or cast Cu		38	2551 Type K soft 4378 Type K hard 1724 Type L soft 2999 Type L hard
				Braze Weld	Wrought or cast Cu		38	2551 Type K soft 1724 Type L soft
	6 to 305	Copper, hard	Type M	Solder[c] Rolled groove (51 to 203) Press-connect (13 to 102) Push connect (13 to 51) Mechanical formed	Wrought or cast Cu		38	2724 Type M hard
				Braze Weld	Wrought or cast Cu		38	1586 Type M soft
	13 to 203	CPVC	Schedule 40,[f] 80		CPVC		99 Limit, 93 operating	
	10 to 25	PEX	SDR-9	Crimp Clamp Expansion Compression Push fit Proprietary	Bronze Brass Copper Engineered plastic		38	1000

Table 5.1 Common Applications of Pipe, Fittings, and Valves for Heating and Air Conditioning [2021F, Ch 22, Tbl 1] *(Continued)*

Application	Size, mm	Material	Type	Joint Type	Fitting Material	Class (When Applicable)	System[g] Temp., °C	Maximum Pressure at Temperature,[a,b] kPa
Potable water, inside building (cont'd)	13 to 152	PE	Schedule 40,[f] 80, SDR	Thermal fusion, compression	PE		49 (60 limit for some applications)	Depends on pipe, grade, schedule, size generally 207 to 758 at 54°C
	13 to 152	PP	Schedule 40,[f] 80, SDR	Thermal fusion, flange, Thread[e]	PP		82	345
Water services, underground	Through 152	Ductile iron	Class 50	Mechanical joint	Cast iron		24	1724
	6 to 305	Copper, hard or soft	Type K or L	Solder[c] Flared (soft) Rolled groove (51 to 203) Press-connect (13 to 102) Push connect (13 to 51) Mechanical formed	Wrought or cast Cu		38	2551 Type K soft 4378 Type K hard 1724 Type L soft 2999 Type L hard
				Braze Weld	Wrought or cast Cu		38	2551 Type K soft 1724 Type L soft
				Flange	Bronze		38	
	6 to 305	Copper, hard	Type M	Solder[c] Rolled groove (51 to 203) Press-connect (13 to 102) Push connect (13 to 51) Mechanical formed	Wrought or cast Cu		38	2724 Type K hard
				Braze Weld	Wrought or cast Cu		38	1586 Type M soft

Table 5.1 Common Applications of Pipe, Fittings, and Valves for Heating and Air Conditioning [2021F, Ch 22, Tbl 1] ***(Continued)***

Application	Size, mm	Material	Type	Joint Type	Fitting Material	Class (When Applicable)	System[g] Temp., °C	Maximum Pressure at Temperature,[a,b] kPa
Water services, underground (cont'd)	10 to 25	PEX	SDR-9	Crimp Clamp Expansion Compression Push fit Proprietary	Bronze Brass Copper Engineered plastic		23	1000
	6 to 508	PVC	Schedule 40, 80, 120, SDR	Solvent weld, thread,[f] thermal weld	PVC		66 limit, 60 operating	545 to 724, depending on schedule and size
Drainage, waste, and vent (DWV)	32 to 203	Copper, hard	DWV	Solder	Wrought or cast Cu		38	1724 DWV hard
	32 to 305	ABS	Schedule DWV, 40,[f] 80, SDR	Solvent weld, thread, flange	ABS		71 limit	Depends on pipe class: approximately 345 at 71°C
	32 to 508	PV	Schedule 40,[f] 80, 120, SDR	Solvent weld, thread, thermal weld	PVC		66 limit, 60 operating	545 to 724, depending on schedule and size

[a]Maximum allowable working pressures have been derated in this table. Higher system pressures can be used for lower temperatures and smaller pipe sizes. Pipe, fittings, joints, and valves must all be considered.
[b]Temperature and pressure relationships can vary based on pipe material composition, size, class, and schedule.
[c]Lead- and antimony-based solders are prohibited for potable water systems. Brazing should be used.
[d]Piping codes typically require thicker-walled pipe for threaded joints to maintain corrosion allowance and pressure ratings.
[e]All plumbing codes require both hot and cold water piping to have a 689 kPa at 82°C rating.
[f]Threads are not recommended on Schedule 40 plastic pipe.
[g]Designer should confirm that all materials are suitably rated for intended operation.

Table 5.2 Properties of Pipe Materials[a] [2021F, Ch 22, Tbl 18]

Material			Tensile Strength, MPa (at 23°C)	Hydrostatic[b] Design Stress, MPa (at 23°C)		Upper Temperature Limit, °C		HDS[b] Upper Limit, MPa	Density, kg/m³	Impact Strength, N (at 23°C)	Modulus of Elasticity, GPa (at 23°C)	Coefficient of Expansion, μm/(m·K)	Thermal Conductivity, W/(m·K)	Relative Pipe Cost[c]
Designation	Type and Grade	Cell No.		Mfr.	ASME B31	Mfr.	ASME B31							
Metals														
Copper	Type L	Drawn	248		62		204	56	8900		117	17.1	33.5	3.5
Steel	A 53 B	ERW	413		88		427	63	7800	1600	190	11.4	3.8	1.3
Stainless steel	304	Drawn or Welded					177		7900		193	17.6	1.2	
Thermoplastics														
PVC 1120	T I,G1	12454-B	52	14	14	60	66	3.0	1400	43	2.90	54	0.159	1.0
PVC 1200	T I,G2	12454-C			14		66				2.83	63		
PVC 2120	T II,G1	14333-D			14		66					54		
CPVC 4120	T IV,G1	23447-B	55	14	14	99	99	2.2	1550	80	2.92	63	0.137	2.9
PE 2306	Gr. P23				4.3		60				0.62	144		
PE 3306	Gr. P34				4.3		70				0.90	126		
PE 3406	Gr. P33				4.3		82				1.03	108		
HDPE 3408	Gr. P34	355434-C	34	11	5.5	60	82	5.5	960	640	0.76	216	0.389	1.1
PP			34	4.9		100	99		910	70	0.83	108	0.187	2.9
ABS	Acrylonitrile copolymer	6-3-3	38			80			1060	450	1.65	101	0.245	3.4
ABS 1210	T I,G2	5-2-2			7		82	4.4			1.72	99		
ABS 1316	T I,G3	3-5-5			11		82	6.9			2.34	72		
ABS 2112	T II,G1	4-4-5			8.6		82	5.5				72		
PVDF			48	8.8		138	135	2.1	1780	200	0.86	142	0.115	28.0

Table 5.2 Properties of Pipe Materials[a] [2021F, Ch 22, Tbl 18] *(Continued)*

Material			Tensile Strength, MPa (at 23°C)	Hydrostatic[b] Design Stress, MPa (at 23°C)		Upper Temperature Limit, °C		HDS[b] Upper Limit, MPa	Density, kg/m^3	Impact Strength, N (at 23°C)	Modulus of Elasticity, GPa (at 23°C)	Coefficient of Expansion, μm/(m·K)	Thermal Conductivity, W/(m·K)	Relative Pipe Cost[c]
Designation	Type and Grade	Cell No.		Mfr.	ASME B31	Mfr.	ASME B31							
Thermosetting														
Epoxy-glass	RTRP-11AF		303	55		99		48			6.90	16 to 23	0.418	
PEX	A,B,C[d]		22	4.3		93	82	0.54	940	200	0.52	162	0.462	0.75
Polyester-glass	RTRP-12EF		303	62		93		34			6.90	16 to 20	0.187	
For Comparison														
Steel	A 53 B	ERW	413		88		427	63	7800	1600	190	11.4	49.6	1.3
Copper	Type L	Drawn	248		62		204	56	8900		117	17.1		3.5

[a] Properties listed are for the specific materials listed; each plastic has other formulations. Consult the manufacturer of the system chosen. These values are for comparative purposes.
[b] Hydrostatic design stress (HDS) is equivalent to allowable design stress.
[c] Based on cost of pipe only, without factoring in fittings, joints, hangers, and labor.
[d] A, B, and C are the three manufacturing processes of PEX pipe. The classifications are not related to a ranking system.

Table 5.3 Steel Pipe Data [2021F, Ch 22, Tbl 16]

U.S. Nominal Size, in.	Nominal Size, mm	Schedule[a]	Wall Thickness t, mm	Inside Diameter d, mm	Surface Area		Cross Section		Mass		Working Pressure[c] ASTM A53 B to 200°C		
					Outside, m^2/m	Inside, m^2/m	Metal Area, mm^2	Flow Area, mm^2	Pipe, kg/m	Water, kg/m	Mfr. Process	Joint Type[b]	kPa (gage)
1/4	8	40 ST	2.24	9.25	0.043	0.029	80.6	67.1	0.631	0.067	CW	T	1296
		80 XS	3.02	7.67	0.043	0.024	101.5	46.2	0.796	0.046	CW	T	6006
3/8	10	40 ST	2.31	12.52	0.054	0.039	107.7	123.2	0.844	0.123	CW	T	1400
		80 XS	3.20	10.74	0.054	0.034	140.2	90.7	1.098	0.091	CW	T	5654
1/2	15	40 ST	2.77	15.80	0.067	0.050	161.5	196.0	1.265	0.196	CW	T	1476
		80 XS	3.73	13.87	0.067	0.044	206.5	151.1	1.618	0.151	CW	T	5192
3/4	20	40 ST	2.87	20.93	0.084	0.066	214.6	344.0	1.68	0.344	CW	T	1496
		80 XS	3.91	18.85	0.084	0.059	279.7	279.0	2.19	0.279	CW	T	4695
1	25	40 ST	3.38	26.64	0.105	0.084	318.6	557.6	2.50	0.558	CW	T	1558
		80 XS	4.55	24.31	0.105	0.076	412.1	464.1	3.23	0.464	CW	T	4427
1 1/4	32	40 ST	3.56	35.05	0.132	0.110	431.3	965.0	3.38	0.965	CW	T	1579
		80 XS	4.85	32.46	0.132	0.102	568.7	827.6	4.45	0.828	CW	T	4096
1 1/2	40	40 ST	3.68	40.89	0.152	0.128	515.5	1 313	4.05	1.313	CW	T	1593
		80 XS	5.08	38.10	0.152	0.120	689.0	1 140	5.40	1.140	CW	T	3972
2	50	40 ST	3.91	52.50	0.190	0.165	690.3	2 165	5.43	2.165	CW	T	1586
		80 XS	5.54	49.25	0.190	0.155	953	1 905	7.47	1.905	CW	T	3799
2 1/2	65	40 ST	5.16	62.71	0.229	0.197	1 099	3 089	8.62	3.089	CW	W	3675
		80 XS	7.01	59.00	0.229	0.185	1 454	2 734	11.40	2.734	CW	W	5757

Table 5.3 Steel Pipe Data [2021F, Ch 22, Tbl 16] ***(Continued)***

U.S. Nominal Size, in.	Nominal Size, mm	Schedule[a]	Wall Thickness *t*, mm	Inside Diameter *d*, mm	Surface Area		Cross Section		Mass		Working Pressure[c] ASTM A53 B to 200°C		
					Outside, m^2/m	Inside, m^2/m	Metal Area, mm^2	Flow Area, mm^2	Pipe, kg/m	Water, kg/m	Mfr. Process	Joint Type[b]	kPa (gage)
3	80	40 ST	5.49	77.93	0.279	0.245	1 438	4 769	11.27	4.769	CW	W	3323
		80 XS	7.62	73.66	0.279	0.231	1 946	4 261	15.25	4.261	CW	W	5288
4	100	40 ST	6.02	102.26	0.359	0.321	2 048	8 213	16.04	8.213	CW	W	2965
		80 XS	8.56	97.18	0.359	0.305	2 844	7 417	22.28	7.417	CW	W	4792
6	150	40 ST	7.11	154.05	0.529	0.484	3 601	18 639	28.22	18.64	ERW	W	4799
		80 XS	10.97	146.33	0.529	0.460	5 423	16 817	42.49	16.82	ERW	W	8336
8	200	30	7.04	205.0	0.688	0.644	4 687	33 000	36.73	33.01	ERW	W	3627
		40 ST	8.18	202.7	0.688	0.637	5 419	32 280	42.45	32.28	ERW	W	4433
		80 XS	12.70	193.7	0.688	0.608	8 234	29 460	64.51	29.46	ERW	W	7626
10	250	30	7.80	257.5	0.858	0.809	6 498	52 060	50.91	52.06	ERW	W	3344
		40 ST	9.27	254.5	0.858	0.800	7 683	50 870	60.23	50.87	ERW	W	4178
		XS	12.70	247.7	0.858	0.778	10 388	48 170	81.39	48.17	ERW	W	6116
		80	15.06	242.9	0.858	0.763	12 208	46 350	95.65	46.35	ERW	W	7453
12	300	30	8.38	307.1	1.017	0.965	8 307	74 060	65.09	74.06	ERW	W	3096
		ST	9.53	304.8	1.017	0.958	9 406	72 970	73.70	72.97	ERW	W	3641
		40	10.31	303.2	1.017	0.953	10 158	72 190	79.59	72.21	ERW	W	4020
		XS	12.70	298.5	1.017	0.938	12 414	69 940	97.28	69.96	ERW	W	5157
		80	17.45	289.0	1.017	0.908	16 797	65 550	131.62	65.57	ERW	W	7419

Table 5.3 Steel Pipe Data [2021F, Ch 22, Tbl 16] ***(Continued)***

U.S. Nominal Size, in.	Nominal Size, mm	Schedule[a]	Wall Thickness t, mm	Inside Diameter d, mm	Surface Area		Cross Section		Mass		Working Pressure[c] ASTM A53 B to 200°C		
					Outside, m^2/m	Inside, m^2/m	Metal Area, mm^2	Flow Area, mm^2	Pipe, kg/m	Water, kg/m	Mfr. Process	Joint Type[b]	kPa (gage)
14	350	30 ST	9.53	336.6	1.117	1.057	10 356	88 970	81.15	88.96	ERW	W	3316
		40	11.10	333.4	1.117	1.047	12 013	87 290	94.13	87.30	ERW	W	3999
		XS	12.70	330.2	1.117	1.037	13 681	85 610	107.21	85.63	ERW	W	4695
		80	19.05	317.5	1.117	0.997	20 142	79 160	157.82	79.17	ERW	W	7453
16	400	30 ST	9.53	387.4	1.277	1.217	11 876	117 800	93.06	117.8	ERW	W	2903
		40 XS	12.70	381.0	1.277	1.197	15 708	114 000	123.09	114.0	ERW	W	4109
18	450	ST	9.53	438.2	1.436	1.376	13 396	150 800	104.98	150.8	ERW	W	2579
		30	11.10	435.0	1.436	1.367	15 556	148 600	121.90	148.6	ERW	W	3110
		XS	12.70	431.8	1.436	1.357	17 735	146 450	138.97	146.4	ERW	W	3654
		40	14.27	428.7	1.436	1.347	19 863	144 300	155.65	144.3	ERW	W	4185
20	500	20 ST	9.53	489.0	1.596	1.536	14 916	187 700	116.88	187.4	ERW	W	2324
		30 XS	12.70	482.6	1.596	1.516	19 762	182 900	154.85	182.9	ERW	W	3289
		40	15.06	477.9	1.596	1.501	23 325	179 400	182.78	179.4	ERW	W	4006

[a]Numbers are schedule numbers per ASME Standard B36.10M; ST = Standard; XS = Extra Strong.
[b]T = Thread; W = Weld.
[c]Working pressures were calculated per ASME B31.9 using furnace butt-weld (continuous weld, CW) pipe through 100 mm and electric resistance weld (ERW) thereafter. The allowance A has been taken as
(1) 12.5% of t for mill tolerance on pipe wall thickness, *plus*
(2) An arbitrary corrosion allowance of 0.64 mm for pipe sizes through NPS 2 and 1.65 mm from NPS 2 1/2 through 20, *plus*
(3) A thread cutting allowance for sizes through NPS 2.
Because the pipe wall thickness of threaded standard pipe is so small after deducting allowance A, the mechanical strength of the pipe is impaired. It is good practice to limit standard threaded pipe pressure to 620 kPa (gage) for steam and 860 kPa (gage) for water.

Table 5.4 Copper Tube Data [2021F, Ch 22, Tbl 17]

U.S. Nominal Size, in.	Type	Wall Thickness t, mm	Diameter		Surface Area		Cross Section		Mass		Working Pressure[a,b,c] ASTM B88 to 120°C MPa (gage)	
			Outside D, mm	Inside d, mm	Outside, m^2/m	Inside, m^2/m	Metal Area, mm^2	Flow Area, mm^2	Tube, kg/m	Water, kg/m	Annealed	Drawn
1/4	K	0.89	9.53	7.75	0.030	0.0244	24	47	0.216	0.047	5.868	11.004
	L	0.76	9.53	8.00	0.030	0.0250	21	50	0.188	0.050	5.033	9.432
3/8	K	1.24	12.70	10.21	0.040	0.0320	45	82	0.400	0.082	6.164	11.556
	L	0.89	12.70	10.92	0.040	0.0344	33	94	0.295	0.094	4.399	8.253
	M	0.64	12.70	11.43	0.040	0.0360	24	103	0.216	0.103	3.144	5.895
1/2	K	1.24	15.88	13.39	0.050	0.0421	57	141	0.512	0.141	4.930	9.246
	L	1.02	15.88	13.84	0.050	0.0436	48	151	0.424	0.151	4.027	7.543
	M	0.71	15.88	14.45	0.050	0.0454	34	164	0.302	0.164	2.820	5.282
5/8	K	1.24	19.05	16.56	0.060	0.0521	70	215	0.622	0.215	4.109	7.702
	L	1.07	19.05	16.92	0.060	0.0530	60	225	0.539	0.225	3.523	6.605
3/4	K	1.65	22.23	18.92	0.070	0.0594	106	281	0.954	0.281	4.668	8.757
	L	1.14	22.23	19.94	0.070	0.0628	75	312	0.677	0.312	3.234	6.061
	M	0.81	22.23	20.60	0.070	0.0646	55	333	0.488	0.333	2.303	4.309
1	K	1.65	28.58	25.27	0.090	0.0792	139	502	1.249	0.502	3.634	6.812
	L	1.27	28.58	26.04	0.090	0.0817	109	532	0.973	0.532	2.792	5.240
	M	0.89	28.58	26.80	0.090	0.0841	77	564	0.691	0.564	1.958	3.668
1 1/4	K	1.65	34.93	31.62	0.110	0.0994	173	785	1.543	0.785	2.972	5.571
	L	1.40	34.93	32.13	0.110	0.1009	147	811	1.316	0.811	2.517	4.716
	M	1.07	34.93	32.79	0.110	0.1030	114	845	1.015	0.845	1.924	3.599
	DWV	1.02	34.93	32.89	0.110	0.1033	108	850	0.967	0.850	1.827	3.427

Table 5.4 Copper Tube Data [2021F, Ch 22, Tbl 17] ***(Continued)***

U.S. Nominal Size, in.	Type	Wall Thickness t, mm	Diameter: Outside D, mm	Diameter: Inside d, mm	Surface Area: Outside, m^2/m	Surface Area: Inside, m^2/m	Cross Section: Metal Area, mm^2	Cross Section: Flow Area, mm^2	Mass: Tube, kg/m	Mass: Water, kg/m	Working Pressure[a,b,c] ASTM B88 to 120°C, MPa (gage): Annealed	Working Pressure[a,b,c] ASTM B88 to 120°C, MPa (gage): Drawn
1 1/2	K	1.83	41.28	37.62	0.130	0.1183	226	1 111	2.025	1.111	2.786	5.226
	L	1.52	41.28	38.23	0.130	0.1201	190	1 148	1.701	1.148	2.324	4.351
	M	1.24	41.28	38.79	0.130	0.1219	157	1 181	1.399	1.182	1.896	3.558
	DWV	1.07	41.28	39.14	0.130	0.1228	135	1 203	1.204	1.203	1.627	3.048
2	K	2.11	53.98	49.76	0.170	0.1564	343	1 945	3.070	1.945	2.455	4.606
	L	1.78	53.98	50.42	0.170	0.1585	292	1 997	2.606	1.997	2.069	3.951
	M	1.47	53.98	51.03	0.170	0.1603	243	2 045	2.171	2.045	1.717	3.220
	DWV	1.07	53.98	51.84	0.170	0.1628	177	2 111	1.585	2.111	1.241	2.331
2 1/2	K	2.41	66.68	61.85	0.209	0.1942	487	3 004	4.35	3.004	2.275	4.268
	L	2.03	66.68	62.61	0.209	0.1966	413	3 079	3.69	3.079	1.917	3.592
	M	1.65	66.68	63.37	0.209	0.1990	337	3 154	3.02	3.154	1.558	2.917
3	K	2.77	79.38	73.84	0.249	0.2320	666	4 282	5.96	4.282	2.193	4.109
	L	2.29	79.38	74.80	0.249	0.2350	554	4 395	4.95	4.395	1.813	3.392
	M	1.83	79.38	75.72	0.249	0.2378	446	4 503	3.98	4.503	1.448	2.717
	DWV	1.14	79.38	77.09	0.249	0.2423	281	4 667	2.51	4.667	0.903	1.696
3 1/2	K	3.05	92.08	85.98	0.289	0.2701	852	5 806	7.62	5.806	2.082	3.903
	L	2.54	92.08	87.00	0.289	0.2733	714	5 944	6.39	5.944	1.738	3.254
	M	2.11	92.08	87.86	0.289	0.2761	596	6 063	5.33	6.063	1.441	2.703
4	K	3.40	104.78	97.97	0.329	0.3078	1084	7 538	9.69	7.538	2.041	3.827
	L	2.79	104.78	99.19	0.329	0.3115	895	7 727	8.00	7.727	1.675	3.144
	M	2.41	104.78	99.95	0.329	0.3139	776	7 846	6.94	7.846	1.448	2.717
	DWV	1.47	104.78	101.83	0.329	0.3200	478	8 144	4.27	8.144	0.883	1.655

Table 5.4 Copper Tube Data [2021F, Ch 22, Tbl 17] *(Continued)*

U.S. Nominal Size, in.	Type	Wall Thickness t, mm	Diameter: Outside D, mm	Diameter: Inside d, mm	Surface Area: Outside, m^2/m	Surface Area: Inside, m^2/m	Cross Section: Metal Area, mm^2	Cross Section: Flow Area, mm^2	Mass: Tube, kg/m	Mass: Water, kg/m	Working Pressure[a,b,c] ASTM B88 to 120°C, MPa (gage): Annealed	Working Pressure[a,b,c] ASTM B88 to 120°C, MPa (gage): Drawn
5	K	4.06	130.18	122.05	0.409	0.3834	1610	11 699	14.39	11.70	1.965	3.682
	L	3.18	130.18	123.83	0.409	0.3889	1266	12 042	11.32	12.04	1.531	2.875
	M	2.77	130.18	124.64	0.409	0.3917	1108	12 201	9.91	12.20	1.338	2.510
	DWV	1.83	130.18	126.52	0.409	0.3975	737	12 572	6.59	12.57	0.883	1.655
6	K	4.88	155.58	145.82	0.489	0.4581	2309	16 701	20.64	16.70	1.972	3.696
	L	3.56	155.58	148.46	0.489	0.4663	1698	17 311	15.18	17.31	1.434	2.696
	M	3.10	155.58	149.38	0.489	0.4694	1484	17 525	13.27	17.53	1.255	2.351
	DWV	2.11	155.58	151.36	0.489	0.4755	1016	17 993	9.09	17.99	0.855	1.600
8	K	6.88	206.38	192.61	0.648	0.6050	4314	29 137	38.56	29.14	2.096	3.930
	L	5.08	206.38	196.22	0.648	0.6163	3212	30 238	28.71	30.24	1.544	2.903
	M	4.32	206.38	197.74	0.648	0.6212	2741	30 710	24.50	30.71	1.317	2.468
	DWV	2.77	206.38	200.84	0.648	0.6309	1771	31 680	15.83	31.62	0.841	1.579
10	K	8.59	257.18	240.00	0.808	0.7541	6705	45 241	59.93	45.15	2.096	3.937
	L	6.35	257.18	244.48	0.808	0.7681	5004	46 942	44.73	46.94	1.551	2.910
	M	5.38	257.18	246.41	0.808	0.7742	4259	47 686	38.07	47.69	1.317	2.468
12	K	10.29	307.98	287.40	0.968	0.9028	9621	64 873	85.99	64.87	2.103	3.937
	L	7.11	307.98	293.75	0.968	0.9229	6722	67 771	60.09	67.77	1.455	2.724
	M	6.45	307.98	295.07	0.968	0.9269	6112	68 382	54.63	68.38	1.317	2.468

[a]When using soldered or brazed fittings, the joint determines the limiting pressure.
[b]Working pressures were calculated using ASME Standard B31.9 allowable stresses. A 5% mill tolerance has been used on the wall thickness. Higher tube ratings can be calculated using the allowable stress for lower temperatures.
[c]If soldered or brazed fittings are used on hard drawn tubing, use the annealed ratings. Full-tube allowable pressures can be used with suitably rated flare or compression-type fittings.

Table 5.5 Thermal Expansion of Metal Pipe [2021F, Ch 22, Tbl 13]

	Saturated Steam Pressure, kPa (gage)	Temperature, °C	Linear Thermal Expansion, mm/m		
			Carbon Steel	Type 304 Stainless Steel	Copper
		−34	−0.16	−0.25	−0.27
		−29	−0.10	−0.17	−0.18
		−23	−0.05	−0.08	−0.09
		−18	0	0	0
		−12	0.07	0.09	0.10
		−7	0.13	0.18	0.20
Vacuum	−100.7	0	0.20	0.30	0.31
Vacuum	−100.7	4	0.25	0.38	0.38
Vacuum	−100.0	10	0.32	0.47	0.47
Vacuum	−99.3	16	0.38	0.56	0.57
Vacuum	−98.6	21	0.44	0.65	0.66
Vacuum	−97.9	27	0.51	0.75	0.75
Vacuum	−96.5	32	0.57	0.84	0.85
Vacuum	−94.5	38	0.63	0.93	0.94
Vacuum	−89.6	49	0.76	1.13	1.14
Vacuum	−81.4	60	0.88	1.31	1.33
Vacuum	−69.0	71	1.02	1.49	1.50
Vacuum	−49.6	82	1.14	1.68	1.71
Vacuum	−22.1	93	1.27	1.87	1.92
	0	100	1.35	1.98	2.03
	17.2	104	1.41	2.07	2.10
	71.0	116	1.54	2.26	2.30
	142.7	127	1.68	2.45	2.49
	238.6	138	1.82	2.64	2.68
	360.6	149	1.96	2.83	2.88
	517.1	160	2.11	3.03	3.08
	712.3	171	2.25	3.23	3.28
	953.6	182	2.40	3.43	3.48
	1 249	193	2.54	3.63	3.68
	1 604	204	2.69	3.83	4.06
	9 039	304	4.11	5.65	5.77
		404	5.67	7.56	7.72
		504	7.31	9.54	9.76

Table 5.6 Suggested Hanger Spacing and Rod Size for Straight Horizontal Runs [2021F, Ch 22, Tbl 11]

Nominal O.D., mm	Hanger Spacing, m			Rod Size, mm
	Standard Steel Pipe*		Copper Tube	
	Water	Steam	Water	
15	2.1	2.4	1.5	6.4
20	2.1	2.7	1.5	6.4
25	2.1	2.7	1.8	6.4
40	2.7	3.7	2.4	10
50	3.0	4.0	2.4	10
65	3.4	4.3	2.7	10
80	3.7	4.6	3.0	10
100	4.3	5.2	3.7	13
150	5.2	6.4	4.3	13
200	5.8	7.3	4.9	16
250	6.1	7.9	5.5	19
300	7.0	9.1	5.8	22
350	7.6	9.8		25
400	8.2	10.7		25
450	8.5	11.3		32
500	9.1	11.9		32

Source: Adapted from MSS Standard SP-69
*Spacing does not apply where span calculations are made or where concentrated loads are placed between supports such as flanges, valves, specialties, etc.

Table 5.7 Capacities of ASTM A36 Steel Threaded Rods [2021F, Ch 22, Tbl 10]

Rod Diameter, mm	Root Area of Coarse Thread, mm^2	Maximum Load,[a] N
6.4	17.4	1 070
10	43.9	2 720
13	81.3	5 030
16	130.3	8 060
19	194.8	12 100
22	270.3	16 800
25	356.1	22 100
32	573.5	35 600

[a] Based on an allowable stress of 83 MPa reduced by 25% using the root area in accordance with ASME Standard B31.1 and MSS Standard SP-58.

6. SERVICE WATER HEATING

Water heating energy use is second only to space conditioning in most residential buildings, and is also significant in many commercial and industrial settings. In some climates and applications, water heating is the largest energy use in a building. Moreover, quick availability of adequate amounts of hot water is an important factor in user satisfaction. Both water and energy waste can be significant in poorly designed service water-heating systems: from over- or undersizing pipes and equipment, from poor building layout, and from poor system design and operating strategies. Good service water-heating system design and operating practices can often reduce first costs as well as operating costs.

System Elements

A service water-heating system has (1) one or more heat energy sources, (2) heat transfer equipment, (3) a distribution system, and (4) terminal hot-water usage devices.

Heat energy sources may be (1) fuel combustion; (2) electrical conversion; (3) solar energy; (4) geothermal, air, or other environmental energy; and/or (5) recovered waste heat from sources such as flue gases, ventilation and air-conditioning systems, refrigeration cycles, and process waste discharge.

Heat transfer equipment is direct, indirect, or a combination of the two. For direct equipment, heat is derived from combustion of fuel or direct conversion of electrical energy into heat and is applied within the water-heating equipment. For indirect heat transfer equipment, heat energy is developed from remote heat sources (e.g., boilers; solar energy collection; air, geothermal, or other environmental source; cogeneration; refrigeration; waste heat) and is then transferred to the water in a separate piece of equipment. Storage tanks may be part of or associated with either type of heat transfer equipment.

Distribution systems transport hot water produced by water-heating equipment to terminal hot-water usage devices. Water consumed must be replenished from the building water service main. For locations where constant supply temperatures are desired, circulation piping or a means of heat maintenance must be provided.

End-use fixtures are plumbing faucets, accessories, and equipment requiring hot water that may have periods of irregular flow, constant flow, and no flow. These patterns and their related water usage vary with different buildings, process applications, and personal preferences. Examples of end-use accessories are prerinse spray valves, faucet aerators, showerheads, washdown sprayers, and hose bibbs. Examples of end-use equipment are dishwashers, clothes washers, and pressure washers.

Legionellosis (Legionnaires' Disease)

Legionnaires' disease (a form of severe pneumonia) is caused by inhaling aerosolized water droplets containing the bacteria *Legionella pneumophila*. Susceptibility to Legionnaire's disease varies among individuals. People with compromised immune systems (e.g., organ transplant patients or others on immunosuppressant drugs, AIDS patients, smokers, elderly, those with other chronic health conditions or injuries) are at greater risk of contracting the disease at lower exposure levels.

Most water supplied to buildings contains some *Legionella* bacteria (and/or other microorganisms), often at levels too low to detect. The concern is that organism colonies can grow (amplify) within the building hot- and cold-water systems under certain conditions. At high *Legionella* concentrations, a hazard may exist. Some examples of conditions potentially conducive to *Legionella* growth are temperatures within a certain range (warm, not too hot or cold), locations of flow stagnation (e.g., pipe dead legs, low flow velocities, other flow stagnation points, intermittent or seasonal use), and inadequate oxidant residual levels. For more specific water system design guidance, refer to ANSI/ASHRAE Standard 188 and ASHRAE Guideline 12.

Hot-Water Load

Methods for sizing storage water heaters vary. Those using recovery versus storage curves are based on extensive research. All methods provide adequate hot water if the designer allows for unusual conditions. To serve a hot-water load adequately, the needs of both the peak energy withdrawal rate and total integrated energy delivery for end uses must be met. Meeting these needs can be done either by providing a heating rate large enough to meet the peak energy withdrawal rate of the system (and modulating that heating input for smaller loads), or by providing a lower heating rate combined with storage (from which the peak rates can be satisfied). Lower costs are usually achieved by using at least some storage. A variety of different heating rate/storage volume combinations can be used to meet the needs of a given water-heating load profile.

Load Diversity

The greatest difficulty in designing water-heating systems comes from uncertainty about design hot-water loads, especially for buildings not yet built. Although it is fairly simple to test maximum flow rates of various hot-water fixtures and appliances, actual flow rates and durations are user-dependent. Moreover, the timing of different hot-water use events varies from day to day, with some overlap, but almost never will all fixtures be used simultaneously. As the number of hot-water-using fixtures and appliances grows, the percent of those fixtures used simultaneously decreases.

Some of the hot-water load information here is based on limited-scale field testing combined with statistical analysis to estimate load demand or **diversity** factors (percent of total possible load that is ever actually used at one time) versus number of end use points, number of people, etc. Much of the work to provide these diversity factors dates from the 1930s to the 1960s; it remains, however, the best information currently available (with a few exceptions, as noted). Of greatest concern is the fact that most of the data from those early studies were for fixtures that used water at much higher flow rates than modern energy-efficient fixtures (e.g., low-flow shower heads and sink aerators, energy-efficient washing machines and dishwashers). Using the older load diversity information usually results in a water-heating system that adequately serves the loads, but often results in substantial oversizing. Oversizing can be a deterrent to using modern high-efficiency water-heating equipment, which may have higher first cost per unit of capacity than less efficient equipment.

Table 6.1 Typical Residential Use of Hot Water [2019A, Ch 51, Tbl 3]

Use	High Flow, Litres/Task	Low Flow (Water Savers Used), Litres/Task	Ultralow Flow, Litres/Task
Food preparation	19	11	11
Hand dish washing	15	15	11
Automatic dishwasher	57	57	11 to 38
Clothes washer	121	80	19 to 57
Shower or bath	76	57	38 to 57
Face and hand washing	15	8	4 to 8

Table 6.2 HUD-FHA Minimum Water Heater Capacities for One- and Two-Family Living Units [2019A, Ch 51, Tbl 4]

Number of Baths	1 to 1.5			2 to 2.5				3 to 3.5			
Number of Bedrooms	**1**	**2**	**3**	**2**	**3**	**4**	**5**	**3**	**4**	**5**	**6**
Gas[a]											
Storage, L	76	114	114	114	150	150	190	150	190	190	190
kW input	7.9	10.5	10.5	10.5	10.5	11.1	13.8	11.1	11.1	13.8	14.6
1 h draw, L	163	227	227	227	265	273	341	273	311	341	350
Recovery, mL/s	24	32	32	32	32	36	42	34	34	42	44
Electric[a]											
Storage, L	76	114	150	150	190	190	250	190	250	250	300
kW input	2.5	3.5	4.5	4.5	5.5	5.5	5.5	5.5	5.5	5.5	5.5
1 h draw, L	114	167	220	220	273	273	334	273	334	334	387
Recovery, mL/s	10	15	19	19	23	23	23	23	23	23	23
Oil[a]											
Storage, L	114	114	114	114	114	114	114	114	114	114	114
kW input	20.5	20.5	20.5	20.5	20.5	20.5	20.5	20.5	20.5	20.5	20.5
1 h draw, L	337	337	337	337	337	337	337	337	337	337	337
Recovery, mL/s	62	62	62	62	62	62	62	62	62	62	62
Tank-Type Indirect[b,c]											
I-W-H-rated draw, L in 3 h, 55 K rise		150	150		250	250[e]	250	250	250	250	250
Manufacturer-rated draw, L in 3 h, 55 K rise		186	186		284	284[e]	284	284	284	284	284
Tank capacity, L		250	250		250	250[e]	310	250	310	310	310
Tankless-Type Indirect[c,d]											
I-W-H-rated draw, mL/s, 55 K rise		170	170		200	200[e]	240	200	240	240	240
Manufacturer-rated draw, L in 5 min, 55 K rise		57	57		95	95[e]	133	95	133	133	133

Note: Applies to tank-type water heaters only

[a]Storage capacity, input, and recovery requirements indicated are typical and may vary with manufacturer. Any combination of requirements to produce stated 1 h draw is satisfactory.

[b]Boiler-connected water heater capacities (82°C boiler water, internal or external connection).

[c]Heater capacities and inputs are minimum allowable. Variations in tank size are permitted when recovery is based on 4.2 mL/(s·kW) at 55 K rise for electrical, AGA recovery ratings for gas, and IBR ratings for steam and hot-water heaters.

[d]Boiler-connected heater capacities (93°C boiler water, internal or external connection).

[e]Also for 1 to 1.5 baths and 4 bedrooms for indirect water heaters.

Table 6.3 Overall (OVL) and Peak Average Hot-Water Use [2019A, Ch 51, Tbl 5]

Group	Average Hot-Water Use, L							
	Hourly		Daily		Weekly		Monthly	
	OVL	Peak	OVL	Peak	OVL	Peak	OVL	Peak
All families	9.8	17.3	236	254	1652	1873	7178	7700
"Typical" families	9.9	21.9	239	252	1673	1981	7270	7866

Table 6.4 Hot-Water Demands and Use for Various Types of Buildings* [2019A, Ch 51, Tbl 6]

Type of Building	Maximum Hourly	Maximum Daily	Average Daily
Men's dormitories	14.4 L/student	83.3 L/student	49.7 L/student
Women's dormitories	19 L/student	100 L/student	46.6 L/student
Motels: Number of units[a]			
20 or less	23 L/unit	132.6 L/unit	75.8 L/unit
60	20 L/unit	94.8 L/unit	53.1 L/unit
100 or more	15 L/unit	56.8 L/unit	37.9 L/unit
Nursing homes	17 L/bed	114 L/bed	69.7 L/bed
Office buildings	1.5 L/person	7.6 L/person	3.8 L/person
Food service establishments:			
Type A: Full-meal restaurants and cafeterias	5.7 L/max meals/h	41.7 L/max meals/day	9.1 L/average meals/day[b]
Type B: Drive-ins, grills, luncheonettes, sandwich and snack shops	2.6 L/max meals/h	22.7 L/max meals/day	2.6 L/average meals/day[b]
Apartment houses: Number of apartments			
20 or less	45.5 L/apartment	303.2 L/apartment	159.2 L/apartment
50	37.9 L/apartment	276.7 L/apartment	151.6 L/apartment
75	32.2 L/apartment	250 L/apartment	144 L/apartment
100	26.5 L/apartment	227.4 L/apartment	140.2 L/apartment
200 or more	19 L/apartment	195 L/apartment	132.7 L/apartment
Elementary schools	2.3 L/student	5.7 L/student	2.3 L/student[b]
Junior and senior high schools	3.8 L/student	13.6 L/student	6.8 L/student[b]

*Data predate modern low-flow fixtures and appliances.
[a]Interpolate for intermediate values.[b]Per day of operation.

Table 6.5 Hot Water Demand per Fixture for Various Types of Buildings [2019A, Ch 51, Tbl 10]
(Litres of water per hour per fixture, calculated at a final temperature of 60°C)

	Apartment House	Club	Gymnasium	Hospital	Hotel	Industrial Plant	Office Building	Private Residence	School	YMCA
1. Basin, private lavatory	7.6	7.6	7.6	7.6	7.6	7.6	7.6	7.6	7.6	7.6
2. Basin, public lavatory	15	23	30	23	30	45.5	23	—	57	30
3. Bathtub[c]	76	76	114	76	76	—	—	76	—	114
4. Dishwasher[a]	57	190-570	—	190-570	190-760	76-380	—	57	76-380	76-380
5. Foot basin	11	11	46	11	11	46	—	11	11	46
6. Kitchen sink	38	76	—	76	114	76	76	38	76	76
7. Laundry, stationary tub	76	106	—	106	106	—	—	76	—	106
8. Pantry sink	19	38	—	38	38	—	38	19	38	38
9. Shower	114	568	850	284	284	850	114	114	850	850
10. Service sink	76	76	—	76	114	76	76	57	76	76
11. Hydrotherapeutic shower				1520						
12. Hubbard bath				2270						
13. Leg bath				380						
14. Arm bath				130						
15. Sitz bath				114						
16. Continuous-flow bath				625						
17. Circular wash sink				76	76	114	76		114	
18. Semicircular wash sink				38	38	57	38		57	
19. DEMAND FACTOR	0.30	0.30	0.40	0.25	0.25	0.40	0.30	0.30	0.40	0.40
20. STORAGE CAPACITY FACTOR[b]	1.25	0.90	1.00	0.60	0.80	1.00	2.00	0.70	1.00	1.00

Note: Data sources predate low-flow fixtures and appliances.
[a]Dishwasher requirements should be taken from this table or from manufacturers' data for model to be used, if known.
[b]Ratio of storage tank capacity to probable maximum demand/h. Storage capacity may be reduced where unlimited supply of steam is available from central street steam system or large boiler plant.
[c]Whirlpool baths require specific consideration based on capacity. They are not included in the bathtub category.

Table 6.6 Hot-Water Demand in Fixture Units (60°C Water) [2019A, Ch 51, Tbl 16]

	Apartments	Club	Gymnasium	Hospital	Hotels and Dormitories	Industrial Plant	Office Building	School	YMCA
Basin, private lavatory	0.75	0.75	0.75	0.75	0.75	0.75	0.75	0.75	0.75
Basin, public lavatory	—	1	1	1	1	1	1	1	1
Bathtub	1.5	1.5	—	1.5	1.5	—	—	—	—
Dishwasher*	1.5	Five fixture units per 250 seating capacity							
Therapeutic bath	—	—	—	5	—	—	—	—	—
Kitchen sink	0.75	1.5	—	3	1.5	3	—	0.75	3
Pantry sink	—	2.5	—	2.5	2.5	—	—	2.5	2.5
Service sink	1.5	2.5	—	2.5	2.5	2.5	2.5	2.5	2.5
Shower	1.5	1.5	1.5	1.5	1.5	3.5	—	1.5	1.5
Circular wash fountain	—	2.5	2.5	2.5	—	4	—	2.5	2.5
Semicircular wash fountain	—	1.5	1.5	1.5	—	3	—	1.5	1.5

Note: Data predate modern low-flow fixtures and appliances.
* Fixture unit: A number, on an arbitrarily chosen scale, that expresses the load-producing effects on the system of different kinds of fixtures.

Table 6.7 Tankless Water Heater Output Heat Rates, kW*
[2019A, Ch 51, Tbl 15]

Flow Rate, mL/s	Temperature Rise						
	6 K	14 K	28 K	31 K	42 K	43 K	56 K
6.3	0.15	0.37	0.74	0.81	1.11	1.14	1.48
31.5	0.74	1.85	3.69	4.06	5.54	5.69	7.39
63.1	1.48	3.69	7.39	8.12	11.1	11.4	14.8
94.6	2.22	5.54	11.1	12.2	16.6	17.1	22.2
126	2.95	7.39	14.8	16.2	22.2	22.8	29.5
158	3.69	9.23	18.5	20.3	27.7	28.4	36.9
189	4.43	11.1	22.2	24.4	33.2	34.1	44.3
221	5.17	12.9	25.8	28.4	38.8	39.8	51.7
252	5.91	14.8	29.5	32.5	44.3	45.5	59.1
284	6.65	16.6	33.2	36.6	49.9	51.2	66.5
315	7.39	18.5	36.9	40.6	55.4	56.9	73.9
379	8.86	22.2	44.3	48.7	66.5	68.2	88.6
442	10.3	25.8	51.7	56.9	77.5	79.6	103.4
505	11.8	29.5	59.1	65.0	88.6	91.0	118.2
568	13.3	33.2	66.5	73.1	99.7	102.4	132.9
631	14.8	36.9	73.9	81.2	110.8	113.7	147.7

*Divide table values by input efficiency to determine required heat input rate.

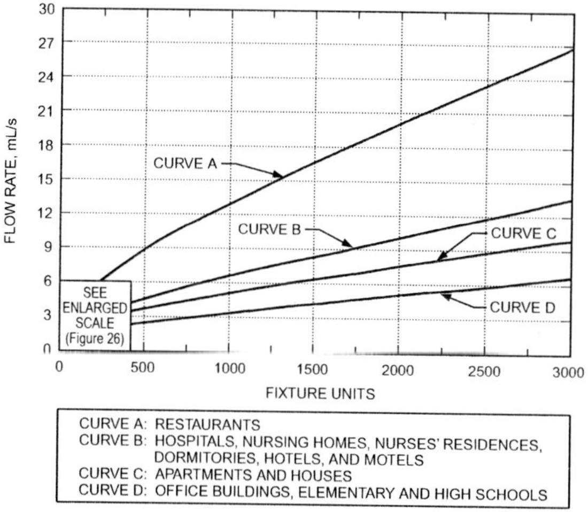

Figure 6.1 Modified Hunter Curve for Calculating Hot-Water Flow Rate [2019A, Ch 51, Fig 27] (Data Predate Modern Low-Flow Fixtures and Appliances)

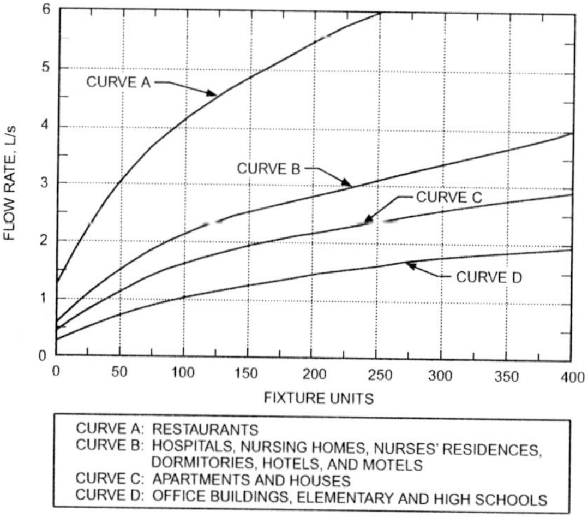

Figure 6.2 Enlarged Section of Modified Hunter Curve [2019A, Ch 51, Fig 28] (Data Predate Modern Low-Flow Fixtures and Appliances)

Temperature Requirement

Typical temperature requirements for some services are shown in Table 6.8. A 60°C water temperature minimizes flue gas condensation in the equipment.

Table 6.8 Representative Hot-Water Temperatures [2019A, Ch 51, Tbl 19]

Use	Temperature, °C
Lavatory	
Hand washing	40
Shaving	45
Showers and tubs	43
Therapeutic baths	35
Commercial or institutional laundry, based on fabric	up to 82
Residential dish washing and laundry	60
Surgical scrubbing	43
Commercial spray-type dish washing[a]	
Single- or multiple-tank hood or rack type	
Wash	65 minimum
Final rinse	82 to 90
Single-tank conveyor type	
Wash	71 minimum
Final rinse	82 to 90
Single-tank rack or door type	
Single-temperature wash and rinse	74 minimum
Chemical sanitizing types[b]	60
Multiple-tank conveyor type	
Wash	65 minimum
Pumped rinse	71 minimum
Final rinse	82 to 90
Chemical sanitizing glass washer	
Wash	60
Rinse	24 minimum

[a]As required by NSF.
[b]See manufacturer for actual temperature required.

Other Safety Concerns

Regulatory agencies differ as to the selection of protective devices and methods of installation. It is therefore essential to check and comply with the manufacturer's instructions and the applicable local codes. In the absence of such instructions and codes, the following recommendations may be used as a guide:

- Water expands when it is heated. Although the water-heating system is initially under service pressure, the pressure rises rapidly if backflow is prevented by devices such as a check valve, pressure-reducing valve, or backflow preventer in the cold-water line or by temporarily shutting off the cold-water valve. When backflow is prevented, the pressure rise during heating may cause the safety relief valve to weep to relieve the pressure. However, if the safety relief valve is inadequate, inoperative, or missing, pressure rise may rupture the tank or cause other damage. Systems having this potential problem must be protected by a properly sized expansion tank located on the cold-water line downstream of and as close as practical to the device preventing backflow.
- Temperature-limiting devices (energy cutoff/high limit) prevent water temperatures from exceeding 99°C by stopping the flow of fuel or energy. These devices should be listed and labeled by a recognized certifying agency.
- Safety relief valves open when pressure exceeds the valve setting. These valves are typically applied to water-heating and hot-water supply boilers. The set pressure should not exceed the maximum allowable working pressure of the boiler. The heat input pressure steam rating (in kW) should equal or exceed the maximum out-put rating for the boiler. The valves should comply with current applicable standards or the ASME *Boiler and Pressure Vessel Code*.
- Temperature and pressure safety relief valves also open if the water temperature reaches 99°C. These valves are typically applied to water heaters and hot-water storage tanks. The heat input temperature/steam rating (in kW) should equal or exceed the heat input rating of the water heater. Combination temperature- and pressure-relief valves should be installed with the temperature-sensitive element located in the top 150 mm of the tank (i.e., where the water is hottest).
- To reduce scald hazards, discharge temperature at fixtures accessible to the occupant should not exceed 50°C. Thermostatically controlled mixing valves can be used to blend hot and cold water to maintain safe service hot-water temperatures.
- A relief valve should be installed in any part of the system containing a heat input device that can be isolated by valves. The heat input device may be solar water-heating panels, desuperheater water heaters, heat recovery devices, or similar equipment.

7. REFRIGERATION CYCLES

Refrigeration cycles transfer thermal energy from a region of low temperature T_R to one of higher temperature. Usually the higher-temperature heat sink is the ambient air or cooling water, at temperature T_0, the temperature of the surroundings.

The first and second laws of thermodynamics can be applied to individual components to determine mass and energy balances and the irreversibility of the components. This procedure is illustrated in later sections in this chapter.

Performance of a refrigeration cycle is usually described by a **coefficient of performance (COP)**, defined as the benefit of the cycle (amount of heat removed) divided by the required energy input to operate the cycle:

$$\text{COP} \equiv \frac{\text{Useful refrigerating effect}}{\text{Net energy supplied from external sources}} \tag{7.1}$$

For a mechanical vapor compression system, the net energy supplied is usually in the form of work, mechanical or electrical, and may include work to the compressor and fans or pumps. Thus,

$$\text{COP} = \frac{Q_{evap}}{W_{net}} \tag{7.2}$$

In an absorption refrigeration cycle, the net energy supplied is usually in the form of heat into the generator and work into the pumps and fans, or

$$\text{COP} = \frac{Q_{evap}}{Q_{gen} + W_{net}} \tag{7.3}$$

In many cases, work supplied to an absorption system is very small compared to the amount of heat supplied to the generator, so the work term is often neglected.

Applying the second law to an entire refrigeration cycle shows that a completely reversible cycle operating under the same conditions has the maximum possible COP. Departure of the actual cycle from an ideal reversible cycle is given by the **refrigerating efficiency**:

$$\eta_R = \frac{\text{COP}}{(\text{COP})_{rev}} \tag{7.4}$$

Theoretical Refrigeration Cycle

$$\text{COP} = \frac{{}_4Q_1}{{}_1W_2} = \frac{h_1 - h_4}{h_2 - h_1} \tag{7.5}$$

where

Heat into evaporator ${}_4Q_1 = m(h_1 - h_4)$, kW

Work of compression ${}_1W_2 = m(h_2 - h_1)$ with s = constant, kW

Heat out to condenser ${}_2Q_3 = m(h_2 - h_3)$, kW

Expansion by throttling flow $h_3 = h_4$

m	=	refrigerant flow rate, kg/s
h	=	enthalpy, kJ/kg
s	=	entropy, kJ/kg·K

Theoretical Compressor Displacement

$$D = m\, v_1 \text{ m}^3\text{/s} \tag{7.6}$$

where v_1 = specific volume at suction, m^3/kg.

For a given cycle, capacity in kW of refrigeration:

$$m = \frac{\text{kW}}{h_1 - h_4} \tag{7.7}$$

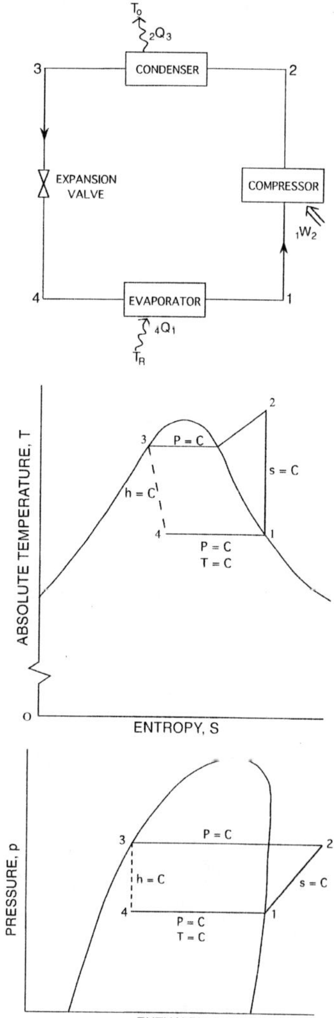

Figure 7.1 Theoretical Single-Stage Vapor Compression Refrigeration Cycle
[2021F, Ch 2, Fig 8]

Actual Refrigeration Systems. Actual systems operating steadily differ from the ideal cycles considered in the previous sections in many respects. Pressure drops occur everywhere in the system except in the compression process. Heat transfers between the refrigerant and its environment in all components.

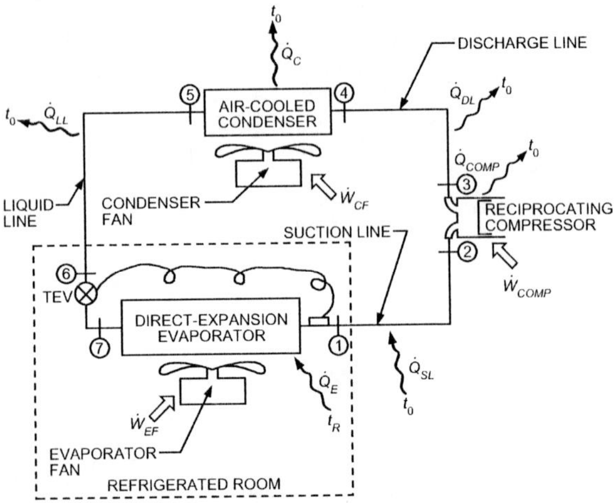

Figure 7.2 Schematic of Real, Direct-Expansion, Single-Stage Mechanical Vapor-Compression Refrigeration System [2021F, Ch 2, Fig 14]

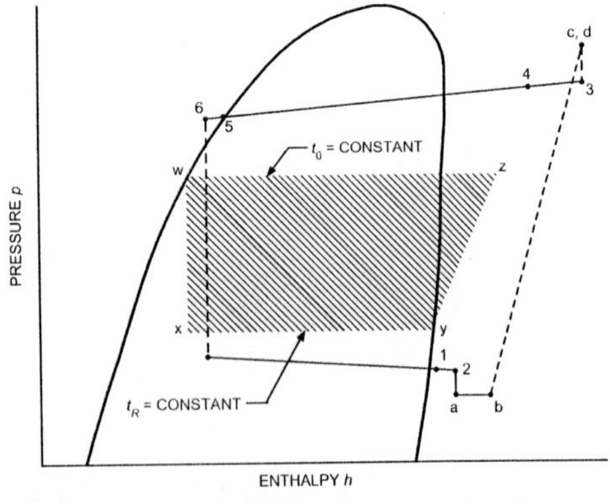

Figure 7.3 Pressure-Enthalpy Diagram of Actual System and Theoretical Single-Stage System Operating Between Same Inlet Air Temperatures t_R and t_0 [2021F, Ch 2, Fig 15]

8. REFRIGERANTS

Table 8.1 Physical Properties of Selected Refrigerants[a] [2021F, Ch 29, Tbl 5]

Refrigerant									
No.	**Chemical Name or Composition (% by Mass)**	**Chemical Formula**	**Molecular Mass**	**Boiling Pt.[f] (NBP) at 101.325 kPa, °C**	**Freezing Point, °C**	**Critical Temp., °C**	**Critical Pressure, kPa**	**Critical Density, kg/m³**	**Refractive Index of Liquid[b,c]**
728	Nitrogen	N_2	28.013	–195.8	–210.0	–146.96	3395.8	313.3	1.205 (83 K) 589.3 nm
729	Air	—	28.959	–194.25	—	–140.59	3789.6	335.94	—
740	Argon	Ar	39.948	–185.85	–189.34	–122.46	4863.0	535.6	1.233 (84 K) 589.3 nm
732	Oxygen	O_2	31.999	–182.96	–218.79	–118.57	5043.0	436.14	1.221 (92 K) 589.3 nm
50	Methane	CH_4	16.043	–161.48	–182.46	–82.586	4599.2	162.66	—
14	Tetrafluoromethane	CF_4	88.005	–128.05	–183.61	–45.64	3750.0	625.66	—
170	Ethane	C_2H_6	30.07	–88.581	–182.8	32.72	4872.2	206.18	—
508A	R-23/116 (39/61)	—	100.1	–87.60	—	10.192	3650.8	567.58	—
508B	R-23/116 (46/54)	—	95.394	–87.6	—	11.205	3771.6	568.45	—
23	Trifluoromethane	CHF_3	70.014	–82.018	–155.13	26.143	4832	526.5	—
13	Chlorotrifluoromethane	$CClF_3$	104.46	–81.48	–181.15	28.85	3879	582.88	1.146 (25)[2]
744	Carbon dioxide	CO_2	44.01	–78.4[d]	–56.558[e]	30.978	7377.3	467.6	1.195 (15)
504	R-32/115 (48.2/51.8)	—	79.249	–57.906	—	62.138	4428.8	504.68	—
32	Difluoromethane	CH_2F_2	52.024	–51.651	–136.81	78.105	5782.0	424	—
410A	R-32/125 (50/50)	—	72.585	–51.446	—	71.358	4902.6	459.53	—
125	Pentafluoroethane	C_2HF_5	120.02	–48.09	–100.63	66.023	3617.7	573.58	—

Table 8.1 Physical Properties of Selected Refrigerants[a] [2021F, Ch 29, Tbl 5] (*Continued*)

Refrigerant									
No.	Chemical Name or Composition (% by Mass)	Chemical Formula	Molecular Mass	Boiling Pt.[f] (NBP) at 101.325 kPa, °C	Freezing Point, °C	Critical Temp., °C	Critical Pressure, kPa	Critical Density, kg/m³	Refractive Index of Liquid[b,c]
1270	Propylene	C_3H_6	42.08	–47.62	–185.2	91.061	4554.8	230.03	1.3640 (–50)[1]
143a	Trifluoroethane	CH_3CF_3	84.041	–47.241	–111.81	72.707	3761.0	431.0	—
507A	R-125/143a (50/50)	—	98.859	–46.741	—	70.617	3705	490.77	—
404A	R-125/143a/134a (44/52/4)	—	97.604	–46.222	—	72.046	3728.9	486.53	—
502	R-22/115 (48.8/51.2)	—	111.63	–45.174	—	80.507	4016.8	568.70	—
407C	R-32/125/134a (23/25/52)	—	86.204	–43.627	—	86.034	4629.8	484.23	—
290	Propane	C_3H_8	44.096	–42.11	–187.62	96.74	4251.2	220.4	1.3397 (–42)
22	Chlorodifluoromethane	$CHClF_2$	86.468	–40.81	–157.42	96.145	4990.0	523.84	1.234 (25)[2]
115	Chloropentafluoroethane	$CClF_2CF_3$	154.47	–39.25	–99.39	79.95	3129.0	614.8	1.221 (25)[2]
500	R-12/152a (73.8/26.2)	—	99.303	–33.603	—	102.09	4168.6	495.1	—
717	Ammonia	NH_3	17.03	–33.327	–77.655	132.25	11 333.0	225.0[d]	1.325 (16.5)
12	Dichlorodifluoromethane	CCl_2F_2	120.91	–29.752	–157.05	111.97	4136.1	565.0	1.288 (25)[2]
1234yf	2,3,3,3-tetrafluoroprop-1-ene	$CF_3CF{=}CH_2$	114.04	–29.45		94.7	3382.2	475.55	
134a	Tetrafluoroethane	CF_3CH_2F	102.03	–26.074	–103.3	101.06	4059.3	511.9	—
152a	Difluoroethane	CHF_2CH_3	66.051	–24.023	–118.59	113.26	4516.8	368	—
1234ze (E)	Trans-1,3,3,3- tetrafluoropropene	$CF_3CH{=}CHF$	114.04	–18.95		109.37	3636.3	489.24	
124	Chlorotetrafluoroethane	$CHClFCF_3$	136.48	–11.963	–199.15	122.28	3624.3	560.0	—
600a	Isobutane	C_4H_{10}	58.122	–11.75	–159.42	134.66	3629.0	225.5	1.3514 (–25)[1]

Table 8.1 Physical Properties of Selected Refrigerants[a] [2021F, Ch 29, Tbl 3] (Continued)

Refrigerant									
No.	Chemical Name or Composition (% by Mass)	Chemical Formula	Molecular Mass	Boiling Pt.[f] (NBP) at 101.325 kPa, °C	Freezing Point, °C	Critical Temp., °C	Critical Pressure, kPa	Critical Density, kg/m^3	Refractive Index of Liquid[b,c]
142b	Chlorodifluoroethane	$CClF_2CH_3$	100.5	–9.15	–130.43	137.11	4055.0	466.0	—
C318	Octafluorocyclobutane	C_4F_8	200.03	–5.975	–39.8	115.23	2777.5	619.97	
600	Butane	C_4H_{10}	58.122	–0.49	–138.27	151.98	3796.0	227.94	1.3562 (–15)[1]
1336mzz (Z)	Cis-1,1,1,4,4,4-hexafluoro-2-butene	$CF_3CH{=}CHCF_3$	164.1	92.1	—	171.3	2903	504.67	—
114	Dichlorotetrafluoroethane	$CClF_2CClF_2$	170.92	3.586	–92.5	145.68	3257.0	579.97	1.294 (25)
1233zd (E)	Trans-1-chloro-3,3,3-trifluoro-1-propene	$CF_3CH{=}CHCl$	130.5	18.1	—	165.6	3580	480.769	—
11	Trichlorofluoromethane	CCl_3F	137.37	23.708	–110.47	197.96	4407.6	554.0	1.362 (25)[2]
123	Dichlorotrifluoroethane	$CHCl_2CF_3$	152.93	27.823	–107.15	183.68	3661.8	550.0	—
141b	Dichlorofluoroethane	CCl_2FCH_3	116.95	32.05	–103.5	204.4	4212.0	458.6	—
113	Trichlorotrifluoroethane	CCl_2FCClF_2	187.38	47.585	–36.22	214.06	3392.2	560.0	1.357 (25)[2]
718[3]	Water	H_2O	18.015	99.974	0.01	373.95	22 064.0	322.0	—

Notes:
[a] Data from NIST (2010) REFPROP v. 9.0.
[b] Temperature of measurement (°C, unless kelvin is noted) shown in parentheses. Data from CRC (1987), unless otherwise noted.
[c] For the sodium D line.
[d] Sublimes.
[e] At 527 kPa.
[f] Bubble point used for blends

References:
[1] Kirk and Othmer (1956).
[2] *Bulletin* B-32A (DuPont).
[3] *Handbook of Chemistry* (1967).

Ozone-depleting substances, including CFCs and HCFCs, are to be phased out of production under the Montreal Protocol. In the United States, production and importation of CFCs were banned completely in 1996. HCFCs are being phased down, with complete phaseout set for 2030.

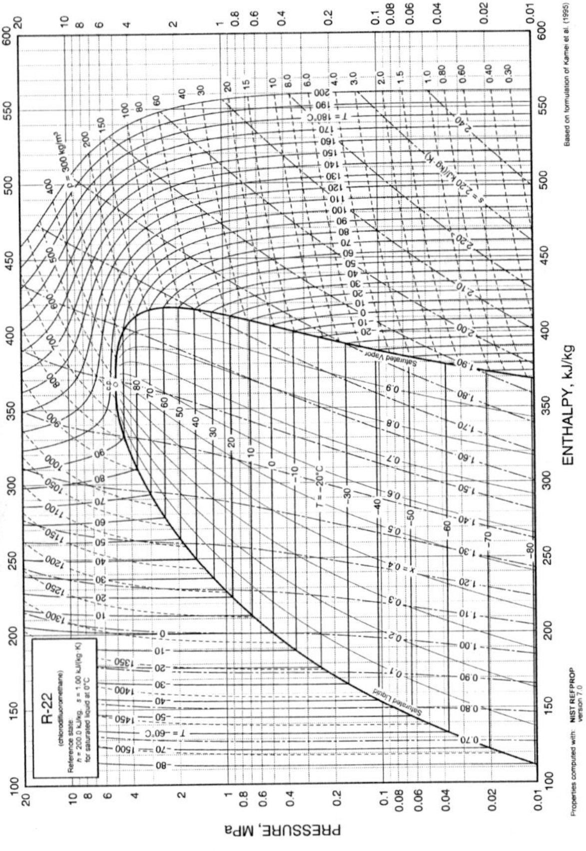

Figure 8.1 Pressure-Enthalpy Diagram for Refrigerant 22
[2021F, Ch 30, Fig 2]

Table 8.2 R-22 (Chlorodifluoromethane)
Properties of Saturated Liquid and Saturated Vapor [2021F, Ch 30, Tbl R-22]

Temp.,* °C	Pressure, MPa	Density, kg/m³ Liquid	Volume, m³/kg Vapor	Enthalpy, kJ/kg		Entropy, kJ/(kg·K)	
				Liquid	Vapor	Liquid	Vapor
−100	0.00201	1571.3	8.26600	90.71	358.97	0.5050	2.0543
−90	0.00481	1544.9	3.64480	101.32	363.85	0.5646	1.9980
−80	0.01037	1518.2	1.77820	111.94	368.77	0.6210	1.9508
−70	0.02047	1491.2	0.94342	122.58	373.70	0.6747	1.9108
−60	0.03750	1463.7	0.53680	133.27	378.59	0.7260	1.8770
−50	0.06453	1435.6	0.32385	144.03	383.42	0.7752	1.8480
−48	0.07145	1429.9	0.29453	146.19	384.37	0.7849	1.8428
−46	0.07894	1424.2	0.26837	148.36	385.32	0.7944	1.8376
−44	0.08705	1418.4	0.24498	150.53	386.26	0.8039	1.8327
−42	0.09580	1412.6	0.22402	152.70	387.20	0.8134	1.8278
−40.81[b]	0.10132	1409.2	0.21260	154.00	387.75	0.8189	1.8250
−40	0.10523	1406.8	0.20521	154.89	388.13	0.8227	1.8231
−38	0.11538	1401.0	0.18829	157.07	389.06	0.8320	1.8186
−36	0.12628	1395.1	0.17304	159.27	389.97	0.8413	1.8141
−34	0.13797	1389.1	0.15927	161.47	390.89	0.8505	1.8098
−32	0.15050	1383.2	0.14682	163.67	391.79	0.8596	1.8056
−30	0.16389	1377.2	0.13553	165.88	392.69	0.8687	1.8015
−28	0.17819	1371.1	0.12528	168.10	393.58	0.8778	1.7975
−26	0.19344	1365.0	0.11597	170.33	394.47	0.8868	1.7937
−24	0.20968	1358.9	0.10749	172.56	395.34	0.8957	1.7899
−22	0.22696	1352.7	0.09975	174.80	396.21	0.9046	1.7862
−20	0.24531	1346.5	0.09268	177.04	397.06	0.9135	1.7826
−18	0.26479	1340.3	0.08621	179.30	397.91	0.9223	1.7791
−16	0.28543	1334.0	0.08029	181.56	398.75	0.9311	1.7757
−14	0.30728	1327.6	0.07485	183.83	399.57	0.9398	1.7723
−12	0.33038	1321.2	0.06986	186.11	400.39	0.9485	1.7690
−10	0.35479	1314.7	0.06527	188.40	401.20	0.9572	1.7658
−8	0.38054	1308.2	0.06103	190.70	401.99	0.9658	1.7627
−6	0.40769	1301.6	0.05713	193.01	402.77	0.9744	1.7596
−4	0.43628	1295.0	0.05352	195.33	403.55	0.9830	1.7566
−2	0.46636	1288.3	0.05019	197.66	404.30	0.9915	1.7536
0	0.49799	1281.5	0.04710	200.00	405.05	1.0000	1.7507
2	0.53120	1274.7	0.04424	202.35	405.78	1.0085	1.7478
4	0.56605	1267.8	0.04159	204.71	406.50	1.0169	1.7450
6	0.60259	1260.8	0.03913	207.09	407.20	1.0254	1.7422
8	0.64088	1253.8	0.03683	209.47	407.89	1.0338	1.7395
10	0.68095	1246.7	0.03470	211.87	408.56	1.0422	1.7368
12	0.72286	1239.5	0.03271	214.28	409.21	1.0505	1.7341
14	0.76668	1232.2	0.03086	216.70	409.85	1.0589	1.7315
16	0.81244	1224.9	0.02912	219.14	410.47	1.0672	1.7289
18	0.86020	1217.4	0.02750	221.59	411.07	1.0755	1.7263
20	0.91002	1209.9	0.02599	224.06	411.66	1.0838	1.7238
22	0.96195	1202.3	0.02457	226.54	412.22	1.0921	1.7212
24	1.01600	1194.6	0.02324	229.04	412.77	1.1004	1.7187
26	1.07240	1186.7	0.02199	231.55	413.29	1.1086	1.7162
28	1.13090	1178.8	0.02082	234.08	413.79	1.1169	1.7136
30	1.19190	1170.7	0.01972	236.62	414.26	1.1252	1.7111
32	1.25520	1162.6	0.01869	239.19	414.71	1.1334	1.7086
34	1.32100	1154.3	0.01771	241.77	415.14	1.1417	1.7061
36	1.38920	1145.8	0.01679	244.38	415.54	1.1499	1.7036
38	1.46010	1137.3	0.01593	247.00	415.91	1.1582	1.7010
40	1.53360	1128.5	0.01511	249.65	416.25	1.1665	1.6985
42	1.60980	1119.6	0.01433	252.32	416.55	1.1747	1.6959
44	1.68870	1110.6	0.01360	255.01	416.83	1.1830	1.6933
46	1.77040	1101.4	0.01291	257.73	417.07	1.1913	1.6906
48	1.85510	1091.9	0.01226	260.47	417.27	1.1997	1.6879
50	1.94270	1082.3	0.01163	263.25	417.44	1.2080	1.6852
52	2.03330	1072.4	0.01104	266.05	417.56	1.2164	1.6824
54	2.12700	1062.3	0.01048	268.89	417.63	1.2248	1.6795
56	2.22390	1052.0	0.00995	271.76	417.66	1.2333	1.6766
58	2.32400	1041.3	0.00944	274.66	417.63	1.2418	1.6736
60	2.42750	1030.4	0.00896	277.61	417.55	1.2504	1.6705
65	2.70120	1001.4	0.00785	285.18	417.06	1.2722	1.6622
70	2.99740	969.7	0.00685	293.10	416.09	1.2945	1.6529
75	3.31770	934.4	0.00595	301.46	414.49	1.3177	1.6424
80	3.66380	893.7	0.00512	310.44	412.01	1.3423	1.6299
85	4.03780	844.8	0.00434	320.38	408.19	1.3690	1.6142
90	4.44230	780.1	0.00356	332.09	401.87	1.4001	1.5922
95	4.88240	662.9	0.00262	349.56	387.28	1.4462	1.5486
96.15[c]	4.99000	523.8	0.00191	366.90	366.90	1.4927	1.4927

*Temperatures on ITS-90 scale [b]Normal boiling point [c]Critical point

Table 8.3 Superheated Vapor Thermodynamic Properties of R-22

Temp., °C	Pressure = 200 kPa Sat. Temp. = −25.20°C			Pressure = 400 kPa Sat. Temp. = −6.57°C		
	ρ	*h*	*s*	ρ	*h*	*s*
−20	8.647	243.8	0.9847			
−10	8.267	250.0	1.009			
0	7.923	256.3	1.032	16.56	252.9	0.9569
10	7.609	262.8	1.056	15.82	259.5	0.9809
20	7.321	269.3	1.078	15.15	266.3	1.004
40	6.810	282.7	1.122	13.99	280.1	1.050
60	6.370	296.4	1.165	13.01	294.1	1.093
Temp., °C	**Pressure = 500 kPa Sat. Temp. = 0.11°C**			**Pressure = 600 kPa Sat. Temp. = 5.84°C**		
10	20.19	257.8	0.9552	24.77	265.1	0.9332
20	19.28	264.7	0.9791	23.58	263.1	0.9576
30	19.46	271.7	1.002	22.53	270.2	0.9814
40	17.73	278.7	1.025	21.58	277.3	1.005
50	17.06	285.8	1.048	20.73	284.5	1.027
60	16.44	293.0	1.069	19.95	291.8	1.049
70	15.87	300.2	1.091	19.24	299.1	1.071
Temp., °C	**Pressure = 800 kPa Sat. Temp. = 15.44°C**			**Pressure = 1000 kPa Sat. Temp. = 23.38°C**		
20	32.77	259.7	0.9214			
40	29.67	274.5	0.9702	38.34	271.4	0.9416
60	27.23	289.3	1.016	34.89	286.8	0.9891
80	25.24	304.4	1.060	32.15	302.2	1.034
100	23.57	319.8	1.102	29.90	317.9	1.077
120	22.14	335.4	1.143	28.00	333.7	1.119
140	20.89	351.3	1.183	26.37	349.9	1.159
Temp., °C	**Pressure = 1200 kPa Sat. Temp. = 30.21°C**			**Pressure = 1400 kPa Sat. Temp. = 36.25°C**		
40	47.73	268.2	0.9163	58.00	364.7	0.8930
60	43.00	284.1	0.9657	51.61	281.3	0.9445
80	39.36	300.0	1.012	46.90	297.7	0.9922
100	36.44	315.9	1.056	43.20	314.0	1.037
120	34.02	332.1	1.098	40.19	330.4	1.080
140	31.96	348.4	1.138	37.66	346.9	1.121
160	30.18	365.0	1.178	35.49	363.7	1.161
Temp., °C	**Pressure = 1600 kPa Sat. Temp. = 41.69°C**			**Pressure = 1800 kPa Sat. Temp. = 46.65°C**		
60	60.82	278.4	0.9249	70.76	275.2	0.9064
80	54.80	295.3	0.9742	63.11	292.8	0.9575
100	50.20	311.9	1.020	57.47	309.9	1.005
120	46.53	328.6	1.064	53.04	326.9	1.049
140	43.48	345.4	1.105	49.43	343.9	1.091
160	40.90	362.4	1.146	46.40	361.1	1.132

ρ = vapor density, kg/m^3; *h* = enthalpy, kJ/kg; *s* = entropy, kJ/(kg·K)

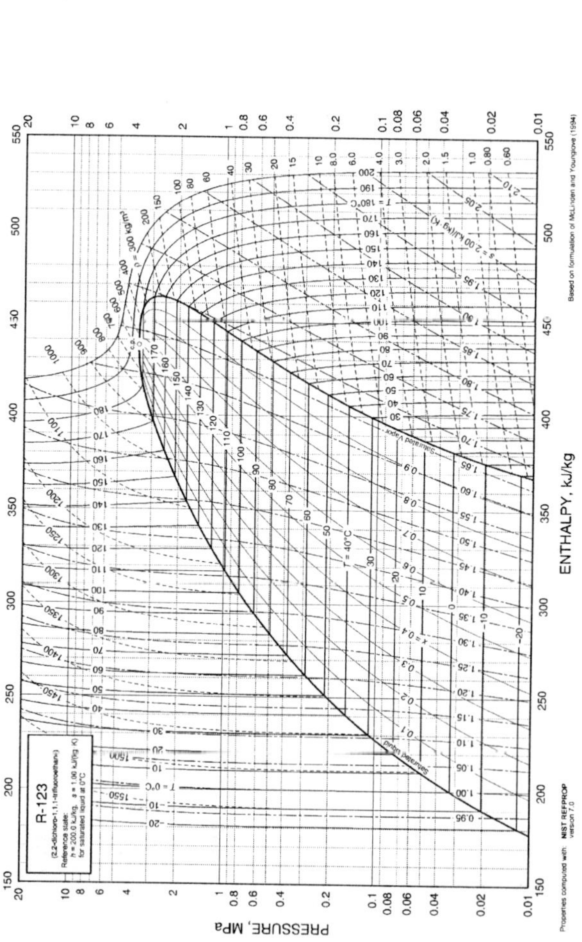

Figure 8.2 **Pressure-Enthalpy Diagram for Refrigerant 123**
[2021F, Ch 30, Fig 5]

Table 8.4 R-123 (2,2-Dichloro-1,1,1-Trifluoroethane) Properties of Saturated Liquid and Saturated Vapor [2021F, Ch 30, Tbl R-123]

Temp.,* °C	Pressure, MPa	Density, kg/m³ Liquid	Volume, m³/kg Vapor	Enthalpy, kJ/kg Liquid	Enthalpy, kJ/kg Vapor	Entropy, kJ/(kg·K) Liquid	Entropy, kJ/(kg·K) Vapor	c_p/c_v Vapor
−80	0.00013	1709.6	83.6670	123.92	335.98	0.6712	1.7691	1.117
−70	0.00034	1687.4	32.8420	133.17	341.25	0.7179	1.7422	1.113
−60	0.00081	1665.1	14.3330	142.46	346.66	0.7625	1.7206	1.110
−50	0.00177	1642.6	6.84600	151.81	352.21	0.8054	1.7034	1.107
−40	0.00358	1620.0	3.53190	161.25	357.88	0.8468	1.6901	1.105
−30	0.00675	1597.0	1.94700	170.78	363.65	0.8868	1.6800	1.103
−20	0.01200	1573.8	1.13640	180.41	369.52	0.9256	1.6726	1.102
−10	0.02025	1550.1	0.69690	190.15	375.45	0.9633	1.6675	1.102
0	0.03265	1526.1	0.44609	200.00	381.44	1.0000	1.6642	1.102
2	0.03574	1521.3	0.40991	201.98	382.64	1.0072	1.6638	1.103
4	0.03907	1516.4	0.37720	203.97	383.84	1.0144	1.6634	1.103
6	0.04264	1511.5	0.34759	205.97	385.05	1.0216	1.6631	1.103
8	0.04647	1506.6	0.32075	207.96	386.25	1.0287	1.6628	1.103
10	0.05057	1501.6	0.29637	209.97	387.46	1.0358	1.6626	1.104
12	0.05495	1496.7	0.27420	211.97	388.66	1.0428	1.6625	1.104
14	0.05963	1491.7	0.25401	213.99	389.87	1.0499	1.6624	1.104
16	0.06463	1486.7	0.23559	216.00	391.08	1.0569	1.6623	1.105
18	0.06995	1481.7	0.21877	218.02	392.29	1.0638	1.6623	1.105
20	0.07561	1476.6	0.20338	220.05	393.49	1.0707	1.6624	1.106
22	0.08163	1471.5	0.18929	222.08	394.70	1.0776	1.6625	1.106
24	0.08802	1466.4	0.17637	224.12	395.91	1.0845	1.6626	1.107
26	0.09480	1461.3	0.16451	226.16	397.12	1.0913	1.6628	1.107
27.82[b]	0.10133	1456.6	0.15453	228.03	398.22	1.0975	1.6630	1.108
28	0.10198	1456.2	0.15360	228.21	398.32	1.0981	1.6630	1.108
30	0.10958	1451.0	0.14356	230.26	399.53	1.1049	1.6633	1.109
32	0.11762	1445.8	0.13431	232.31	400.73	1.1116	1.6635	1.109
34	0.12611	1440.6	0.12577	234.38	401.93	1.1183	1.6639	1.110
36	0.13507	1435.4	0.11789	236.44	403.14	1.1250	1.6642	1.111
38	0.14452	1430.1	0.11060	238.51	404.34	1.1317	1.6646	1.112
40	0.15447	1424.8	0.10385	240.59	405.54	1.1383	1.6651	1.113
42	0.16495	1419.4	0.09759	242.67	406.73	1.1449	1.6655	1.114
44	0.17597	1414.1	0.09179	244.76	407.93	1.1515	1.6660	1.115
46	0.18755	1408.7	0.08641	246.86	409.12	1.1581	1.6665	1.116
48	0.19971	1403.3	0.08140	248.95	410.31	1.1646	1.6670	1.117
50	0.21246	1397.8	0.07674	251.06	411.50	1.1711	1.6676	1.119
52	0.22584	1392.3	0.07240	253.17	412.69	1.1776	1.6682	1.120
54	0.23985	1386.8	0.06836	255.28	413.87	1.1840	1.6688	1.121
56	0.25451	1381.2	0.06458	257.41	415.05	1.1905	1.6694	1.123
58	0.26985	1375.6	0.06106	259.53	416.23	1.1969	1.6701	1.124
60	0.28589	1370.0	0.05777	261.67	417.40	1.2033	1.6707	1.126
62	0.30264	1364.3	0.05469	263.81	418.57	1.2096	1.6714	1.127
64	0.32013	1358.6	0.05180	265.95	419.73	1.2160	1.6721	1.129
66	0.33838	1352.8	0.04910	268.10	420.89	1.2223	1.6728	1.131
68	0.35740	1347.0	0.04656	270.26	422.05	1.2286	1.6735	1.133
70	0.37722	1341.2	0.04418	272.42	423.20	1.2349	1.6743	1.135
72	0.39787	1335.3	0.04195	274.60	424.35	1.2411	1.6750	1.137
74	0.41936	1329.3	0.03985	276.77	425.50	1.2474	1.6758	1.139
76	0.44171	1323.4	0.03787	278.96	426.63	1.2536	1.6766	1.142
78	0.46494	1317.3	0.03601	281.15	427.77	1.2598	1.6774	1.144
80	0.48909	1311.2	0.03426	283.35	428.89	1.2660	1.6781	1.147
82	0.51416	1305.1	0.03261	285.55	430.01	1.2722	1.6789	1.150
84	0.54019	1298.9	0.03105	287.77	431.13	1.2783	1.6797	1.152
86	0.56720	1292.6	0.02958	289.99	432.23	1.2845	1.6806	1.156
88	0.59520	1286.3	0.02819	292.22	433.33	1.2906	1.6814	1.159
90	0.62423	1279.9	0.02687	294.45	434.43	1.2967	1.6822	1.162
92	0.65430	1273.5	0.02563	296.70	435.51	1.3028	1.6830	1.166
94	0.68544	1266.9	0.02445	298.95	436.59	1.3089	1.6838	1.169
96	0.71768	1260.3	0.02334	301.21	437.66	1.3150	1.6846	1.173
98	0.75103	1253.7	0.02228	303.49	438.72	1.3211	1.6854	1.177
100	0.78553	1246.9	0.02128	305.77	439.77	1.3271	1.6862	1.182
110	0.97603	1211.9	0.01697	317.32	444.88	1.3572	1.6902	1.208
120	1.19900	1174.4	0.01361	329.15	449.67	1.3872	1.6938	1.243
130	1.45780	1133.6	0.01094	341.32	454.07	1.4173	1.6969	1.294
140	1.75630	1088.3	0.00879	353.92	457.94	1.4475	1.6992	1.369
150	2.09870	1036.8	0.00703	367.10	461.05	1.4782	1.7003	1.493
160	2.49010	975.7	0.00555	381.13	463.01	1.5101	1.6991	1.726
170	2.93720	896.9	0.00425	396.61	462.89	1.5443	1.6939	2.309
180	3.45060	765.9	0.00292	416.22	456.82	1.5867	1.6763	6.158
183.68[c]	3.66180	550.0	0.00182	437.39	437.39	1.6325	1.6325	∞

*Temperatures on ITS-90 scale

[b]Normal boiling point

[c]Critical point

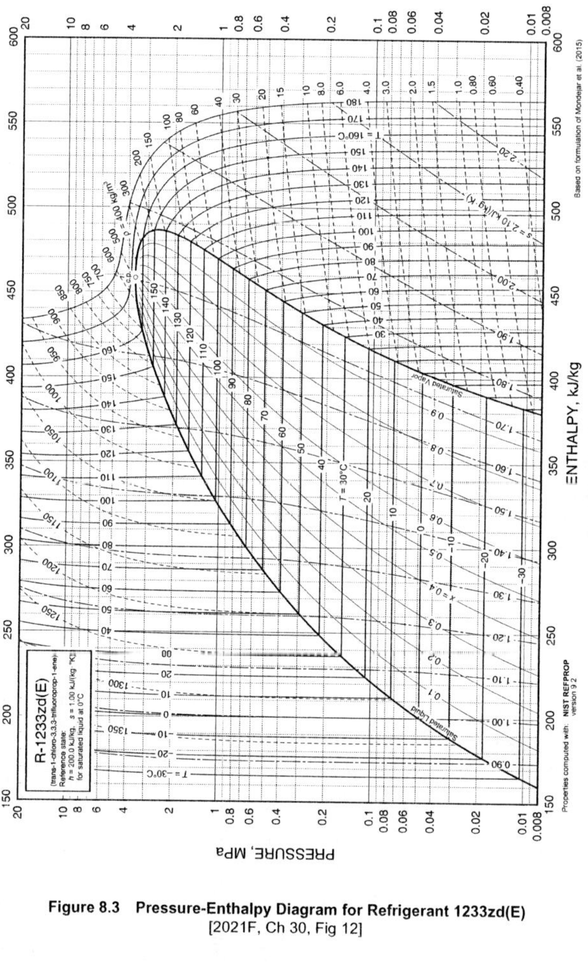

Figure 8.3 Pressure-Enthalpy Diagram for Refrigerant 1233zd(E)
[2021F, Ch 30, Fig 12]

Table 8.5 Refrigerant 1233zd(E) (trans-1-chloro-3,3,3-trifluoroprop-1-ene) Properties of Saturated Liquid and Saturated Vapor [2021F, Ch 30, Tbl R-1234zd(E)]

Temp., °C	Pressure, MPa	Density, kg/m³ Liquid	Volume, m³/kg Vapor	Enthalpy, kJ/kg Liquid	Enthalpy, kJ/kg Vapor	Entropy, kJ/(kg·K) Liquid	Entropy, kJ/(kg·K) Vapor	Specific Heat c_p, kJ/(kg·K) Liquid	Specific Heat c_p, kJ/(kg·K) Vapor	c_p/c_v Vapor
−50	0.00289	1429.8	4.9048	142.87	369.48	0.7694	1.7849	1.109	0.660	1.108
−45	0.00410	1419.3	3.5348	148.43	372.76	0.7940	1.7773	1.115	0.672	1.107
−40	0.00570	1408.7	2.5918	154.02	376.08	0.8182	1.7707	1.122	0.683	1.105
−35	0.00781	1398.1	1.9310	159.65	379.43	0.8421	1.7650	1.128	0.694	1.104
−30	0.01052	1387.4	1.4601	165.31	382.82	0.8656	1.7602	1.135	0.704	1.104
−25	0.01398	1376.6	1.1193	171.00	386.24	0.8888	1.7561	1.142	0.715	1.103
−20	0.01833	1365.7	0.86907	176.73	389.68	0.9116	1.7528	1.149	0.726	1.103
−18	0.02035	1361.3	0.78804	179.03	391.06	0.9207	1.7517	1.152	0.730	1.102
−16	0.02256	1357.0	0.71589	181.34	392.45	0.9297	1.7506	1.154	0.734	1.102
−14	0.02495	1352.6	0.65150	183.65	393.83	0.9386	1.7497	1.157	0.739	1.102
−12	0.02755	1348.1	0.59392	185.97	395.22	0.9475	1.7488	1.160	0.743	1.102
−10	0.03036	1343.7	0.54235	188.29	396.62	0.9564	1.7480	1.163	0.747	1.103
−8	0.03340	1339.3	0.49606	190.62	398.01	0.9652	1.7473	1.166	0.751	1.103
−6	0.03668	1334.8	0.45444	192.96	399.41	0.9740	1.7467	1.169	0.756	1.103
−4	0.04021	1330.3	0.41695	195.30	400.80	0.9827	1.7462	1.171	0.760	1.103
−2	0.04402	1325.8	0.38313	197.65	402.20	0.9914	1.7458	1.174	0.764	1.103
0	0.04811	1321.3	0.35257	200.00	403.60	1.0000	1.7454	1.177	0.769	1.104
2	0.05250	1316.7	0.32490	202.36	405.00	1.0086	1.7451	1.180	0.773	1.104
4	0.05720	1312.2	0.29981	204.72	406.40	1.0171	1.7448	1.183	0.778	1.104
6	0.06224	1307.6	0.27702	207.10	407.80	1.0257	1.7447	1.186	0.782	1.105
8	0.06762	1303.0	0.25630	209.47	409.21	1.0341	1.7445	1.189	0.787	1.105
10	0.07337	1298.3	0.23743	211.86	410.61	1.0426	1.7445	1.192	0.791	1.106
12	0.07949	1293.7	0.22022	214.25	412.01	1.0510	1.7445	1.195	0.796	1.106
14	0.08601	1289.0	0.20449	216.64	413.41	1.0593	1.7445	1.198	0.801	1.107
16	0.09295	1284.3	0.19011	219.04	414.81	1.0676	1.7447	1.201	0.805	1.108
18	0.10032	1279.6	0.17694	221.45	416.20	1.0759	1.7448	1.205	0.810	1.108
18.26[b]	0.10132	1278.9	0.17529	221.77	416.39	1.0770	1.7448	1.205	0.811	1.108
20	0.10815	1274.8	0.16486	223.87	417.60	1.0841	1.7450	1.208	0.815	1.109
22	0.11644	1270.0	0.15376	226.29	419.00	1.0924	1.7453	1.211	0.820	1.110
24	0.12522	1265.2	0.14356	228.72	420.39	1.1005	1.7456	1.214	0.825	1.111
26	0.13452	1260.4	0.13417	231.15	421.78	1.1087	1.7459	1.218	0.830	1.112
28	0.14434	1255.5	0.12552	233.59	423.17	1.1168	1.7463	1.221	0.835	1.113
30	0.15471	1250.6	0.11753	236.04	424.56	1.1249	1.7467	1.224	0.840	1.114
32	0.16566	1245.7	0.11015	238.50	425.94	1.1329	1.7472	1.228	0.845	1.115
34	0.17719	1240.7	0.10333	240.96	427.32	1.1409	1.7477	1.231	0.850	1.116
36	0.18934	1235.7	0.09702	243.43	428.70	1.1489	1.7482	1.235	0.856	1.117
38	0.20212	1230.7	0.09116	245.91	430.08	1.1569	1.7488	1.238	0.861	1.119
40	0.21555	1225.6	0.08573	248.39	431.45	1.1648	1.7493	1.242	0.866	1.120
42	0.22966	1220.5	0.08069	250.88	432.81	1.1727	1.7500	1.246	0.872	1.122
44	0.24447	1215.4	0.07600	253.38	434.18	1.1805	1.7506	1.250	0.878	1.123
46	0.26000	1210.2	0.07163	255.89	435.54	1.1884	1.7513	1.254	0.884	1.125
48	0.27628	1205.0	0.06757	258.40	436.89	1.1962	1.7520	1.258	0.889	1.127
50	0.29332	1199.7	0.06378	260.93	438.24	1.2040	1.7527	1.262	0.895	1.128
52	0.31116	1194.4	0.06024	263.46	439.59	1.2117	1.7534	1.266	0.902	1.130
54	0.32981	1189.1	0.05693	266.00	440.93	1.2195	1.7542	1.270	0.908	1.132
56	0.34929	1183.7	0.05385	268.55	442.26	1.2272	1.7550	1.274	0.914	1.134
58	0.36964	1178.3	0.05096	271.10	443.59	1.2349	1.7558	1.279	0.921	1.137
60	0.39088	1172.8	0.04825	273.67	444.91	1.2426	1.7566	1.283	0.927	1.139
62	0.41303	1167.2	0.04571	276.24	446.23	1.2502	1.7574	1.288	0.934	1.142
64	0.43612	1161.7	0.04333	278.83	447.54	1.2578	1.7582	1.293	0.941	1.144
66	0.46017	1156.0	0.04110	281.42	448.84	1.2655	1.7591	1.298	0.948	1.147
68	0.48521	1150.3	0.03900	284.03	450.14	1.2730	1.7600	1.303	0.955	1.150
70	0.51126	1144.6	0.03702	286.64	451.43	1.2806	1.7608	1.308	0.963	1.153
72	0.53835	1138.8	0.03517	289.26	452.70	1.2882	1.7617	1.313	0.970	1.156
74	0.56651	1132.9	0.03342	291.90	453.98	1.2957	1.7626	1.319	0.978	1.159
76	0.59576	1127.0	0.03177	294.54	455.24	1.3032	1.7635	1.324	0.986	1.163
78	0.62613	1121.0	0.03021	297.20	456.49	1.3107	1.7644	1.330	0.995	1.167
80	0.65765	1114.9	0.02875	299.86	457.73	1.3182	1.7653	1.336	1.003	1.171
90	0.83345	1083.4	0.02253	313.38	463.78	1.3555	1.7697	1.370	1.050	1.195
100	1.0423	1049.7	0.01778	327.24	469.49	1.3926	1.7739	1.412	1.107	1.227
110	1.2880	1013.1	0.01408	341.51	474.77	1.4297	1.7775	1.465	1.179	1.273
120	1.5750	972.6	0.01116	356.29	479.44	1.4671	1.7803	1.538	1.277	1.342
130	1.9080	926.9	0.00881	371.74	483.27	1.5050	1.7816	1.643	1.423	1.455
140	2.2928	873.1	0.00687	388.12	485.81	1.5441	1.7805	1.819	1.675	1.661
150	2.7364	805.4	0.00522	406.03	486.16	1.5856	1.7750	2.193	2.232	2.141
160	3.2485	704.0	0.00368	427.43	481.49	1.6340	1.7588	3.751	4.545	4.191
166.45[c]	3.6237	480.2	0.00208	458.27	458.27	1.7032	1.7032	∞	∞	∞

[b]Normal boiling point

[c]Critical point

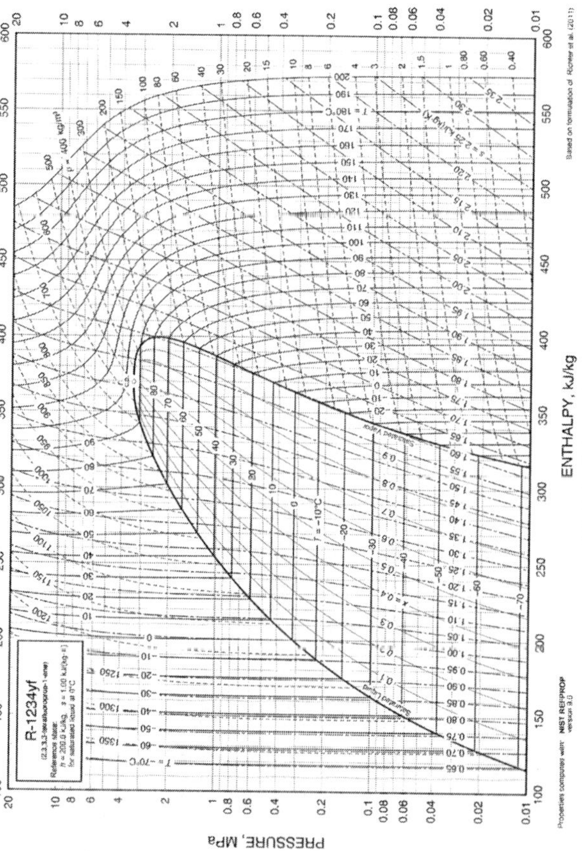

Figure 8.4 Pressure-Enthalpy Diagram for Refrigerant 1234yf
[2021F, Ch 30, Fig 13]

Table 8.6 Refrigerant 1234yf (2,3,3,3-tetrafluoroprop-1-ene) Properties of Saturated Liquid and Saturated Vapor [2021F, Ch 30, Tbl R-1234yf]

Temp., °C	Pressure, MPa	Density, kg/m³	Volume, m³/kg	Enthalpy, kJ/kg		Entropy, kJ/(kg·K)		Specific Heat c_p, kJ/(kg·K)		c_p/c_v
		Liquid	Vapor	Liquid	Vapor	Liquid	Vapor	Liquid	Vapor	Vapor
−50	0.03742	1318.4	0.42472	139.63	329.85	0.7573	1.6098	1.128	0.746	1.124
−45	0.04862	1305.2	0.33259	145.31	333.21	0.7825	1.6060	1.143	0.762	1.125
−40	0.06237	1291.9	0.26354	151.07	336.58	0.8074	1.6031	1.157	0.778	1.126
−35	0.07904	1278.3	0.21110	156.90	339.95	0.8321	1.6007	1.173	0.794	1.127
−30	0.09906	1264.5	0.17079	162.81	343.32	0.8566	1.5990	1.188	0.811	1.129
−29.49[b]	0.10132	1263.1	0.16719	163.42	343.67	0.8591	1.5988	1.190	0.813	1.129
−28	0.10810	1259.0	0.15732	165.20	344.67	0.8663	1.5984	1.195	0.818	1.130
−26	0.11778	1253.4	0.14512	167.60	346.02	0.8760	1.5980	1.201	0.825	1.131
−24	0.12812	1247.7	0.13404	170.01	347.36	0.8857	1.5976	1.207	0.832	1.132
−22	0.13916	1242.0	0.12398	172.43	348.71	0.8954	1.5973	1.214	0.839	1.133
−20	0.15092	1236.3	0.11482	174.87	350.05	0.9050	1.5970	1.220	0.847	1.135
−18	0.16344	1230.5	0.10647	177.32	351.39	0.9146	1.5968	1.227	0.854	1.136
−16	0.17676	1224.7	0.09884	179.79	352.73	0.9242	1.5967	1.234	0.862	1.137
−14	0.19090	1218.8	0.09187	182.27	354.06	0.9338	1.5967	1.240	0.869	1.139
−12	0.20590	1212.9	0.08548	184.76	355.39	0.9433	1.5967	1.247	0.877	1.141
−10	0.22178	1207.0	0.07962	187.26	356.72	0.9528	1.5968	1.254	0.885	1.142
−8	0.23860	1200.9	0.07424	189.78	358.04	0.9623	1.5969	1.261	0.893	1.144
−6	0.25637	1194.9	0.06930	192.31	359.36	0.9717	1.5970	1.268	0.901	1.146
−4	0.27514	1188.7	0.06474	194.86	360.68	0.9812	1.5973	1.275	0.909	1.149
−2	0.29495	1182.5	0.06054	197.42	361.99	0.9906	1.5975	1.282	0.918	1.151
0	0.31582	1176.3	0.05667	200.00	363.29	1.0000	1.5978	1.289	0.926	1.153
2	0.33780	1170.0	0.05309	202.59	364.59	1.0094	1.5981	1.297	0.935	1.156
4	0.36092	1163.6	0.04977	205.20	365.88	1.0187	1.5985	1.304	0.944	1.159
6	0.38523	1157.2	0.04670	207.82	367.16	1.0281	1.5989	1.312	0.953	1.162
8	0.41075	1150.6	0.04385	210.45	368.44	1.0374	1.5993	1.320	0.962	1.165
10	0.43753	1144.0	0.04121	213.10	369.70	1.0467	1.5998	1.327	0.972	1.168
12	0.46561	1137.4	0.03875	215.77	370.96	1.0560	1.6003	1.335	0.982	1.172
14	0.49503	1130.6	0.03646	218.45	372.21	1.0653	1.6008	1.344	0.992	1.176
16	0.52583	1123.8	0.03433	221.15	373.45	1.0746	1.6013	1.352	1.002	1.180
18	0.55804	1116.9	0.03235	223.87	374.67	1.0838	1.6018	1.361	1.013	1.185
20	0.59172	1109.9	0.03049	226.60	375.89	1.0931	1.6024	1.369	1.024	1.190
22	0.62690	1102.8	0.02876	229.34	377.09	1.1023	1.6029	1.378	1.035	1.195
24	0.66363	1095.5	0.02714	232.11	378.28	1.1115	1.6034	1.387	1.047	1.201
26	0.70194	1088.2	0.02562	234.89	379.45	1.1208	1.6040	1.397	1.060	1.207
28	0.74189	1080.8	0.02420	237.69	380.61	1.1300	1.6045	1.407	1.073	1.213
30	0.78351	1073.3	0.02287	240.51	381.75	1.1392	1.6051	1.417	1.086	1.220
32	0.82686	1065.7	0.02162	243.35	382.87	1.1484	1.6056	1.427	1.101	1.228
34	0.87197	1057.9	0.02044	246.21	383.98	1.1576	1.6061	1.438	1.116	1.236
36	0.91890	1050.0	0.01934	249.09	385.06	1.1668	1.6066	1.449	1.132	1.245
38	0.96769	1042.0	0.01830	251.98	386.13	1.1759	1.6071	1.461	1.149	1.254
40	1.0184	1033.8	0.01732	254.90	387.17	1.1851	1.6075	1.473	1.167	1.265
42	1.0711	1025.5	0.01639	257.84	388.19	1.1943	1.6079	1.486	1.186	1.276
44	1.1257	1017.0	0.01552	260.81	389.18	1.2035	1.6083	1.500	1.207	1.289
46	1.1825	1008.3	0.01469	263.80	390.14	1.2128	1.6087	1.515	1.229	1.302
48	1.2413	999.4	0.01392	266.81	391.08	1.2220	1.6089	1.531	1.252	1.317
50	1.3023	990.4	0.01318	269.85	391.98	1.2312	1.6092	1.548	1.277	1.333
52	1.3656	981.1	0.01248	272.92	392.85	1.2405	1.6094	1.566	1.305	1.351
54	1.4311	971.6	0.01182	276.02	393.68	1.2498	1.6095	1.586	1.335	1.371
56	1.4989	961.8	0.01119	279.15	394.48	1.2592	1.6095	1.607	1.367	1.393
58	1.5692	951.7	0.01059	282.32	395.23	1.2685	1.6095	1.631	1.403	1.418
60	1.6419	941.3	0.01002	285.53	395.93	1.2779	1.6093	1.656	1.442	1.445
62	1.7171	930.6	0.00948	288.77	396.58	1.2874	1.6091	1.685	1.485	1.477
64	1.7949	919.5	0.00897	292.06	397.18	1.2969	1.6087	1.717	1.534	1.512
66	1.8754	907.9	0.00848	295.39	397.71	1.3065	1.6082	1.752	1.589	1.553
68	1.9586	895.8	0.00801	298.78	398.18	1.3162	1.6076	1.792	1.651	1.600
70	2.0445	883.2	0.00756	302.22	398.57	1.3260	1.6068	1.837	1.724	1.655
72	2.1334	870.0	0.00712	305.72	398.87	1.3359	1.6058	1.890	1.808	1.719
74	2.2252	856.1	0.00671	309.30	399.07	1.3459	1.6045	1.951	1.907	1.797
76	2.3201	841.4	0.00631	312.95	399.16	1.3561	1.6030	2.025	2.028	1.891
78	2.4181	825.7	0.00592	316.69	399.11	1.3664	1.6011	2.114	2.176	2.009
80	2.5194	809.0	0.00555	320.54	398.90	1.3770	1.5989	2.227	2.364	2.158
85	2.7879	760.4	0.00464	330.81	397.40	1.4049	1.5909	2.702	3.168	2.803
90	3.0803	694.1	0.00372	342.79	393.32	1.4370	1.5762	4.186	5.688	4.820
94.70[c]	3.3822	475.6	0.00210	369.55	369.55	1.5087	1.5087	∞	∞	∞

[b]Normal boiling point

[c]Critical point

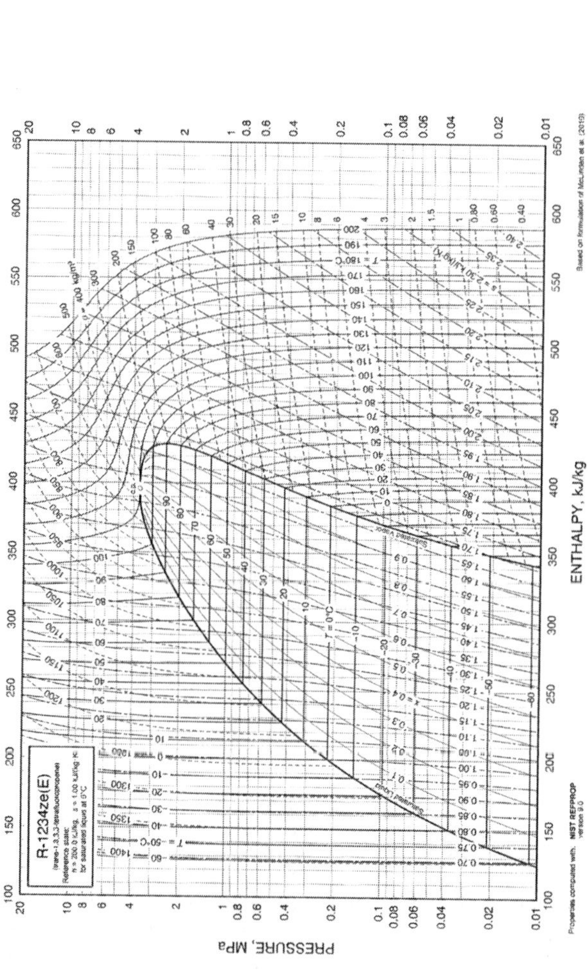

Figure 8.5 Pressure-Enthalpy Diagram for Refrigerant 1234ze(E)
[2021F, Ch 30, Fig 14]

Table 8.7 Refrigerant 1234ze(E) (Trans-1,3,3,3-tetrafluoropropene) Properties of Saturated Liquid and Saturated Vapor [2021F, Ch 30, Tbl R-1234ze(E)]

Temp., °C	Pressure, MPa	Density, kg/m³	Volume, m³/kg	Enthalpy, kJ/kg		Entropy, kJ/(kg·K)		Specific Heat c_p, kJ/(kg·K)		c_p/c_v
		Liquid	Vapor	Liquid	Vapor	Liquid	Vapor	Liquid	Vapor	Vapor
−50	0.02092	1375.3	0.76762	136.20	348.81	0.7429	1.6956	1.240	0.755	1.115
−45	0.02793	1362.4	0.58586	142.42	352.37	0.7704	1.6907	1.246	0.766	1.115
−40	0.03675	1349.5	0.45337	148.66	355.94	0.7975	1.6865	1.251	0.777	1.115
−35	0.04768	1336.5	0.35533	154.94	359.51	0.8241	1.6831	1.258	0.789	1.116
−30	0.06109	1323.3	0.28176	161.25	363.09	0.8503	1.6803	1.265	0.801	1.117
−25	0.07735	1309.9	0.22584	167.60	366.65	0.8761	1.6782	1.272	0.814	1.118
−20	0.09687	1296.4	0.18282	173.99	370.20	0.9015	1.6766	1.280	0.826	1.120
−18.97[b]	0.10132	1293.6	0.17525	175.31	370.93	0.9067	1.6763	1.282	0.829	1.121
−18	0.10569	1290.9	0.16843	176.56	371.62	0.9116	1.6761	1.284	0.832	1.121
−16	0.11512	1285.4	0.15540	179.13	373.03	0.9216	1.6756	1.287	0.837	1.122
−14	0.12520	1279.9	0.14357	181.71	374.44	0.9316	1.6752	1.291	0.843	1.123
−12	0.13596	1274.3	0.13281	184.30	375.85	0.9415	1.6749	1.294	0.848	1.125
−10	0.14744	1268.7	0.12302	186.90	377.25	0.9513	1.6747	1.298	0.854	1.126
−8	0.15965	1263.1	0.11409	189.50	378.64	0.9612	1.6745	1.302	0.860	1.127
−6	0.17263	1257.4	0.10593	192.11	380.04	0.9709	1.6744	1.306	0.865	1.128
−4	0.18642	1251.7	0.09847	194.73	381.42	0.9807	1.6743	1.310	0.871	1.130
−2	0.20105	1245.9	0.09163	197.36	382.80	0.9904	1.6743	1.314	0.877	1.132
0	0.21655	1240.1	0.08537	200.00	384.18	1.0000	1.6743	1.318	0.884	1.133
2	0.23296	1234.3	0.07961	202.65	385.55	1.0096	1.6743	1.323	0.890	1.135
4	0.25031	1228.4	0.07431	205.30	386.91	1.0192	1.6744	1.327	0.897	1.137
6	0.26863	1222.4	0.06944	207.97	388.27	1.0287	1.6746	1.332	0.903	1.139
8	0.28797	1216.4	0.06494	210.64	389.62	1.0382	1.6748	1.337	0.910	1.141
10	0.30836	1210.4	0.06079	213.32	390.96	1.0476	1.6750	1.342	0.917	1.144
12	0.32983	1204.3	0.05695	216.02	392.29	1.0571	1.6752	1.347	0.924	1.146
14	0.35243	1198.1	0.05340	218.72	393.61	1.0664	1.6755	1.353	0.932	1.149
16	0.37619	1191.9	0.05012	221.44	394.93	1.0758	1.6758	1.358	0.939	1.152
18	0.40114	1185.6	0.04707	224.16	396.23	1.0851	1.6761	1.364	0.947	1.155
20	0.42734	1179.3	0.04423	226.90	397.53	1.0944	1.6765	1.370	0.955	1.158
22	0.45482	1172.8	0.04160	229.65	398.81	1.1037	1.6768	1.376	0.963	1.161
24	0.48362	1166.4	0.03915	232.41	400.09	1.1129	1.6772	1.382	0.971	1.165
26	0.51377	1159.8	0.03687	235.19	401.35	1.1221	1.6776	1.389	0.980	1.169
28	0.54533	1153.2	0.03475	237.98	402.60	1.1313	1.6780	1.396	0.989	1.173
30	0.57833	1146.4	0.03276	240.78	403.83	1.1405	1.6784	1.403	0.999	1.177
32	0.61281	1139.6	0.03091	243.59	405.06	1.1497	1.6788	1.410	1.008	1.181
34	0.64882	1132.8	0.02918	246.42	406.27	1.1588	1.6792	1.418	1.018	1.186
36	0.68640	1125.8	0.02755	249.27	407.46	1.1679	1.6796	1.426	1.029	1.192
38	0.72560	1118.7	0.02603	252.13	408.64	1.1770	1.6801	1.434	1.039	1.197
40	0.76645	1111.5	0.02461	255.00	409.80	1.1861	1.6805	1.443	1.051	1.203
42	0.80901	1104.2	0.02327	257.89	410.95	1.1952	1.6809	1.452	1.063	1.210
44	0.85332	1096.8	0.02202	260.80	412.07	1.2043	1.6813	1.462	1.075	1.217
46	0.89943	1089.3	0.02084	263.73	413.18	1.2134	1.6816	1.472	1.088	1.224
48	0.94738	1081.6	0.01972	266.68	414.26	1.2224	1.6820	1.482	1.101	1.232
50	0.99722	1073.8	0.01868	269.64	415.33	1.2315	1.6823	1.493	1.116	1.241
52	1.0490	1065.9	0.01769	272.63	416.37	1.2405	1.6826	1.505	1.131	1.250
54	1.1028	1057.8	0.01676	275.63	417.38	1.2496	1.6829	1.517	1.147	1.260
56	1.1586	1049.5	0.01588	278.66	418.37	1.2587	1.6831	1.530	1.165	1.272
58	1.2165	1041.1	0.01505	281.71	419.33	1.2677	1.6833	1.544	1.184	1.284
60	1.2766	1032.5	0.01427	284.78	420.26	1.2768	1.6835	1.558	1.204	1.297
62	1.3388	1023.7	0.01353	287.88	421.16	1.2859	1.6836	1.574	1.225	1.312
64	1.4033	1014.7	0.01282	291.01	422.03	1.2950	1.6836	1.591	1.249	1.328
66	1.4702	1005.5	0.01216	294.16	422.86	1.3041	1.6836	1.609	1.274	1.345
68	1.5393	996.0	0.01152	297.34	423.65	1.3133	1.6835	1.628	1.302	1.365
70	1.6110	986.2	0.01092	300.56	424.40	1.3225	1.6834	1.649	1.333	1.386
72	1.6851	976.2	0.01035	303.80	425.10	1.3317	1.6831	1.672	1.367	1.411
74	1.7617	965.9	0.00980	307.09	425.76	1.3410	1.6828	1.698	1.404	1.438
76	1.8410	955.2	0.00929	310.41	426.36	1.3503	1.6824	1.726	1.445	1.468
78	1.9230	944.1	0.00879	313.77	426.90	1.3596	1.6818	1.757	1.492	1.503
80	2.0077	932.7	0.00832	317.19	427.38	1.3691	1.6811	1.792	1.545	1.542
85	2.2321	901.9	0.00722	325.95	428.25	1.3930	1.6787	1.901	1.712	1.671
90	2.4755	867.2	0.00623	335.12	428.52	1.4177	1.6749	2.062	1.963	1.868
95	2.7395	826.9	0.00531	344.89	427.93	1.4435	1.6691	2.327	2.386	2.206
100	3.0260	777.3	0.00444	355.59	425.95	1.4715	1.6600	2.859	3.258	2.905
105	3.3378	708.1	0.00356	368.23	421.12	1.5040	1.6438	4.573	6.122	5.179
109.36[c]	3.6349	489.2	0.00204	395.54	395.54	1.5744	1.5744	∞	∞	∞

[b]Normal boiling point

[c]Critical point

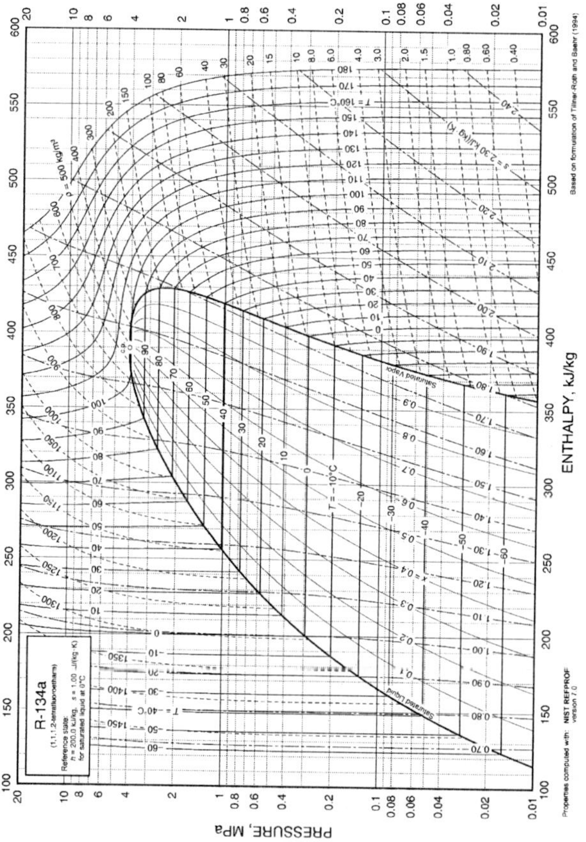

Figure 8.6 Pressure-Enthalpy Diagram for Refrigerant 134a
[2021F, Ch 30, Fig 8]

Table 8.8 R-134a (1,1,1,2-Tetrafluoroethane) Properties of Saturated Liquid and Saturated Vapor [2021F, Ch 30, Tbl R-134a]

Temp.,* °C	Pressure, MPa	Density, kg/m^3 Liquid	Volume, m^3/kg Vapor	Enthalpy, kJ/kg Liquid	Enthalpy, kJ/kg Vapor	Entropy, kJ/(kg·K) Liquid	Entropy, kJ/(kg·K) Vapor	c_p/c_v Vapor
−103.30[a]	0.00039	1591.1	35.4960	71.46	334.94	0.4126	1.9639	1.164
−100	0.00056	1582.4	25.1930	75.36	336.85	0.4354	1.9456	1.162
−90	0.00152	1555.8	9.7698	87.23	342.76	0.5020	1.8972	1.156
−80	0.00367	1529.0	4.2682	99.16	348.83	0.5654	1.8580	1.151
−70	0.00798	1501.9	2.0590	111.20	355.02	0.6262	1.8264	1.148
−60	0.01591	1474.3	1.0790	123.36	361.31	0.6846	1.8010	1.146
−50	0.02945	1446.3	0.60620	135.67	367.65	0.7410	1.7806	1.146
−40	0.05121	1417.7	0.36108	148.14	374.00	0.7956	1.7643	1.148
−30	0.08438	1388.4	0.22594	160.79	380.32	0.8486	1.7515	1.152
−28	0.09270	1382.4	0.20680	163.34	381.57	0.8591	1.7492	1.153
−26.07[b]	0.10133	1376.7	0.19018	165.81	382.78	0.8690	1.7472	1.154
−26	0.10167	1376.5	0.18958	165.90	382.82	0.8694	1.7471	1.154
−24	0.11130	1370.4	0.17407	168.47	384.07	0.8798	1.7451	1.155
−22	0.12165	1364.4	0.16006	171.05	385.32	0.8900	1.7432	1.156
−20	0.13273	1358.3	0.14739	173.64	386.55	0.9002	1.7413	1.158
−18	0.14460	1352.1	0.13592	176.23	387.79	0.9104	1.7396	1.159
−16	0.15728	1345.9	0.12551	178.83	389.02	0.9205	1.7379	1.161
−14	0.17082	1339.7	0.11605	181.44	390.24	0.9306	1.7363	1.163
−12	0.18524	1333.4	0.10744	184.07	391.46	0.9407	1.7348	1.165
−10	0.20060	1327.1	0.09959	186.70	392.66	0.9506	1.7334	1.167
−8	0.21693	1320.8	0.09242	189.34	393.87	0.9606	1.7320	1.169
−6	0.23428	1314.3	0.08587	191.99	395.06	0.9705	1.7307	1.171
−4	0.25268	1307.9	0.07987	194.65	396.25	0.9804	1.7294	1.174
−2	0.27217	1301.4	0.07436	197.32	397.43	0.9902	1.7282	1.176
0	0.29280	1294.8	0.06931	200.00	398.60	1.0000	1.7271	1.179
2	0.31462	1288.1	0.06466	202.69	399.77	1.0098	1.7260	1.182
4	0.33766	1281.4	0.06039	205.40	400.92	1.0195	1.7250	1.185
6	0.36198	1274.7	0.05644	208.11	402.06	1.0292	1.7240	1.189
8	0.38761	1267.9	0.05280	210.84	403.20	1.0388	1.7230	1.192
10	0.41461	1261.0	0.04944	213.58	404.32	1.0485	1.7221	1.196
12	0.44301	1254.0	0.04633	216.33	405.43	1.0581	1.7212	1.200
14	0.47288	1246.9	0.04345	219.09	406.53	1.0677	1.7204	1.204
16	0.50425	1239.8	0.04078	221.87	407.61	1.0772	1.7196	1.209
18	0.53718	1232.6	0.03830	224.66	408.69	1.0867	1.7188	1.214
20	0.57171	1225.3	0.03600	227.47	409.75	1.0962	1.7180	1.219
22	0.60789	1218.0	0.03385	230.29	410.79	1.1057	1.7173	1.224
24	0.64578	1210.5	0.03186	233.12	411.82	1.1152	1.7166	1.230
26	0.68543	1202.9	0.03000	235.97	412.84	1.1246	1.7159	1.236
28	0.72688	1195.2	0.02826	238.84	413.84	1.1341	1.7152	1.243
30	0.77020	1187.5	0.02664	241.72	414.82	1.1435	1.7145	1.249
32	0.81543	1179.6	0.02513	244.62	415.78	1.1529	1.7138	1.257
34	0.86263	1171.6	0.02371	247.54	416.72	1.1623	1.7131	1.265
36	0.91185	1163.4	0.02238	250.48	417.65	1.1717	1.7124	1.273
38	0.96315	1155.1	0.02113	253.43	418.55	1.1811	1.7118	1.282
40	1.0166	1146.7	0.01997	256.41	419.43	1.1905	1.7111	1.292
42	1.0722	1138.2	0.01887	259.41	420.28	1.1999	1.7103	1.303
44	1.1301	1129.5	0.01784	262.43	421.11	1.2092	1.7096	1.314
46	1.1903	1120.6	0.01687	265.47	421.92	1.2186	1.7089	1.326
48	1.2529	1111.5	0.01595	268.53	422.69	1.2280	1.7081	1.339
50	1.3179	1102.3	0.01509	271.62	423.44	1.2375	1.7072	1.354
52	1.3854	1092.9	0.01428	274.74	424.15	1.2469	1.7064	1.369
54	1.4555	1083.2	0.01351	277.89	424.83	1.2563	1.7055	1.386
56	1.5282	1073.4	0.01278	281.06	425.47	1.2658	1.7045	1.405
58	1.6036	1063.2	0.01209	284.27	426.07	1.2753	1.7035	1.425
60	1.6818	1052.9	0.01144	287.50	426.63	1.2848	1.7024	1.448
62	1.7628	1042.2	0.01083	290.78	427.14	1.2944	1.7013	1.473
64	1.8467	1031.2	0.01024	294.09	427.61	1.3040	1.7000	1.501
66	1.9337	1020.0	0.00969	297.44	428.02	1.3137	1.6987	1.532
68	2.0237	1008.3	0.00916	300.84	428.36	1.3234	1.6972	1.567
70	2.1168	996.2	0.00865	304.28	428.65	1.3332	1.6956	1.607
72	2.2132	983.8	0.00817	307.78	428.86	1.3430	1.6939	1.653
74	2.3130	970.8	0.00771	311.33	429.00	1.3530	1.6920	1.705
76	2.4161	957.3	0.00727	314.94	429.04	1.3631	1.6899	1.766
78	2.5228	943.1	0.00685	318.63	428.98	1.3733	1.6876	1.838
80	2.6332	928.2	0.00645	322.39	428.81	1.3836	1.6850	1.924
85	2.9258	887.2	0.00550	332.22	427.76	1.4104	1.6771	2.232
90	3.2442	837.8	0.00461	342.93	425.42	1.4390	1.6662	2.820
95	3.5912	772.7	0.00374	355.25	420.67	1.4715	1.6492	4.369
100	3.9724	651.2	0.00268	373.30	407.68	1.5188	1.6109	20.81
101.06[c]	4.0593	511.9	0.00195	389.64	389.64	1.5621	1.5621	∞

*Temperatures on ITS-90 scale [a]?Triple point [b]?Normal boiling point [c]?Critical point

Table 8.9 Superheated Vapor Thermodynamic Properties of R-134a

Temp, °C	Pressure = 0.101325 MPa Sat. temp. = −26.07°C			Pressure = 0.200 MPa Sat. temp. = −10.07°C		
	V	*h*	*s*	*V*	*h*	*s*
−20.00	5.11	387.68	1.7667	−10.00	10.01	392.77
−10.00	4.89	395.65	1.7976	0.00	9.54	401.21
0.00	4.69	403.74	1.8278	10.00	9.13	409.73
10.00	4.50	411.97	1.8574	20.00	8.76	418.35
20.00	4.34	420.34	1.8864	30.00	8.42	427.07
30.00	4.18	428.85	1.9150	40.00	8.12	435.90
40.00	4.04	437.52	1.9431	50.00	7.83	444.87
50.00	3.91	446.33	1.9708	60.00	7.57	453.97
60.00	3.78	455.30	1.9981			

Temp, °C	Pressure = 0.400 MPa Sat. temp. = 8.94°C			Pressure = 1.000 MPa Sat. temp. = 39.39°C		
	V	*h*	*s*	*V*	*h*	*s*
10.00	19.41	404.78	1.7263	70.00	41.21	452.05
20.00	18.45	414.00	1.7583	80.00	39.36	462.47
30.00	17.61	423.21	1.7892	90.00	37.74	472.86
40.00	16.87	432.46	1.8192	100.00	36.29	483.26
50.00	16.20	441.76	1.8485	110.00	34.99	493.69
60.00	15.60	451.15	1.8771	120.00	33.80	504.19
				130.00	32.71	514.75

Temp, °C	Pressure = 1.400 MPa Sat. temp. = 52.43°C			Pressure = 1.600 MPa Sat. temp. = 57.91°C		
	V	*h*	*s*	*V*	*h*	*s*
60.00	66.61	433.69	1.7347	60.00	80.74	428.99
70.00	62.25	445.31	1.7691	70.00	74.43	441.47
80.00	58.74	456.56	1.8014	80.00	69.61	453.30
90.00	55.79	467.60	1.8322	90.00	65.71	464.76
100.00	53.24	478.53	1.8619	100.00	62.43	476.01
110.00	51.03	489.39	1.8906	110.00	59.62	487.13
120.00	49.05	500.25	1.9186	120.00	57.14	498.19
130.00	47.28	511.11	1.9459	130.00	54.95	509.23

Temp, °C	Pressure = 2.000 MPa Sat. temp. = 67.49°C		
	V	*h*	*s*
70.00	104.37	432.22	1.7091
80.00	94.85	445.86	1.7483
90.00	87.97	458.49	1.7835
100.00	82.58	470.57	1.8164
110.00	78.17	482.32	1.8474
120.00	74.44	493.86	1.8772
130.00	71.18	505.30	1.9059
140.00	68.33	516.68	1.9338

V = vapor volume, kg/m^3
h = enthalpy, kJ/kg
s = entropy, kJ/kg·K

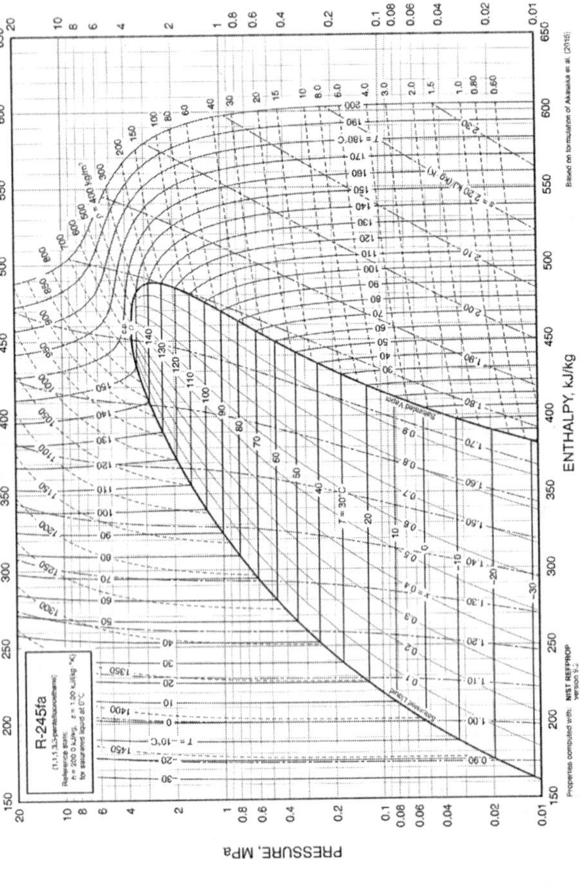

Figure 8.7 Pressure-Enthalpy Diagram for Refrigerant 245fa
[2021F, Ch 30, Fig 11]

Table 8.10 Refrigerant 245fa (1,1,1,3,3-pentafluoroethane) Properties of Saturated Liquid and Saturated Vapor [2021F, Ch 30, Tbl R-245fa]

Temp., °C	Pressure, MPa	Density, kg/m³	Volume, m³/kg	Enthalpy, kJ/kg		Entropy, kJ/(kg·K)		Specific Heat c_p, kJ/(kg·K)		c_p/c_v
		Liquid	Vapor	Liquid	Vapor	Liquid	Vapor	Liquid	Vapor	Vapor
−60	0.00131	1548.2	10.041	127.42	361.18	0.7006	1.7973	1.166	0.695	1.099
−50	0.00289	1525.0	4.7815	139.14	368.14	0.7543	1.7805	1.178	0.716	1.097
−40	0.00584	1501.5	2.4619	150.99	375.25	0.8063	1.7681	1.192	0.738	1.095
−30	0.01103	1477.7	1.3552	162.99	382.49	0.8566	1.7594	1.207	0.760	1.094
−20	0.01961	1453.5	0.79014	175.14	389.83	0.9056	1.7537	1.224	0.783	1.093
−10	0.03307	1429.0	0.48405	187.48	397.25	0.9533	1.7505	1.242	0.807	1.094
0	0.05327	1403.9	0.30947	200.00	404.72	1.0000	1.7495	1.261	0.833	1.095
2	0.05831	1398.8	0.28429	202.53	406.22	1.0092	1.7495	1.265	0.838	1.096
4	0.06372	1393.7	0.26153	205.07	407.72	1.0184	1.7496	1.270	0.843	1.096
6	0.06953	1388.6	0.24093	207.61	409.22	1.0275	1.7497	1.274	0.849	1.097
8	0.07576	1383.4	0.22226	210.17	410.72	1.0366	1.7500	1.278	0.854	1.097
10	0.08242	1378.3	0.20530	212.73	412.22	1.0457	1.7502	1.282	0.860	1.098
12	0.08954	1373.0	0.18987	215.30	413.72	1.0547	1.7506	1.287	0.865	1.098
14	0.09714	1367.8	0.17582	217.88	415.22	1.0637	1.7510	1.291	0.871	1.099
15.05[b]	0.10132	1365.1	0.16896	219.24	416.01	1.0684	1.7512	1.293	0.874	1.100
16	0.10525	1362.5	0.16301	220.47	416.72	1.0727	1.7514	1.296	0.877	1.100
18	0.11388	1357.2	0.15131	223.07	418.22	1.0816	1.7519	1.300	0.883	1.101
20	0.12306	1351.9	0.14061	225.68	419.72	1.0905	1.7524	1.305	0.888	1.102
22	0.13282	1346.6	0.13080	228.29	421.22	1.0994	1.7530	1.309	0.894	1.102
24	0.14317	1341.2	0.12182	230.92	422.71	1.1082	1.7537	1.314	0.900	1.103
26	0.15415	1335.7	0.11356	233.56	424.21	1.1170	1.7544	1.319	0.907	1.104
28	0.16578	1330.3	0.10598	236.20	425.70	1.1258	1.7551	1.324	0.913	1.106
30	0.17808	1324.8	0.09899	238.86	427.19	1.1346	1.7559	1.329	0.919	1.107
32	0.19108	1319.3	0.09256	241.52	428.68	1.1433	1.7567	1.334	0.926	1.108
34	0.20481	1313.7	0.08662	244.20	430.17	1.1520	1.7575	1.339	0.932	1.109
36	0.21930	1308.1	0.08113	246.89	431.65	1.1607	1.7584	1.344	0.939	1.111
38	0.23457	1302.4	0.07606	249.58	433.13	1.1694	1.7593	1.349	0.946	1.112
40	0.25065	1296.7	0.07137	252.29	434.61	1.1780	1.7602	1.355	0.952	1.114
42	0.26757	1291.0	0.06702	255.01	436.08	1.1866	1.7612	1.360	0.959	1.115
44	0.28535	1285.2	0.06298	257.74	437.55	1.1952	1.7622	1.366	0.967	1.117
46	0.30403	1279.3	0.05923	260.48	439.01	1.2038	1.7632	1.371	0.974	1.119
48	0.32364	1273.4	0.05575	263.23	440.47	1.2123	1.7642	1.377	0.981	1.121
50	0.34421	1267.5	0.05251	266.00	441.93	1.2209	1.7653	1.383	0.989	1.123
52	0.36576	1261.5	0.04949	268.77	443.38	1.2294	1.7664	1.389	0.996	1.125
54	0.38833	1255.4	0.04667	271.56	444.82	1.2379	1.7675	1.395	1.004	1.128
56	0.41195	1249.3	0.04405	274.36	446.26	1.2463	1.7686	1.401	1.012	1.130
58	0.43665	1243.2	0.04159	277.17	447.69	1.2548	1.7697	1.407	1.021	1.133
60	0.46246	1236.9	0.03930	279.99	449.11	1.2632	1.7709	1.414	1.029	1.135
62	0.48941	1230.6	0.03715	282.83	450.53	1.2717	1.7720	1.420	1.038	1.138
64	0.51754	1224.3	0.03514	285.68	451.94	1.2801	1.7732	1.427	1.046	1.141
66	0.54689	1217.8	0.03326	288.54	453.34	1.2885	1.7744	1.434	1.056	1.145
68	0.57747	1211.3	0.03149	291.42	454.73	1.2968	1.7755	1.441	1.065	1.148
70	0.60933	1204.7	0.02984	294.31	456.11	1.3052	1.7767	1.448	1.075	1.152
72	0.64250	1198.0	0.02828	297.22	457.48	1.3136	1.7779	1.456	1.084	1.156
74	0.67702	1191.3	0.02681	300.13	458.84	1.3219	1.7791	1.464	1.095	1.160
76	0.71292	1184.4	0.02543	303.07	460.19	1.3303	1.7803	1.472	1.105	1.165
78	0.75024	1177.5	0.02414	306.02	461.53	1.3386	1.7815	1.480	1.116	1.169
80	0.78901	1170.5	0.02291	309.08	462.86	1.3469	1.7826	1.488	1.128	1.175
82	0.82927	1163.4	0.02176	311.96	464.17	1.3552	1.7838	1.497	1.140	1.180
84	0.87105	1156.1	0.02067	314.96	465.46	1.3636	1.7850	1.506	1.152	1.186
86	0.91440	1148.8	0.01964	317.97	466.74	1.3719	1.7861	1.516	1.165	1.192
88	0.95935	1141.3	0.01867	321.01	468.01	1.3802	1.7872	1.525	1.178	1.199
90	1.0059	1133.7	0.01774	324.06	469.26	1.3885	1.7883	1.536	1.192	1.206
92	1.0542	1126.0	0.01687	327.12	470.48	1.3968	1.7894	1.546	1.207	1.214
94	1.1042	1118.2	0.01605	330.21	471.69	1.4051	1.7905	1.557	1.223	1.222
96	1.1559	1110.2	0.01526	333.32	472.88	1.4134	1.7915	1.569	1.239	1.231
98	1.2095	1102.0	0.01452	336.45	474.04	1.4217	1.7925	1.582	1.257	1.241
100	1.2649	1093.7	0.01381	339.60	475.19	1.4301	1.7934	1.595	1.275	1.252
105	1.4117	1072.1	0.01220	347.57	477.92	1.4509	1.7956	1.630	1.327	1.283
110	1.5711	1049.3	0.01077	355.71	480.46	1.4719	1.7975	1.673	1.389	1.323
115	1.7437	1024.9	0.00950	364.02	482.77	1.4930	1.7990	1.723	1.466	1.374
120	1.9304	998.6	0.00836	372.54	484.79	1.5144	1.7999	1.786	1.563	1.442
125	2.1320	970.0	0.00734	381.31	486.45	1.5360	1.8001	1.867	1.693	1.535
130	2.3495	938.4	0.00641	390.39	487.64	1.5581	1.7993	1.977	1.875	1.670
135	2.5841	902.7	0.00555	399.87	488.22	1.5809	1.7973	2.137	2.148	1.877
140	2.8370	861.1	0.00476	409.90	487.91	1.6046	1.7934	2.398	2.606	2.232
145	3.1099	810.0	0.00399	420.80	486.22	1.6300	1.7865	2.927	3.536	2.961
150	3.4050	738.4	0.00321	433.55	481.69	1.6594	1.7732	4.734	6.570	5.361
153.86[c]	3.6510	519.4	0.00193	459.00	459.00	1.7183	1.7183	∞	∞	∞

[b]Normal boiling point   [c]Critical point

Refrigerants

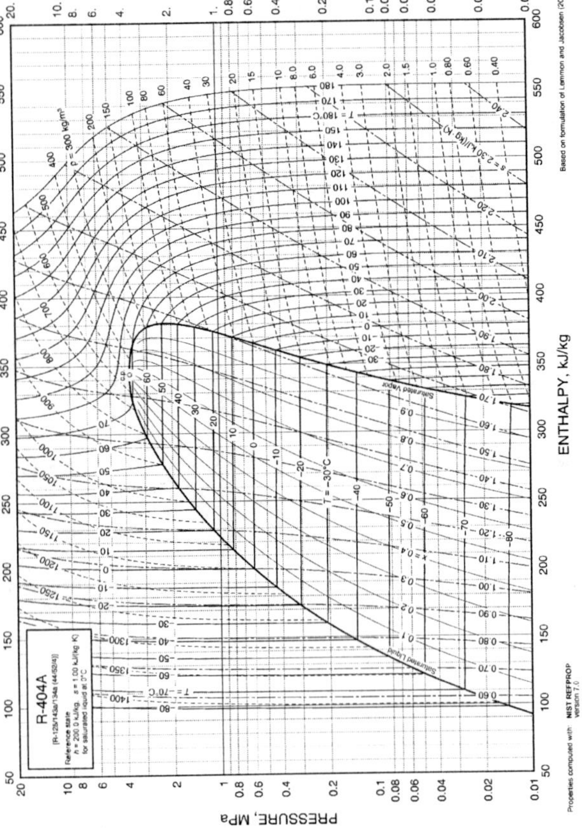

Figure 8.8 Pressure-Enthalpy Diagram for Refrigerant 404A
[2021F, Ch 30, Fig 15]

Table 8.11 R-404A [R-125/143a/134a (44/52/4)]
Properties of Liquid on Bubble Line and Vapor on Dew Line [2021F, Ch 30, Tbl R-404A]

Pressure, MPa	Temperature,* °C		Density, kg/m³	Volume, m³/kg	Enthalpy, kJ/kg		Entropy, kJ/(kg·K)		Specific Heat c_p, kJ/(kg·K)		c_p/c_v
	Bubble	Dew	Liquid	Vapor	Liquid	Vapor	Liquid	Vapor	Liquid	Vapor	Vapor
0.005	−93.70	−92.50	1447.1	3.05794	81.16	311.61	0.4716	1.7532	1.220	0.640	1.163
0.006	−91.48	−90.32	1440.6	2.57690	83.85	312.92	0.4865	1.7450	1.218	0.646	1.162
0.007	−89.56	−88.42	1434.9	2.22992	86.19	314.06	0.4993	1.7382	1.216	0.651	1.161
0.008	−87.86	−86.74	1429.9	1.96748	88.26	315.07	0.5106	1.7324	1.215	0.655	1.161
0.009	−86.32	−85.22	1425.4	1.76182	90.13	315.99	0.5206	1.7273	1.214	0.660	1.160
0.01	−84.93	−83.84	1421.3	1.59620	91.83	316.83	0.5296	1.7229	1.214	0.663	1.160
0.02	−75.05	−74.08	1392.4	0.83425	103.81	322.78	0.5917	1.6953	1.215	0.691	1.159
0.04	−63.85	−62.97	1359.4	0.43619	117.48	329.58	0.6587	1.6707	1.225	0.725	1.159
0.06	−56.57	−55.75	1337.7	0.29837	126.44	334.00	0.7007	1.6578	1.234	0.749	1.161
0.08	−51.03	−50.25	1321.0	0.22779	133.31	337.36	0.7320	1.6494	1.243	0.767	1.163
0.1	−46.50	−45.74	1307.1	0.18467	138.97	340.08	0.7571	1.6434	1.251	0.784	1.166
0.10132[b]	−46.22	−45.47	1306.3	0.18240	139.31	340.25	0.7586	1.6430	1.252	0.785	1.166
0.12	−42.63	−41.90	1295.1	0.15551	143.83	342.40	0.7783	1.6387	1.259	0.798	1.169
0.14	−39.24	−38.53	1284.5	0.13443	148.12	344.41	0.7967	1.6349	1.266	0.811	1.171
0.16	−36.20	−35.51	1275.0	0.11846	151.97	346.20	0.8130	1.6318	1.273	0.823	1.174
0.18	−33.45	−32.78	1266.2	0.10592	155.49	347.81	0.8277	1.6292	1.279	0.834	1.177
0.2	−30.93	−30.27	1258.0	0.09581	158.73	349.28	0.8411	1.6270	1.285	0.844	1.179
0.22	−28.59	−27.94	1250.4	0.08748	161.75	350.63	0.8534	1.6250	1.291	0.855	1.182
0.24	−26.42	−25.78	1243.3	0.08049	164.57	351.88	0.8649	1.6233	1.297	0.864	1.185
0.26	−24.37	−23.75	1236.5	0.07454	167.23	353.04	0.8755	1.6217	1.303	0.873	1.188
0.28	−22.45	−21.83	1230.1	0.06941	169.75	354.13	0.8855	1.6203	1.308	0.882	1.190
0.3	−20.62	−20.02	1223.9	0.06494	172.14	355.15	0.8950	1.6190	1.313	0.891	1.193
0.32	−18.89	−18.29	1218.0	0.06101	174.43	356.12	0.9039	1.6179	1.319	0.899	1.196
0.34	−17.24	−16.65	1212.4	0.05752	176.61	357.03	0.9125	1.6168	1.324	0.907	1.199
0.36	−15.66	−15.08	1206.9	0.05441	178.71	357.90	0.9206	1.6158	1.329	0.915	1.202
0.38	−14.15	−13.57	1201.6	0.05162	180.73	358.72	0.9283	1.6149	1.334	0.923	1.205
0.4	−12.69	−12.12	1196.5	0.04909	182.68	359.51	0.9358	1.6141	1.339	0.931	1.208
0.42	−11.29	−10.73	1191.6	0.04680	184.56	360.26	0.9429	1.6133	1.344	0.938	1.211
0.44	−9.94	−9.39	1186.7	0.04471	186.38	360.98	0.9498	1.6125	1.349	0.946	1.214
0.46	−8.64	−8.09	1182.0	0.04279	188.15	361.67	0.9564	1.6118	1.353	0.953	1.217
0.48	−7.37	−6.83	1177.5	0.04103	189.86	362.33	0.9628	1.6112	1.358	0.960	1.220
0.5	−6.15	−5.61	1173.0	0.03940	191.53	362.96	0.9690	1.6105	1.363	0.967	1.223
0.55	−3.24	−2.72	1162.3	0.03584	195.51	364.45	0.9837	1.6091	1.374	0.984	1.231
0.6	−0.53	−0.02	1152.0	0.03284	199.26	365.81	0.9973	1.6078	1.386	1.001	1.239
0.65	2.02	2.52	1142.3	0.03029	202.81	367.06	1.0101	1.6066	1.397	1.018	1.247
0.7	4.42	4.91	1132.9	0.02809	206.18	368.21	1.0222	1.6055	1.409	1.034	1.256
0.75	6.70	7.18	1123.8	0.02618	209.41	369.28	1.0336	1.6044	1.420	1.051	1.264
0.8	8.87	9.34	1115.1	0.02449	212.49	370.27	1.0444	1.6035	1.432	1.067	1.274
0.85	10.94	11.40	1106.5	0.02300	215.46	371.19	1.0547	1.6025	1.443	1.084	1.283
0.9	12.92	13.37	1098.2	0.02166	218.32	372.05	1.0646	1.6016	1.455	1.100	1.293
0.95	14.81	15.26	1090.2	0.02046	221.09	372.85	1.0741	1.6007	1.466	1.117	1.303
1.0	16.64	17.08	1082.2	0.01937	223.77	373.59	1.0832	1.5999	1.478	1.134	1.313
1.1	20.09	20.52	1066.9	0.01749	228.89	374.94	1.1005	1.5982	1.503	1.169	1.336
1.2	23.32	23.73	1052.0	0.01590	233.75	376.12	1.1166	1.5965	1.528	1.206	1.360
1.3	26.35	26.75	1037.5	0.01455	238.37	377.14	1.1318	1.5949	1.554	1.244	1.386
1.4	29.22	29.60	1023.4	0.01338	242.81	378.02	1.1462	1.5932	1.582	1.285	1.414
1.5	31.93	32.30	1009.5	0.01236	247.07	378.78	1.1599	1.5914	1.611	1.329	1.445
1.6	34.51	34.87	995.7	0.01146	251.19	379.42	1.1730	1.5896	1.643	1.376	1.478
1.7	36.97	37.32	982.1	0.01066	255.17	379.95	1.1856	1.5878	1.676	1.426	1.515
1.8	39.33	39.67	968.6	0.00994	259.05	380.38	1.1977	1.5858	1.712	1.481	1.556
1.9	41.58	41.91	955.1	0.00930	262.83	380.70	1.2095	1.5838	1.751	1.541	1.601
2.0	43.75	44.07	941.6	0.00871	266.52	380.92	1.2208	1.5817	1.794	1.607	1.652
2.1	45.84	46.15	928.1	0.00817	270.14	381.05	1.2319	1.5794	1.841	1.681	1.709
2.2	47.85	48.15	914.4	0.00768	273.70	381.08	1.2427	1.5770	1.893	1.763	1.774
2.3	49.80	50.08	900.6	0.00723	277.20	381.01	1.2532	1.5745	1.952	1.856	1.847
2.4	51.68	51.95	886.5	0.00680	280.66	380.83	1.2635	1.5718	2.019	1.962	1.932
2.5	53.50	53.76	872.2	0.00641	284.09	380.55	1.2737	1.5689	2.095	2.085	2.032
2.6	55.26	55.51	857.5	0.00604	287.50	380.15	1.2837	1.5658	2.183	2.229	2.149
2.7	56.97	57.21	842.4	0.00569	290.89	379.62	1.2937	1.5624	2.288	2.401	2.289
2.8	58.63	58.86	826.8	0.00536	294.29	378.96	1.3036	1.5587	2.414	2.609	2.459
2.9	60.24	60.46	810.5	0.00505	297.70	378.14	1.3135	1.5547	2.569	2.868	2.672
3.0	61.81	62.01	793.4	0.00475	301.15	377.15	1.3234	1.5503	2.765	3.197	2.944
3.2	64.82	64.99	755.6	0.00417	308.25	374.49	1.3438	1.5397	3.381	4.233	3.797
3.4	67.67	67.81	709.8	0.00361	315.97	370.45	1.3657	1.5255	4.771	6.536	5.689
3.729[c]	72.05	72.05	486.5	0.00206	343.92	343.92	1.4455	1.4455	—	—	—

*Temperatures on ITS-90 scale

[b]Bubble and dew points at one standard atmosphere

Refrigerants

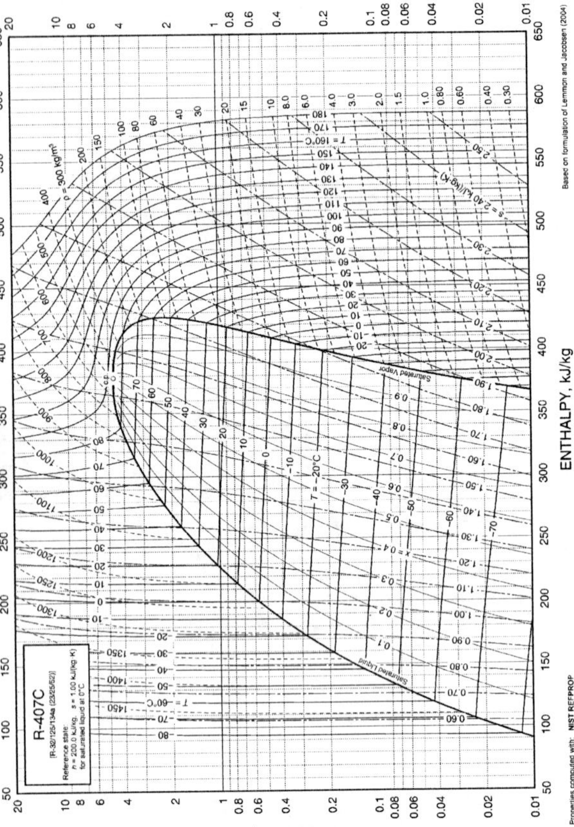

Figure 8.9 Pressure-Enthalpy Diagram for Refrigerant 407C
[2021F, Ch 30, Fig 16]

Table 8.12 R-407C [R-32/125/134a (23/25/52)]
Properties of Liquid on Bubble Line and Vapor on Dew Line [2021F, Ch 30, Tbl R-407C]

Pressure, MPa	Temperature,* °C		Density, kg/m³	Volume, m³/kg	Enthalpy, kJ/kg		Entropy, kJ/(kg·K)		Specific Heat c_p, kJ/(kg·K)		c_p/c_v
	Bubble	Dew	Liquid	Vapor	Liquid	Vapor	Liquid	Vapor	Liquid	Vapor	Vapor
0.01	−82.45	−74.81	1495.5	1.89703	90.48	366.78	0.5259	1.9471	1.281	0.668	1.182
0.02	−72.50	−65.02	1466.7	0.99017	103.24	372.75	0.5910	1.9104	1.283	0.694	1.181
0.04	−61.25	−53.95	1433.7	0.51705	117.72	379.47	0.6612	1.8761	1.291	0.727	1.182
0.06	−53.96	−46.79	1412.0	0.35346	127.17	383.77	0.7050	1.8573	1.299	0.750	1.184
0.08	−48.42	−41.34	1395.3	0.26975	134.39	386.99	0.7374	1.8445	1.306	0.769	1.187
0.1	−43.90	−36.90	1381.5	0.21865	140.31	389.59	0.7635	1.8349	1.312	0.786	1.190
0.10132[b]	−43.63	−36.63	1380.7	0.21595	140.67	389.75	0.7650	1.8343	1.312	0.787	1.190
0.12	−40.05	−33.11	1369.7	0.18411	145.39	391.78	0.7854	1.8273	1.318	0.800	1.193
0.14	−36.67	−29.79	1359.1	0.15916	149.86	393.68	0.8043	1.8210	1.324	0.813	1.196
0.16	−33.65	−26.83	1349.7	0.14025	153.86	395.36	0.8211	1.8156	1.329	0.825	1.199
0.18	−30.92	−24.15	1341.0	0.12542	157.51	396.86	0.8362	1.8110	1.334	0.837	1.201
0.2	−28.41	−21.69	1333.0	0.11347	160.87	398.22	0.8499	1.8069	1.339	0.848	1.204
0.22	−26.09	−19.41	1325.5	0.10362	163.99	399.47	0.8625	1.8033	1.344	0.858	1.207
0.24	−23.93	−17.29	1318.4	0.09536	166.91	400.62	0.8742	1.8000	1.349	0.868	1.210
0.26	−21.90	−15.31	1311.8	0.08833	169.65	401.69	0.8851	1.7970	1.354	0.877	1.213
0.28	−19.99	−13.43	1305.5	0.08227	172.24	402.69	0.8954	1.7942	1.358	0.886	1.216
0.3	−18.19	−11.66	1299.5	0.07699	174.71	403.62	0.9050	1.7917	1.362	0.895	1.219
0.32	−16.47	−9.98	1293.7	0.07235	177.06	404.49	0.9141	1.7894	1.367	0.903	1.222
0.34	−14.83	−8.38	1288.2	0.06824	179.30	405.32	0.9228	1.7872	1.371	0.911	1.224
0.36	−13.27	−6.85	1282.9	0.06457	181.45	406.10	0.9310	1.7851	1.375	0.919	1.227
0.38	−11.77	−5.38	1277.8	0.06127	183.52	406.85	0.9389	1.7832	1.379	0.927	1.230
0.4	−10.33	−3.97	1272.8	0.05830	185.52	407.55	0.9465	1.7814	1.383	0.934	1.233
0.42	−8.94	−2.61	1268.0	0.05559	187.44	408.23	0.9537	1.7796	1.387	0.942	1.236
0.44	−7.61	−1.31	1263.4	0.05313	189.30	408.87	0.9607	1.7780	1.391	0.949	1.239
0.46	−6.31	−0.04	1258.8	0.05087	191.11	409.48	0.9674	1.7764	1.395	0.956	1.242
0.48	−5.06	1.18	1254.4	0.04879	192.86	410.07	0.9739	1.7750	1.399	0.963	1.245
0.5	−3.85	2.36	1250.1	0.04687	194.56	410.64	0.9801	1.7735	1.403	0.970	1.248
0.55	−0.98	5.17	1239.8	0.04267	198.61	411.95	0.9950	1.7702	1.413	0.987	1.255
0.6	1.70	7.79	1230.0	0.03915	202.42	413.15	1.0087	1.7672	1.422	1.004	1.262
0.65	4.22	10.24	1220.7	0.03615	206.02	414.25	1.0216	1.7644	1.432	1.020	1.270
0.7	6.60	12.56	1211.7	0.03356	209.44	415.25	1.0338	1.7618	1.441	1.036	1.278
0.75	8.85	14.76	1203.1	0.03131	212.71	416.18	1.0452	1.7594	1.451	1.052	1.286
0.8	11.00	16.85	1194.9	0.02933	215.83	417.03	1.0561	1.7571	1.460	1.067	1.294
0.85	13.04	18.84	1186.8	0.02757	218.83	417.83	1.0665	1.7550	1.469	1.082	1.302
0.9	15.00	20.74	1179.1	0.02600	221.71	418.57	1.0764	1.7529	1.479	1.098	1.310
0.95	16.88	22.56	1171.5	0.02460	224.50	419.25	1.0859	1.7509	1.488	1.113	1.319
1.0	18.69	24.32	1164.1	0.02332	227.19	419.89	1.0950	1.7491	1.498	1.128	1.327
1.1	22.11	27.63	1149.9	0.02111	232.34	421.03	1.1122	1.7455	1.517	1.159	1.346
1.2	25.30	30.73	1136.2	0.01926	237.20	422.03	1.1283	1.7421	1.537	1.190	1.365
1.3	28.30	33.63	1123.0	0.01768	241.82	422.89	1.1434	1.7389	1.557	1.222	1.385
1.4	31.14	36.37	1110.2	0.01631	246.24	423.63	1.1577	1.7358	1.578	1.255	1.406
1.5	33.83	38.97	1097.7	0.01512	250.48	424.27	1.1713	1.7328	1.600	1.289	1.428
1.6	36.39	41.43	1085.5	0.01408	254.57	424.80	1.1843	1.7298	1.622	1.324	1.452
1.7	38.84	43.78	1073.5	0.01315	258.51	425.25	1.1967	1.7269	1.645	1.361	1.477
1.8	41.18	46.03	1061.7	0.01231	262.33	425.61	1.2080	1.7241	1.669	1.400	1.504
1.9	43.43	48.18	1050.0	0.01157	266.05	425.89	1.2200	1.7212	1.695	1.440	1.533
2.0	45.59	50.25	1038.5	0.01089	269.66	426.10	1.2311	1.7184	1.722	1.483	1.564
2.1	47.67	52.24	1027.1	0.01027	273.19	426.23	1.2418	1.7155	1.750	1.529	1.597
2.2	49.68	54.15	1015.7	0.00971	276.64	426.29	1.2522	1.7126	1.780	1.577	1.633
2.3	51.63	56.00	1004.4	0.00919	280.02	426.28	1.2624	1.7097	1.813	1.629	1.671
2.4	53.51	57.79	993.1	0.00871	283.34	426.20	1.2723	1.7068	1.847	1.684	1.713
2.5	55.34	59.51	981.8	0.00827	286.60	426.06	1.2819	1.7038	1.884	1.744	1.758
2.6	57.11	61.19	970.5	0.00786	289.82	425.85	1.2914	1.7007	1.924	1.810	1.808
2.7	58.83	62.81	959.0	0.00747	292.99	425.57	1.3006	1.6976	1.968	1.881	1.863
2.8	60.51	64.38	947.5	0.00711	296.12	425.21	1.3097	1.6944	2.016	1.958	1.923
2.9	62.14	65.91	935.9	0.00677	299.23	424.79	1.3187	1.6911	2.069	2.044	1.990
3.0	63.73	67.40	924.1	0.00645	302.31	424.29	1.3276	1.6877	2.128	2.139	2.065
3.2	66.80	70.25	899.9	0.00587	308.43	423.06	1.3450	1.6805	2.268	2.365	2.243
3.4	69.73	72.94	874.6	0.00533	314.54	421.46	1.3622	1.6726	2.451	2.657	2.475
3.6	72.53	75.50	847.6	0.00484	320.71	419.45	1.3795	1.6639	2.701	3.050	2.789
3.8	75.22	77.92	818.1	0.00439	327.02	416.91	1.3970	1.6540	3.065	3.613	3.239
4.0	77.82	80.21	785.1	0.00395	333.64	413.66	1.4152	1.6424	3.647	4.486	3.935
4.2	80.32	82.37	746.0	0.00352	340.83	409.34	1.4348	1.6281	4.726	6.029	5.159
4.63[c]	86.03	86.03	484.2	0.00207	378.48	378.48	1.5384	1.5384	—	—	—

*Temperatures on ITS-90 scale [b]Bubble and dew points at one standard atmosphere [c]Critical point

Refrigerants

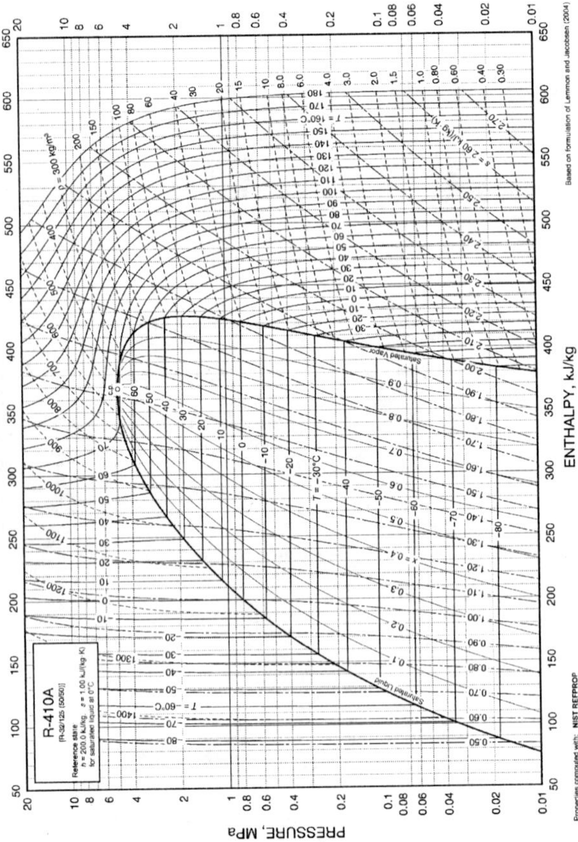

Figure 8.10 Pressure-Enthalpy Diagram for Refrigerant 410A
[2021F, Ch 30, Fig 17]

Table 8.13 R-410A [R-32/125 (50/50)]
Properties of Liquid on Bubble Line and Vapor on Dew Line [2021F, Ch 30, Tbl R-410A]

Pressure, MPa	Temperature,* °C		Density, kg/m³	Volume, m³/kg	Enthalpy, kJ/kg		Entropy, kJ/(kg·K)		Specific Heat c_p, kJ/(kg·K)		c_p/c_v
	Bubble	Dew	Liquid	Vapor	Liquid	Vapor	Liquid	Vapor	Liquid	Vapor	Vapor
0.01	–88.23	–88.14	1460.6	2.09888	76.56	378.76	0.4588	2.0927	1.344	0.668	1.227
0.02	–78.79	–78.70	1432.9	1.09659	89.26	384.25	0.5258	2.0432	1.345	0.696	1.228
0.04	–68.12	–68.04	1401.1	0.57309	103.64	390.29	0.5978	1.9956	1.351	0.734	1.231
0.06	–61.22	–61.14	1380.0	0.39193	113.00	394.10	0.6426	1.9687	1.358	0.762	1.235
0.08	–55.98	–55.90	1363.9	0.29918	120.14	396.92	0.6758	1.9500	1.364	0.785	1.239
0.1	–51.70	–51.62	1350.5	0.24256	125.99	399.17	0.7024	1.9358	1.369	0.805	1.243
0.10132[b]	–51.44	–51.36	1349.7	0.23957	126.34	399.31	0.7040	1.9350	1.370	0.807	1.244
0.12	–48.06	–47.98	1339.0	0.20427	130.99	401.05	0.7247	1.9243	1.375	0.823	1.247
0.14	–44.87	–44.79	1328.8	0.17661	135.39	402.67	0.7441	1.9147	1.380	0.839	1.251
0.16	–42.02	–41.94	1319.6	0.15565	139.34	404.09	0.7612	1.9065	1.385	0.854	1.255
0.18	–39.44	–39.36	1311.2	0.13921	142.93	405.36	0.7766	1.8993	1.390	0.868	1.259
0.2	–37.07	–36.99	1303.4	0.12595	146.23	406.50	0.7905	1.8928	1.395	0.881	1.263
0.22	–34.89	–34.80	1296.2	0.11503	149.29	407.53	0.8034	1.8871	1.399	0.893	1.266
0.24	–32.85	–32.76	1289.4	0.10587	152.15	408.49	0.8153	1.8818	1.404	0.904	1.270
0.26	–30.94	–30.85	1283.0	0.09807	154.84	409.36	0.8264	1.8770	1.408	0.916	1.274
0.28	–29.14	–29.05	1276.9	0.09135	157.38	410.18	0.8368	1.8726	1.413	0.926	1.277
0.3	–27.44	–27.35	1271.1	0.08550	159.80	410.94	0.8466	1.8685	1.417	0.936	1.281
0.32	–25.82	–25.73	1265.5	0.08035	162.10	411.65	0.8558	1.8647	1.421	0.946	1.285
0.34	–24.28	–24.19	1260.2	0.07579	164.29	412.32	0.8646	1.8611	1.426	0.956	1.288
0.36	–22.81	–22.72	1255.0	0.07172	166.40	412.95	0.8703	1.8577	1.430	0.965	1.292
0.38	–21.40	–21.31	1250.1	0.06806	168.43	413.54	0.8810	1.8545	1.434	0.975	1.295
0.4	–20.04	–19.95	1245.3	0.06476	170.38	414.10	0.8887	1.8514	1.438	0.983	1.299
0.42	–18.74	–18.65	1240.6	0.06176	172.26	414.64	0.8960	1.8486	1.443	0.992	1.303
0.44	–17.48	–17.39	1236.1	0.05902	174.08	415.14	0.9031	1.8458	1.447	1.001	1.306
0.46	–16.27	–16.18	1231.8	0.05652	175.84	415.63	0.9099	1.8432	1.451	1.009	1.310
0.48	–15.10	–15.00	1227.5	0.05421	177.55	416.09	0.9165	1.8407	1.455	1.017	1.313
0.5	–13.96	–13.86	1223.3	0.05209	179.21	416.53	0.9228	1.8383	1.459	1.025	1.317
0.55	–11.26	–11.16	1213.4	0.04743	183.17	417.54	0.9379	1.8326	1.469	1.045	1.326
0.6	–8.74	–8.64	1203.9	0.04352	186.89	418.46	0.9518	1.8275	1.479	1.064	1.335
0.65	–6.38	–6.28	1194.9	0.04019	190.40	419.28	0.9649	1.8227	1.489	1.083	1.344
0.7	–4.15	–4.05	1186.3	0.03732	193.74	420.03	0.9772	1.8183	1.499	1.101	1.354
0.75	–2.04	–1.93	1178.1	0.03482	196.92	420.71	0.9888	1.8141	1.509	1.119	1.363
0.8	–0.03	0.08	1170.1	0.03262	199.96	421.33	0.9998	1.8102	1.519	1.136	1.373
0.85	1.89	1.99	1162.4	0.03068	202.88	421.89	1.0103	1.8065	1.529	1.154	1.382
0.9	3.72	3.83	1154.9	0.02894	205.69	422.41	1.0204	1.8030	1.540	1.171	1.392
0.95	5.48	5.58	1147.6	0.02738	208.40	422.88	1.0300	1.7996	1.550	1.188	1.402
1.0	7.17	7.27	1140.5	0.02596	211.02	423.31	1.0392	1.7964	1.560	1.205	1.413
1.1	10.36	10.47	1126.8	0.02351	216.03	424.07	1.0567	1.7903	1.581	1.239	1.434
1.2	13.34	13.46	1113.7	0.02145	220.76	424.68	1.0730	1.7846	1.603	1.274	1.457
1.3	16.15	16.26	1101.0	0.01970	225.26	425.19	1.0883	1.7792	1.624	1.31	1.481
1.4	18.79	18.91	1088.8	0.01819	229.56	425.59	1.1027	1.7741	1.647	1.347	1.506
1.5	21.30	21.41	1076.9	0.01687	233.68	425.89	1.1165	1.7691	1.670	1.385	1.532
1.6	23.68	23.80	1065.2	0.01571	237.65	426.11	1.1296	1.7644	1.694	1.424	1.560
1.7	25.96	26.07	1053.8	0.01468	241.48	426.25	1.1421	1.7597	1.719	1.465	1.590
1.8	28.13	28.25	1042.6	0.01376	245.19	426.31	1.1542	1.7552	1.745	1.509	1.621
1.9	30.22	30.34	1031.6	0.01293	248.79	426.31	1.1657	1.7508	1.772	1.555	1.655
2.0	32.22	32.34	1020.7	0.01218	252.29	426.24	1.1769	1.7464	1.800	1.603	1.690
2.1	34.16	34.28	1009.9	0.0115	255.71	426.10	1.1878	1.7421	1.830	1.655	1.728
2.2	36.02	36.14	999.2	0.01088	259.05	425.90	1.1983	1.7379	1.861	1.709	1.769
2.3	37.82	37.94	988.6	0.01031	262.32	425.64	1.2085	1.7336	1.894	1.768	1.813
2.4	39.56	39.68	978.0	0.00978	265.52	425.33	1.2185	1.7294	1.929	1.831	1.860
2.5	41.25	41.37	967.5	0.00929	268.67	424.95	1.2282	1.7251	1.967	1.898	1.911
2.6	42.89	43.00	957.0	0.00883	271.77	424.51	1.2377	1.7209	2.008	1.971	1.966
2.7	44.48	44.59	946.4	0.00841	274.82	424.02	1.2470	1.7166	2.052	2.050	2.026
2.8	46.02	46.14	935.8	0.00802	277.84	423.47	1.2561	1.7123	2.100	2.136	2.091
2.9	47.53	47.64	925.2	0.00764	280.82	422.85	1.2651	1.7079	2.153	2.230	2.163
3.0	48.99	49.10	914.5	0.00729	283.78	422.18	1.2740	1.7035	2.211	2.333	2.243
3.2	51.81	51.91	892.6	0.00665	289.62	420.62	1.2913	1.6944	2.348	2.575	2.429
3.4	54.49	54.59	870.0	0.00607	295.43	418.78	1.3085	1.6849	2.522	2.879	2.663
3.6	57.05	57.15	846.3	0.00555	301.26	416.60	1.3254	1.6747	2.752	3.276	2.970
3.8	59.50	59.59	821.0	0.00506	307.16	414.03	1.3425	1.6638	3.070	3.815	3.386
4.0	61.85	61.93	793.5	0.00460	313.24	410.97	1.3600	1.6517	3.541	4.596	3.987
4.2	64.10	64.17	762.6	0.00417	319.65	407.24	1.3783	1.6380	4.306	5.826	4.929
4.903[c]	71.36	71.36	459.5	0.00218	368.55	368.55	1.5181	1.5181	—	—	—

*Temperatures on ITS-90 scale   [b]Bubble and dew points at one standard atmosphere   [c]Critical point

Refrigerants

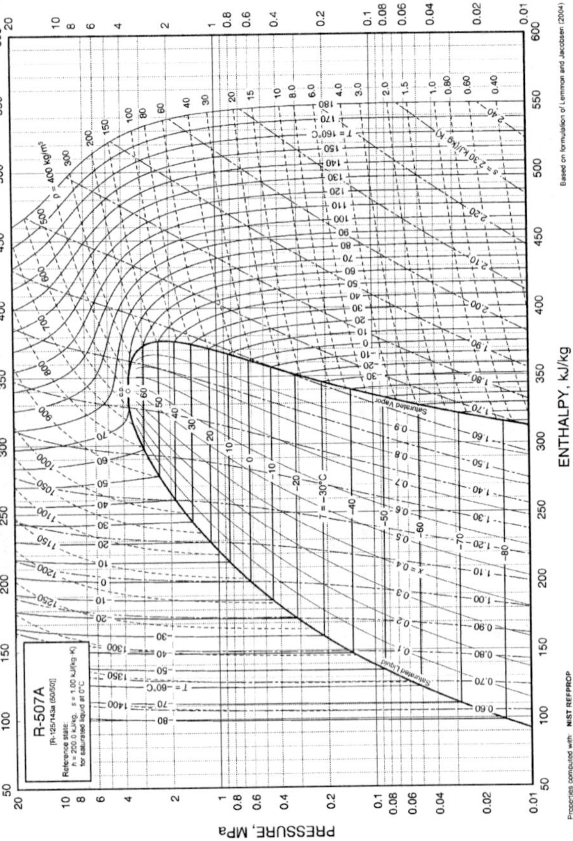

Figure 8.11 Pressure-Enthalpy Diagram for Refrigerant 507A
[2021F, Ch 30, Fig 18]

Table 8.14 R-507A [R-125/143a (50/50)]
Properties of Saturated Liquid and Saturated Vapor [2021F, Ch 30, Tbl R-507A]

Temp.,* °C	Pressure,** MPa	Density, kg/m³ Liquid	Volume, m³/kg Vapor	Enthalpy, kJ/kg Liquid	Enthalpy, kJ/kg Vapor	Entropy, kJ/(kg·K) Liquid	Entropy, kJ/(kg·K) Vapor	Specific Heat c_p, kJ/(kg·K) Liquid	Specific Heat c_p, kJ/(kg·K) Vapor	c_p/c_v Vapor
–100	0.00295	1476.9	4.92920	74.41	303.90	0.4323	1.7579	1.219	0.618	1.164
–95	0.00458	1461.7	3.25360	80.48	306.85	0.4669	1.7377	1.210	0.631	1.162
–90	0.00693	1446.8	2.20850	86.51	309.83	0.5003	1.7197	1.205	0.644	1.161
–85	0.01019	1431.9	1.53750	92.53	312.83	0.5327	1.7036	1.203	0.658	1.159
–80	0.01464	1417.1	1.09510	98.54	315.85	0.5642	1.6893	1.203	0.672	1.159
–75	0.02058	1402.3	0.79638	104.57	318.88	0.5950	1.6766	1.205	0.686	1.158
–70	0.02836	1387.4	0.59012	110.60	321.92	0.6250	1.6652	1.208	0.701	1.158
–65	0.03837	1372.5	0.44482	116.66	324.96	0.6545	1.6552	1.213	0.716	1.159
–60	0.05105	1357.4	0.34056	122.74	328.00	0.6833	1.6463	1.220	0.732	1.160
–55	0.06688	1342.3	0.26444	128.87	331.03	0.7116	1.6384	1.227	0.749	1.161
–50	0.08638	1326.9	0.20801	135.03	334.05	0.7395	1.6314	1.235	0.766	1.164
–48	0.09533	1320.7	0.18960	137.51	335.25	0.7505	1.6288	1.239	0.773	1.165
–46.74[b]	0.10132	1316.8	0.17902	139.07	336.01	0.7574	1.6273	1.241	0.777	1.166
–46	0.10499	1314.5	0.17313	139.99	336.45	0.7615	1.6264	1.243	0.780	1.166
–44	0.11541	1308.2	0.15836	142.48	337.65	0.7724	1.6241	1.247	0.787	1.167
–42	0.12662	1301.9	0.14510	144.99	338.84	0.7832	1.6219	1.251	0.795	1.169
–40	0.13867	1295.6	0.13317	147.49	340.03	0.7940	1.6198	1.255	0.803	1.170
–38	0.15159	1289.2	0.12240	150.01	341.21	0.8047	1.6178	1.259	0.810	1.172
–36	0.16542	1282.8	0.11268	152.54	342.38	0.8153	1.6159	1.264	0.818	1.174
–34	0.18022	1276.3	0.10388	155.08	343.55	0.8260	1.6141	1.269	0.826	1.176
–32	0.19602	1269.7	0.09590	157.63	344.72	0.8365	1.6123	1.274	0.835	1.178
–30	0.21287	1263.2	0.08865	160.18	345.88	0.8470	1.6107	1.279	0.843	1.180
–28	0.23081	1256.5	0.08205	162.75	347.03	0.8575	1.6092	1.284	0.852	1.183
–26	0.24989	1249.8	0.07604	165.33	348.17	0.8679	1.6077	1.289	0.861	1.186
–24	0.27016	1243.1	0.07055	167.92	349.30	0.8783	1.6063	1.295	0.870	1.188
–22	0.29167	1236.3	0.06553	170.52	350.43	0.8886	1.6049	1.301	0.879	1.191
–20	0.31446	1229.4	0.06094	173.13	351.54	0.8989	1.6037	1.307	0.888	1.195
–18	0.33858	1222.5	0.05673	175.76	352.65	0.9091	1.6024	1.313	0.898	1.198
–16	0.36408	1215.4	0.05286	178.39	353.75	0.9193	1.6013	1.319	0.908	1.202
–14	0.39102	1208.4	0.04931	181.04	354.83	0.9295	1.6001	1.326	0.918	1.206
–12	0.41945	1201.2	0.04603	183.71	355.91	0.9397	1.5991	1.333	0.929	1.210
–10	0.44941	1193.9	0.04301	186.39	356.97	0.9498	1.5980	1.340	0.940	1.214
–8	0.48096	1186.6	0.04023	189.08	358.02	0.9599	1.5971	1.348	0.951	1.219
–6	0.51416	1179.2	0.03765	191.78	359.06	0.9699	1.5961	1.355	0.962	1.224
–4	0.54906	1171.7	0.03527	194.51	360.08	0.9800	1.5952	1.363	0.974	1.230
–2	0.58571	1164.0	0.03306	197.25	361.08	0.9900	1.5943	1.372	0.987	1.236
0	0.62417	1156.3	0.03101	200.00	362.07	1.0000	1.5934	1.381	0.999	1.242
2	0.66450	1148.5	0.02910	202.77	363.05	1.0100	1.5925	1.390	1.012	1.249
4	0.70676	1140.5	0.02733	205.56	364.00	1.0199	1.5917	1.399	1.026	1.256
6	0.75099	1132.4	0.02568	208.37	364.94	1.0299	1.5908	1.410	1.040	1.264
8	0.79728	1124.2	0.02415	211.20	365.85	1.0398	1.5900	1.420	1.055	1.272
10	0.84566	1115.9	0.02271	214.04	366.75	1.0498	1.5891	1.431	1.071	1.282
12	0.89622	1107.4	0.02138	216.91	367.61	1.0597	1.5883	1.443	1.088	1.291
14	0.94900	1098.7	0.02012	219.80	368.46	1.0696	1.5874	1.455	1.105	1.302
16	1.00410	1089.9	0.01895	222.71	369.28	1.0796	1.5865	1.468	1.124	1.314
18	1.06150	1080.9	0.01785	225.65	370.07	1.0895	1.5856	1.482	1.144	1.327
20	1.12140	1071.7	0.01683	228.61	370.83	1.0995	1.5846	1.497	1.165	1.341
22	1.18370	1062.4	0.01586	231.60	371.55	1.1094	1.5836	1.513	1.188	1.356
24	1.24860	1052.8	0.01495	234.61	372.23	1.1194	1.5826	1.530	1.212	1.372
26	1.31610	1043.0	0.01410	237.66	372.91	1.1294	1.5815	1.548	1.239	1.391
28	1.38640	1032.9	0.01329	240.73	373.52	1.1394	1.5804	1.568	1.268	1.411
30	1.45940	1022.6	0.01253	243.84	374.10	1.1495	1.5792	1.589	1.299	1.433
32	1.53520	1011.9	0.01182	246.98	374.63	1.1595	1.5779	1.612	1.333	1.458
34	1.61400	1001.0	0.01114	250.16	375.11	1.1697	1.5765	1.637	1.371	1.485
36	1.69580	989.7	0.01050	253.39	375.54	1.1799	1.5750	1.664	1.413	1.516
38	1.78070	978.1	0.00989	256.65	375.91	1.1901	1.5734	1.695	1.459	1.551
40	1.86880	966.0	0.00932	259.96	376.22	1.2004	1.5717	1.729	1.511	1.591
42	1.96020	953.5	0.00877	263.33	376.46	1.2108	1.5698	1.767	1.570	1.636
44	2.05490	940.5	0.00825	266.74	376.61	1.2213	1.5678	1.811	1.638	1.689
46	2.15310	926.9	0.00776	270.23	376.68	1.2320	1.5655	1.860	1.716	1.750
48	2.25480	912.7	0.00728	273.78	376.66	1.2427	1.5631	1.918	1.807	1.823
50	2.36030	897.7	0.00683	277.41	376.52	1.2536	1.5603	1.985	1.915	1.910
55	2.64090	856.2	0.00578	286.91	375.54	1.2818	1.5519	2.225	2.304	2.228
60	2.94760	806.1	0.00480	297.28	373.26	1.3120	1.5401	2.677	3.060	2.855
65	3.28380	739.1	0.00384	309.30	368.44	1.3465	1.5215	3.940	5.190	4.625
70	3.65570	599.6	0.00260	328.32	353.47	1.4007	1.4740	31.960	44.630	36.780
70.62[c]	3.70500	490.8	0.00204	340.45	340.45	1.4358	1.4358	∞	∞	∞

*Temperatures on ITS-90 scale

**Small deviations from azeotropic behavior occur at some conditions; tabulated pressures are average of bubble and dew-point pressures

[b]Normal boiling point

[c]Critical point

Refrigerants

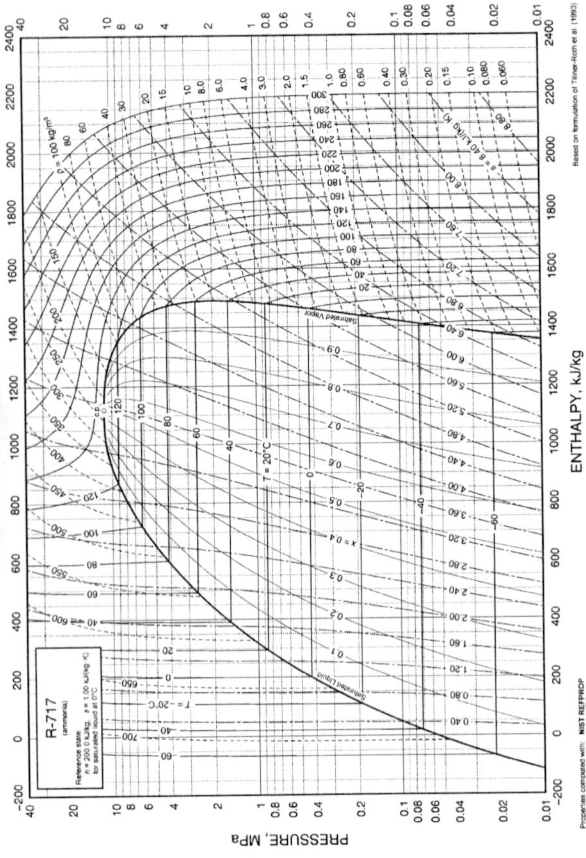

Figure 8.12 Pressure-Enthalpy Diagram for Refrigerant 717 (Ammonia)
[2021F, Ch 30, Fig 18]

Table 8.15 R-717 (Ammonia)
Properties of Saturated Liquid and Saturated Vapor [2021F, Ch 30, Tbl R-717]

Temp.,* °C	Pressure, MPa	Density, kg/m³ Liquid	Volume, m³/kg Vapor	Enthalpy, kJ/kg		Entropy, kJ/(kg·K)		c_p/c_v Vapor
				Liquid	Vapor	Liquid	Vapor	
−77.65[a]	0.00609	732.9	15.602	−143.15	1341.23	−0.4716	7.1213	1.325
−70	0.01094	724.7	9.0079	−110.81	1355.55	−0.3094	6.9088	1.327
−60	0.02189	713.6	4.7057	−68.06	1373.73	−0.1040	6.6602	1.330
−50	0.04084	702.1	2.6277	−24.73	1391.19	0.0945	6.4396	1.335
−40	0.07169	690.2	1.5533	19.17	1407.76	0.2867	6.2425	1.342
−38	0.07971	687.7	1.4068	28.01	1410.96	0.3245	6.2056	1.343
−36	0.08845	685.3	1.2765	36.88	1414.11	0.3619	6.1694	1.345
−34	0.09795	682.8	1.1604	45.77	1417.23	0.3992	6.1339	1.347
−33.33[b]	0.10133	682.0	1.1242	48.76	1418.26	0.4117	6.1221	1.348
−32	0.10826	680.3	1.0567	54.67	1420.29	0.4362	6.0992	1.349
−30	0.11943	677.8	0.96396	63.60	1423.31	0.4730	6.0651	1.351
−28	0.13151	675.3	0.88082	72.55	1426.28	0.5096	6.0317	1.353
−26	0.14457	672.8	0.80614	81.52	1429.21	0.5460	5.9989	1.355
−24	0.15864	670.3	0.73896	90.51	1432.08	0.5821	5.9667	1.358
−22	0.17379	667.7	0.67840	99.52	1434.91	0.6180	5.9351	1.360
−20	0.19008	665.1	0.62373	108.55	1437.68	0.6538	5.9041	1.363
−18	0.20756	662.6	0.57428	117.60	1440.39	0.6893	5.8736	1.365
−16	0.22630	660.0	0.52949	126.67	1443.06	0.7246	5.8437	1.368
−14	0.24637	657.3	0.48885	135.76	1445.66	0.7597	5.8143	1.371
−12	0.26782	654.7	0.45192	144.88	1448.21	0.7946	5.7853	1.375
−10	0.29071	652.1	0.41830	154.01	1450.70	0.8293	5.7569	1.378
−8	0.31513	649.4	0.38767	163.16	1453.14	0.8638	5.7289	1.382
−6	0.34114	646.7	0.35970	172.34	1455.51	0.8981	5.7013	1.385
−4	0.36880	644.0	0.33414	181.54	1457.81	0.9323	5.6741	1.389
−2	0.39819	641.3	0.31074	190.76	1460.06	0.9662	5.6474	1.393
0	0.42938	638.6	0.28930	200.00	1462.24	1.0000	5.6210	1.398
2	0.46246	635.8	0.26962	209.27	1464.35	1.0336	5.5951	1.402
4	0.49748	633.1	0.25153	218.55	1466.40	1.0670	5.5695	1.407
6	0.53453	630.3	0.23489	227.87	1468.37	1.1003	5.5442	1.412
8	0.57370	627.5	0.21956	237.20	1470.28	1.1334	5.5192	1.417
10	0.61505	624.6	0.20543	246.57	1472.11	1.1664	5.4946	1.422
12	0.65866	621.8	0.19237	255.95	1473.88	1.1992	5.4703	1.428
14	0.70463	618.9	0.18031	265.37	1475.56	1.2318	5.4463	1.434
16	0.75303	616.0	0.16914	274.81	1477.17	1.2643	5.4226	1.440
18	0.80395	613.1	0.15879	284.28	1478.70	1.2967	5.3991	1.446
20	0.85748	610.2	0.14920	293.78	1480.16	1.3289	5.3759	1.453
22	0.91369	607.2	0.14029	303.31	1481.53	1.3610	5.3529	1.460
24	0.97268	604.3	0.13201	312.87	1482.82	1.3929	5.3301	1.468
26	1.03450	601.3	0.12431	322.47	1484.02	1.4248	5.3076	1.475
28	1.09930	598.2	0.11714	332.09	1485.14	1.4565	5.2853	1.484
30	1.16720	595.2	0.11046	341.76	1486.17	1.4881	5.2631	1.492
32	1.23820	592.1	0.10422	351.45	1487.11	1.5196	5.2412	1.501
34	1.31240	589.0	0.09840	361.19	1487.95	1.5509	5.2194	1.510
36	1.39000	585.8	0.09296	370.96	1488.70	1.5822	5.1978	1.520
38	1.47090	582.6	0.08787	380.78	1489.36	1.6134	5.1763	1.530
40	1.55540	579.4	0.08310	390.64	1489.91	1.6446	5.1549	1.541
42	1.64350	576.2	0.07863	400.54	1490.36	1.6756	5.1337	1.553
44	1.73530	572.9	0.07445	410.48	1490.70	1.7065	5.1126	1.565
46	1.83100	569.6	0.07052	420.48	1490.94	1.7374	5.0915	1.577
48	1.93050	566.3	0.06682	430.52	1491.06	1.7683	5.0706	1.591
50	2.03400	562.9	0.06335	440.62	1491.07	1.7990	5.0497	1.605
55	2.31110	554.2	0.05554	466.10	1490.57	1.8758	4.9977	1.643
60	2.61560	545.2	0.04880	491.97	1489.27	1.9523	4.9458	1.687
65	2.94910	536.0	0.04296	518.26	1487.09	2.0288	4.8939	1.739
70	3.31350	526.3	0.03787	545.04	1483.94	2.1054	4.8415	1.799
75	3.71050	516.2	0.03342	572.37	1479.72	2.1823	4.7885	1.870
80	4.14200	505.7	0.02951	600.34	1474.31	2.2596	4.7344	1.955
85	4.61000	494.5	0.02606	629.04	1467.53	2.3377	4.6789	2.058
90	5.11670	482.8	0.02300	658.61	1459.19	2.4168	4.6213	2.187
95	5.66430	470.2	0.02027	689.19	1449.01	2.4973	4.5612	2.349
100	6.25530	456.6	0.01782	721.00	1436.63	2.5797	4.4975	2.562
105	6.89230	441.9	0.01561	754.35	1421.57	2.6647	4.4291	2.851
110	7.57830	425.6	0.01360	789.68	1403.08	2.7533	4.3542	3.26
115	8.31700	407.2	0.01174	827.74	1379.99	2.8474	4.2702	3.91
120	9.11250	385.5	0.00999	869.92	1350.23	2.9502	4.1719	5.04
125	9.97002	357.8	0.00828	919.68	1309.12	3.0702	4.0483	7.62
130	10.89770	312.3	0.00638	992.02	1239.32	3.2437	3.8571	20.66
132.25[c]	11.33300	225.0	0.00444	1119.22	1119.22	3.5542	3.5542	∞

*Temperatures on ITS-90 scale [a]Triple point [b]Normal boiling point [c]Critical point

Refrigerants

Table 8.16 Comparative Refrigerant Performance per Kilowatt of Refrigeration [2021F, Ch 29, Tbl 8]

Refrigerant No.	Chemical Name or Composition (% by mass)	Evaporator Pressure, MPa	Condenser Pressure, MPa	Compression Ratio	Net Refrigerating Effect, kJ/kg	Refrigerant Circulated, g/s	Liquid Circulated, L/s	Specific Volume of Suction Gas, m^3/kg	Compressor Displacement, L/s	Power Consumption, kW	Coefficient of Performance	Compressor Discharge Temp., °C
Evaporator –31.7°C/Condenser 30°C												
744	Carbon dioxide	1.349	7.213	5.35	132.1	7.57	0.0128	0.0285	0.2160	0.5892	1.698	91.3
170	Ethane	1.012	4.655	4.6	153.6	6.51	0.0236	0.0548	0.3567	0.5947	1.681	57.9
1270	Propylene	0.199	1.305	6.57	269.1	3.72	0.0075	0.2266	0.8422	0.3471	2.88	49.1
507A	R-125/143a (50/50)	0.199	1.460	7.34	101.1	9.89	0.0097	0.0949	0.9360	0.3887	2.573	38.1
404A	R-125/143a/134a (44/52/4)	0.190	1.421	7.46	104.9	9.54	0.0093	0.1005	0.9565	0.3853	2.595	38.9
502	R-22/115 (48.8/51.2)	0.183	1.304	7.14	97.8	10.22	0.0086	0.0924	0.9470	0.3651	2.739	41.3
22	Chlorodifluoromethane	0.152	1.192	7.81	155.3	6.44	0.0055	0.1448	0.9326	0.3369	2.967	65.4
717	Ammonia	0.110	1.167	10.61	1079.1	0.93	0.0016	1.0425	0.9643	0.3327	3.007	140.9
Evaporator –6.7°C/Condenser 30°C												
744	Carbon dioxide	2.909	7.213	2.48	129.5	7.72	0.0130	0.0127	0.0977	0.2845	3.514	61.3
170	Ethane	2.024	4.655	2.3	163.1	6.13	0.0222	0.0263	0.1612	0.2786	3.588	46.6
32	Difluoromethane	0.653	1.928	2.95	258.6	3.87	0.0041	0.0563	0.2178	0.1690	5.924	59.7
410A	R-32/125 (50/50)	0.643	1.886	2.94	170.9	5.85	0.0057	0.0406	0.2381	0.1728	5.78	46.6
507A	R-125/143a (50/50)	0.503	1.460	2.9	114.9	8.70	0.0085	0.0385	0.3349	0.1798	5.564	34.2
404A	R-125/143a/134a (44/52/4)	0.486	1.421	2.92	118.8	8.42	0.0083	0.0405	0.3410	0.1785	5.598	34.6
1270	Propylene	0.476	1.305	2.74	294.4	3.40	0.0068	0.0986	0.3359	0.1675	5.975	39.3
502	R-22/115 (48.8/51.2)	0.457	1.304	2.86	109.5	9.13	0.0077	0.0386	0.3527	0.1724	5.799	35.4
22	Chlorodifluoromethane	0.399	1.192	2.99	165.9	6.03	0.0051	0.0584	0.3520	0.1637	6.105	47.8
407C	R-32/125/134a (23/25/52)	0.396	1.267	3.19	167.1	5.98	0.0053	0.0588	0.3518	0.1686	5.93	43.9
290	Propane	0.385	1.079	2.8	288.6	3.47	0.0072	0.1180	0.4093	0.1669	5.987	34.9
717	Ammonia	0.332	1.167	3.51	1113.0	0.90	0.0015	0.3689	0.3313	0.1599	6.254	82.1
1234yf	2,3,3,3-tetrafluoropropene*	0.250	0.783	3.13	120.5	8.30	0.0077	0.0718	0.5954	0.1715	5.835	30.0
134a	Tetrafluoroethane	0.228	0.770	3.37	153.0	6.54	0.0055	0.0880	0.5745	0.1650	6.063	34.8
1234ze(E)	trans-1,3,3,3-tetrafluoropropene*	0.168	0.578	3.44	139.6	7.16	0.0063	0.1086	0.7798	0.1658	6.03	30.0
600a	Isobutane*	0.123	0.405	3.29	278.0	3.60	0.0066	0.2984	1.0723	0.1620	6.171	30.0

Table 8.16 Comparative Refrigerant Performance per Kilowatt of Refrigeration [2021F, Ch 29, Tbl 8] ***(Continued)***

Refrigerant No.	Chemical Name or Composition (% by mass)	Evaporator Pressure, MPa	Condenser Pressure, MPa	Compression Ratio	Net Refrigerating Effect, kJ/kg	Refrigerant Circulated, g/s	Liquid Circulated, L/s	Specific Volume of Suction Gas, m^3/kg	Compressor Displacement, L/s	Power Consumption, kW	Coefficient of Performance	Compressor Discharge Temp., °C
Evaporator 7.2°C/Condenser 30°C												
32	Difluoromethane	1.018	1.928	1.89	261.1	3.83	0.0040	0.0360	0.1381	0.0944	10.602	46.9
410A	R-32/125 (50/50)	1.000	1.886	1.89	175.0	5.71	0.0055	0.0260	0.1484	0.0965	10.379	39.8
502	R-22/115 (48.8/51.2)	0.703	1.304	1.85	115.3	8.67	0.0073	0.0252	0.2187	0.0956	10.474	33.2
407C	R-32/125/134a (23/25/52)	0.640	1.267	1.98	173.7	5.76	0.0051	0.0367	0.2112	0.0939	10.655	39.3
22	Chlorodifluoromethane	0.626	1.192	1.9	171.0	5.85	0.0050	0.0377	0.2205	0.0918	10.885	40.3
290	Propane	0.588	1.079	1.84	303.9	3.29	0.0068	0.0787	0.2580	0.0931	10.743	32.6
717	Ammonia	0.558	1.167	2.09	1127.8	0.89	0.0015	0.2254	0.1998	0.0893	11.186	58.6
500	R-12/152a (73.8/26.2)	0.458	0.880	1.92	150.4	6.65	0.0059	0.0453	0.3010	0.0916	10.925	34.6
1234yf	2,3,3,3-tetrafluoropropene*	0.401	0.783	1.96	129.0	7.75	0.0072	0.0453	0.3514	0.0941	10.623	30.0
12	Dichlorodifluoromethane	0.388	0.744	1.92	126.9	7.88	0.0061	0.0449	0.3536	0.0910	11.004	33.1
134a	Tetrafluoroethane	0.377	0.770	2.04	161.0	6.21	0.0052	0.0542	0.3364	0.0918	10.903	32.6
1234ze(E)	trans-1,3,3,3-tetrafluoropropene*	0.280	0.578	2.06	149.1	6.71	0.0059	0.0668	0.4483	0.0918	10.899	30.0
600a	Isobutane*	0.201	0.405	2.01	296.3	3.37	0.0062	0.1879	0.6332	0.0901	11.084	30.0
600	Butane*	0.134	0.283	2.11	326.9	3.06	0.0054	0.2853	0.8725	0.0891	11.226	30.0
123	Dichlorotrifluoroethane	0.045	0.110	2.44	155.5	6.43	0.0044	0.3309	2.1269	0.0878	11.397	30.0
113	Trichlorotrifluoroethane*	0.021	0.054	2.57	137.6	7.27	0.0047	0.5874	4.2686	0.0876	11.409	30.0

*Superheat required

Data from NIST CYCLE_D 4.0, zero subcool, zero superheat unless noted, no line losses, 100% efficiencies, average temperatures.

Table 8.17 Suction, Discharge, and Liquid Line Capacities in Kilowatts for Refrigerant 22 (Single- or High-Stage Applications) [2018R, Ch 1, Tbl 3]

Nominal Line OD, mm	Suction Lines ($\Delta t = 0.04$ K/m) Saturated Suction Temperature, °C					Discharge Lines ($\Delta t = 0.02$ K/m, $\Delta p = 74.90$) Saturated Suction Temperature, °C			Liquid Lines See note a	
	−40	−30	−20	−5	5					
	Corresponding Δp, Pa/m								Velocity = 0.5 m/s	Δt = 0.02 K/m $\Delta p = 749$
	196	277	378	572	731	−40	−20	5		
TYPE L COPPER LINE										
12	0.32	0.50	0.75	1.28	1.76	2.30	2.44	2.60	7.08	11.24
15	0.61	0.95	1.43	2.45	3.37	4.37	4.65	4.95	11.49	21.54
18	1.06	1.66	2.49	4.26	5.85	7.59	8.06	8.59	17.41	37.49
22	1.88	2.93	4.39	7.51	10.31	13.32	14.15	15.07	26.66	66.18
28	3.73	5.82	8.71	14.83	20.34	26.24	27.89	29.70	44.57	131.0
35	6.87	10.70	15.99	27.22	37.31	48.03	51.05	54.37	70.52	240.7
42	11.44	17.80	26.56	45.17	61.84	79.50	84.52	90.00	103.4	399.3
54	22.81	35.49	52.81	89.69	122.7	157.3	167.2	178.1	174.1	794.2
67	40.81	63.34	94.08	159.5	218.3	279.4	297.0	316.3	269.9	1415.0
79	63.34	98.13	145.9	247.2	337.9	431.3	458.5	488.2	376.5	2190.9
105	136.0	210.3	312.2	527.8	721.9	919.7	977.6	1041.0	672.0	4697.0
STEEL LINE										
10	0.47	0.72	1.06	1.78	2.42	3.04	3.23	3.44	10.66	15.96
15	0.88	1.35	1.98	3.30	4.48	5.62	5.97	6.36	16.98	29.62
20	1.86	2.84	4.17	6.95	9.44	11.80	12.55	13.36	29.79	62.55
25	3.52	5.37	7.87	13.11	17.82	22.29	23.70	25.24	48.19	118.2
32	7.31	11.12	16.27	27.11	36.79	46.04	48.94	52.11	83.56	244.4
40	10.98	16.71	24.45	40.67	55.21	68.96	73.31	78.07	113.7	366.6

Table 8.17 Suction, Discharge, and Liquid Line Capacities in Kilowatts for Refrigerant 22 (Single- or High-Stage Applications) [2018R, Ch 1, Tbl 3] ***(Continued)***

Nominal Line OD, mm	Suction Lines ($\Delta t = 0.04$ K/m) Saturated Suction Temperature, °C					Discharge Lines ($\Delta t = 0.02$ K/m, $\Delta p = 74.90$) Saturated Suction Temperature, °C			Liquid Lines See note a	
	−40	−30	−20	−5	5				Velocity = 0.5 m/s	$\Delta t =$ 0.02 K/m $\Delta p = 749$
	Corresponding Δp, Pa/m									
	196	277	378	572	731	−40	−20	5		
STEEL LINE *(Continued)*										
50	21.21	32.23	47.19	78.51	106.4	132.9	141.3	150.5	187.5	707.5
65	33.84	51.44	75.19	124.8	169.5	211.4	224.7	239.3	267.3	1127.3
80	59.88	90.95	132.8	220.8	299.5	373.6	397.1	422.9	412.7	1991.3
100	122.3	185.6	270.7	450.1	610.6	761.7	809.7	862.2	711.2	4063.2

Notes:

1. Table capacities are in kilowatts of refrigeration.

   Δp = pressure drop per unit equivalent length of line, Pa/m

   Δt = corresponding change in saturation temperature, K/m

2. Line capacity for other saturation temperatures Δt and equivalent lengths L_e

$$\text{Line capacity} = \text{Table capacity}\left(\frac{\text{Table } L_e}{\text{Actual } L_e} \times \frac{\text{Actual } \Delta t}{\text{Table } \Delta t}\right)^{0.55}$$

3. Saturation temperature Δt for other capacities and equivalent lengths L_e

$$\Delta t = \text{Table } \Delta t \left(\frac{\text{Actual } L_e}{\text{Table } L_e}\right)\left(\frac{\text{Actual capacity}}{\text{Table capacity}}\right)^{1.8}$$

4. Values based on 40°C condensing temperature. Multiply table capacities by the following factors for other condensing temperatures.

Condensing Temperature, °C	Suction Line	Discharge Line
20	1.18	0.80
30	1.10	0.88
40	1.00	1.00
50	0.91	1.11

[a] Sizing is recommended where any gas generated in receiver must return up condensate line to condenser without restricting condensate flow. Water-cooled condensers, where receiver ambient temperature may be higher than refrigerant condensing temperature, fall into this category.

[b] Line pressure drop Δp is conservative; if subcooling is substantial or line is short, a smaller size line may be used. Applications with very little subcooling or very long lines may require a larger line.

[c] System working pressures may exceed calculated allowable pressure in some listed type L annealed copper tubes at certain saturated condensing temperatures. Review maximum working pressure allowances for the pipe material used before selecting pipe sizes to ensure the pipe is properly rated for system working and design pressures.

Table 8.18 Suction, Discharge, and Liquid Line Capacities in Kilowatts for Refrigerant 134a (Single- or High-Stage Applications)
[2018R, Ch 1, Tbl 5]

Nominal Line OD, mm	Suction Lines (Δt = 0.04 K/m)					Discharge Lines (Δt = 0.02 K/m, Δp = 538 Pa/m)			Liquid Lines	
	Saturated Suction Temperature, °C					Saturated Suction Temperature, °C			See note a	
	−10	−5	0	5	10				Velocity = 0.5 m/s	Δt = 0.02 K/m Δp = 538 Pa/m
	Corresponding Δp, Pa/m									
	318	368	425	487	555	−10	0	10		
TYPE L COPPER LINE										
12	0.62	0.76	0.92	1.11	1.33	1.69	1.77	1.84	6.51	8.50
15	1.18	1.45	1.76	2.12	2.54	3.23	3.37	3.51	10.60	16.30
18	2.06	2.52	3.60	3.69	4.42	5.61	5.85	6.09	16.00	28.40
22	3.64	4.45	5.40	6.50	7.77	9.87	10.30	10.70	24.50	50.10
28	7.19	8.80	10.70	12.80	15.30	19.50	20.30	21.10	41.00	99.50
35	13.20	16.10	19.50	23.50	28.10	35.60	37.20	38.70	64.90	183.00
42	21.90	26.80	32.40	39.00	46.50	59.00	61.60	64.10	95.20	304.00
54	43.60	53.20	64.40	77.30	92.20	117.00	122.00	127.00	160.00	605.00
67	77.70	94.60	115.00	138.00	164.00	208.00	217.00	226.00	248.00	1080.00
79	120.00	147.00	177.00	213.00	253.00	321.00	335.00	349.00	346.00	1670.00
105	257.00	313.00	379.00	454.00	541.00	686.00	715.00	744.00	618.00	3580.00
STEEL LINE										
10	0.87	1.06	1.27	1.52	1.80	2.28	2.38	2.47	9.81	12.30
15	1.62	1.96	2.36	2.81	3.34	4.22	4.40	4.58	15.60	22.80
20	3.41	4.13	4.97	5.93	7.02	8.88	9.26	9.64	27.40	48.20
25	6.45	7.81	9.37	11.20	13.30	16.70	17.50	18.20	44.40	91.00
32	13.30	16.10	19.40	23.10	27.40	34.60	36.10	37.50	76.90	188.00
40	20.00	24.20	29.10	34.60	41.00	51.90	54.10	56.30	105.00	283.00

Table 8.18 Suction, Discharge, and Liquid Line Capacities in Kilowatts for Refrigerant 134a (Single- or High-Stage Applications) [2018R, Ch 1, Tbl 5] *(Continued)*

Nominal Line OD, mm	Suction Lines ($\Delta t = 0.04$ K/m) Saturated Suction Temperature, °C					Discharge Lines ($\Delta t = 0.02$ K/m, $\Delta p = 538$ Pa/m) Saturated Suction Temperature, °C			Liquid Lines (See note a)	
	−10	−5	0	5	10				Velocity = 0.5 m/s	$\Delta t = 0.02$ K/m $\Delta p = 538$ Pa/m
	Corresponding Δp, Pa/m									
	318	368	425	487	555	−10	0	10		
STEEL LINE *(Continued)*										
50	38.60	46.70	56.00	66.80	79.10	100.00	104.00	108.00	173.00	546.00
65	61.50	74.30	89.30	106.00	126.00	159.00	166.00	173.00	246.00	871.00
80	109.00	131.00	158.00	288.00	223.00	281.00	294.00	306.00	380.00	1540.00
100	222.00	268.00	322.00	383.00	454.00	573.00	598.00	622.00	655.00	3140.00

Notes:

1. Table capacities are in kilowatts of refrigeration.

Δp = pressure drop per equivalent line length, Pa/m

Δt = corresponding change in saturation temperature, K/m

2. Line capacity for other saturation temperatures Δt and equivalent lengths L_e

$$\text{Line capacity} = \text{Table capacity}\left(\frac{\text{Table } L_e}{\text{Actual } L_e} \times \frac{\text{Actual } \Delta t}{\text{Table } \Delta t}\right)^{0.55}$$

3. Saturation temperature Δt for other capacities and equivalent lengths L_e

$$\Delta t = \text{Table } \Delta t \left(\frac{\text{Actual } L_e}{\text{Table } L_e}\right)\left(\frac{\text{Actual capacity}}{\text{Table capacity}}\right)^{1.8}$$

4. Values based on 40°C condensing temperature. Multiply table capacities by the following factors for other condensing temperatures.

Condensing Temperature, °C	Suction Line	Discharge Line
20	1.239	0.682
30	1.120	0.856
40	1.0	1.0
50	0.888	1.110

[a] Sizing is recommended where any gas generated in receiver must return up condensate line to condenser without restricting condensate flow. Water-cooled condensers, where receiver ambient temperature may be higher than refrigerant condensing temperature, fall into this category.

[b] Line pressure drop Δp is conservative; if subcooling is substantial or line is short, a smaller size line may be used. Applications with very little subcooling or very long lines may require a larger line.

[c] System working pressures may exceed calculated allowable pressure in some listed type L annealed copper tubes at certain saturated condensing temperatures. Review maximum working pressure allowances for the pipe material used before selecting pipe sizes to ensure the pipe is properly rated for system working and design pressures.

Table 8.19 Suction, Discharge, and Liquid Line Capacities in Kilowatts for Refrigerant 404A (Single- or High-Stage Applications)
[2018R, Ch 1, Tbl 6]

Line Size	Suction Lines ($\Delta t = 0.04$ K/m)						Discharge Lines ($\Delta t = 0.02$ K/m, $\Delta p = 74.90$)[c]						Liquid Lines (40°C)[a,c]		
	Saturated Suction Temperature, °C						Saturated Suction Temperature, °C								
Type L Copper, OD, mm	−50	−40	−30	−20	−5	5	−50	−40	−30	−20	−5	5	Velocity = 0.5 m/s	Δt = 0.02 K/m Drop	Δt = 0.05 K/m Drop
	Corresponding Δp, Pa/m						Corresponding Δp, Pa/m								
	165.5	240.6	337.2	455.1	679.1	863.2	875.6	875.6	875.6	875.6	875.6	875.6		Δp = 875.6	Δp = 2189.1
12	0.16	0.27	0.43	0.67	1.19	1.69	1.73	1.87	2.00	2.13	2.31	2.42	4.1	8.0	13.3
15	0.30	0.52	0.83	1.28	2.27	3.22	3.29	3.55	3.81	4.05	4.40	4.61	6.7	15.3	25.2
18	0.53	0.90	1.45	2.22	3.94	5.57	5.71	6.16	6.59	7.02	7.62	7.99	10.1	26.6	43.7
22	0.94	1.59	2.55	3.91	6.93	9.79	10.00	10.79	11.56	12.30	13.36	14.01	15.5	46.8	76.7
28	1.86	3.14	5.04	7.72	13.66	19.25	19.68	21.23	22.74	24.21	26.29	27.57	26.0	92.5	151.1
35	3.43	5.78	9.26	14.15	25.00	35.17	35.96	38.78	41.54	44.23	48.03	50.37	41.1	169.3	276.3
42	5.71	9.61	15.36	23.46	41.32	58.16	59.48	64.15	68.72	73.16	79.45	83.32	60.3	280.4	456.2
54	11.37	19.12	30.50	46.57	81.90	114.98	117.62	126.86	135.89	144.67	157.11	164.76	101.4	556.9	903.2
67	20.31	34.10	54.30	82.75	145.45	203.96	208.67	225.07	241.08	256.66	278.73	292.29	157.3	989.8	1601.8
79	31.54	52.78	84.12	128.09	224.52	314.97	321.69	346.97	371.66	395.67	429.70	450.60	219.3	1529.9	2473.4
105	67.66	113.08	179.89	273.26	478.70	670.69	685.09	738.92	791.51	842.65	915.11	959.63	391.5	3264.9	5265.6

Table 8.19 Suction, Discharge, and Liquid Line Capacities in Kilowatts for Refrigerant 404A (Single- or High-Stage Applications) [2018R, Ch 1, Tbl 6] ***(Continued)***

Line Size		Suction Lines (Δt = 0.04 K/m)						Discharge Lines (Δt = 0.02 K/m, Δp = 74.90)[c]						Liquid Lines (40°C)[a,c]		
		Saturated Suction Temperature, °C						Saturated Suction Temperature, °C						Velocity = 0.5 m/s	Δt = 0.02 K/m Drop	Δt = 0.05 K/m Drop
		−50	−40	−30	−20	−5	5	−50	−40	−30	−20	−5	5			
		Corresponding Δp, Pa/m						Corresponding Δp, Pa/m								
		165.5	240.6	337.2	455.1	679.1	863.2	875.6	875.6	875.6	875.6	875.6	875.6		Δp = 875.6	Δp = 2189.1
Steel																
mm	SCH															
10	80	0.16	0.26	0.40	0.61	1.05	1.46	1.49	1.61	1.72	1.83	1.99	2.09	4.6	7.2	11.5
15	80	0.31	0.51	0.80	1.20	2.07	2.88	2.94	3.17	3.39	3.61	3.92	4.12	7.6	14.3	22.7
20	80	0.70	1.15	1.80	2.70	4.66	6.48	6.61	7.13	7.64	8.14	8.84	9.27	14.1	32.1	51.1
25	80	1.37	2.25	3.53	5.30	9.13	12.68	12.95	13.97	14.96	15.93	17.30	18.14	23.4	63.0	100.0
32	80	2.95	4.83	7.57	11.35	19.57	27.20	27.72	29.90	32.03	34.10	37.03	38.83	41.8	134.9	214.0
40	80	4.49	7.38	11.55	17.29	29.81	41.42	42.22	45.54	48.78	51.94	56.40	59.14	57.5	205.7	326.5
50	40	10.47	17.16	26.81	40.20	69.20	96.18	98.04	105.75	113.27	120.59	130.96	137.33	109.2	477.6	758.2
65	40	16.68	27.33	42.72	63.93	110.18	152.98	155.95	168.20	180.17	191.81	208.31	218.44	155.7	761.1	1205.9
80	40	29.51	48.38	75.47	112.96	194.49	270.35	275.59	297.25	318.40	338.98	368.13	386.03	240.5	1344.9	2131.2
100	40	60.26	98.60	153.84	230.29	396.56	550.03	560.67	604.72	647.76	689.61	748.91	785.34	414.3	2735.7	4335.6
125	40	108.75	177.97	277.71	415.78	714.27	991.91	1012.44	1091.99	1169.71	1245.28	1352.37	1418.15	650.6	4939.2	7819.0
150	40	176.25	287.77	449.08	671.57	1155.17	1604.32	1635.36	1763.85	1889.38	2011.45	2184.43	2290.69	940.3	7988.0	12 629.7
200	40	360.41	589.35	918.60	1373.79	2363.28	3277.89	3341.30	3603.84	3860.32	4109.73	4463.15	4680.25	1628.2	16 342.0	25 838.1
250	40	652.69	1065.97	1661.62	2485.16	4275.41	5930.04	6044.77	6519.73	6983.73	7434.94	8074.30	8467.06	2566.4	29 521.7	46 743.9
300	ID[b]	1044.01	1705.26	2658.28	3970.05	6830.36	9488.03	9671.59	10 431.52	11 173.92	11 895.85	12 918.83	13 547.24	3680.9	47 161.0	74 677.7
350	30	1351.59	2207.80	3436.53	5140.20	8843.83	12 266.49	12 503.79	13 486.26	14 446.06	15 379.40	16 701.95	17 514.38	4487.7	61 061.2	96 691.3
400	30	1947.52	3176.58	4959.92	7407.49	12 725.25	17 677.86	18 019.86	19 435.74	20 818.96	22 164.04	24 070.04	25 240.87	5944.7	87 994.9	139 346.8

Table 8.19 Suction, Discharge, and Liquid Line Capacities in Kilowatts for Refrigerant 404A (Single- or High-Stage Applications) [2018R, Ch 1, Tbl 6] ***(Continued)***

[a] Sizing shown is recommended where any gas generated in receiver must return up condensate line to condenser without restricting condensate flow. Water-cooled condensers, where receiver ambient temperature may be higher than refrigerant condensing temperature, fall into this category.

[b] Pipe inside diameter is same as nominal pipe size.

[c] System working pressures may exceed calculated allowable pressure in some listed type L annealed copper tubes at certain saturated condensing temperatures. Review maximum working pressure allowances for the pipe material used before selecting pipe sizes to ensure the pipe is properly rated for system working and design pressures.

Notes:

1. Table capacities are in kilowatts of refrigeration.
   Δp = pressure drop per unit equivalent length of line, Pa/m
   Δt = corresponding change in saturation temperature, K/m
2. Line capacity for other saturation temperatures Δt and equivalent lengths L_e

$$\text{Line capacity} = \text{Table capacity}\left(\frac{\text{Table } L_e}{\text{Actual } L_e} \times \frac{\text{Actual } \Delta t}{\text{Table } \Delta t}\right)^{0.55}$$

3. Saturation temperature Δt for other capacities and equivalent lengths L_e

$$\Delta t = \text{Table } \Delta t \left(\frac{\text{Actual } L_e}{\text{Table } L_e}\right)\left(\frac{\text{Actual capacity}}{\text{Table capacity}}\right)^{1.8}$$

4. Capacity (kW) based on standard refrigerant cycle of 40°C liquid and saturated evaporator outlet temperature. Liquid capacity (kW) based on –5°C evaporator temperature.
5. Thermophysical properties and viscosity data based on calculations from NIST REFPROP program Version 6.01.
6. For brazed Type L copper tubing larger than 28 mm OD for discharge or liquid service, see Safety Requirements section.
7. Values are based on 40°C condensing temperature. Multiply table capacities by the following factors for other condensing temperatures.

Cond. Temp., °C	Suction Line	Discharge Line
20	1.344	0.812
30	1.177	0.906
40	1.000	1.000
50	0.809	1.035

Table 8.20 Suction, Discharge, and Liquid Line Capacities in Kilowatts for Refrigerant 407C (Single- or High-Stage Applications)
[2018R, Ch 1, Tbl 9]

Line Size	Suction Lines (Δt = 0.04 K/m)						Discharge Lines (Δt = 0.02 K/m, Δp = 74.90)[c]						Liquid Lines (40°C)[a,c]		
Type L Copper, OD, mm	Saturated Suction Temperature, °C						Saturated Suction Temperature, °C						Velocity = 0.5 m/s	Δt = 0.02 K/m Drop	Δt = 0.05 K/m Drop
	−50	−40	−30	−20	−5	5	−50	−40	−30	−20	−5	5			
	Corresponding Δp, Pa/m						Corresponding Δp, Pa/m								
	173.7	251.7	350.3	471.6	700.5	882.5	896.3	896.3	896.3	896.3	896.3	896.3		Δp = 896.3	Δp = 2240.8
12	0.16	0.28	0.44	0.68	1.21	1.70	1.72	1.86	2.00	2.13	2.32	2.43	4.0	7.9	13.0
15	0.31	0.53	0.85	1.30	2.31	3.24	3.27	3.54	3.80	4.05	4.41	4.63	6.5	15.0	24.7
18	0.55	0.92	1.47	2.26	4.00	5.61	5.66	6.12	6.57	7.01	7.63	8.01	9.8	26.1	42.8
22	0.97	1.63	2.60	3.98	7.02	9.85	9.93	10.73	11.52	12.29	13.37	14.04	15.0	45.9	75.1
28	1.91	3.22	5.14	7.85	13.83	19.38	19.53	21.12	22.67	24.18	26.31	27.63	25.1	90.5	147.8
35	3.52	5.91	9.42	14.37	25.28	35.40	35.68	38.58	41.42	44.17	48.07	50.47	39.7	165.6	270.0
42	5.86	9.82	15.65	23.83	41.86	58.55	59.03	63.82	68.52	73.07	79.52	83.50	58.2	274.8	447.1
54	11.68	19.55	31.07	47.24	82.83	115.76	116.74	126.22	135.51	144.51	157.26	165.12	98.0	544.0	883.9
67	20.86	34.83	55.25	84.08	147.12	205.36	206.75	223.53	239.99	255.92	278.52	292.43	151.9	967.0	1567.7
79	32.31	54.01	85.61	129.94	227.12	317.17	319.34	345.26	370.68	395.29	430.19	451.67	211.9	1497.3	2420.9
105	69.31	115.54	182.78	277.24	484.29	675.47	678.77	733.87	787.90	840.21	914.39	960.06	378.2	3189.5	5154.4

Table 8.20 Suction, Discharge, and Liquid Line Capacities in Kilowatts for Refrigerant 407C (Single- or High-Stage Applications) [2018R, Ch 1, Tbl 9] *(Continued)*

Line Size		Suction Lines (Δt = 0.04 K/m)						Discharge Lines (Δt = 0.02 K/m, Δp = 74.90)[c]						Liquid Lines (40°C)[a,c]		
		Saturated Suction Temperature, °C						Saturated Suction Temperature, °C						Velocity = 0.5 m/s	Δt = 0.02 K/m Drop	Δt = 0.05 K/m Drop
		−50	−40	−30	−20	−5	5	−50	−40	−30	−20	−5	5			
		Corresponding Δp, Pa/m						Corresponding Δp, Pa/m							Δp =	Δp =
		173.7	251.7	350.3	471.6	700.5	882.5	896.3	896.3	896.3	896.3	896.3	896.3		896.3	2240.8
Steel																
mm	SCH															
10	80	0.16	0.26	0.41	0.62	1.06	1.47	1.48	1.60	1.72	1.83	1.99	2.09	4.4	7.1	11.3
15	80	0.31	0.52	0.81	1.21	2.09	2.90	2.91	3.15	3.38	3.60	3.92	4.12	7.4	13.9	22.2
20	80	0.71	1.17	1.83	2.74	4.71	6.52	6.55	7.09	7.61	8.11	8.83	9.27	13.6	31.4	49.9
25	80	1.40	2.29	3.58	5.36	9.23	12.77	12.83	13.87	14.89	15.88	17.28	18.15	22.6	61.6	97.7
32	80	3.01	4.93	7.68	11.50	19.76	27.33	27.47	29.70	31.88	34.00	37.00	38.85	40.3	132.0	209.4
40	80	4.59	7.52	11.72	17.54	30.09	41.63	41.83	45.23	48.56	51.78	56.35	59.17	55.6	201.0	319.0
50	40	10.69	17.50	27.25	40.71	69.87	96.67	97.14	105.02	112.76	120.24	130.86	137.39	105.5	466.6	740.7
65	40	17.06	27.88	43.32	64.81	111.37	153.76	154.51	167.05	179.35	191.26	208.14	218.54	150.4	743.5	1178.1
80	40	30.20	49.26	76.63	114.52	196.37	271.72	273.05	295.22	316.95	338.00	367.84	386.21	232.3	1313.9	2082.0
100	40	61.60	100.39	156.20	233.20	400.40	552.81	555.50	600.59	644.81	687.62	748.33	785.70	400.3	2675.6	4235.5
125	40	111.17	181.20	281.64	421.03	721.18	998.16	1003.06	1084.49	1164.33	1241.63	1351.25	1418.74	628.6	4825.1	7638.5
150	40	179.98	292.99	455.44	680.92	1166.35	1612.43	1620.28	1751.80	1880.77	2005.64	2182.72	2291.73	908.5	7803.5	12 338.1
200	40	368.55	600.02	931.61	1393.04	2386.16	3294.46	3310.49	3579.22	3842.72	4097.86	4459.65	4682.37	1573.2	15 964.7	25 241.5
250	40	666.52	1085.29	1685.18	2516.51	4316.82	5960.02	5989.03	6475.19	6951.89	7413.46	8067.98	8470.90	2479.7	28 840.0	45 664.6
300	ID[b]	1067.53	1736.16	2695.93	4020.13	6896.51	9535.99	9582.41	10 360.26	11 122.98	11 861.49	12 908.71	13 553.39	3556.5	46 140.3	72 953.4
350	30	1380.23	2247.80	3485.20	5205.04	8929.47	12 328.49	12 388.50	13 394.13	14 380.20	15 334.97	16 688.86	17 522.33	4336.1	59 651.3	94 458.7
400	30	1991.54	3239.15	5030.17	7500.91	12 848.49	17 767.21	17 853.70	19 302.97	20 724.05	22 100.02	24 051.18	25 252.33	5743.9	85 963.1	136 129.3

Table 8.20 Suction, Discharge, and Liquid Line Capacities in Kilowatts for Refrigerant 407C (Single- or High-Stage Applications) [2018R, Ch 1, Tbl 9] *(Continued)*

[a] Sizing shown is recommended where any gas generated in receiver must return up condensate line to condenser without restricting condensate flow. Water-cooled condensers, where receiver ambient temperature may be higher than refrigerant condensing temperature, fall into this category.

[b] Pipe inside diameter is same as nominal pipe size.

[c] System working pressures may exceed calculated allowable pressure in some listed type L annealed copper tubes at certain saturated condensing temperatures. Review maximum working pressure allowances for the pipe material used before selecting pipe sizes to ensure the pipe is properly rated for system working and design pressures.

Notes:

1. Table capacities are in kilowatts of refrigeration.
   Δp = pressure drop per unit equivalent length of line, Pa/m
   Δt = corresponding change in saturation temperature, K/m
2. Line capacity for other saturation temperatures Δt and equivalent lengths L_e

$$\text{Line capacity} = \text{Table capacity}\left(\frac{\text{Table } L_e}{\text{Actual } L_e} \times \frac{\text{Actual } \Delta t}{\text{Table } \Delta t}\right)^{0.55}$$

3. Saturation temperature Δt for other capacities and equivalent lengths L_e

$$\Delta t = \text{Table } \Delta t \left(\frac{\text{Actual } L_e}{\text{Table } L_e}\right)\left(\frac{\text{Actual capacity}}{\text{Table capacity}}\right)^{1.8}$$

4. Capacity (kW) based on standard refrigerant cycle of 40°C liquid and saturated evaporator outlet temperature. Liquid capacity (kW) based on −5°C evaporator temperature.
5. Thermophysical properties and viscosity data based on calculations from NIST REFPROP program Version 6.01.
6. For brazed Type L copper tubing larger than 28 mm OD for discharge or liquid service, see Safety Requirements section.
7. Values are based on 40°C condensing temperature. Multiply table capacities by the following factors for other condensing temperatures.

Cond. Temp., °C	Suction Line	Discharge Line
20	1.357	0.765
30	1.184	0.908
40	1.000	1.000
50	0.801	1.021

Table 8.21 Suction, Discharge, and Liquid Line Capacities in Kilowatts for Refrigerant 410A (Single- or High-Stage Applications)
[2018R, Ch 1, Tbl 8]

Line Size	Suction Lines ($\Delta t = 0.04$ K/m)						Discharge Lines ($\Delta t = 0.02$ K/m, $\Delta p = 74.90$)[c]						Liquid Lines (40°C)[a,c]		
	Saturated Suction Temperature, °C						Saturated Suction Temperature, °C								
	−50	−40	−30	−20	−5	5	−50	−40	−30	−20	−5	5	Velocity = 0.5 m/s	$\Delta t = 0.02$ K/m Drop	$\Delta t = 0.05$ K/m Drop
Type L Copper, OD, mm	Corresponding Δp, Pa/m						Corresponding Δp, Pa/m								
	218.6	317.2	443.3	599.1	894.2	1137.6	1172.1	1172.1	1172.1	1172.1	1172.1	1172.1		$\Delta p =$ 1179	$\Delta p =$ 2935.8
12	0.32	0.52	0.80	1.20	2.05	2.83	3.47	3.60	3.73	3.84	4.00	4.07	6.2	14.3	23.5
15	0.61	0.99	1.54	2.29	3.90	5.37	6.60	6.85	7.09	7.31	7.60	7.75	10.1	27.2	44.6
18	1.06	1.72	2.68	3.98	6.76	9.30	11.43	11.87	12.29	12.67	13.16	13.42	15.4	47.3	77.2
22	1.87	3.04	4.72	7.00	11.89	16.32	20.04	20.81	21.54	22.20	23.08	23.53	23.5	83.0	135.3
28	3.72	6.03	9.32	13.82	23.43	32.11	39.44	40.95	42.39	43.70	45.42	46.31	39.3	163.7	266.4
35	6.84	11.07	17.11	25.33	42.82	58.75	72.05	74.82	77.46	79.84	82.98	84.62	62.2	299.6	486.0
42	11.39	18.39	28.38	42.00	70.89	97.02	119.01	123.57	127.93	131.87	137.06	139.76	91.3	495.7	804.1
54	22.70	36.61	56.35	83.26	140.29	191.84	235.35	244.38	253.00	260.80	271.06	276.39	153.7	982.0	1590.3
67	40.48	65.21	100.35	147.94	249.16	340.33	417.58	433.60	448.89	462.73	480.93	490.40	238.2	1746.4	2816.7
79	62.89	101.10	155.22	229.02	384.65	525.59	643.78	668.47	692.05	713.37	741.44	756.03	332.2	2695.2	4350.8
105	134.69	216.27	331.96	488.64	820.20	1119.32									

Table 8.21 Suction, Discharge, and Liquid Line Capacities in Kilowatts for Refrigerant 410A (Single- or High-Stage Applications) [2018R, Ch 1, Tbl 8] *(Continued)*

Line Size		Suction Lines (Δt = 0.04 K/m)						Discharge Lines (Δt = 0.02 K/m, Δp = 74.90)[c]						Liquid Lines (40°C)[a,c]		
		Saturated Suction Temperature, °C						Saturated Suction Temperature, °C								
		−50	−40	−30	−20	−5	5	−50	−40	−30	−20	−5	5	Velocity = 0.5 m/s	Δt = 0.02 K/m Drop	Δt = 0.05 K/m Drop
		Corresponding Δp, Pa/m						Corresponding Δp, Pa/m								
		218.6	317.2	443.3	599.1	894.2	1137.6	1172.1	1172.1	1172.1	1172.1	1172.1	1172.1		Δp = 1179	Δp = 2935.8
Steel																
mm	SCH															
10	80	0.31	0.49	0.74	1.08	1.80	2.44	2.98	3.10	3.21	3.31	3.44	3.50	6.9	12.7	20.1
15	80	0.61	0.97	1.47	2.14	3.54	4.80	5.87	6.09	6.31	6.50	6.76	6.89	11.5	25.0	39.6
20	80	1.39	2.19	3.32	4.82	7.98	10.81	13.21	13.72	14.20	14.64	15.22	15.52	21.3	56.2	89.1
25	80	2.72	4.30	6.50	9.45	15.63	21.16	25.86	26.85	27.80	28.66	29.79	30.37	35.5	110.2	174.5
32	80	5.86	9.24	13.95	20.26	33.47	45.30	55.37	57.50	59.53	61.36	63.77	65.03	63.2	235.9	
40	80	8.94	14.09	21.28	30.91	50.97	68.99	84.33	87.57	90.66	93.45	97.13	99.04	87.1	359.8	568.9
50	40	20.81	32.75	49.39	71.75	118.34	160.19	195.83	203.34	210.51	217.00	225.54	229.98	165.4	835.4	1320.9
65	40	33.22	52.18	78.69	114.11	188.61	254.80	311.49	323.43	334.84	345.16	358.74	365.80	235.8	1328.6	2101.0
80	40	58.79	92.36	139.17	201.84	332.58	450.29	550.47	571.59	591.74	609.98	633.98	646.46	364.2	2347.8	3713.1
100	40	119.78	188.24	283.69	411.01	678.11	916.08	1121.21	1164.22	1205.28	1242.42	1291.30	1316.72	627.6	4787.0	7562.8
125	40	216.38	339.76	511.52	742.06	1221.40	1654.16	2022.16	2099.73	2173.77	2240.77	2328.92	2374.75	985.4	8622.2	13 639.9
150	40	350.32	549.37	827.18	1200.12	1975.34	2672.01	3266.45	3391.75	3511.36	3619.58	3761.97	3836.01	1424.2	13 944.5	22 032.9
200	40	717.23	1125.10	1692.00	2451.89	4041.21	5459.36	6673.89	6929.91	7174.29	7395.39	7586.32	7837.60	2466.2	28 528.0	45 016.9
250	40	1297.30	2035.01	3060.66	4435.35	7310.97	9876.55	12 073.76	12 536.92	12 979.03	13 379.03	13 905.35	14 179.04	3887.3	51 535.6	81 440.3
300	ID[b]	2075.09	3255.45	4896.39	7085.49	11 679.95	15 802.42	19 317.94	20 059.00	20 766.37	21 406.37	22 248.47	22 686.37	5575.3	82 451.9	130 304.0
350	30	2686.45	4214.83	6329.87	9173.88	15 122.98	20 429.97	24 974.96	25 933.02	26 847.53	27 674.95	28 763.66	29 329.79	6797.4	106 757.2	168 461.9
400	30	3870.92	6064.31	9135.88	13 220.36	21 760.24	29 442.67	35 992.70	37 373.41	38 691.36	39 883.80	41 452.79	42 268.67	9004.3	153 611.4	242 779.1

Table 8.21 Suction, Discharge, and Liquid Line Capacities in Kilowatts for Refrigerant 410A (Single- or High-Stage Applications) [2018R, Ch 1, Tbl 8] *(Continued)*

[a] Sizing shown is recommended where any gas generated in receiver must return up condensate line to condenser without restricting condensate flow. Water-cooled condensers, where receiver ambient temperature may be higher than refrigerant condensing temperature, fall into this category.

[b] Pipe inside diameter is same as nominal pipe size.

[c] System working pressures may exceed calculated allowable pressure in some listed type L annealed copper tubes at certain saturated condensing temperatures. Review maximum working pressure allowances for the pipe material used before selecting pipe sizes to ensure the pipe is properly rated for system working and design pressures.

Notes:

1. Table capacities are in kilowatts of refrigeration.
   Δp = pressure drop per unit equivalent length of line, Pa/m
   Δt = corresponding change in saturation temperature, K/m
2. Line capacity for other saturation temperatures Δt and equivalent lengths L_e

$$\text{Line capacity} = \text{Table capacity}\left(\frac{\text{Table } L_e}{\text{Actual } L_e} \times \frac{\text{Actual } \Delta t}{\text{Table } \Delta t}\right)^{0.55}$$

3. Saturation temperature Δt for other capacities and equivalent lengths L_e

$$\Delta t = \text{Table } \Delta t \left(\frac{\text{Actual } L_e}{\text{Table } L_e}\right)\left(\frac{\text{Actual capacity}}{\text{Table capacity}}\right)^{1.8}$$

4. Capacity (kW) based on standard refrigerant cycle of 40°C liquid and saturated evaporator outlet temperature. Liquid capacity (kW) based on −5°C evaporator temperature.
5. Thermophysical properties and viscosity data based on calculations from NIST REFPROP program Version 6.01.
6. For brazed Type L copper tubing larger than 15 mm OD for discharge or liquid service, see Safety Requirements section.
7. Values are based on 40°C condensing temperature. Multiply table capacities by the following factors for other condensing temperatures.

Cond. Temp., °C	Suction Line	Discharge Line
20	1.238	0.657
30	1.122	0.866
40	1.000	1.000
50	0.867	1.117

Table 8.22 Suction, Discharge, and Liquid Line Capacities in Kilowatts for Refrigerant 507A (Single- or High-Stage Applications)

[2018R, Ch 1, Tbl 7]

Line Size	Suction Lines ($\Delta t = 0.04$ K/m)						Discharge Lines ($\Delta t = 0.02$ K/m, $\Delta p = 74.90$)[c]						Liquid Lines (40°C)[a,c]		
	Saturated Suction Temperature, °C						Saturated Suction Temperature, °C								
Type L Copper, OD, mm	–50	–40	–30	–20	–5	5	–50	–40	–30	–20	–5	5	Velocity = 0.5 m/s	$\Delta t = 0.02$ K/m Drop	$\Delta t = 0.05$ K/m Drop
	Corresponding Δp, Pa/m						Corresponding Δp, Pa/m								
	173.7	251.7	350.3	471.6	700.5	882.5	896.3	896.3	896.3	896.3	896.3	896.3		$\Delta p = 896.3$	$\Delta p = 2240.8$
12	0.16	0.28	0.44	0.68	1.21	1.70	1.72	1.86	2.00	2.13	2.32	2.43	4.0	7.9	13.0
15	0.31	0.53	0.85	1.30	2.31	3.24	3.27	3.54	3.80	4.05	4.41	4.63	6.5	15.0	24.7
18	0.55	0.92	1.47	2.26	4.00	5.61	5.66	6.12	6.57	7.01	7.63	8.01	9.8	26.1	42.8
22	0.97	1.63	2.60	3.98	7.02	9.85	9.93	10.73	11.52	12.29	13.37	14.04	15.0	45.9	75.1
28	1.91	3.22	5.14	7.85	13.83	19.38	19.53	21.12	22.67	24.18	26.31	27.63	25.1	90.5	147.8
35	3.52	5.91	9.42	14.37	25.28	35.40	35.68	38.58	41.42	44.17	48.07	50.47	39.7	165.6	270.0
42	5.86	9.82	15.65	23.83	41.86	58.55	59.03	63.82	68.52	73.07	79.52	83.50	58.2	274.8	447.1
54	11.68	19.55	31.07	47.24	82.83	115.76	116.74	126.22	135.51	144.51	157.26	165.12	98.0	544.0	883.9
67	20.86	34.83	55.25	84.08	147.12	205.36	206.75	223.53	239.99	255.92	278.52	292.43	151.9	967.0	1567.7
79	32.31	54.01	85.61	129.94	227.12	317.17	319.34	345.26	370.68	395.29	430.19	451.67	211.9	1497.3	2420.9
105	69.31	115.54	182.78	277.24	484.29	675.47									

Table 8.22 Suction, Discharge, and Liquid Line Capacities in Kilowatts for Refrigerant 507A (Single- or High-Stage Applications) [2018R, Ch 1, Tbl 7] *(Continued)*

Line Size		Suction Lines ($\Delta t = 0.04$ K/m)						Discharge Lines ($\Delta t = 0.02$ K/m, $\Delta p = 74.90$)[c]						Liquid Lines (40°C)[a,c]		
		Saturated Suction Temperature, °C						Saturated Suction Temperature, °C								
		−50	−40	−30	−20	−5	5	−50	−40	−30	−20	−5	5	Velocity = 0.5 m/s	$\Delta t = 0.02$ K/m Drop	$\Delta t = 0.05$ K/m Drop
		Corresponding Δp, Pa/m						Corresponding Δp, Pa/m								
		173.7	251.7	350.3	471.6	700.5	882.5	896.3	896.3	896.3	896.3	896.3	896.3		$\Delta p =$ 896.3	$\Delta p =$ 2240.8
Steel																
mm	SCH															
10	80	0.16	0.26	0.41	0.62	1.06	1.47	1.48	1.60	1.72	1.83	1.99	2.09	4.4	7.1	11.3
15	80	0.31	0.52	0.81	1.21	2.09	2.90	2.91	3.15	3.38	3.60	3.92	4.12	7.4	13.9	22.2
20	80	0.71	1.17	1.83	2.74	4.71	6.52	6.55	7.09	7.61	8.11	8.83	9.27	13.6	31.4	49.9
25	80	1.40	2.29	3.58	5.36	9.23	12.77	12.83	13.87	14.89	15.88	17.28	18.15	22.6	61.6	97.7
32	80	3.01	4.93	7.68	11.50	19.76	27.33	27.47	29.70	31.88	34.00	37.00	38.85	40.3	132.0	209.4
40	80	4.59	7.52	11.72	17.54	30.09	41.63	41.83	45.23	48.56	51.78	56.35	59.17	55.6	201.0	319.0
50	40	10.69	17.50	27.25	40.71	69.87	96.67	97.14	105.02	112.76	120.24	130.86	137.39	105.5	466.6	740.7
65	40	17.06	27.88	43.32	64.81	111.37	153.76	154.51	167.05	179.35	191.26	208.14	218.54	150.4	743.5	1178.1
80	40	30.20	49.26	76.63	114.52	196.37	271.72	273.05	295.22	316.95	338.00	367.84	386.21	232.3	1313.9	2082.0
100	40	61.60	100.39	156.20	233.20	400.40	552.81	555.50	600.59	644.81	687.62	748.33	785.70	400.3	2675.6	4235.5
125	40	111.17	181.20	281.64	421.03	721.18	998.16	1003.06	1084.49	1164.33	1241.63	1351.25	1418.74	628.6	4825.1	7638.5
150	40	179.98	292.99	455.44	680.92	1166.35	1612.43	1620.28	1751.80	1880.77	2005.64	2182.72	2291.73	908.5	7803.5	12 338.1
200	40	368.55	600.02	931.61	1393.04	2386.16	3294.46	3310.49	3579.22	3842.72	4097.86	4459.65	4682.37	1573.2	15 964.7	25 241.5
250	40	666.52	1085.29	1685.18	2516.51	4316.82	5960.02	5989.03	6475.19	6951.89	7413.46	8067.98	8470.90	2479.7	28 840.0	45 664.6
300	ID[b]	1067.53	1736.16	2695.93	4020.13	6896.51	9535.99	9582.41	10 360.26	11 122.98	11 861.49	12 908.71	13 553.39	3556.5	46 140.3	72 953.4
350	30	1380.23	2247.80	3485.20	5205.04	8929.47	12 328.49	12 388.50	13 394.13	14 380.20	15 334.97	16 688.86	17 522.33	4336.1	59 651.3	94 458.7
400	30	1991.54	3239.15	5030.17	7500.91	12 848.49	17 767.21	17 853.70	19 302.97	20 724.05	22 100.02	24 051.18	25 252.33	5743.9	85 963.1	136 129.3

Table 8.22 Suction, Discharge, and Liquid Line Capacities in Kilowatts for Refrigerant 507A (Single- or High-Stage Applications [2018R, Ch 1, Tbl 7] *(Continued)*

[a]Sizing shown is recommended where any gas generated in receiver must return up condensate line to condenser without restricting condensate flow. Water-cooled condensers, where receiver ambient temperature may be higher than refrigerant condensing temperature, fall into this category.

[b]Pipe inside diameter is same as nominal pipe size.

[c]System working pressures may exceed calculated allowable pressure in some listed type L annealed copper tubes at certain saturated condensing temperatures. Review maximum working pressure allowances for the pipe material used before selecting pipe sizes to ensure the pipe is properly rated for system working and design pressures.

Notes:

1. Table capacities are in kilowatts of refrigeration.
   Δp = pressure drop per unit equivalent length of line, Pa/m
   Δt = corresponding change in saturation temperature, K/m
2. Line capacity for other saturation temperatures Δt and equivalent lengths L_e

$$\text{Line capacity} = \text{Table capacity}\left(\frac{\text{Table } L_e}{\text{Actual } L_e} \times \frac{\text{Actual } \Delta t}{\text{Table } \Delta t}\right)^{0.55}$$

3. Saturation temperature Δt for other capacities and equivalent lengths L_e

$$\Delta t = \text{Table } \Delta t \left(\frac{\text{Actual } L_e}{\text{Table } L_e}\right)\left(\frac{\text{Actual capacity}}{\text{Table capacity}}\right)^{1.8}$$

4. Capacity (kW) based on standard refrigerant cycle of 40°C liquid and saturated evaporator outlet temperature. Liquid capacity (kW) based on –5°C evaporator temperature.
5. Thermophysical properties and viscosity data based on calculations from NIST REFPROP program Version 6.01.
6. For brazed Type L copper tubing larger than 28 mm OD for discharge or liquid service, see Safety Requirements section.
7. Values are based on 40°C condensing temperature. Multiply table capacities by the following factors for other condensing temperatures.

Cond. Temp., °C	Suction Line	Discharge Line
20	1.357	0.765
30	1.184	0.908
40	1.000	1.000
50	0.801	1.021

Table 8.23 Suction, Discharge, and Liquid Capacities in Kilowatts for Ammonia (Single- or High-Stage Applications) [2018R, Ch 2, Tbl 2]

Steel Nominal Line Size, mm	Suction Lines ($\Delta t = 0.02$ K/m) Saturated Suction Temperature, °C					Discharge Lines $\Delta t = 0.02$ K/m, $\Delta p = 684.0$ Pa/m Saturated Suction Temp., °C			Steel Nominal Line Size, mm	Liquid Lines	
	−40 $\Delta p = 76.9$	−30 $\Delta p = 116.3$	−20 $\Delta p = 168.8$	−5 $\Delta p = 276.6$	+5 $\Delta p = 370.5$	−40	−20	+5		Velocity = 0.5 m/s	$\Delta p = 450.0$
10	0.8	1.2	1.9	3.5	4.9	8.0	8.3	8.5	10	3.9	63.8
15	1.4	2.3	3.6	6.5	9.1	14.9	15.3	15.7	15	63.2	118.4
20	3.0	4.9	7.7	13.7	19.3	31.4	32.3	33.2	20	110.9	250.2
25	5.8	9.4	14.6	25.9	36.4	59.4	61.0	62.6	25	179.4	473.4
32	12.1	19.6	30.2	53.7	75.4	122.7	126.0	129.4	32	311.0	978.0
40	18.2	29.5	45.5	80.6	113.3	184.4	189.4	194.5	40	423.4	1469.4
50	35.4	57.2	88.1	155.7	218.6	355.2	364.9	374.7	50	697.8	2840.5
65	56.7	91.6	140.6	248.6	348.9	565.9	581.4	597.0	65	994.8	4524.8
80	101.0	162.4	249.0	439.8	616.9	1001.9	1029.3	1056.9	80	1536.3	8008.8
100	206.9	332.6	509.2	897.8	1258.6	2042.2	2098.2	2154.3	—	—	—
125	375.2	601.8	902.6	1622.0	2271.4	3682.1	3783.0	3884.2	—	—	—
150	608.7	975.6	1491.4	2625.4	3672.5	5954.2	6117.4	6281.0	—	—	—
200	1252.3	2003.3	3056.0	5382.5	7530.4	12 195.3	12 529.7	12 864.8	—	—	—
250	2271.0	3625.9	5539.9	9733.7	13619.6	22 028.2	22 632.2	23 237.5	—	—	—
300	3640.5	5813.5	8873.4	15568.9	21787.1	35 239.7	36 206.0	37 174.3	—	—	—

Notes:

1. Table capacities are in kilowatts of refrigeration.

   Δp = pressure drop due to line friction, Pa/m

   Δt = corresponding change in saturation temperature, K/m

2. Line capacity for other saturation temperatures Δt and equivalent lengths L_e

$$\text{Line capacity} = \text{Table capacity}\left(\frac{\text{Table } L_e}{\text{Actual } L_e} \times \frac{\text{Actual } \Delta t}{\text{Table } \Delta t}\right)^{0.55}$$

3. Saturation temperature Δt for other capacities and equivalent lengths L_e

$$\Delta t = \text{Table } \Delta t \left(\frac{\text{Actual } L_e}{\text{Table } L_e}\right)\left(\frac{\text{Actual capacity}}{\text{Table capacity}}\right)^{1.8}$$

4. Values are based on 30°C condensing temperature. Multiply table capacities by the following factors for other condensing temperatures:

Condensing Temperature, °C	Suction Lines	Discharge Lines
20	1.04	0.86
30	1.00	1.00
40	0.96	1.24
50	0.91	1.43

5. Liquid line capacities based on −5°C suction.

Table 8.24 Liquid Ammonia Line Capacities in Kilowatts [2018R, Ch 2, Tbl 3]

Nominal Size, mm	Pumped Liquid Overfeed Ratio			High-Pressure Liquid at 21 kPa[a]	Hot-Gas Defrost[a]	Equalizer High Side[b]	Thermosiphon Lubricant Cooling Lines Gravity Flow[c]		
	3:1	4:1	5:1				Supply	Return	Vent
15	35	26	21	106					
20	77	58	46	243	32	176			
25	151	114	92	472	56	352			
32	329	246	197	1007	99	528			
40	513	387	308	1544	106	791	59	35	60
50	1175	879	703	3573	176	1055	138	88	106
65	1875	1407	1125	5683	324	1759	249	155	187
80	2700	2026	1620	10 150	570	3517	385	255	323
100	4800	3600	2880	—	1154	7034	663	413	586
125	—	—	—	—	2089	—	1041	649	1062
150	—	—	—	—	3411	—	1504	938	1869
200	—	—	—	—	—	—	2600	1622	3400

Source: Wile (1977).
[a]Rating for hot-gas branch lines under 30 m with minimum inlet pressure of 724 kPa (gage), defrost pressure of 483 kPa (gage), and −29°C evaporators designed for a 5.6 K temperature differential
[b]Line sizes based on experience using total system evaporator kilowatts.
[c]From Frick Co. (2004). Values for line sizes above 100 mm are extrapolated.

Refrigerants

Table 8.25 Minimum Refrigeration Capacity in Kilowatts for Oil Entrainment up Suction Risers (Copper Tubing, ASTM B 88M Type B, Metric Size) [2018R, Ch 1, Tbl 19]

Refrigerant	Saturated Temp., °C	Suction Gas Temp., °C	Tubing Nominal OD, mm											
			12	15	18	22	28	35	42	54	67	79	105	130
22	–40	–35	0.182	0.334	0.561	0.956	1.817	3.223	5.203	9.977	14.258	26.155	53.963	93.419
		–25	0.173	0.317	0.532	0.907	1.723	3.057	4.936	9.464	16.371	24.811	51.189	88.617
		–15	0.168	0.307	0.516	0.880	1.672	2.967	4.791	9.185	15.888	24.080	49.681	86.006
	–20	–15	0.287	0.527	0.885	1.508	2.867	5.087	8.213	15.748	27.239	41.283	85.173	147.449
		–5	0.273	0.501	0.841	1.433	2.724	4.834	7.804	14.963	25.882	39.226	80.929	140.102
		5	0.264	0.485	0.815	1.388	2.638	4.680	7.555	14.487	25.058	37.977	78.353	135.642
	–5	0	0.389	0.713	1.198	2.041	3.879	6.883	11.112	21.306	36.854	55.856	115.240	199.499
		10	0.369	0.676	1.136	1.935	3.678	6.526	10.535	20.200	34.940	52.954	109.254	189.136
		20	0.354	0.650	1.092	1.861	3.537	6.275	10.131	19.425	33.600	50.924	105.065	181.884
	5	10	0.470	0.862	1.449	2.468	4.692	8.325	13.441	25.771	44.577	67.560	139.387	241.302
		20	0.440	0.807	1.356	2.311	4.393	7.794	12.582	24.126	41.731	63.246	130.488	225.896
		30	0.422	0.774	1.301	2.217	4.213	7.476	12.069	23.141	40.027	60.665	125.161	216.675

Table 8.25 Minimum Refrigeration Capacity in Kilowatts for Oil Entrainment up Suction Risers (Copper Tubing, ASTM B 88M Type B, Metric Size) [2018R, Ch 1, Tbl 19] *(Continued)*

Refrigerant	Saturated Temp., °C	Suction Gas Temp., °C	Tubing Nominal OD, mm: 12	15	18	22	28	35	42	54	67	79	105	130
134a	−10	−5	0.274	0.502	0.844	1.437	2.732	4.848	7.826	15.006	25.957	39.340	81.164	140.509
		5	0.245	0.450	0.756	1.287	2.447	4.342	7.010	13.440	23.248	35.235	72.695	125.847
		15	0.238	0.436	0.732	1.247	2.370	4.206	6.790	13.019	22.519	34.129	70.414	121.898
	−5	0	0.296	0.543	0.913	1.555	2.956	5.244	8.467	16.234	28.081	42.559	87.806	152.006
		10	0.273	0.500	0.840	1.431	2.720	4.827	7.792	14.941	25.843	39.168	80.809	139.894
		20	0.264	0.484	0.813	1.386	2.634	4.674	7.546	14.468	25.026	37.929	78.254	135.471
	5	10	0.357	0.655	1.100	1.874	3.562	6.321	10.204	19.565	33.843	51.292	105.823	183.197
		20	0.335	0.615	1.033	1.761	3.347	5.938	9.586	18.380	31.792	48.184	99.412	172.098
		30	0.317	0.582	0.978	1.667	3.168	5.621	9.075	17.401	30.099	45.617	94.115	162.929
	10	15	0.393	0.721	1.211	2.063	3.921	6.957	11.232	21.535	37.250	56.456	116.479	201.643
		25	0.370	0.679	1.141	1.944	3.695	6.555	10.583	20.291	35.098	53.195	109.749	189.993
		35	0.358	0.657	1.104	1.881	3.576	6.345	10.243	19.640	33.971	51.486	106.224	183.891

Notes:
1. Refrigeration capacity in kilowatts is based on saturated evaporator as shown in table and condensing temperature of 40 °C. For other liquid line temperatures, use correction factors in table at right.
2. Values computed using ISO 32 mineral oil for R-22 and R-502. R-134a computed using ISO 32 ester-based oil.

Refrigerant	Liquid Temperature, °C: 20	30	50
22	1.17	1.08	0.91
134a	1.20	1.10	0.89

Table 8.26 Minimum Refrigeration Capacity in Kilowatts for Oil Entrainment up Hot-Gas Risers (Copper Tubing, ASTM B 88M Type B, Metric Size) [2018R, Ch 1, Tbl 20]

Refrigerant	Saturated Discharge Temp., °C	Discharge Gas Temp., °C	Tubing Diameter, Nominal OD, mm											
			12	15	18	22	28	35	42	54	67	79	105	130
22	20	60	0.563	0.032	0.735	2.956	5.619	9.969	16.094	30.859	43.377	80.897	116.904	288.938
		70	0.549	1.006	1.691	2.881	5.477	9.717	15.687	30.078	52.027	48.851	162.682	281.630
		80	0.535	0.982	1.650	2.811	5.343	9.480	15.305	29.346	50.761	76.933	158.726	173.780
	30	70	0.596	1.092	1.836	3.127	5.945	10.547	17.028	32.649	56.474	85.591	176.588	305.702
		80	0.579	1.062	1.785	3.040	5.779	10.254	16.554	31.740	54.901	83.208	171.671	297.190
		90	0.565	0.035	1.740	2.964	5.635	9.998	16.140	30.948	53.531	81.131	167.386	289.773
	40	80	0.618	1.132	1.903	3.242	6.163	10.934	17.653	33.847	58.546	88.732	183.069	316.922
		90	0.601	1.103	1.853	3.157	6.001	10.647	17.189	32.959	47.009	86.403	178.263	308.603
		100	0.584	1.071	1.800	3.067	5.830	10.343	16.698	32.018	55.382	83.936	173.173	299.791
	50	90	0.630	1.156	1.943	3.310	6.291	11.162	18.020	34.552	59.766	90.580	186.882	323.523
		100	0.611	1.121	1.884	3.209	6.100	10.823	17.473	33.503	57.951	87.831	181.209	313.702
		110	0.595	1.092	1.834	3.125	5.941	10.540	17.016	32.627	56.435	85.532	176.467	305.493

Table 8.26 Minimum Refrigeration Capacity in Kilowatts for Oil Entrainment up Hot-Gas Risers (Copper Tubing, ASTM B 88M Type B, Metric Size) [2018R, Ch 1, Tbl 20] *(Continued)*

Refrigerant	Saturated Discharge Temp., °C	Discharge Gas Temp., °C	Tubing Diameter, Nominal OD, mm											
			12	15	18	22	28	35	42	54	67	79	105	130
134a	20	60	0.469	0.860	1.445	2.462	4.681	8.305	13.408	25.709	44.469	67.396	139.050	240.718
		70	0.441	0.808	1.358	2.314	4.399	7.805	12.600	24.159	41.788	63.334	130.668	226.207
		80	0.431	0.790	1.327	2.261	4.298	7.626	12.311	23.605	40.830	61.881	127.671	221.020
	30	70	0.493	0.904	1.519	2.587	4.918	8.726	14.087	27.011	46.722	70.812	145.096	252.916
		80	0.463	0.849	1.426	2.430	4.260	8.196	13.232	25.371	43.885	66.512	137.225	237.560
		90	0.452	0.829	1.393	2.374	4.513	8.007	12.926	24.785	42.870	64.974	134.052	232.066
	40	80	0.507	0.930	1.563	2.662	5.061	8.979	14.496	27.794	48.075	72.863	150.328	260.242
		90	0.477	0.874	1.469	2.502	4.756	8.439	13.624	26.122	45.184	68.480	141.285	244.588
		100	0.465	0.852	1.432	2.439	4.637	8.227	13.281	25.466	44.048	66.759	137.735	238.443
	50	90	0.510	0.936	1.573	2.679	5.093	9.037	14.589	27.973	48.385	73.332	151.296	261.918
		100	0.479	0.878	1.476	2.514	4.779	8.480	13.690	25.248	45.402	68.811	141.969	245.772
		110	0.467	0.857	1.441	2.454	4.665	8.278	13.364	25.624	44.322	67.173	138.590	239.921

Notes:
1. Refrigeration capacity in kilowatts is based on saturated evaporator at −5°C, and condensing temperature as shown in table. For other liquid line temperatures, use correction factors in table at right.
2. Values computed using ISO 32 mineral oil for R-22, and ISO 32 ester-based oil for R-134a.

Refrigerant	Saturated Suction Temperature, °C						
	−50	−40	−30	−20	0	5	10
22	0.87	0.90	0.93	0.96	—	1.02	—
134a	—	—	—	—	1.02	1.04	1.06

Lubricants In Refrigerant Systems

Oil in refrigerant compressors lubricates, acts as coolant, and seals the suction from the discharge side. Oil mixes well with hydrocarbon refrigerants at higher temperatures; miscibility is reduced as temperature lowers. Oil leaves the compressor and dissolves into the refrigerant in the condenser, and passes through the liquid line to the evaporator where it separates. In higher temperature systems, it returns by gravity or is dragged by the returning vapor. Low temperature halocarbon systems need an oil separator at the compressor discharge. Oil return up vertical piping requires significant refrigerant velocity. Lubricants are generally not miscible with ammonia and separate easily out of the liquid. Oil separators at the discharge of compressors are essential. Oil must be periodically or continuously removed and returned to the compressor.

There is no ideal lubricant. For halocarbon refrigerants, there are mineral lubricants, both naphthenic and paraffinic, and synthetic lubricants, ester and glycol. Viscosity grades required vary with the temperature and the solubility of the refrigerant in the lubricant. Additives are used to enhance lubricant properties or impact new characteristics. They may be polar compounds, polymers, or compounds containing active elements such as sulfur or phosphorus. Lubricants should be dry; normally almost all hydrocarbon lubricants have a moisture content of about 30 ppm. Synthetic lubricants polyalkylene glycols (PAGs) are used commonly in automobile R-134a systems; polyalphaolefins (PAOs) are mainly used an immiscible oil in ammonia systems; polyol esters are used with HFC refrigerants in all types of compressors. Low pour point is essential for oils in ammonia systems.

When retrofitting to a new refrigerant, follow the recommendations of the equipment manufacturer on the lubricants that are suitable for use.

Mixing lubricants can cause serious problems; to extend equipment life, it is important to use lubricants approved or specified by the system or compressor manufacturer.

Table 8.27 Secondary Coolant Performance Comparisons [2018R, Ch 13, Tbl 1]

Secondary Coolant	Concentration (by Weight), %	Freeze Point, °C	L/(s·kW)[a]	Pressure Drop,[b] kPa	Heat Transfer Coefficient[c] h_i, W/(m^2·K
Propylene glycol	39	–20.6	0.0459	20.064	1164
Ethylene glycol	38	–21.6	0.0495	16.410	2305
Methanol	26	–20.7	0.0468	14.134	2686
Sodium chloride	23	–20.6	0.0459	15.858	3169
Calcium chloride	22	–22.1	0.0500	16.685	3214
Aqua ammonia	14	–21.7	0.0445	16.823	3072
Trichloroethylene	100	–86.1	0.1334	14.548	2453
d-Limonene	100	–96.7	0.1160	10.204	1823
Methylene chloride	100	–96.7	0.1146	12.824	3322
R-11	100	–111.1	0.1364	14.341	2430

[a]Based on inlet secondary coolant temperature at pump of 3.9°C.
[b]Based on one length of 4.9 m tube with 26.8 mm ID and use of Moody Chart (1944) for an average velocity of 2.13 m/s. Input/output losses equal $V^2\rho/2$ for 2.13 m/s velocity. Evaluations are at a bulk temperature of –6.7°C and a temperature range of 5.6 K.
[c]Based on curve fit equation for Kern's (1950) adaptation of Sieder and Tate's (1936) heat transfer equation using 4.9 m tube for L/D = 181 and film temperature of 2.8°C lower than average bulk temperature with 2.134 m/s velocity.

Table 8.28 Relative Pumping Energy Required* [2018R, Ch 13, Tbl 3]

Secondary Coolant	Energy Factor
Aqua ammonia	1.000
Methanol	1.078
Propylene glycol	1.142
Ethylene glycol	1.250
Sodium chloride	1.295
Calcium chloride	1.447
d-Limonene	2.406
Methylene chloride	3.735
Trichloroethylene	4.787
R-11	5.022

*Based on same pump pressure, refrigeration load, –6.7°C average temperature, 6 K range, and freezing point (for water-based secondary coolants) 11 to 13 K below lowest secondary coolant temperature.

9. REFRIGERANT SAFETY

(For more guidance, see ANSI/ASHRAE Standard 15, *Safety Standard for Refrigeration Systems*, and ANSI/ASHRAE Standard 34, *Designation and Safety Classification of Refrigerants*)

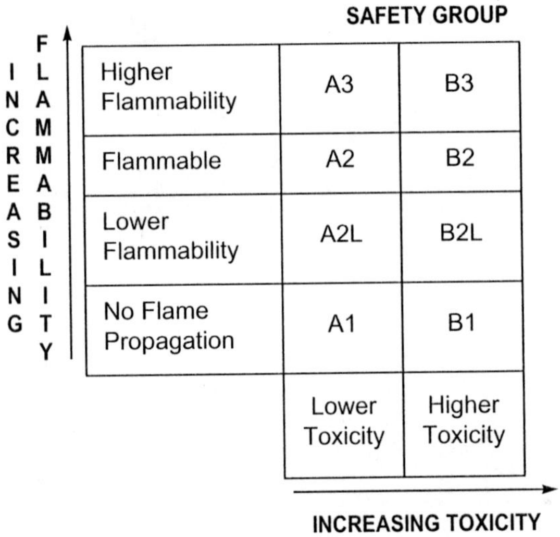

Figure 9.1 Refrigerant Safety Group Classification [Std 34-2019, Fig 6-1]

Table 9.1 Refrigerant Data and Safety Classifications [Std 34-2019, Tbl 4-1]

Refrigerant No.	Chemical Name[a,b]	Chemical Formula[a]	OEL[f], ppm v/v	Safety Group	RCL[c] (ppm v/v)	RCL[c] (lb/Mcf)	RCL[c] (g/m^3)	Highly Toxic or Toxic[d] Under Code Classification
Methane Series								
11	trichlorofluoromethane	CCl_3F	1000	A1	1100	0.39	6.2	Neither
12	dichlorodifluoromethane	CCl_2F_2	1000	A1	18,000	5.6	90	Neither
12B1	bromochlorodifluoromethane	$CBrClF_2$						Neither
13	chlorotrifluoromethane	$CClF_3$	1000	A1				Neither
13B1	bromotrifluoromethane	$CBrF_3$	1000	A1				Neither
14[e]	tetrafluoromethane (carbon tetrafluoride)	CF_4	1000	A1	110,000	25	400	Neither
21	dichlorofluoromethane	$CHCl_2F$		B1				Toxic
22	chlorodifluoromethane	$CHClF_2$	1000	A1	59,000	13	210	Neither
23	trifluoromethane	CHF_3	1000	A1	41,000	7.3	120	Neither
30	dichloromethane (methylene chloride)	CH_2Cl_2		B1				Neither
31	chlorofluoromethane	CH_2ClF						Neither
32	difluoromethane (methylene fluoride)	CH_2F_2	1000	A2L	35,000	4.8	77	Neither
40	chloromethane (methyl chloride)	CH_3Cl		B2				Toxic
41	fluoromethane (methyl fluoride)	CH_3F						Neither
50	methane	CH_4	1000	A3				Neither
Ethane Series								
113	1,1,2-trichloro-1,2,2-trifluoroethane	CCl_2FCClF_2	1000	A1	2600	1.2	20	Neither
114	1,2-dichloro-1,1,2,2-tetrafluoroethane	$CClF_2CClF_2$	1000	A1	20,000	8.7	140	Neither
115[g]	chloropentafluoroethane	$CClF_2CF_3$	1000	A1	120,000	47	760	Neither
116[e]	hexafluoroethane	CF_3CF_3	1000	A1	97,000	34	550	Neither
123	2,2-dichloro-1,1,1-trifluoroethane	$CHCl_2CF_3$	50	B1	9100	3.5	57	Neither

Table 9.1 Refrigerant Data and Safety Classifications [Std 34-2019, Tbl 4-1] ***(Continued)***

Refrigerant No.	Chemical Name[a,b]	Chemical Formula[a]	OEL[f], ppm v/v	Safety Group	RCL[c] (ppm v/v)	RCL[c] (lb/Mcf)	RCL[c] (g/m^3)	Highly Toxic or Toxic[d] Under Code Classification
124	2-chloro-1,1,1,2-tetrafluoroethane	$CHClFCF_3$	1000	A1	10,000	3.5	56	Neither
125[e]	pentafluoroethane	CHF_2CF_3	1000	A1	75,000	23	370	Neither
134a	1,1,1,2-tetrafluoroethane	CH_2FCF_3	1000	A1	50,000	13	210	Neither
141b	1,1-dichloro-1-fluoroethane	CH_3CCl_2F	500		2600	0.78	12	Neither
142b	1-chloro-1,1-difluoroethane	CH_3CClF_2	1000	A2	20,000	5.1	83	Neither
143a	1,1,1-trifluoroethane	CH_3CF_3	1000	A2L	21,000	4.5	70	Neither
152a	1,1-difluoroethane	CH_3CHF_2	1000	A2	12,000	2.0	32	Neither
170	ethane	CH_3CH_3	1000	A3	7000	0.54	8.7	Neither
Ethers								
E170	methoxymethane (dimethyl ether)	CH_3OCH_3	1000	A3	8500	1.0	16	Neither
Propane								
218[e]	octafluoropropane	$CF_3CF_2CF_3$	1000	A1	90,000	43	690	Neither
227ea[e]	1,1,1,2,3,3,3-heptafluoropropane	CF_3CHFCF_3	1000	A1	84,000	36	580	Neither
236fa	1,1,1,3,3,3-hexafluoropropane	$CF_3CH_2CF_3$	1000	A1	55,000	21	340	Neither
245fa	1,1,1,3,3-pentafluoropropane	$CHF_2CH_2CF_3$	300	B1	34,000	12	190	Neither
290	propane	$CH_3CH_2CH_3$	1000	A3	5300	0.56	9.5	Neither
Cyclic Organic Compounds								
C318	octafluorocyclobutane	$-(CF_2)_4-$	1000	A1	80,000	41	660	Neither
Miscellaneous Organic Compounds								
hydrocarbons								
600	butane	$CH_3CH_2CH_2CH_3$	1000	A3	1000	0.15	2.4	Neither
600a	2-methylpropane (isobutane)	$CH(CH_3)_2CH_3$	1000	A3	4000	0.59	9.6	Neither

Table 9.1 Refrigerant Data and Safety Classifications [Std 34-2019, Tbl 4-1] ***(Continued)***

Refrigerant No.	Chemical Name[a,b]	Chemical Formula[a]	OEL[f], ppm v/v	Safety Group	RCL[c] (ppm v/v)	RCL[c] (lb/Mcf)	RCL[c] (g/m³)	Highly Toxic or Toxic[d] Under Code Classification
601	pentane	$CH_3CH_2CH_2CH_2CH_3$	600	A3	1000	0.18	2.9	Neither
601a	2-methylbutane (isopentane)	$(CH_3)_2CHCH_2CH_3$	600	A3	1000	0.18	2.9	Neither
oxygen compounds								
610	ethoxyethane (ethyl ether)	$CH_3CH_2OCH_2CH_3$	400					Neither
611	methyl formate	$HCOOCH_3$	100	B2				Neither
sulfur compounds								
620	(Reserved for future assignment)							
Nitrogen Compounds								
630	methanamine (methyl amine)	CH_3NH_2						Toxic
631	ethanamine (ethyl amine)	$CH_3CH_2(NH_2)$						Neither
Inorganic Compounds								
702	hydrogen	H_2		A3				Neither
704	helium	He		A1				Neither
717	ammonia	NH_3	25	B2L	320	0.014	0.22	Neither
718	water	H_2O		A1				Neither
720	neon	Ne		A1				Neither
728	nitrogen	N_2		A1				Neither
732	oxygen	O_2						Neither
740	argon	Ar		A1				Neither
744	carbon dioxide	CO_2	5000	A1	30,000	3.4	54	Neither
744A	nitrous oxide	N_2O						Neither
764	sulfur dioxide	SO_2		B1				Neither

Table 9.1 Refrigerant Data and Safety Classifications [Std 34-2019, Tbl 4-1] ***(Continued)***

Refrigerant No.	Chemical Name[a,b]	Chemical Formula[a]	OEL[f], ppm v/v	Safety Group	RCL[c] (ppm v/v)	RCL[c] (lb/Mcf)	RCL[c] (g/m^3)	Highly Toxic or Toxic[d] Under Code Classification
Unsaturated Organic Compounds								
1130(E)	trans-1,2-dichloroethene	CHCl=CHCl	200	B1	100	0.25	4	Neither
1132a	1,1-difluoroethylene	$CF_2=CH_2$	500	A2	13,000	2.0	33	Neither
1150	ethene (ethylene)	$CH_2=CH_2$	200	A3				Neither
1224yd(Z)	(Z)-1-chloro-2,3,3,3-tetrafluoropropene	CF3CF=CHCl	1000	A1	60,000	23	360	Neither
1233zd(E)	trans-1-chloro-3,3,3-trifluoro-1-propene	$CF_3CH=CHCl$	800	A1	16,000	5.3	85	Neither
1234yf	2,3,3,3-tetrafluoro-1-propene	$CF_3CF=CH_2$	500	A2L	16,000	4.7	75	Neither
1234ze(E)	trans-1,3,3,3-tetrafluoro-1-propene	$CF_3CH=CFH$	800	A2L	16,000	4.7	75	Neither
1270	propene (propylene)	$CH_3CH=CH_2$	500	A3	1000	0.11	1.7	Neither
1336mzz(E)	trans-1,1,1,4,4,4-hexafluoro-2-butene	$CF_3CH=CHCF_3$	400	A1	7200	3.0	48	Neither
1336mzz(Z)	cis-1,1,1,4,4,4-hexaflouro-2-butene	$CF_3CHCHCF_3$	500	A1	13,000	5.4	87	Neither

a. The chemical name and chemical formula are not part of this standard. Chemical names conform to IUPAC nomenclature[6,7] except where shortened, unambiguous names are used following ANSI/ASHRAE Standard 34 convention.
b. The preferred chemical name is followed by the popular name in parentheses.
c. Data taken from J.M. Calm, "ARTI Refrigerant Database," Air- Conditioning and Refrigeration Technology Institute (ARTI), Arlington, VA, July 2001; J.M. Calm, "Toxicity Data to Determine Refrigerant Concentration Limits," Report DE/CE 23810-110, Air- Conditioning and Refrigeration Technology Institute (ARTI), Arlington, VA, September 2000; J.M. Calm, "The Toxicity of Refrigerants," *Proceedings of the 1996 International Refrigeration Conference*, Purdue University, West Lafayette, IN, pp. 157–62, 1996; D.P. Wilson and R.G. Richard, "Determination of Refrigerant Lower Flammability Limits (LFLs) in Compliance with Proposed Addendum p to ANSI/ASHRAE Standard 34-1992 (1073-RP)," *ASHRAE Transactions* 2002, 108(2); D.W. Coombs, "HFC-32 Assessment of Anesthetic Potency in Mice by Inhalation," Huntingdon Life Sciences Ltd., Huntingdon, Cambridgeshire, England, February 2004 and amendment February 2006; D.W. Coombs, "HFC-22 An Inhalation Study to Investigate the Cardiac Sensitization Potential in the Beagle Dog," Huntingdon Life Sciences Ltd., Huntingdon, Cambridgeshire, England, August 2005; and other toxicity studies.
d. *Highly toxic*, *toxic*, or *neither*, where *highly toxic* and *toxic* are as defined in the *International Fire Code*, *Uniform Fire Code*, and OSHA regulations, and *neither* identifies those refrigerants having lesser toxicity than either of those groups[1,2,3].
e. At locations with altitudes higher than 1500 m, the ODL and RCL shall be 69,100 ppm.
f. The OELs are eight-hour TWAs, as defined in Section 3, unless otherwise noted; a "C" designation denotes a ceiling limit.
g. At locations with altitudes higher than 1000 m but below or equal to 1500 m, the ODL and RCL shall be 112,000 ppm, and at altitudes higher than 1500 m, the ODL and RCL shall be 69,100 ppm.

Table 9.2 Data and Safety Classifications for Refrigerant Blends [Std 34-2019, Tbl 4-2]

Refrigerant No.	Composition (Mass%)	Composition Tolerances	OEL[h], ppm v/v	Safety Group	RCL[a] (ppm v/v)	RCL[a] (lb/Mcf)	RCL[a] (g/m^3)	Highly Toxic or Toxic[f] Under Code Classification
Zeotropes								
400	R-12/114 (must be specified)			A1				Neither
	(50.0/50.0)		1000	A1	28,000	10	160	
	(60.0/40.0)		1000	A1	30,000	11	170	
401A	R-22/152a/124 (53.0/13.0/34.0)	(±2.0/+0.5, −1.5/±1.0)	1000	A1	27,000	6.6	110	Neither
401B	R-22/152a/124 (61.0/11.0/28.0)	(±2.0/+0.5, −1.5/±1.0)	1000	A1	30,000	7.2	120	Neither
401C	R-22/152a/124 (33.0/15.0/52.0)	(±2.0/+0.5, −1.5/±1.0)	1000	A1	20,000	5.2	84	Neither
402A	R-125/290/22 (60.0/2.0/38.0)	(±2.0/+0.1, −1.0/±2.0)	1000	A1	66,000	17	270	Neither
402B	R-125/290/22 (38.0/2.0/60.0)	(±2.0/+0.1, −1.0/±2.0)	1000	A1	63,000	15	240	Neither
403A	R-290/22/218 (5.0/75.0/20.0)	(+0.2, −2.0/±2.0/±2.0)	1000	A2	33,000	7.6	120	Neither
403B[g]	R-290/22/218 (5.0/56.0/39.0)	(+0.2, −2.0/±2.0/±2.0)	1000	A1	70,000	18	290	Neither
404A[i]	R-125/143a/134a (44.0/52.0/4.0)	(±2.0/±1.0/±2.0)	1000	A1	130,000	31	500	Neither
405A	R-22/152a/142b/C318 (45.0/7.0/5.5/42.5)	Individual components = (±2.0/±1.0/±1.0/±2.0); sum of R-152a and R-142b = (+0.0, −2.0)	1000		57,000	16	260	Neither
406A	R-22/600a/142b (55.0/4.0/41.0)	(±2.0/±1.0/±1.0)	1000	A2	21,000	4.7	25	Neither
407A[g]	R-32/125/134a (20.0/40.0/40.0)	(±2.0/±2.0/±2.0)	1000	A1	83,000	19	300	Neither
407B[g]	R-32/125/134a (10.0/70.0/20.0)	(±2.0/±2.0/±2.0)	1000	A1	79,000	21	330	Neither
407C[g]	R-32/125/134a (23.0/25.0/52.0)	(±2.0/±2.0/±2.0)	1000	A1	81,000	18	290	Neither
407D	R-32/125/134a (15.0/15.0/70.0)	(±2.0/±2.0/±2.0)	1000	A1	68,000	16	250	Neither
407E[g]	R-32/125/134a (25.0/15.0/60.0)	(±2.0/±2.0/±2.0)	1000	A1	80,000	17	280	Neither
407F	R-32/125/134a (30.0/30.0/40.0)	(±2.0/±2.0/±2.0)	1000	A1	95,000	20	320	Neither
407G	R-32/125/134a (2.5/2.5/95.0)	(±0.5/±0.5/±1.0)	1000	A1	52,000	13	210	Neither
407H	R-32/125/134a (32.5/15.0/52.5)	(±1.0/±1.0/±2.0)	1000	A1	92,000	19	300	Neither

Table 9.2 Data and Safety Classifications for Refrigerant Blends [Std 34-2019, Tbl 4-2] ***(Continued)***

Refrigerant No.	Composition (Mass%)	Composition Tolerances	OEL[h], ppm v/v	Safety Group	RCL[a] (ppm v/v)	RCL[a] (lb/Mcf)	RCL[a] (g/m³)	Highly Toxic or Toxic[f] Under Code Classification
407I	R-32/125/134a (19.5/8.5/72.0)	(+1.0, –2.0/+2.0, –1.0/±2.0)	1000	A1	71,100	16	250	Neither
408A[g]	R-125/143a/22 (7.0/46.0/47.0)	(±2.0/±1.0/±2.0)	1000	A1	95,000	21	340	Neither
409A	R-22/124/142b (60.0/25.0/15.0)	(±2.0/±2.0/±1.0)	1000	A1	29,000	7.1	110	Neither
409B	R-22/124/142b (65.0/25.0/10.0)	(±2.0/±2.0/±1.0)	1000	A1	30,000	7.3	120	Neither
410A[i]	R-32/125 (50.0/50.0)	(+0.5, –1.5/+1.5, –0.5)	1000	A1	140,000	26	420	Neither
410B[i]	R-32/125 (45.0/55.0)	(±1.0/±1.0)		A1	140,000	27	430	Neither
411A[e]	R-1270/22/152a (1.5/87.5/11.0)	(+0.0, –1.0/+2.0, –0.0/+0.0, –1.0)	990	A2	14,000	2.9	46	Neither
411B[e]	R-1270/22/152a (3.0/94.0/3.0)	(+0.0, –1.0/+2.0, –0.0/+0.0, –1.0)	980	A2	13,000	2.8	45	Neither
412A	R-22/218/142b (70.0/5.0/25.0)	(±2.0/±2.0/±1.0)	1000	A2	22,000	5.1	82	Neither
413A	R-218/134a/600a (9.0/88.0/3.0)	(±1.0/±2.0/+0.0, –1.0)	1000	A2	22,000	5.8	94	Neither
414A	R-22/124/600a/142b (51.0/28.5/4.0/16.5)	(±2.0/±2.0/±0.5/+0.5, –1.0)	1000	A1	26,000	6.4	100	Neither
414B	R-22/124/600a/142b (50.0/39.0/1.5/9.5)	(±2.0/±2.0/±0.5/+0.5, –1.0)	1000	A1	23,000	6.0	95	Neither
415A	R-22/152a (82.0/18.0)	(±1.0/±1.0)	1000	A2	14,000	2.9	47	Neither
415B	R-22/152a (25.0/75.0)	(±1.0/±1.0)	1000	A2	12,000	2.1	34	Neither
416A[e]	R-134a/124/600 (59.0/39.5/1.5)	(+0.5, –1.0/+1.0, –0.5/+1.0, –0.2)	1000	A1	14,000	3.9	62	Neither
417A[e]	R-125/134a/600 (46.6/50.0/3.4)	(±1.1/±1.0/+0.1, –0.4)	1000	A1	13,000	3.5	56	Neither
417B	R-125/134a/600 (79.0/18.3/2.7)	(±1.0/±1.0/+0.1, –0.5)	1000	A1	15,000	4.3	70	Neither
417C	R-125/134a/600 (19.5/78.8/1.7)	(±1.0/±1.0/+0.1, –0.5)	1000	A1	21,000	5.4	87	Neither
418A	R-290/22/152a (1.5/96.0/2.5)	(±0.5/±1.0/±0.5)	1000	A2	22,000	4.8	77	Neither
419A[g]	R-125/134a/E170 (77.0/19.0/4.0)	(±1.0/±1.0/±1.0)	1000	A2	15,000	4.2	67	Neither
419B	R-125/134a/E170 (48.5/48.0/3.5)	(±1.0/±1.0/±0.5)	1000	A2	17,000	4.6	74	Neither
420A	R-134a/142b (88.0/12.0)	(+1.0, –0.0/+0.0, –1.0)	1000	A1	45,000	12	190	Neither
421A	R-125/134a (58.0/42.0)	(±1.0/±1.0)	1000	A1	61,000	17	280	Neither

Table 9.2 Data and Safety Classifications for Refrigerant Blends [Std 34-2019, Tbl 4-2] ***(Continued)***

Refrigerant No.	Composition (Mass%)	Composition Tolerances	OEL[h], ppm v/v	Safety Group	RCL[a] (ppm v/v)	RCL[a] (lb/Mcf)	RCL[a] (g/m³)	Highly Toxic or Toxic[f] Under Code Classification
421B	R-125/134a (85.0/15.0)	(±1.0/±1.0)	1000	A1	69,000	21	330	Neither
422A	R-125/134a/600a (85.1/11.5/3.4)	(±1.0/±1.0/+0.1, −0.4)	1000	A1	63,000	18	290	Neither
422B	R-125/134a/600a (55.0/42.0/3.0)	(±1.0/±1.0/+0.1, −0.5)	1000	A1	56,000	16	250	Neither
422C	R-125/134a/600a (82.0/15.0/3.0)	(±1.0/±1.0/+0.1, −0.5)	1000	A1	62,000	18	290	Neither
422D	R-125/134a/600a (65.1/31.5/3.4)	+0.9, −1.1/±1.0/+0.1, −0.4)	1000	A1	58,000	16	260	Neither
422E	R-125/134a/600a (58.0/39.3/2.7)	(±1.0/+1.7, −1.3/+0.3, −0.2)	1000	A1	57,000	16	260	Neither
423A	R-134a/227ea (52.5/47.5)	(±1.0/±1.0)	1000	A1	59,000	19	310	Neither
424A[e]	R-125/134a/600a/600/601a (50.5/47.0/0.9/1.0/0.6)	(±1.0/±1.0/+0.1, −0.2/+0.1, +0.2/+0.1, −0.2)	970	A1	23,000	6.2	100	Neither
425A	R-32/134a/227ea (18.5/69.5/12.0)	(±0.5/±0.5/±0.5)	1000	A1	72,000	16	260	Neither
426A[e]	R-125/134a/600/601a (5.1/93.0/1.3/0.6)	(±1.0/±1.0/+0.1, −0.2/+0.1, −0.2)	990	A1	20,000	5.2	83	Neither
427A	R-32/125/143a/134a (15.0/25.0/10.0/50.0)	(±2.0/±2.0/±2.0/±2.0)	1000	A1	79,000	18	290	Neither
428A	R-125/143a/290/600a (77.5/20.0/0.6/1.9)	(±1.0/±1.0/+0.1, −0.2/+0.1, −0.2)	1000	A1	83,000	23	370	Neither
429A	R-E170/152a/600a (60.0/10.0/30.0)	(±1.0/±1.0/±1.0)	1000	A3	6300	0.81	13	Neither
430A	R-152a/600a (76.0/24.0)	(±1.0/±1.0)	1000	A3	8000	1.3	21	Neither
431A	R-290/152a (71.0/29.0)	(±1.0/±1.0)	1000	A3	5500	0.69	11	Neither
432A	R-1270/E170 (80.0/20.0)	(±1.0/±1.0)	700	A3	1200	0.13	2.1	Neither
433A	R-1270/290 (30.0/70.0)	(±1.0/±1.0)	880	A3	3100	0.34	5.5	Neither
433B	R-1270/290 (5.0/95.0)	(±1.0/±1.0)	950	A3	4500	0.51	8.1	Neither
433C	R-1270/290 (25.0/75.0)	(±1.0/±1.0)	790	A3	3600	0.41	6.6	Neither
434A[g]	R-125/143a/134a/600a (63.2/18.0/16.0/2.8)	(±1.0/±1.0/±1.0/+0.1, −0.2)	1000	A1	73,000	20	320	Neither
435A	R-E170/152a (80.0/20.0)	(±1.0/±1.0)	1000	A3	8500	1.1	17	Neither
436A	R-290/600a (56.0/44.0)	(±1.0/±1.0)	1000	A3	4000	0.50	8.1	Neither

Table 9.2 Data and Safety Classifications for Refrigerant Blends [Std 34-2019, Tbl 4-2] ***(Continued)***

Refrig-erant No.	Composition (Mass%)	Composition Tolerances	OEL[h], ppm v/v	Safety Group	RCL[a] (ppm v/v)	RCL[a] (lb/Mcf)	RCL[a] (g/m³)	Highly Toxic or Toxic[f] Under Code Classification
436B	R-290/600a (52.0/48.0)	(±1.0/±1.0)	1000	A3	4000	0.51	8.2	Neither
436C	R-290/600a (95.0/5.0)	(±1.2/±1.2)	990	A3	5000	0.57	9.1	Neither
437A	R-125/134a/600/601 (19.5/78.5/1.4/0.6)	(+0.5, –1.8/+1.5, –0.7/+0.1, –0.2/+0.1, –0.2)	990	A1	19,000	5.0	82	Neither
438A	R-32/125/134a/600/601a (8.5/45.0/44.2/1.7/0.6)	(+0.5, –1.5/±1.5/±1.5/+0.1, –0.2/+0.1, –0.2)	990	A1	20,000	4.9	79	Neither
439A	R-32/125/600a (50.0/47.0/3.0)	(±1.0/±1.0/±0.5)	990	A2	26,000	4.7	76	Neither
440A	R-290/134a/152a (0.6/1.6/97.8)	(±0.1/±0.6/±0.5)	1000	A2	12,000	1.9	31	Neither
441A	R-170/290/600a/600 (3.1/54.8/6.0/36.1)	(±0.3/±2.0/±0.6/±2.0)	1000	A3	3200	0.39	6.3	Neither
442A	R-32/125/134a/152a/227ea (31.0/31.0/30.0/3.0/5.0)	(±1.0/±1.0±/1.0/±0.5/±1.0)	1000	A1	100,000	21	330	Neither
443A	R-1270/290/600a (55.0/40.0/5.0)	(±2.0/±2.0/±1.2)	580	A3	1700	0.19	3.1	Neither
444A	R-32/152a/1234ze(E) (12.0/5.0/83.0)	(±1.0/±1.0/±2.0)	850	A2L	21,000	5.1	81	Neither
444B	R-32/152a/1234ze(E) (41.5/10.0/48.5)	(±1.0/±1.0/±1.0)	890	A2L	23,000	4.3	69	Neither
445A	R-744/134a/1234ze(E) (6.0/9.0/85.0)	(±1.0/±1.0/±2.0)	930	A2L	16,000	4.2	67	Neither
446A	R-32/1234ze(E)/600 (68.0/29.0/3.0)	(+0.5, –1.0/+2.0, –0.6/+0.1, –1.0)	960	A2L	16,000	2.5	39	Neither
447A	R-32/125/1234ze(E) (68.0/3.5/28.5)	(+1.5, –0.5/+1.5, –0.5/+1.0, –1.0)	900	A2L	16,000	2.6	42	Neither
447B	R-32/125/1234ze(E) (68.0/8.0/24.0)	(+1.0, –2.0/+2.0, –1.0/+1.0, –2.0)	970	A2L	30,000	23	360	Neither
448A	R-32/125/1234yf/134a/1234ze(E) (26.0/26.0/20.0/21.0/7.0)	(+0.5, –2.0/+2.0, –0.5/+0.5, –2.0/+2.0, –1.0/+0.5, –2.0)	890	A1	110,000	24	390	Neither
449A	R-32 /125 /1234yf /134a (24.3/24.7/25.3/25.7)	(+0.2, –1.0/+1.0, –0.2/+0.2, –1.0/+1.0, –0.2)	830	A1	100,000	23	370	Neither
449B	R-32/125/1234yf/134a (25.2/24.3/23.2/27.3)	(+0.3, –1.5/+1.5, –0.3/+0.3, –1.5/+1.5, –0.3)	850	A1	100,000	23	370	Neither
449C	R-32/125/1234yf/134a (20.0/20.0/31.0/29.0)	(+0.5, –1.5/+1.5, –0.5/+0.5, –1.5/+1.5, –0.5)	800	A1	98,000	23	360	Neither

Table 9.2 Data and Safety Classifications for Refrigerant Blends [Std 34-2019, Tbl 4-2] *(Continued)*

Refrigerant No.	Composition (Mass%)	Composition Tolerances	OEL[h], ppm v/v	Safety Group	RCL[a] (ppm v/v)	RCL[a] (lb/Mcf)	RCL[a] (g/m^3)	Highly Toxic or Toxic[f] Under Code Classification
450A	R-134a/1234ze(E) (42.0/58.0)	(±2.0/±2.0)	880	A1	72,000	20	320	Neither
451A	R-1234yf/134a (89.8/10.2)	(±0.2/±0.2)	520	A2L	18,000	5.3	81	Neither
451B	R-1234yf/134a (88.8/11.2)	(±0.2/±0.2)	530	A2L	18,000	5.3	81	Neither
452A	R-32/125/1234yf (11.0/59.0/30.0)	(±1.7/±1.8/+0.1, −1.0)	780	A1	10,000	27	440	Neither
452B	R-32/125/1234yf (67.0/7.0/26.0)	(±2.0/±1.5/±2.0)	870	A2L	30,000	23	360	Neither
452C	R-32/125/1234yf (12.5/61.0/26.5)	(+0.5, −1.5/±1.0/+0.5, −1.5)	800	A1	100,000	27	430	Neither
453A	R-32/125/134a/227ea/600/601a (20.0/20.0/53.8/5.0/0.6/0.6)	(±1.0/±1.0/±1.0/±0.5/+0.1, −0.2/+0.1, −0.2)	1000	A1	34,000	7.8	120	Neither
454A	R-32/1234yf (35.0/65.0)	(+2.0/−2.0, +2.0/−2.0)	690	A2L	16,000	28	450	Neither
454B	R-32/1234yf (68.9/31.1)	(+1.0/−1.0, +1.0/−1.0)	850	A2L	19,000	22	360	Neither
454C	R-32/1234yf (21.5/78.5)	(±2.0/±2.0)	620	A2L	19,000	29	460	Neither
455A	R-744/32/1234yf (3.0/21.5/75.5)	(+2.0, −1.0/+1.0, −2.0/±2.0)	650	A2L	30,000	23	380	Neither
456A	R-32/134a/1234ze(E) (6.0/45.0/49.0)	(±1.0/±1.0/±1.0)	900	A1	77,000	20	320	Neither
457A	R-32/1234yf/152a (18.0/70.0/12.0)	(+0.5, −1.5/+0.5, −1.5/+0.1, −1.9)	650	A2L	15,000	25	400	Neither
458A	R-32/125/134a/227ea/236fa (20.5/4.0/61.4/13.5/0.6)	(±0.5/±0.5/±0.5/±0.5/±0.1)	1000	A1	76,000	18	280	Neither
459A	R-32/1234yf/1234ze(E) (68.0/26.0/6.0)	(+0.5, −1.5/±2.0/+1.5, −0.5)	870	A2L	27,000	23	360	Neither
459B	R-32/1234yf/1234ze(E) (21.0/69.0/10.0)	(+0.5, −1.0/±2.0/±1.0)	640	A2L	16,000	30	470	Neither
460A	R-32/125/134a/1234ze(E) (12.0/52.0/14.0/22.0)	(±1.0/±1.0/±1.0/±1.0)	650	A1	92,000	24	380	Neither
460B	R-32/125/134a/1234ze(E) (28.0/25.0/20.0/27.0)	(±1.0/±1.0/±1.0/±1.0)	950	A1	120,000	25	400	Neither
460C	R-32/125/134a/1234ze(E) (2.5/2.5/46.0/49.0)	(±0.5/±0.5/±1.0/±1.0)	900	A1	73,000	20	310	Neither

Table 9.2 Data and Safety Classifications for Refrigerant Blends [Std 34-2019, Tbl 4-2] ***(Continued)***

Refrigerant No.	Composition (Mass%)	Composition Tolerances	OEL[h], ppm v/v	Safety Group	RCL[a] (ppm v/v)	RCL[a] (lb/Mcf)	RCL[a] (g/m^3)	Highly Toxic or Toxic[f] Under Code Classification
461A	R-125/143a/134a/227ea/600a (55.0/5.0/32.0/5.0/3.0)	(±1.0/ ±0.5/±1.0/±0.5/+0.1, –0.4)	1000	A1	61,000	17	270	Neither
462A	R-32/125/143a/134a/600 (9.0/42.0/2.0/44.0/3.0)	(+1.5, –1.0/±2.0/±1.0/±2.0/±1.0)	1000	A2	16,000	3.9	62	Neither
463A	R-744/32/125/1234yf/134a (6.0/36.0/30.0/14.0/14.0)	(+2.0, –1.0/±2.0/±2.0/±2.0/±2.0)	990	A1	98,000	19	300	Neither
464A	R-32/125/1234ze(E)/227ea (27.0/27.0/40.0/6.0)	(±1.0/±1.0/±1.0/±0.5)	930	A1	120,000	27	430	Neither
465A	R-32/290/1234yf (21.0/7.9/71.1)	(+0.5, –1.5/+0.1, –0.9/±1.0)	660	A2	12,000	2.5	40	Neither
Azeotropes[b]								
500	R-12/152a (73.8/26.2)		1000	A1	30,000	7.6	120	Neither
501	R-22/12 (75.0/25.0)[c]		1000	A1	54,000	13	210	Neither
502[g]	R-22/115 (48.8/51.2)		1000	A1	73,000	21	330	Neither
503	R-23/13 (40.1/59.9)		1000					Neither
504[i]	R-32/115 (48.2/51.8)		1000		140,000	28	450	Neither
505	R-12/31 (78.0/22.0)[c]							Neither
506	R-31/114 (55.1/44.9)							Neither
507A[d,i]	R-125/143a (50.0/50.0)		1000	A1	130,000	32	520	Neither
508A[d]	R-23/116 (39.0/61.0)		1000	A1	55,000	14	220	Neither
508B	R-23/116 (46.0/54.0)		1000	A1	52,000	13	200	Neither
509A[d,g]	R-22/218 (44.0/56.0)		1000	A1	75,000	24	390	Neither
510A	R-E170/600a (88.0/12.0)	(±0.5/±0.5)	1000	A3	7300	0.87	14	Neither
511A	R-290/E170 (95.0/5.0)	(±1.0/±1.0)	1000	A3	5300	0.59	9.5	Neither

Table 9.2 Data and Safety Classifications for Refrigerant Blends [Std 34-2019, Tbl 4-2] ***(Continued)***

Refrigerant No.	Composition (Mass%)	Composition Tolerances	OEL[h], ppm v/v	Safety Group	RCL[a] (ppm v/v)	RCL[a] (lb/Mcf)	RCL[a] (g/m^3)	Highly Toxic or Toxic[f] Under Code Classification
512A	R-134a/152a (5.0/95.0)	(±1.0/±1.0)	1000	A2	11,000	1.9	31	Neither
513A	R-1234yf/134a (56.0/44.0)	(±1.0/±1.0)	650	A1	72,000	20	320	Neither
513B	R-1234yf/134a (58.5/41.5)	(±0.5/±0.5)	640	A1	74,000	21	330	Neither
514A	R-1336mzz(Z)/1130 (E) (74.7/25.3)	(+1.5, –0.5/+0.5, –1.5)	320	B1	2400	0.86	14	Neither
515A	R-1234ze(E)/227ea (88.0/12.0)	(+1.0, –2.0/+2.0, –1.0)	810	A1	62,000	19	300	Neither
516A	R-1234yf/134a/152a (77.5/8.5/14.0)	(±1.4/+0.5, –1.5/+0.1, –1.9)	590	A2L	27,000	7.0	110	Neither

a. Data taken from J.M. Calm, "ARTI Refrigerant Database," Air- Conditioning and Refrigeration Technology Institute (ARTI), Arlington, VA, July 2001; J.M. Calm, "Toxicity Data to Determine Refrigerant Concentration Limits," Report DE/CE 23810-110, Air- Conditioning and Refrigeration Technology Institute (ARTI), Arlington, VA, September 2000; J.M. Calm, "The Toxicity of Refrigerants," *Proceedings of the 1996 International Refrigeration Conference*, Purdue University, West Lafayette, IN, pp. 157–62, 1996; D.P. Wilson and R.G. Richard, "Determination of Refrigerant Lower Flammability Limits (LFLs) in Compliance with Proposed Addendum p to ANSI/ASHRAE Standard 34-1992 (1073-RP)," *ASHRAE Transactions* 2002, 108(2); D.W. Coombs, "HFC-32 Assessment of Anesthetic Potency in Mice by Inhalation," Huntingdon Life Sciences Ltd., Huntingdon, Cambridgeshire, England, February 2004 and amendment February 2006; D.W. Coombs, "HFC-22 An Inhalation Study to Investigate the Cardiac Sensitization Potential in the Beagle Dog," Huntingdon Life Sciences Ltd., Huntingdon, Cambridgeshire, England, August 2005; and other toxicity studies.

b. Azeotropic refrigerants exhibit some segregation of components at conditions of temperature and pressure other than those at which they were formulated. The extent of segregation depends on the particular azeotrope and hardware system configuration.

c. The exact composition of this azeotrope is in question, and additional experimental studies are needed.

d. R-507, R-508, and R-509 are allowed alternative designations for R-507A, R-508A, and R-509A due to a change in designations after assignment of R-500 through R-509. Corresponding changes were not made for R-500 through R-506.

e. The RCL values for these refrigerant blends are approximated in the absence of adequate data for a component comprising less than 4% m/m of the blend and expected to have only a small influence in an acute, accidental release.

f. *Highly toxic*, *toxic*, or *neither*, where *highly toxic* and *toxic* are as defined in the *International Fire Code*, *Uniform Fire Code*, and OSHA regulations, and *neither* identifies those refrigerants having lesser toxicity than either of those groups [1,2,3].

g. At locations with altitudes higher than 1500 m, the ODL and RCL shall be 69,100 ppm.

h. The OELs are eight-hour TWAs as defined in Section 3 unless otherwise noted; a "C" designation denotes a ceiling limit.

i. At locations with altitudes higher than 1000 m but below or equal to 1500 m, the ODL and RCL shall be 112,000 ppm, and at altitudes higher than 1500 m, the ODL and RCL shall be 69,100 ppm.

Refrigerating Machinery Rooms

When required, the machinery room must

- Be dimensioned so parts are accessible with space for service and maintenance.
- Have tight-fitting doors opening outward, self-closing if they open into the building, with no openings permitting passage of escaping refrigerant into the building except gasketed access panels of ductwork and air-handling equipment.
- Contain a leak detector located where refrigerant from a leak will concentrate that actuates visual and audible alarms inside the room and outside each entrance, and activates the mechanical ventilation.
- On alarm, exhaust of $Q_{L/s} = 100 \times G^{0.5}$ where G is kilograms of refrigerant in the largest system, with openings for inlet air to avoid recirculation; multiple fans or multi-speed fans to operate to reduce airflow for normal operation to at least 2.54 L/s per m^2 or 9.44 L/s per person, and operable when occupied to limit temperature rise to 10°C above inlet air or a maximum of 50°C.
- Combustion air can be used for equipment in the machinery room only if ducted from outside the room and sealed from refrigerant entry, and a refrigerant detector is employed to shut off combustion on refrigerant leaks. (Exceptions: CO_2 or water refrigerant; or ammonia only driven by internal combustion engine.)
- Be no airflow to or from an occupied space through a machinery room unless ducted and sealed against refrigerant leaks.
- Restrict access to authorized personnel with clear signage at each entrance. If system is in an open enclosure outdoors more than 6.1 m from building openings, mechanical or natural ventilation may be used; free-opening area for natural ventilation must be $F_{sqm} = G^{0.5}$.

The total of Group A2, B2, A3, or B3 refrigerants except R-717 (ammonia) must not exceed 500 kg without approval of the authority having jurisdiction. Special requirements in Section 7.5 of ANSI/Standard 15 apply relative to recovered, reclaimed, or recycled refrigerants, or mixing of refrigerants, refrigerant, or lubricant conversion. Group A2, A3, B1, B2, or B3 refrigerants must not be used in high-probability systems (where a refrigerant leak can enter occupied space) for human comfort.

Refrigerant Piping

Piping must not be installed in elevators or other shafts that have moving objects or open into living quarters. It must not penetrate floors except the top floor to the roof, or the basement to the first floor, unless enclosed in a gastight, fire-resistive shaft. The piping must be enclosed in a pipe duct if inside floors.

Pressure Relief Protection

Refrigerating systems must be protected by pressure relief devices. ANSI/ASHRAE Standard 15 covers required location and sizing. All pressure relief valves must be marked "UV" or "VR," and all rupture members marked with data required by paragraph UG 127 of Section VIII, Division 1, of the *ASME Boiler and Pressure Vessel Code*; and fusible plugs must be marked with the melting temperature. Generally, pressure relief devices and fusible plugs must discharge to the atmosphere not less than 4.57 m above ground and 6.1 m from a window, ventilation opening, pedestrian walkway, or exit in any building.

10. REFRIGERATION LOADS AND EQUIPMENT

Transmission Load

Sensible heat gain through walls, floor, and ceiling is calculated at steady state as

$$q = UA\,\Delta t \tag{10.1}$$

where

q	=	heat gain, W
A	=	outer area of section, m^2
Δt	=	difference between outdoor air temperature and air temperature of refrigerated space, K

The overall coefficient of heat transfer U of the wall, floor, or ceiling can be calculated by the following equation:

$$U = \frac{1}{1/h_i + x/k + 1/h_o} \tag{10.2}$$

where

U	=	overall heat transfer coefficient, $W/(m^2 \cdot K)$
x	=	wall thickness, m
k	=	thermal conductivity of wall material, $W/(m \cdot K)$
h_i	=	inner surface conductance, $W/(m^2 \cdot K)$
h_o	=	outer surface conductance, $W/(m^2 \cdot K)$

A value of 9.1 for h_i and h_o is frequently used for still air. If the outer surface is exposed to 25 km/h wind, h_o is increased to 34.

Latent heat gain due to moisture transmission through walls, floors, and ceilings of modern refrigerated facilities is negligible.

Table 10.1 Thermal Conductivity of Cold Storage Insulation [2018R, Ch 24, Tbl 1]

Insulation	Thermal Conductivity[a] k, W/(m·K)
Polyurethane board (R-11 expanded)	0.023 to 0.026
Polyisocyanurate, cellular (R-141b expanded)	0.027
Polystyrene, extruded (R-142b)	0.035
Polystyrene, expanded (R-142b)	0.037
Corkboard[b]	0.043
Foam glass[c]	0.044

[a]Values are for a mean temperature of 24°C, and insulation is aged 180 days.
[b]Seldom used. Data are only for reference.
[c]Virtually no effects from aging.

Table 10.2 Minimum Insulation Thickness [2018R, Ch 24, Tbl 2]

Storage Temperature, °C	Expanded Polyisocyanurate Thickness	
	Northern U.S., mm	Southern U.S., mm
10 to 16	50	50
4 to 10	50	50
–4 to 4	50	75
–9 to –4	75	75
–18 to –9	75	100
–26 to –18	100	100
–40 to –26	125	125

Table 10.3 Allowance for Sun Effect [2018R, Ch 24, Tbl 3]

Typical Surface Types	East Wall, K	South Wall, K	West Wall, K	Flat Roof, K
Dark-colored surfaces				
Slate roofing	5	3	5	11
Tar roofing				
Black paint				
Medium-colored surfaces				
Unpainted wood	4	3	4	9
Brick				
Red tile				
Dark cement				
Red, gray, or green paint				
Light-colored surfaces				
White stone	3	2	3	5
Light-colored cement				
White paint				

Note: Add to the normal temperature difference for heat leakage calculations to compensate for sun effect. Do not use for air-conditioning design.

Product Load

1. Heat removed to cool from initial temperature to some lower temperature above freezing:

$$Q_1 = mc_1(t_1 - t_2) \tag{10.3}$$

2. Heat removed in cooling from the initial temperature to a freezing point of the product:

$$Q_2 = mc_1(t_1 - t_f) \tag{10.4}$$

3. Heat removed to freeze the product:

$$Q_3 = m\,h_{if} \tag{10.5}$$

4. Heat removed in cooling from the freezing point to the final temperature below the freezing point:

$$Q_4 = mc_2\,(t_f - t_3) \tag{10.6}$$

where

Q_1, Q_2, Q_3, Q_4	=	heat removed, kJ
m	=	weight of the product, kg
c_1	=	specific heat of the product above freezing, kJ/(kg·K)
t_1	=	initial temperature of the product above freezing, °C
t_2	=	lower temperature of the product above freezing, °C
t_f	=	freezing temperature of the product, °C
h_{if}	=	latent heat of fusion of the product, kJ/kg
c_2	=	specific heat of the product below freezing, kJ/(kg·K)
t_3	=	final temperature of the product below freezing, °C

Specific heats above and below freezing for many products are given in Table 3 of Chapter 19 of the 2018 *ASHRAE Handbook—Refrigeration.*

Refrigeration system capacity for products brought into refrigerated spaces is determined from the time allotted for heat removal and assumes that the product is properly exposed to remove the heat in that time. The calculation is:

$$q = \frac{Q_2 + Q_3 + Q_4}{3600n} \tag{10.7}$$

where

q	=	product cooling load, kW
n	=	allotted time period, h

A product's latent heat of fusion is related to its water content and can be estimated by multiplying the product's percent of water (expressed as a decimal) by the water's latent heat of fusion, 334 kJ/kg. Most food products freeze in the range of –3°C to –0.5°C. When the exact freezing temperature is not known, assume that it is –2.2°C.

Table 10.4 Heat Gain from Typical Electric Motors [2018R, Ch 24, Tbl 6]

				Location of Motor and Driven Equipment with Respect to Conditioned Space or Airstream		
				A	B	C
Motor Rated, kW	Motor Type	Nominal rpm	Full Load Motor Efficiency, %	Motor in, Driven Equipment in, W	Motor out, Driven Equipment in, W	Motor in, Driven Equipment out, W
0.04	Shaded pole	1500	35	105	35	70
0.06			35	170	59	110
0.09			35	264	94	173
0.12			35	340	117	223
0.19	Split phase	1750	54	346	188	158
0.25			56	439	246	194
0.37			60	621	372	249
0.56	3-Phase	1750	72	776	557	217
0.75			75	993	747	249

Table 10.5 Heat Equivalent of Occupancy [2018R, Ch 24, Tbl 7]

Refrigerated Space Temperature, °C	Heat Equivalent/Person, W
10	210
5	240
0	270
–5	300
–10	330
–15	360
–20	390

Note: Heat equivalent may be estimated by the following equation:
$q_p = 272 - 6t$
where t = temperature of refrigerated space, °C

Packaging Related Load

Cardboard and wood used as part of product packaging adsorb or desorb moisture, depending on air temperature and relative humidity. This moisture sorption represents a conversion between sensible and latent heat: the latent heat of sorption is countered by sensible heat transfer between the packaging and the air by convection. The heat load q_t from the i packaging components is

$$q_t = \frac{\sum m_i c_i (t_1 - t_3)}{3600n} \quad (10.8)$$

$$q_l = \frac{\left[\frac{m(X_1 - X_3)L}{1 + X_1}\right]_{wood} + \left[\frac{m(X_1 - X_3)L}{1 + X_1}\right]_{cardboard}}{3600n} \quad (10.9)$$

where

q_l	=	total heat load, kJ
m_i	=	mass of ith packaging component, kg
c_i	=	specific heat of ith packaging component, kJ/(kg·K)
X_1	=	entering packaging component moisture content, kg/kg dry basis
X_3	=	packaging component moisture content after time n, kg/kg dry basis
L	=	latent heat of sorption, kJ/kg (2500 kJ/kg)

Typical values of c_i are given in Table 10.6. X_1 can be measured or estimated using moisture sorption isotherms based on the air temperature and relative humidity the packaging experienced before entering the refrigerated space. The moisture sorption isotherms for wood and cardboard are given in Figures 10.1 and 10.2. X_3 can be estimated using Figure 10.3 or Figure 10.4 to get Y and

$$X_3 = X^* + (X_1 - X^*)Y \quad (10.10)$$

where

Y	=	fractional unaccomplished moisture change (Figure 10.4)
X^*	=	equilibrium packaging component moisture content, kg/kg dry basis

X^* can be estimated using the moisture sorption isotherms based on the air temperature and relative humidity in the refrigerated space using Figures 10.1 and 10.2 for wood and cardboard, respectively, where equilibrium moisture content (EMC) is plotted against equilibrium relative humidity (ERH).

Table 10.6 Typical Specific Heat Capacities of Common Packaging Materials [2018R, Ch 24, Tbl 8]

Material	Specific Heat Capacity c, kJ/(kg·K)
Wood	1.7
Cardboard	1.4
Plastic	1.6
Aluminum	0.85
Steel	0.5

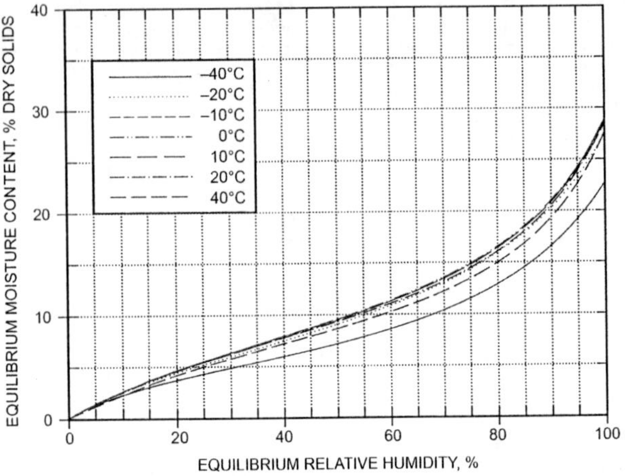

Figure 10.1 Moisture Sorption Isotherms for Wood as Function of Air Temperature and Relative Humidity [2018R, Ch 24, Fig 3]

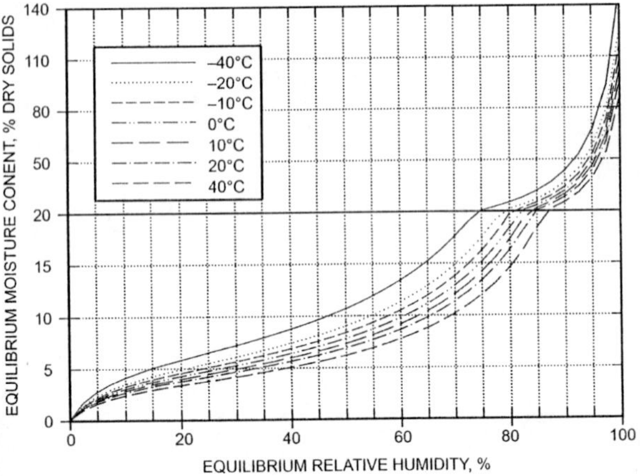

Figure 10.2 Moisture Sorption Isotherms for Cardboard as Function of Air Temperature and Relative Humidity [2018R, Ch 24, Fig 4]

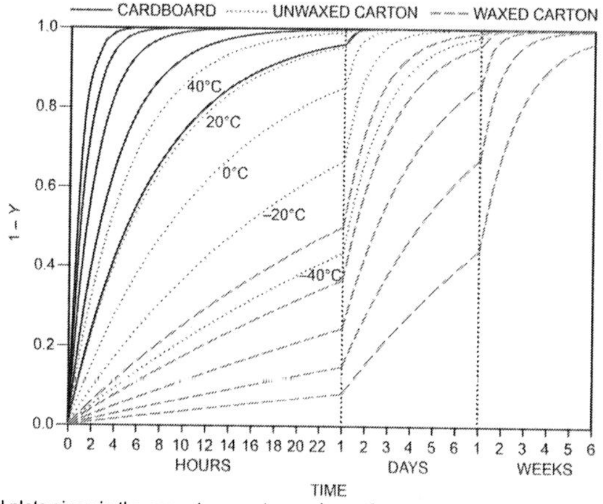

(All plots given in the same temperature order as for unwaxed carton)

Figure 10.3 Fractional Unaccomplished Moisture Change as Function of Time and Temperature for Sheets of Cardboard, Unwaxed Cartons, and Waxed Cartons [2018R, Ch 24, Fig 5]

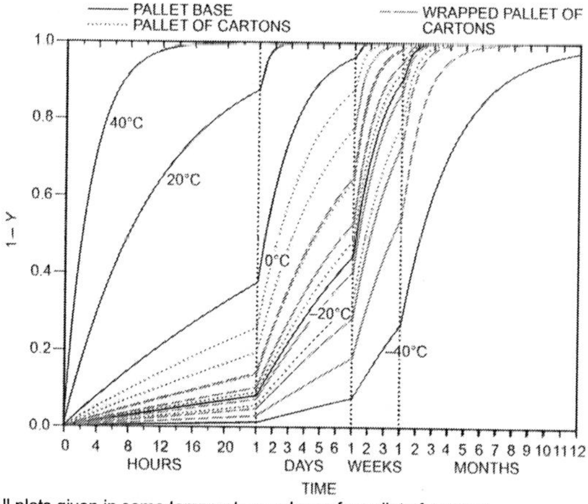

(All plots given in same temperature order as for pallet of cartons)

Figure 10.4 Fractional Unaccomplished Moisture Change as Function of Time and Temperature for Wooden Pallet Bases, Unwrapped Pallets of Cartons, and Wrapped Pallets of Cartons [2018R, Ch 24, Fig 6]

Infiltration Air Load

Heat gain through doorways from air exchange is:

$$q_t = qD_tD_f(1 - E) \tag{10.11}$$

where

q_t	=	average heat gain for the 24 h or other period, kW
q	=	sensible and latent refrigeration load for fully established flow, kW
D_t	=	doorway open-time factor
D_f	=	doorway flow factor
E	=	effectiveness of doorway protective device

$$q = 0.5773790\ WH^{1.5}\ (Q_s/A)(1/R_s) \tag{10.12}$$

where

Q_s/A	=	sensible heat load of infiltration air per square foot of doorway opening as read from Figure 10.5, kW/m^2
W	=	doorway width, m
H	=	doorway height, m
R_s	=	sensible heat ratio of the infiltration air heat gain, from a psychrometric chart

Doorway open-time factor D_t can be calculated as follows:

$$D_t = \frac{(P\theta_p + 60\theta_o)}{3600\theta_d} \tag{10.13}$$

where

D_t	=	decimal portion of time doorway is open
P	=	number of doorway passages
θ_p	=	door open-close time, seconds per passage
θ_o	=	time door simply stands open, min
θ_d	=	the daily (or other) time period, h

Equipment-Related Load

Equipment heat gain is often small at space temperatures above approximately −1°C. Where reheat or other artificial loads are not imposed, total equipment heat gain is about 5% or less of the total load. However, equipment heat gain becomes a major portion of the total load at freezer temperatures. For example, at −30°C the theoretical contribution to total refrigeration load from fan power and coil defrosting alone can exceed, for many cases, 15% of the total load (assuming proper control of defrosting so that the space is not heated excessively).

Safety Factor

The calculated load is typically increased by a safety factor (10% or more is often used) to allow for possible discrepancies between design criteria and actual operation. This factor should be selected in consultation with the facility user and should be applied individually to the first four heat load segments.

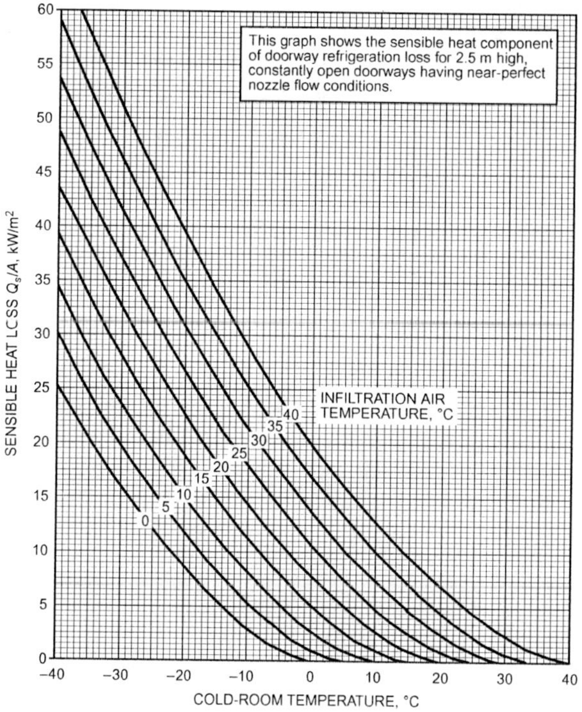

Figure 10.5 Sensible Heat Gain by Air Exchange for Continuously Open Door with Fully Established Flow [2010R, Ch 24, Fig 9]

Refrigeration Equipment

Liquid Coolers

A liquid cooler (hereafter called a cooler) is a heat exchanger in which refrigerant is evaporated, thereby cooling a fluid (usually water or brine) circulating through the cooler.

Various types of liquid coolers and their characteristics are listed in Table 10.7 and Figures 10.6 through 10.12.

Heat transfer for liquid coolers can be expressed by the following steady-state heat transfer equation:

$$q = UA\Delta t_m \tag{10.14}$$

where

q = total heat transfer rate, W
Δt_m = mean temperature difference, K
A = heat transfer surface area associated with U, m^2
U = overall heat transfer coefficient, $W/(m^2 \cdot K)$

The area A can be calculated if the geometry of the cooler is known. The mean temperature difference is

$$\Delta t_m = (\Delta t_1 - \Delta t_2)/\ln(\Delta t_1/\Delta t_2) \tag{10.15}$$

where Δt_1 and Δt_2 are temperature differences between the fluids at each end of the heat exchanger.

U may be calculated by one of the following equations.

Based on inside surface area:

$$U = \frac{1}{1/h_i + [A_i/(A_o h_o)] + (t/k)(A_i/A_m) + r_{fi}} \tag{10.16}$$

Based on outside surface area:

$$U = \frac{1}{[A_o/(A_i h_i)] + 1/h_o + (t/k)(A_o/A_m) + r_{fo}} \tag{10.17}$$

where

h_i = inside heat transfer coefficient based on inside surface area, $W/(m^2 \cdot K)$
h_o = outside heat transfer coefficient based on outside surface area, $W/(m^2 \cdot K)$
A_o = outside heat transfer surface area, m^2
A_i = inside heat transfer surface area, m^2
A_m = mean heat transfer area of metal wall, m^2
k = thermal conductivity of heat transfer material, $W/(m \cdot K)$
t = thickness of heat transfer surface (tube wall thickness), m
r_{fi} = fouling factor of fluid side based on inside surface area, $(m^2 \cdot K)/W$
r_{fo} = fouling factor of fluid side based on outside surface area, $(m^2 \cdot K)/W$

Note: If fluid is on inside, multiply r_{fi} by A_o/A_i to find r_{fo}.
If fluid is on outside, multiply r_{fo} by A_i/A_o to find r_{fi}.

These equations can be applied to incremental sections of the heat exchanger to include local effects on the value of U, and then the increments summed to obtain a more accurate design.

Over time, most fluids foul the fluid-side heat transfer surface, reducing the cooler's overall heat transfer coefficient. If fouling is expected to be a problem, a mechanically cleanable cooler should be used, such as a flooded, Baudelot, or cleanable direct-expansion tube-in-tube cooler. Direct-expansion shell-and-tube, shell-and-coil, and brazed-plate coolers can be cleaned chemically. Flooded coolers and direct-expansion tube-in-tube coolers with enhanced fluid-side heat transfer surfaces tend to be self-cleaning because of high fluid turbulence, so a smaller fouling factor can probably be used for these coolers. Research shows that negligible fouling occurs in closed-loop evaporator tubes at 1 to 1.5 m/s and 2.1 m/s water velocities. AHRI Standard 480 discusses fouling calculations.

Table 10.7 Types of Coolers [2020S, Ch 42, Tbl 1]

Type of Cooler	Subtype	Usual Refrigerant Feed Device	Usual Capacity Range, kW	Commonly Used Refrigerants
Direct-expansion	Shell-and-tube	Thermal expansion valve	7 to 1800	12, 22, 134a, 404A, 407C, 410A, 500, 502, 507A, 717
		Electronic modulation valve	7 to 1800	
	Tube-in-tube	Thermal expansion valve	18 to 90	12, 22, 134a, 717
	Brazed-plate	Thermal expansion valve	2 to 700	12, 22, 134a, 404A 407C, 410A, 500, 502, 507A, 508B, 717, 744
	Semiwelded plate	Thermal expansion valve	175 to 7000	12, 22, 134a, 500, 502, 507A, 717, 744
Flooded	Shell-and-tube	Low-pressure float	90 to 7000	11, 12, 22, 113, 114
		High-pressure float	90 to 21 100	123, 134a, 500, 502, 507A, 717
		Fixed orifice(s)	90 to 21 100	
		Weir	90 to 21 100	
	Spray shell-and-tube	Low-pressure float	180 to 35 000	11, 12, 13B1, 22
		High-pressure float	180 to 35 000	113, 114, 123, 134a
	Brazed-plate	Low-pressure float	2 to 700	12, 22, 134a, 500, 502, 507A, 717, 744
	Semiwelded plate	Low-pressure float	175 to 7000	12, 22, 134a, 500, 502, 507A, 717, 744
Baudelot	Flooded	Low-pressure float	35 to 350	22, 717
	Direct-expansion	Thermal expansion valve	18 to 90	12, 22, 134a, 717
Shell-and-coil	—	Thermal expansion valve	7 to 35	12, 22, 134a, 717

The refrigerant side of the cooler is not subject to fouling, and a fouling factor need not be included for that side.

Typically, the t/k term in Equations 10.16 and 10.17 may be negligible for material with high thermal conductivity. However, with low-thermal-conductivity material or thick-walled tubing, it may become significant. Refer to Chapter 4 of the 2021 *ASHRAE Handbook—Fundamentals* and to Chapter 39 of the 2020 *ASHRAE Handbook—HVAC Systems and Equipment* for further details.

Pressure drop is usually minimal in Baudelot and shell-and-coil coolers but must be considered in direct-expansion and flooded coolers. Both direct-expansion and flooded coolers rely on turbulent fluid flow to improve heat transfer. This turbulence is obtained at the expense of pressure drop.

For air-conditioning, pressure drop is commonly limited to 70 kPa to keep pump size and energy cost reasonable. For flooded coolers, see Chapter 39 of the 2020 *ASHRAE Handbook—HVAC Systems and Equipment* for a discussion of pressure drop for flow in tubes. Pressure drop for fluid flow in shell-and-tube direct-expansion coolers depends greatly on tube and baffle geometry. The following equation projects the change in pressure drop caused by a change in flow:

$$\text{New pressure drop} = \text{Original pressure drop}\left[\frac{\text{New rate}}{\text{Original rate}}\right]^{1.8} \quad (10.18)$$

The refrigerant-side pressure drop must be considered for direct-expansion, shell-and-coil, brazed-plate, and (sometimes) Baudelot coolers. When there is a pressure drop on the refrigerant side, the refrigerant inlet and outlet pressures and corresponding saturated temperature are different. This difference changes the mean temperature difference, which affects the total heat transfer rate. If pressure drop is high, expansion valve operation may be affected because of reduced pressure drop across the valve. This pressure drop varies, depending on the refrigerant used, operating temperature, and type of tubing.

When the fluid being cooled is electrically conductive, the system must be grounded to prevent electrochemical corrosion.

The constant superheat thermal expansion valve is the most common control used, located directly upstream of the cooler.

In flooded coolers, an orifice is often used as the throttling device between condenser and cooler.

Freeze prevention must be considered for coolers operating near the fluid's freezing point. In some coolers, freezing causes extensive damage. Two methods can be used for freeze protection: (1) hold saturated suction pressure above the fluid freezing point or (2) shut the system off if fluid temperature approaches the freezing point.

If the cooler is used only when ambient temperature is above freezing, drain the fluid from the cooler for cold weather. Alternatively, if the cooler is used year-round, the following methods can be used to prevent freezing:

- Heat tape or other heating device to keep cooler above freezing
- For water, adding an appropriate amount of ethylene glycol
- Continuous pump operation

Most compressors discharge a small percentage of oil in the discharge gas. This oil mixes with condensed refrigerant in the condenser and flows to the cooler. Because the oil is nonvolatile, it does not evaporate and may collect in the cooler.

In direct-expansion coolers, gas velocity in the tubes and suction gas header is usually sufficient to carry oil from the cooler into the suction line. From there, with proper piping design, it can be carried back to the compressor. At light load and low temperature, oil may gather in the superheat section of the cooler, detracting from performance. For this reason, operating refrigerant circuits at light load for long periods should be avoided, especially under low-temperature conditions.

In flooded coolers, vapor velocity above the tube bundle is usually insufficient to return oil up the suction line, and oil tends to accumulate in the cooler. With time, depending on the compressor oil loss rate, oil concentration in the cooler may become large. When concentration exceeds about 1%, heat transfer performance may be adversely affected if enhanced tubing is used.

It is common in flooded coolers to take some oil-rich liquid and return it to the compressor on a continuing basis, to establish a rate of return equal to the compressor oil loss rate.

Cooler maintenance centers around (1) safety and (2) cleaning the fluid side. Periodically inspect the cooler for any weakening of its pressure boundaries. The manufacturer or a service organization experienced in cooler maintenance should have details for cleaning.

A cooler operating at a saturated suction temperature lower than the ambient-air dew point should be insulated to prevent condensation.

Forced-Circulation Air Coolers

A cooling coil and a motor-driven fan are the basic components, and coil defrosting means are added for low-temperature operations where coil frosting might impede performance. Blow-through direct-drive propeller fans are most common, but for long throws, draw-through configuration is preferred. For loads above 0°C, coil spacing is usually 240 to 320 fins per metre; below 0°C a maximum of 160 fins per metre is preferred. Even distribution of halocarbon refrigerant is usually attained in direct-expansion coils by refrigerant distributors. Units in larger refrigeration systems are often liquid-pumped recirculating types with orifice disks.

Defrost for coils and drain pans of low-temperature units may be hot-gas, electric, or water. Usually defrosting is done with the fan off. Control of defrost is usually by microprocessor, with a thermostat mounted within the coil. Usually a rise to 7°C returns the unit to the operating cycle. Drain lines should be well-pitched, insulated, and trapped outside the refrigerated space.

Capacities of air coolers are usually based on the temperature difference between inlet air and refrigerant in the coil. The higher the TD, the lower the space relative humidity. Between 5°C and 9°C TD is usual, except for packaged products and workrooms where TD of 14°C is common. Low-temperature units generally have TD below 8°C for system economics and limiting defrost frequency.

Most frequent control of refrigerant flow is an expansion valve, most frequently thermostatic type. Electric expansion valves, requiring a valve, controller, and control sensor, are also available.

Large refrigerating systems more frequently have flooded evaporators, most often low-side float valves. Refrigerant valves opening or closing flow are usually solenoid valves. Larger flows may require pilot-operated solenoid valves. When it is desired to limit compressor motor load during pulldown, an evaporator pressure regulating valve may be used to limit compressor suction pressure.

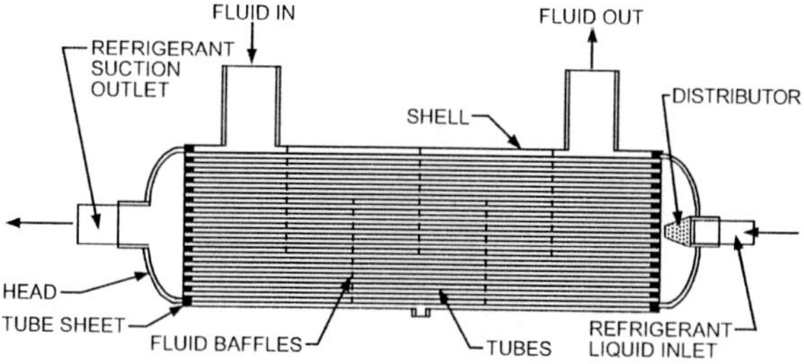

Figure 10.6 Direct-Expansion Shell-and-Tube Cooler [2020S, Ch 42, Fig 1]

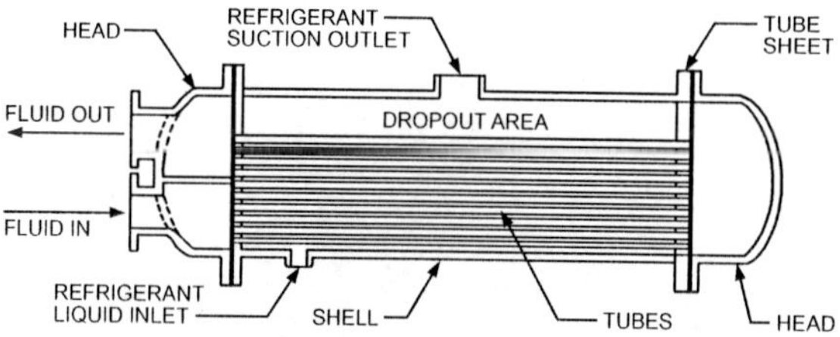

Figure 10.7 Flooded Shell-and-Tube Cooler [2020S, Ch 42, Fig 2]

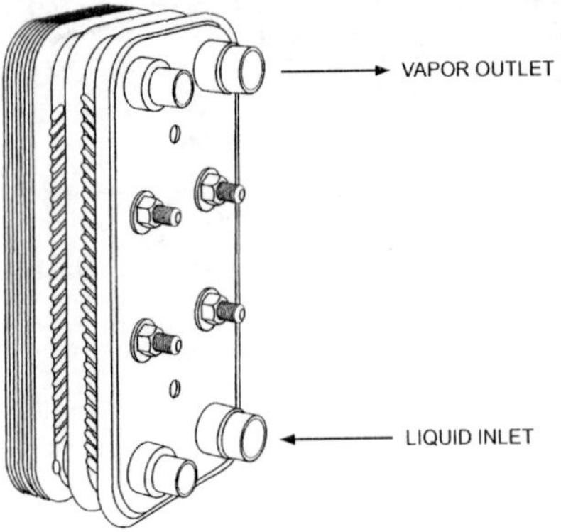

Figure 10.8 Flooded Plate Cooler [2020S, Ch 42, Fig 3]

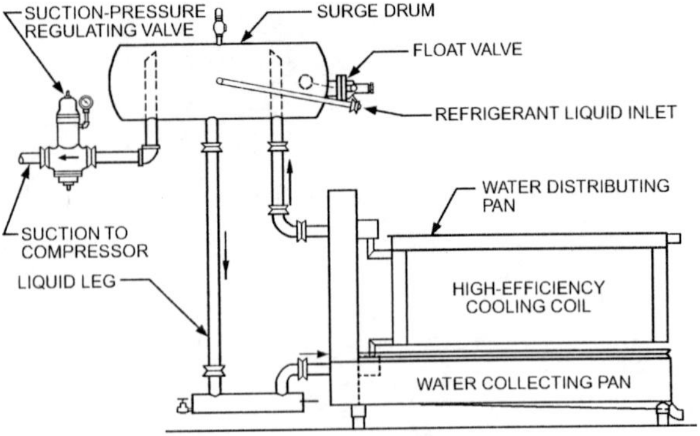

Figure 10.9 Baudelot Cooler [2020S, Ch 42, Fig 4]

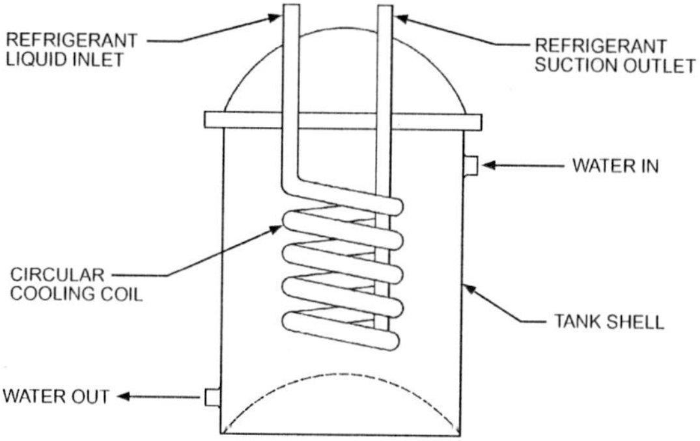

Figure 10.10 Shell-and-Coil Cooler [2020S, Ch 42, Fig 5]

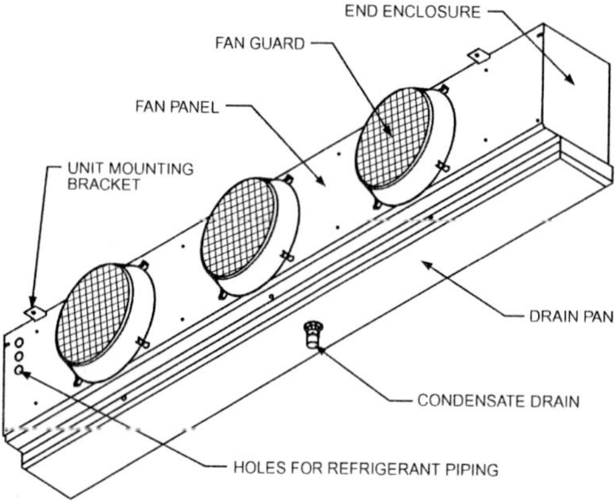

Figure 10.11 Low-Profile Cooler [2018R, Ch 14, Fig 3]

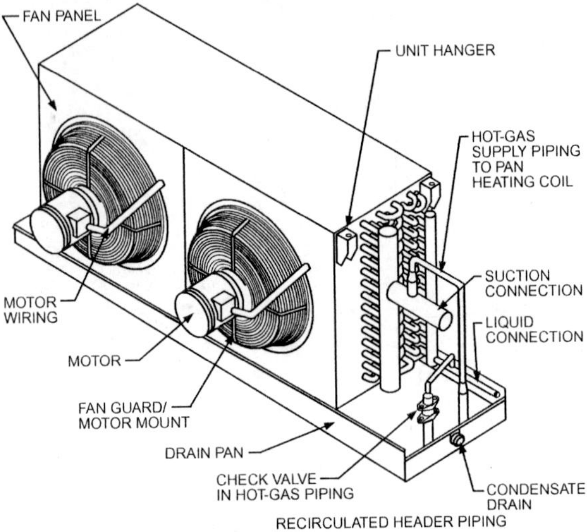

Figure 10.12 Liquid Overfeed Type Unit Cooler [2018R, Ch 14, Fig 4]

11. AIR-CONDITIONING LOAD DATA

Cooling Loads

Cooling loads result from many conduction, convection, and radiation heat transfer processes through the building envelope and from internal sources and system components. Building components or contents that may affect cooling loads include the following:

- **External:** Walls, roofs, windows, skylights, doors, partitions, ceilings, and floors
- **Internal:** Lights, people, appliances, and equipment
- **Infiltration:** Air leakage and moisture migration
- **System:** Outdoor air, duct leakage and heat gain, reheat, fan and pump energy, and energy recovery

ASHRAE provides climatic design information for 9237 locations in the United States, Canada, and around the world. Chapter 14 of the 2021 *ASHRAE Handbook—Fundamentals* provides background discussion on these design data. The complete data tables for all 9237 stations are included with both the PDF version of the Handbook chapter and the Handbook Online version.

ANSI/ASHRAE/ACCA Standard 183 sets the minimum standards for nonresidential load calculations.

Currently there are two ASHRAE cooling load calculation methods. The first is the Heat Balance (HB) method, whose equations are coded in a generic computer program linked to a user interface program. The source code for these programs is in the ASHRAE Load Calculation Toolkit.

The second method is the Radiant Time Series (RTS) method, a simplification of the heat balance method, still requiring a complex computer program for a multiroom building.

Due to the variation in heat transfer coefficients, precision of construction, and manner of actual building operation, a cooling load calculation can never be more than a good estimate of the actual load.

To design and size components of central air-conditioning systems, more than the cooling load is needed. Type of system, fan energy and location, direct heat loss and gain, duct leakage, heat extracted from lights, and type of return system must all be considered.

Heating Loads

Techniques for estimating design heating load for commercial, institutional, and industrial applications are essentially the same as for those estimating design cooling loads for such uses, with the following exceptions:

- Temperatures of outdoor conditioned spaces are generally lower than maintained space temperatures.
- Credit for solar or internal heat gains is not included.
- Thermal storage effect of building structure or content is ignored.

Thermal bridging effects on wall and roof conduction are greater for heating loads than for cooling loads, and greater care must be taken to account for bridging effects on U-factors used in heating load calculations.

Heat losses (negative heat gains) are thus considered to be instantaneous, heat transfer essentially conductive, and latent heat treated only as a function of replacing space humidity lost to the exterior environment.

This simplified approach is justified because it evaluates worst-case conditions that can reasonably occur during a heating season. Therefore, the near-worst-case load is based on the following:

- Design interior and exterior conditions
- Including infiltration and/or ventilation
- No solar effect (at night or on cloudy winter days)
- Before the periodic presence of people, lights, and appliances has an offsetting effect

Typical commercial and retail spaces have nighttime unoccupied periods at a setback temperature where little to no ventilation is required, building lights and equipment are off, and heat loss is primarily through conduction and infiltration. Before being occupied, buildings are warmed to the occupied temperature. During occupied time, building lights, equipment, and people cooling loads can offset conduction heat loss, although some perimeter heat may be required, leaving infiltration and ventilation as the primary heating loads. Ventilation heat load may be offset with heat recovery equipment. These loads (conduction loss, warm-up load, and ventilation load) may not be additive when sizing building heating equipment, and it is prudent to analyze each load and their interactions to arrive at final equipment sizing for heating.

Previous Cooling Load Calculation Methods

Procedures described in Chapters 17 and 18 of the 2021 *ASHRAE Handbook–Fundamentals* are the most current and scientifically derived means for estimating cooling load for a defined building space, but methods in earlier editions of the ASHRAE Handbook are valid for many applications. These earlier procedures are simplifications of the heat balance principles, and their use requires experience to deal with atypical or unusual circumstances. In fact, any cooling or heating load estimate is no better than the assumptions used to define conditions and parameters such as physical makeup of the various envelope surfaces, conditions of occupancy and use, and ambient weather conditions. Experience of the practitioner can never be ignored.

The primary difference between the HB and RTS methods and the older methods is the newer methods' direct approach, compared to the simplifications necessitated by the limited computer capability available previously.

The **transfer function method (TFM)**, for example, required many calculation steps. It was originally designed for energy analysis with emphasis on daily, monthly, and annual energy use, and thus was more oriented to average hourly cooling loads than peak design loads.

The **total equivalent temperature differential method with time averaging (TETD/TA)** has been a highly reliable (if subjective) method of load estimating since its initial presentation in the 1967 *Handbook of Fundamentals*. Originally intended as a manual method of calculation, it proved suitable only as a computer application because of the need to calculate an extended profile of hourly heat gain values, from which radiant components had to be averaged over a time representative of the general mass of the building involved. Because perception of thermal storage characteristics of a given building is almost entirely subjective, with little specific information for the user to judge variations, the TETD/TA method's primary usefulness has always been to the experienced engineer.

The **cooling load temperature differential method with solar cooling load factors (CLTD/CLF)** attempted to simplify the two-step TFM and TETD/TA methods into a single-step technique that proceeded directly from raw data to cooling load without intermediate conversion of radiant heat gain to cooling load. A series of factors were taken from cooling load calculation results (produced by more sophisticated methods) as "cooling load temperature differences" and "cooling load factors" for use in traditional conduction ($q = UA\Delta t$) equations. The results are approximate cooling load values rather than simple heat gain values. The simplifications and assumptions used in the original work to derive those factors limit this method's applicability to those building types and conditions for which the CLTD/CLF factors were derived; the method should not be used beyond the range of applicability.

The TFM, TETD/TA, and CLTD/CLF procedures have not been invalidated or discredited. Experienced engineers have successfully used them in millions of buildings around the world. The accuracy of cooling load calculations in practice depends primarily on the availability of accurate information and the design engineer's judgment in the assumptions made in interpreting the available data. Those factors have much greater influence on a project's success than does the choice of a particular cooling load calculation method.

The primary benefit of HB and RTS calculations is their somewhat reduced dependency on purely subjective input (e.g., determining a proper time-averaging period for TETD/TA; ascertaining appropriate safety factors to add to the rounded-off TFM results; determining whether CLTD/CLF factors are applicable to a specific unique application). However, using the most up-to-date techniques in real-world design still requires judgment on the part of the design engineer and care in choosing appropriate assumptions, just as in applying older calculation methods.

The information that follows includes data on material properties that apply generally to any of the above-mentioned calculation approaches as well as information specific to the CLTD/CLF approach.

Table 11.1 Summary of Load Sources and Equations for Estimating Space Design Cooling Load

[Adapted from *Cooling and Heating Load Calculation Manual*, Second Edition, Tbl 8.1]

Load Source	Equation	Reference, Table, Description
External		
Roof	$q = UA(\text{CLTD})$	Design heat transmission coefficients, Table 11.7 Areas calculated from plans; CLTD from Table 11.8
Walls	$q = UA(\text{CLTD})$	Design heat transmission coefficients, Table 11.7 Areas calculated from plans; CLTD from Table 11.10
Glass Conduction	$q = UA(\text{CLTD})$	Glass area calculated from plans U-factors, page 206 CLTD for conduction load through glass, page 206
Glass Solar	$q = A(\text{SC})\text{SCL}$	Solar cooling load (SCL) factors, Table 11.11 Net glass area from plans Shading coefficients for combination of glass and internal shading, Table 11.12 Compute shaded area from building projections Externally shaded glass: use north orientation data
Partitions, Ceilings, Floors	$q = UA(\text{TD})$	Design heat transmission coefficients, Table 11.7 Area calculated from plans
Internal		
Lights	$q = \text{INPUT}$	Input rating from electrical plans or lighting fixture data, Table 11.14
People		
Sensible	$q_s = \text{No. (Sens. H.G.)}$	Number of people in space Sensible heat gain from occupants, Table 11.13
Latent	$q_l = \text{No. (Lat. H.G.)}$	Latent heat gain from occupants
Equipment and Appliances	$q_s = \text{HEAT GAIN}$	Recommended rate of heat gain, Tables 11.15 through 11.33
Power	$q = \text{HEAT GAIN}$	page 236
Infiltration Air	$Q = \text{L/s}$	
Sensible	$q_s = 120Q\,\Delta t$	Inside-outside air temperature difference, °C
Latent	$q_l = 3.0Q\,\Delta W$	Inside-outside air humidity ratio difference, g/kW
Total	$q = 1.20Q\,\Delta h$	Inside-outside air enthalpy difference, kJ/kg

CAUTION: Approximate data—Use for preliminary computations only. See *Load Calculation Applications Manual*, Second Edition, for more details.

Heat Flow Q Through Building Materials

In addition to heat flow through building materials the resistance of surfaces and air spaces must be included in calculating U-factors.

$$Q\ (\mathrm{W}) = U \times \text{Area}\ (\mathrm{m}^2) \times \text{temperature difference}\ (\mathrm{K}) \qquad (11.1)$$

where U = overall coefficient of heat transmission, W/(m²·K), of materials + interior and exterior resistances:

$$1/U = \Sigma R\ (\text{resistance of components}) \qquad (11.2)$$

For multiple layers of homogeneous materials, R values are added in series:

$$1/U = R_{cold\ surface} + R_1 + R_2 + R_n ... + R_{warm\ surface} \qquad (11.3)$$

For wood stud walls, studs 400 mm on center (series and parallel):

$$1/U = R_{cold\ surface} + \left\{ \frac{+\ 0.25\ R_{stud}}{+\ 0.75\ R_{stud\ space}} \right\} + R_{warm\ surface} \qquad (11.4)$$

Plus, in series, $R_{insulation}$, R_{siding}, $R_{wallboard}$, etc.

For metal framed construction, heat flow through the metal causes thermal bridging, increasing the U-factor significantly. For details, see Normative Appendix A of ANSI/ASHRAE/IES Standard 90.1.

Conductive Heat Flow Through Glazing

Solar radiation gain through glazing is usually more significant in cooling load calculations than conductive heat gain. Solar heat gain is neglected in heating load calculations.

Conductive heat flow through glazing including surface resistance (approximate data)

Single glazing	$U = 6.2$
Double glazing	$U = 3.1$
Triple glazing	$U = 1.9$

Table 11.2 Thermal Resistances of Plane Air Spaces,[a,b,c] (m²·K)/W [2021F, Ch 26, Tbl 3]

		Air Space		Effective Emittance ε_{eff}[d,e]									
				13 mm Air Space[c]					20 mm Air Space[c]				
Position of Air Space	Direction of Heat Flow	Mean Temp.[d], °C	Temp. Diff.,[d] K	0.03	0.05	0.2	0.5	0.82	0.03	0.05	0.2	0.5	0.82
		32.2	5.6	0.37	0.36	0.27	0.17	0.13	0.41	0.39	0.28	0.18	0.13
	Up	10.0	16.7	0.29	0.28	0.23	0.17	0.13	0.30	0.29	0.24	0.17	0.14
		10.0	5.6	0.37	0.36	0.28	0.20	0.15	0.40	0.39	0.30	0.20	0.15
Horiz.	↑	−17.8	11.1	0.30	0.30	0.26	0.20	0.16	0.32	0.32	0.27	0.20	0.16
		−17.8	5.6	0.37	0.36	0.30	0.22	0.18	0.39	0.38	0.31	0.23	0.18
		−45.6	11.1	0.30	0.29	0.26	0.22	0.18	0.31	0.31	0.27	0.22	0.19
		−45.6	5.6	0.36	0.35	0.31	0.25	0.20	0.38	0.37	0.32	0.26	0.21
		32.2	5.6	0.43	0.41	0.29	0.19	0.14	0.62	0.57	0.37	0.21	0.15
	Horiz.	10.0	16.7	0.45	0.43	0.32	0.22	0.16	0.51	0.49	0.35	0.23	0.17
		10.0	5.6	0.47	0.45	0.33	0.22	0.16	0.65	0.61	0.41	0.25	0.18
Vertical		−17.8	11.1	0.50	0.48	0.38	0.26	0.20	0.55	0.53	0.41	0.28	0.21
	→	−17.8	5.6	0.52	0.50	0.39	0.27	0.20	0.66	0.63	0.46	0.30	0.22
		−45.6	11.1	0.51	0.50	0.41	0.31	0.24	0.51	0.50	0.42	0.31	0.24
		−45.6	5.6	0.56	0.55	0.45	0.33	0.26	0.65	0.63	0.51	0.36	0.27
		32.2	5.6	0.44	0.41	0.29	0.19	0.14	0.62	0.58	0.37	0.21	0.15
	Down	10.0	16.7	0.47	0.45	0.33	0.22	0.16	0.66	0.62	0.42	0.25	0.18
		10.0	5.6	0.47	0.45	0.33	0.22	0.16	0.68	0.63	0.42	0.26	0.18
Horiz.	↓	−17.8	11.1	0.52	0.50	0.39	0.27	0.20	0.74	0.70	0.50	0.32	0.23
		−17.8	5.6	0.52	0.50	0.39	0.27	0.20	0.75	0.71	0.51	0.32	0.23
		−45.6	11.1	0.57	0.55	0.45	0.33	0.26	0.81	0.78	0.59	0.40	0.30
		−45.6	5.6	0.58	0.56	0.46	0.33	0.26	0.83	0.79	0.60	0.40	0.30

[a]See Chapter 25 of the 2021 *ASHRAE Handbook—Fundamentals*. Thermal resistance values were determined from $R = 1/C$, where $C = h_c + \varepsilon_{eff} h_r$; h_c is conduction/convection coefficient, $\varepsilon_{eff} h_r$ is radiation coefficient $\approx 0.227\varepsilon_{eff}[(t_m + 273)/100]^3$, and t_m is mean temperature of air space.

[b]Values apply for ideal conditions (i.e., air spaces of uniform thickness bounded by plane, smooth, parallel surfaces with no air leakage to or from the space). **This table should not be used for hollow siding or profiled cladding.**

[c]A single resistance value cannot account for multiple air spaces; each air space requires a separate resistance calculation that applies only for established boundary conditions. Resistances of horizontal spaces with heat flow downward are substantially independent of temperature difference.

[d]Interpolation is permissible for other values of mean temperature, temperature difference, and effective emittance ε_{eff}. Interpolation and moderate extrapolation for air spaces greater than 90 mm are also permissible.

[e]Effective emittance ε_{eff} of air space is given by $1/\varepsilon_{eff} = 1/\varepsilon_1 + 1/\varepsilon_2 - 1$, where ε_1 and ε_2 are emittances of surfaces of air space (see Table 2 of Chapter 26 of the 2021 *ASHRAE Handbook—Fundamentals*). **Also, oxidation, corrosion, and accumulation of dust and dirt can dramatically increase surface emittance. Emittance values of 0.05 should only be used where the highly reflective surface can be maintained over the service life of the assembly.**

Table 11.3 Surface Film Coefficients/Resistances [2021F, Ch 26, Tbl 10]

Position of Surface	Direction of Heat Flow	Surface Emittance, ε					
		Nonreflective		Reflective			
		ε = 0.90		ε = 0.20		ε = 0.05	
Indoor		h_i	R_i	h_i	R_i	h_i	R_i
Horizontal	Upward	9.26	0.11	5.17	0.19	4.32	0.23
Sloping at 45°	Upward	9.09	0.11	5.00	0.20	4.15	0.24
Vertical	Horizontal	8.29	0.12	4.20	0.24	3.35	0.30
Sloping at 45°	Downward	7.50	0.13	3.41	0.29	2.56	0.39
Horizontal	Downward	6.13	0.16	2.10	0.48	1.25	0.80
Outdoor (any position)		h_o	R_o				
Wind (for winter) at 6.7 m/s	Any	34.0	0.030	—	—	—	—
Wind (for summer) at 3.4 m/s	Any	22.7	0.044	—	—	—	—

Notes:
1. Surface conductance h_i and h_o measured in W/(m^2·K); resistance R_i and R_o in (m^2·K)/W.
2. No surface has both an air space resistance value and a surface resistance value.
3. Conductances are for surfaces of the stated emittance facing virtual blackbody surroundings at same temperature as ambient air. Values based on surface/air temperature difference of 5.6 K and surface temperatures of 21°C.
4. See Chapter 4 of the 2021 *ASHRAE Handbook—Fundamentals* for more detailed information.
5. Condensate can have significant effect on surface emittance (see Table 2 of Chapter 26 of the 2021 *ASHRAE Handbook—Fundamentals*). Also, oxidation, corrosion, and accumulation of dust and dirt can dramatically increase surface emittance. Emittance values of 0.05 should only be used where highly reflective surface can be maintained over the service life of the assembly.

Table 11.4 European Surface Film Coefficients/Resistances [2021F, Ch 26, Tbl 11]

Position of Surface	Direction of Heat Flow	h, W/(m^2·K)	R, (m^2·K)/W
Indoors			
Horizontal, sloping till 45°	Upward	10	0.1
	Downward	6	0.17
Vertical, sloping beyond 45°	Any direction	7.7	0.13
Outdoors		25	0.04

Table 11.5 Emissivity of Various Surfaces and Effective Emittances of Facing Air Spaces[a] [2021F, Ch 26, Tbl 2]

Surface	Average Emissivity ε	Effective Emittance ε_{eff} of Air Space: One Surface's Emittance ε; Other, 0.9	Effective Emittance ε_{eff} of Air Space: Both Surfaces' Emittance ε
Aluminum foil, bright	0.05	0.05	0.03
Aluminum foil, with condensate just visible (>0.5 g/m^2)	0.30[b]	0.29	—
Aluminum foil, with condensate clearly visible (>2.0 g/m^2)	0.70[b]	0.65	—
Aluminum sheet	0.12	0.12	0.06
Aluminum-coated paper, polished	0.20	0.20	0.11
Brass, nonoxidized	0.04	0.038	0.02
Copper, black oxidized	0.74	0.41	0.59
Copper, polished	0.04	0.038	0.02
Iron and steel, polished	0.2	0.16	0.11
Iron and steel, oxidized	0.58	0.35	0.41
Lead, oxidized	0.27	0.21	0.16
Nickel, nonoxidized	0.06	0.056	0.03
Silver, polished	0.03	0.029	0.015
Steel, galvanized, bright	0.25	0.24	0.15
Tin, nonoxidized	0.05	0.047	0.026
Aluminum paint	0.50	0.47	0.35
Building materials: wood, paper, masonry, nonmetallic paints	0.90	0.82	0.82
Regular glass	0.84	0.77	0.72

[a]Values apply in 4 to 40 μm range of electromagnetic spectrum. Also, oxidation, corrosion, and accumulation of dust and dirt can dramatically increase surface emittance. Emittance values of 0.05 should only be used where the highly reflective surface can be maintained over the service life of the assembly. Except as noted, data from VDI (1999).

[b]Values based on data in Bassett and Trethowen (1984).

Table 11.6 Effective Thermal Resistance of Ventilated Attics[a] (Summer Condition)

Ventilation Air Temp., °C	Sol-Air[e] Temp., °C	Not Ventilation[b]		Natural Ventilation		Power Ventilation[c]					
		Ventilation Rate, L/s per m^2									
		0		0.5		2.5		5.1		7.6	
		Ceiling Resistance R^d, $K \cdot m^2/W$									
		1.8	3.5	1.8	3.5	1.8	3.5	1.8	3.5	1.8	3.5
Part A. Nonreflective Surfaces											
27	49	0.33	0.33	0.49	0.60	1.11	1.64	1.69	2.82	1.94	3.52
	60	0.33	0.33	0.49	0.62	1.14	1.76	1.72	2.99	2.11	3.70
	71	0.33	0.33	0.49	0.63	1.18	1.94	1.76	3.17	2.29	3.87
32	49	0.33	0.33	0.44	0.49	0.81	1.18	1.07	1.76	1.21	2.29
	60	0.33	0.33	0.46	0.55	0.92	1.39	1.34	2.11	1.51	2.64
	71	0.33	0.33	0.48	0.60	1.02	1.58	1.50	2.46	1.76	2.99
38	49	0.33	0.33	0.39	0.40	0.58	0.77	0.70	1.06	0.72	1.21
	60	0.33	0.33	0.42	0.48	0.74	1.07	1.02	1.53	1.14	1.76
	71	0.33	0.33	0.46	0.56	0.88	1.34	1.27	1.94	1.46	2.29
Part B. Reflective Surfaces[f]											
27	49	1.14	1.14	1.43	1.55	2.29	2.99	2.99	4.40	3.34	5.28
	60	1.14	1.14	1.44	1.58	2.46	3.17	3.17	4.58	3.52	5.46
	71	1.14	1.14	1.46	1.62	2.64	3.17	3.34	4.75	3.70	5.63
32	49	1.14	1.14	1.32	1.41	1.76	2.29	2.11	2.99	2.29	3.34
	60	1.14	1.14	1.36	1.46	2.11	2.64	2.46	3.52	2.82	3.87
	71	1.14	1.14	1.39	1.51	2.29	2.82	2.82	3.87	3.17	4.40
38	49	1.14	1.14	1.23	1.30	1.41	1.76	1.50	2.11	1.55	2.11
	60	1.14	1.14	1.28	1.37	1.76	2.11	1.94	2.64	2.11	2.82
	71	1.14	1.14	1.34	1.44	1.94	2.46	2.29	3.17	2.64	3.52

[a]Although the term effective resistance is commonly used when there is attic ventilation, this table includes values for situations with no ventilation. The effective resistance of the attic added to the resistance (1/U) of the ceiling yields the effective resistance of this combination based on sol-air and room temperatures. These values apply to wood frame construction with a roof deck and roofing that has a conductance of 5.7 $W/(m^2 \cdot K)$.

[b]This condition cannot be achieved in the field unless extreme measures are taken to tightly seal the attic.

[c]Based on air discharging outward from attic.

[d]When determining ceiling resistance, do not add the effect of a reflective surface facing the attic, as it is accounted for in Part B of this table.

[e]Roof surface temperature rather than sol-air temperature can be used if0.04 is subtracted from the attic resistance shown.

[f]Surfaces with effective emittance $\epsilon_{eff} = 0.05$ between ceiling joists facing attic space.

Table 11.7 Building and Insulating Materials: Design Values[a] [2021F, Ch 26, Tbl 1]

Description	Density, kg/m³	Conductivity[b] k, W/(m·K)	Resistance R, (m²·K)/W	Specific Heat, kJ/(kg·K)
Insulating Materials				
Blanket and batt[c,d]				
Glass-fiber batts				0.8
	7.5 to 8.2	0.046 to 0.048	—	—
	9.8 to 12	0.040 to 0.043	—	—
	13 to 14	0.037 to 0.039	—	—
	22	0.033	—	—
Rock and slag wool batts	—	—	—	0.8
	32 to 37	0.036 to 0.037	—	—
	45	0.033 to 0.035	—	—
Mineral wool, felted	16 to 48	0.040	—	—
	16 to 130	0.035	—	—
Board and slabs				
Cellular glass	120	0.042	—	0.8
Cement fiber slabs, shredded wood with Portland cement binder	400 to 430	0.072 to 0.076	—	—
with magnesia oxysulfide binder	350	0.082	—	1.3
Glass fiber board	—	—	—	0.8
	24 to 96	0.033 to 0.035	—	—
Expanded rubber (rigid)	64	0.029	—	1.7
Extruded polystyrene, smooth skin	—	—	—	1.5
aged per CAN/ULC Standard S770-2003	22 to 58	0.026 to 0.029	—	—
aged 180 days	22 to 58	0.029		
European product	30	0.030		
aged 5 years at 24°C	32 to 35	0030	—	—
blown with low global warming potential (GWP) (<5) blowing agent		0.035 to 0.036	—	—

Table 11.7 Building and Insulating Materials: Design Values[a] [2021F, Ch 26, Tbl 1] (*Continued*)

Description	Density, kg/m^3	Conductivity[b] k, $W/(m \cdot K)$	Resistance R, $(m^2 \cdot K)/W$	Specific Heat, $kJ/(kg \cdot K)$
Expanded polystyrene, molded beads	—	—	—	1.5
	16 to 24	0.035 to 0.037	—	—
	29	0.033	—	—
Mineral fiberboard, wet felted	160	0.037	—	0.8
Rock wool board	—	—	—	0.8
floors and walls	64 to 130	0.033 to 0.036	—	—
roofing	160 to 180.	0.039 to 0.042	—	0.8
Acoustical tile[e]	340 to 370	0.052 to 0.053	—	0.6 to 0.8
Perlite board	140	0.052	—	—
Polyisocyanurate	—	—	—	1.5
unfaced, aged per CAN/ULC Standard S770-2003	26 to 37	0.023 to 0.025	—	—
with foil facers, aged 180 days	—	0.022 to 0.023	—	—
Phenolic foam board with facers, aged	—	0.020 to 0.023	—	—
Loose fill				
Cellulose fiber, loose fill	—	—	—	1.4
attic application up to 100 mm	16 to 19	0.045 to 0.046	—	—
attic application > 100 mm	19 to 26	0.039 to 0.040	—	—
wall application, dense packed	56	0.039 to 0.040	—	—
Perlite, expanded	32 to 64	0.039 to 0.045	—	1.1
	64 to 120	0.045 to 0.052	—	—
	120 to 180	0.052 to 0.061	—	—
Glass fiber[d]				
attics, ~100 to 600 mm	6.4 to 8.0	0.052 to 0.055	—	—
attics, ~600 to 1100 mm	8 to 9.6	0.049 to 0.052	—	—
closed attic or wall cavities	29 to 37	0.035 to 0.036	—	—

Table 11.7 Building and Insulating Materials: Design Values[a] [2021F, Ch 26, Tbl 1] (*Continued*)

Description	Density, kg/m^3	Conductivity[b] k, $W/(m \cdot K)$	Resistance R, $(m^2 \cdot K)/W$	Specific Heat, $kJ/(kg \cdot K)$
Rock and slag wool[d]				
attics, ~90 to 115 mm	24 to 26	0.049	—	—
attics, ~125 to 430 mm	24 to 29	0.046 to 0.048	—	—
closed attic or wall cavities	64	0.039 to 0.042	—	—
Vermiculite, exfoliated	112 to 131	0.068	—	1.3
	64 to 96	0.063	—	—
Spray-applied				
Cellulose, sprayed into open wall cavities	26 to 42	0.039 to 0.040	—	—
Glass fiber, sprayed into open wall or attic cavities	16	0.039 to 0.042	—	—
	29 to 37	0.033 to 0.037	—	—
Polyurethane foam	—	—	—	1.5
low density, open cell	7.2 to 10	0.037 to 0.042	—	—
medium density, closed cell, aged 180 days	30 to 51	0.020 to 0.029	—	—
Building Board and Siding				
Board				
Asbestos/cement board	1900	0.57	—	1.00
Cement board	1150	0.25	—	0.84
Fiber/cement board	1400	0.25	—	0.84
	1000	0.19	—	0.84
	400	0.07	—	1.88
	300	0.06	—	1.88
Gypsum or plaster board	640	0.16	—	1.15
Oriented strand board (OSB) 9 to 11 mm	650	—	0.11	1.88
12.7 mm	650	—	0.12	1.88

Table 11.7 Building and Insulating Materials: Design Values[a] [2021F, Ch 26, Tbl 1] (*Continued*)

Description		Density, kg/m^3	Conductivity[b] k, W/(m·K)	Resistance R, (m^2·K)/W	Specific Heat, kJ/(kg·K)
Plywood (douglas fir)	12.7 mm	460	—	0.14	1.88
	15.9 mm	540	—	0.15	1.88
Plywood/wood panels	19.0 mm	450	—	0.19	1.88
Vegetable fiber board		650	—	0.11	1.88
sheathing, regular density	12.7 mm	290	—	0.23	1.30
intermediate density	12.7 mm	350	—	0.19	1.30
nail-based sheathing	12.7 mm	400	—	0.19	1.30
shingle backer	9.5 mm	290	—	0.17	1.30
sound deadening board	12.7 mm	240	—	0.24	1.26
tile and lay-in panels, plain or acoustic		290	0.058	—	0.59
laminated paperboard		480	0.072	—	1.38
homogeneous board from repulped paper		480	0.072	—	1.17
Hardboard					
medium density		800	0.105	—	1.30
high density, service-tempered grade and service grade		880	0.12	—	1.34
high density, standard-tempered grade		1010	0.144	—	1.34
Particleboard					
low density		590	0.102	—	1.30
medium density		800	0.135	—	1.30
high density		1000	1.18	—	—
underlayment	15.9 mm	640	—	1.22	1.21
Waferboard		700	0.072	—	1.88
Shingles					
Asbestos/cement		1900	—	0.037	—
Wood, 400 mm, 190 mm exposure		—	—	0.15	1.30

Table 11.7 Building and Insulating Materials: Design Values[a] [2021F, Ch 26, Tbl 1] (*Continued*)

Description	Density, kg/m^3	Conductivity[b] k, $W/(m \cdot K)$	Resistance R, $(m^2 \cdot K)/W$	Specific Heat, $kJ/(kg \cdot K)$
Wood, double, 400 mm, 300 mm exposure	—	—	0.21	1.17
Wood, plus ins. backer board ... 8 mm	—	—	0.25	1.30
Siding				—
Asbestos/cement, lapped ... 6.4 mm	—	—	0.037	1.01
Asphalt roll siding	—	—	0.026	1.47
Siding				
Asphalt insulating siding (12.7 mm bed)	—	—	0.26	1.47
Hardboard siding ... 11 mm		—	—	0.12
Wood, drop, 200 mm ... 25 mm	—	—	0.14	1.17
Wood, bevel				
200 mm, lapped ... 13 mm	—	—	0.14	1.17
250 mm, lapped ... 19 mm	—	—	0.18	1.17
Wood, plywood, lapped ... 9.5 mm	—	—	0.10	1.22
Aluminum, steel, or vinyl,[h, i] over sheathing			—	
hollow-backed	—	—	0.11	1.22[i]
insulating-board-backed ... 9.5 mm	—	—	0.32	1.34
foil-backed ... 9.5 mm	—	—	0.52	—
insulated vinyl siding ... 19 to 32 mm			0.35 to 0.48[m]	
Architectural (soda-lime float) glass	2500	1.0	—	0.84
Building Membrane				
Vapor-permeable felt	—	—	0.011	—
Vapor: seal, 2 layers of mopped 0.73 kg/m^2 felt	—	—	0.21	—
Vapor: seal, plastic film	—	—	Negligible	—
Finish Flooring Materials				
Carpet and rebounded urethane pad ... 19 mm	110	—	0.42	—

Table 11.7 Building and Insulating Materials: Design Values[a] [2021F, Ch 26, Tbl 1] (*Continued*)

Description		Density, kg/m³	Conductivity[b] k, W/(m·K)	Resistance R, (m²·K)/W	Specific Heat, kJ/(kg·K)
Carpet and rubber pad (one-piece)	9.5 mm	320	—	0.12	—
Pile carpet with rubber pad	9.5 to 12.7 mm	290	—	0.28	—
Linoleum/cork tile	6.4 mm	465	—	0.09	—
PVC/rubber floor covering		—	0.40	—	—
rubber tile	25 mm	1900	—	0.06	—
terrazzo	25 mm	—	—	0.014	0.80
Metals (See 2021F, Ch 33, Tbl 3)					
Roofing					
Asbestos/cement shingles		1920	—	0.037	1.00
Asphalt (bitumen with inert fill)		1600	0.43	—	—
		1900	0.58	—	—
		2300	1.15	—	—
Asphalt roll roofing		920	—	0.027	1.51
Asphalt shingles		920	—	0.078	1.26
Built-up roofing	10 mm	920	—	0.059	1.47
Mastic asphalt (heavy, 20% grit)		950	0.19	—	—
Reed thatch		270	0.09	—	—
Roofing felt		2250	1.20	—	—
Slate	13 mm	—	—	0.009	1.26
Straw thatch		240	0.07	—	—
Wood shingles, plain and plastic-film-faced		—	—	0.166	1.30
Plastering Materials					
Cement plaster, sand aggregate		1860	0.72	—	0.84
Sand aggregate	10 mm	—	—	0.013	0.84
	20 mm	—	—	0.026	0.84

Table 11.7 Building and Insulating Materials: Design Values[a] [2021F, Ch 26, Tbl 1] (*Continued*)

Description		Density, kg/m^3	Conductivity[b] k, $W/(m \cdot K)$	Resistance R, $(m^2 \cdot K)/W$	Specific Heat, $kJ/(kg \cdot K)$
Gypsum plaster		1120	0.38	—	—
		1280	0.46	—	—
Lightweight aggregate	13 mm		720	—	0.056
	16 mm		720	—	0.066
on metal lath	19 mm	—	—	0.083	—
Perlite aggregate		720	0.22	—	1.34
Sand aggregate		1680	0.81	—	0.84
on metal lath	19 mm	—	—	0.023	—
Vermiculite aggregate		480	0.14	—	—
		600	0.20	—	—
		720	0.25	—	—
		840	0.26	—	—
		960	0.30	—	—
Perlite plaster		400	0.08	—	—
		600	0.19	—	—
Pulpboard or paper plaster		600	0.07	—	—
Sand/cement plaster, conditioned		1560	0.63	—	—
Sand/cement/lime plaster, conditioned		1440	0.48	—	—
Sand/gypsum (3:1) plaster, conditioned		1550	0.65	—	—
Masonry Materials					
Masonry units					
Brick, fired clay		2400	1.21 to 1.47	—	—
		2240	1.07 to 1.30	—	—
		2080	0.92 to 1.12	—	—
		1920	0.81 to 0.98	—	0.80

Table 11.7 Building and Insulating Materials: Design Values[a] [2021F, Ch 26, Tbl 1] (*Continued*)

Description	Density, kg/m^3	Conductivity[b] k, $W/(m \cdot K)$	Resistance R, $(m^2 \cdot K)/W$	Specific Heat, $kJ/(kg \cdot K)$
	1760	0.71 to 0.85	—	—
	1600	0.61 to 0.74	—	—
	1440	0.52 to 0.62	—	—
	1280	0.43 to 0.53	—	—
	1120	0.36 to 0.45	—	—
Clay tile, hollow				
1 cell deep .... 75 mm	—	—	0.14	0.88
.... 100 mm	—	—	0.20	—
2 cells deep .... 150 mm	—	—	0.27	—
.... 200 mm	—	—	0.33	—
.... 250 mm	—	—	0.39	—
3 cells deep .... 300 mm	—	—	0.44	—
Lightweight brick	800	0.20	—	—
	770	0.22	—	—
Concrete blocks[f, g]				
Limestone aggregate				
~200 mm, 16.3 kg, 2200 kg/m^3 concrete, 2 cores	—	—	—	—
with perlite-filled cores	—	—	0.37	—
~300 mm, 25 kg, 2200 kg/m^3 concrete, 2 cores	—		—	—
with perlite-filled cores	—	—	0.65	—
Normal-weight aggregate (sand and gravel)				
~200 mm, 16 kg, 2100 kg/m^3 concrete, 2 or 3 cores...	—	—	0.20 to 0.17	0.92
with perlite-filled cores	—	—	0.35	—
with vermiculite-filled cores	—	—	0.34 to 0.24	—
~300 mm, 22.7 kg, 2000 kg/m^3 concrete, 2 cores	—	—	0.217	0.92

Table 11.7 Building and Insulating Materials: Design Values[a] [2021F, Ch 26, Tbl 1] (*Continued*)

Description	Density, kg/m^3	Conductivity[b] k, $W/(m \cdot K)$	Resistance R, $(m^2 \cdot K)/W$	Specific Heat, $kJ/(kg \cdot K)$
Medium-weight aggregate (combinations of normal and lightweight aggregate)				
~200 mm, 13 kg, 1550 to 1800 kg/m^3 concrete, 2 or 3 cores		—	0.30 to 0.22	—
with perlite-filled cores	—	—	0.65 to 0.41	—
with vermiculite-filled cores	—	—	0.58	—
with molded-EPS-filled (beads) cores	—	—	0.56	—
with molded EPS inserts in cores	—	—	0.47	—
Low-mass aggregate (expanded shale, clay, slate or slag, pumice)				
~150 mm, 7 1/2 kg, 1400 kg/m^2 concrete, 2 or 3 cores	—	—	0.34 to 0.29	—
with perlite-filled cores	—	—	0.74	—
with vermiculite-filled cores	—	—	0.53	—
200 mm, 8 to 10 kg, 1150 to 1380 kg/m^2 concrete	—	—	0.56 to 0.33	0.88
with perlite-filled cores	—	—	1.20 to 0.77	—
with vermiculite-filled cores	—	—	0.93 to 0.69	—
with molded-EPS-filled (beads) cores	—	—	0.85	—
with UF foam-filled cores	—	—	0.79	—
with molded EPS inserts in cores	—	—	0.62	—
300 mm, 16 kg, 1400 kg/m^3, concrete, 2 or 3 cores	—	—	0.46 to 0.40	—
with perlite-filled cores	—	—	1.6 to 1.1	—
with vermiculite-filled cores	—	—	1.0	—
Stone, lime, or sand	2880	10.4	—	—
Quartzitic and sandstone	2560	6.2	—	—
	2240	3.46	—	—
	1920	1.88	—	0.88
Calcitic, dolomitic, limestone, marble, and granite	2880	4.33	—	—
	2560	3.17	—	—

Table 11.7 Building and Insulating Materials: Design Values[a] [2021F, Ch 26, Tbl 1] (*Continued*)

Description	Density, kg/m³	Conductivity[b] k, W/(m·K)	Resistance R, (m²·K)/W	Specific Heat, kJ/(kg·K)
	2240	2.31	—	—
	1920	1.59	—	0.88
	1600	1.15	—	—
Gypsum partition tile				
75 by 300 by 760 mm, solid	—	—	0.222	0.79
4 cells	—	—	0.238	—
100 by 300 by 760 mm, 3 cells	—	—	0.294	—
Limestone	2400	0.57	—	0.84
	2600	0.93	—	0.84
Concretes[i]				
Sand and gravel or stone aggregate concretes	2400	1.4 to 2.9	—	—
(concretes with >50% quartz or quartzite sand have conductivities	2240	1.3 to 2.6	—	0.80 to 1.00
in higher end of range)	2080	1.0 to 1.9	—	—
Low-mass aggregate or limestone concretes	1920	0.9 to 1.3	—	—
expanded shale, clay, or slate; expanded slags; cinders;	1600	0.68 to 0.89	—	0.84
pumice (with density up to 1600 kg/m³); scoria (sanded	1280	0.48 to 0.59	—	0.84
concretes have conductivities in higher end of range)	960	0.30 to 0.36	—	—
	640	0.18	—	—
Gypsum/fiber concrete (87.5% gypsum, 12.5% wood chips)	800	0.24	—	0.84
Cement/lime, mortar, and stucco	1920	1.40	—	—
	1600	0.97	—	—
	1280	0.65	—	—
Perlite, vermiculite, and polystyrene beads	800	0.26 to 0.27	—	—
	640	0.20 to 0.22	—	0.63 to 0.96
	480	0.16	—	—

Table 11.7 Building and Insulating Materials: Design Values[a] [2021F, Ch 26, Tbl 1] (*Continued*)

Description	Density, kg/m^3	Conductivity[b] k, $W/(m \cdot K)$	Resistance R, $(m^2 \cdot K)/W$	Specific Heat, $kJ/(kg \cdot K)$
	320	0.12	—	—
Foam concretes	1920	0.75	—	—
	1600	0.60	—	—
	1280	0.44	—	—
	1120	0.36	—	—
Foam concretes and cellular concretes	960	0.30	—	—
	640	0.20	—	—
	320	0.12	—	—
Aerated concrete (oven-dried)	430 to 800	0.20	—	0.84
Polystyrene concrete (oven-dried)	255 to 800	0.37	—	0.84
Polymer concrete	1950	1.64	—	—
	2200	1.03	—	—
Polymer cement	1870	0.78	—	—
Slag concrete	960	0.22	—	—
	1280	0.32	—	—
	1600	0.43	—	—
	2000	1.23	—	—
Woods (12% moisture content)[j]				
Hardwoods	—	—	—	*1.63[n]*
Oak	660 to 750	0.16 to 0.18	—	—
Birch	680 to 725	0.17 to 0.18	—	—
Maple	635 to 700	0.16 to 0.17	—	—
Ash	615 to 670	0.15 to 0.16	—	—
Softwoods	—	—	—	*1.63[n]*
Southern pine	570 to 660	0.14 to 0.16	—	—

Table 11.7 Building and Insulating Materials: Design Values[a] [2021F, Ch 26, Tbl 1] (*Continued*)

Description	Density, kg/m^3	Conductivity[b] k, $W/(m \cdot K)$	Resistance R, $(m^2 \cdot K)/W$	Specific Heat, $kJ/(kg \cdot K)$
Southern yellow pine	500	0.13	—	—
Eastern white pine	400	0.10	—	—
Douglas fir/larch	535 to 580	0.14 to 0.15	—	—
Southern cypress	500 to 515	0.13	—	—
Hem/fir, spruce/pine/fir	390 to 500	0.11 to 0.13	—	—
Spruce	400	0.09	—	—
Western red cedar	350	0.09	—	—
West coast woods, cedars	350 to 500	0.10 to 0.13	—	—
Eastern white cedar	360	0.10	—	—
California redwood	390 to 450	0.11 to 0.12	—	—
Pine (oven-dried)	370	0.092	—	1.88
Spruce (oven-dried)	395	0.10	—	1.88

Notes for Table 11.7:

[a]Values are for mean temperature of 24°C. Representative values for dry materials are intended as design (not specification) values for materials in normal use. Thermal values of insulating materials may differ from design values depending on in situ properties (e.g., density and moisture content, orientation, etc.) and manufacturing variability. For properties of specific product, use values supplied by manufacturer or unbiased tests.

[b]Symbol λ also used to represent thermal conductivity.

[c]Does not include paper backing and facing, if any. Where insulation forms boundary (reflective or otherwise) of airspace, see Tables 2 and 3 of Chapter 26 of the 2021 *ASHRAE Handbook—Fundamentals* for insulating value of airspace with appropriate effective emittance and temperature conditions of space.

[d]Conductivity varies with fiber diameter (see Chapter 25 of the 2021 *ASHRAE Handbook—Fundamentals*). Batt, blanket, and loose-fill mineral fiber insulations are manufactured to achieve specified R-values, the most common of which are listed in the table. Because of differences in manufacturing processes and materials, the product thicknesses, densities, and thermal conductivities vary over considerable ranges for a specified R-value.

[e]Values are for aged products with gas-impermeable facers on the two major surfaces. An aluminum foil facer of 25 μm thickness or greater is generally considered impermeable to gases. For change in conductivity with age of expanded polyisocyanurate, see SPI Bulletin U108.

[f]Cellular phenolic insulation may no longer be manufactured.

[g]Insulating values of acoustical tile vary, depending on density of board and on type, size, and depth of perforations.

[h]Values for fully grouted block may be approximated using values for concrete with similar unit density.

[i]Values for concrete block and concrete are at moisture contents representative of normal use.

[j]Values for metal or vinyl siding applied over flat surfaces vary widely, depending on ventilation of the airspace beneath the siding; whether airspace is reflective or nonreflective; and on thickness, type, and application of insulating backing-board used. Values are averages for use as design guides, and were obtained from several guarded hot box tests (ASTM Standard C1363) on hollow-backed types and types made using backing of wood fiber, foamed plastic, and glass fiber. Departures of ±50% or more from these values may occur.

[k]Vinyl specific heat = 1.0 kJ/(kg·K)

[l]See Adams (1971), MacLean (1941), and Wilkes (1979). Conductivity values listed are for heat transfer across the grain. Thermal conductivity of wood varies linearly with density, and density ranges listed are those normally found for wood species given. If density of wood species is not known, use mean conductivity value. For extrapolation to other moisture contents, the following empirical equation developed by Wilkes (1979) may be used:

$$k = 0.1791 + \frac{(1.874 \times 10^{-2} + 5.753 \times 10^{-4} M)\rho}{1 + 0.01M}$$

where ρ is density of moist wood in kg/m³, and M is moisture content in percent.

[m]Dimension referenced is taken at the maximum siding profile thickness. The range of R values and associated thicknesses represent values for products tested to ASTM Standard D7793, which requires applying 6.7 m/s airstream perpendicular to surface of siding during testing.

[n]From Wilkes (1979), an empirical equation for specific heat of moist wood at 24°C is as follows:

$$c_p = \frac{(0.299 + 0.01M)}{(1 + 0.01M)} + \Delta c_p$$

where Δc_p accounts for heat of sorption and is denoted by $\Delta c_p = M(1.921 \times 10^{-3} - 3.168 \times 10^{-5} M)$ where M is moisture content in percent by mass.

[o]Blank space in reference column indicates historical values from previous volumes of ASHRAE Handbook. Source of information could not be determined.

Cooling Load Temperature Differences (CLTDs)

Table 11.8 CLTDs (°C) for Flat Roofs—40°N Latitude, July

Roof No.	Solar time, h											
	1	2	3	4	5	6	7	8	9	10	11	12
1	0	−1	−2	−3	−3	−3	0	7	16	25	33	41
2	1	0	−1	−2	−3	−3	−2	2	9	18	27	34
3	7	4	3	1	0	−1	0	3	7	13	19	26
4	9	6	4	2	1	−1	−2	−2	0	4	9	16
5	12	9	7	4	3	2	1	1	3	7	12	17
8	16	13	12	9	8	7	6	6	7	9	12	16
9	18	14	12	9	7	5	3	2	2	4	7	11
10	21	18	15	13	11	8	7	6	5	6	7	9
13	19	17	16	14	12	11	10	9	9	9	11	13
14	19	18	17	15	14	13	12	11	11	11	12	13

Roof No.	Solar time, h											
	13	14	15	16	17	18	19	20	21	22	23	24
1	46	49	49	46	41	33	24	14	8	5	3	1
2	41	46	48	47	44	39	31	22	14	8	5	3
3	32	37	40	41	41	37	33	27	21	17	13	9
4	23	30	36	41	43	43	41	37	31	25	19	13
5	23	28	33	37	38	38	36	33	28	23	19	15
8	19	23	27	29	31	32	31	29	27	24	21	18
9	15	20	25	29	33	35	36	35	32	29	25	21
10	13	17	21	24	28	31	32	32	31	29	26	23
13	16	18	21	23	26	27	27	27	26	24	22	21
14	16	18	20	22	23	24	25	25	24	23	22	21

CAUTION: Approximate data—Use for preliminary computations only. Also, see notes on next page.

Notes for CLTD Data for Flat Roofs

1. Data apply directly to (1) dark surface, (2) indoor temperature is 25.5°C, (3) outdoor maximum temperature of 35°C with mean temperature of 29.4°C and daily range of 11.6°C, (4) solar radiation typical of clear day on 21st day of month, (5) outside surface film resistance of 0.059 $m^2 \cdot K/W$, and (6) inside surface resistance of 0.121 $m^2 \cdot K/W$.

2. Adjustments to design temperatures

$$\text{Corr. CLTD} = \text{CLTD} + (25.5 - t_r) + (t_m - 25.5) \qquad (11.5)$$

where t_r = inside temperature and t_m = mean outdoor temperature, or t_m = maximum outdoor temperature − (daily range)/2.

No adjustment recommended for color or for ventilation of air space above a ceiling.

For design purposes, the data suffice for plus or minus 2 weeks from the 21st day of given month.

Table 11.9 Roof Classifications for Use with CLTD Tables for Flat Roofs

Mass Location	Suspended Ceiling	*R*, $m^2 \cdot K/W$	Wood 25 mm	50 mm Heavy Concrete	Steel Deck	Attic Ceiling Comb.
Mass inside insul.	Without	0 to 1.76	*	2	*	*
		1.76 to 3.52	*	4	*	*
		3.52 to 4.40	*	5	*	*
	With	0 to 0.88	*	5	*	*
		0.88 to 1.76	*	8	*	*
		1.76 to 3.52	*	13	*	*
		3.52 to 4.40	*	14	*	*
Mass evenly placed	Without	0 to 0.88	1	2	1	1
		0.88 to 2.64	2	*	1	2
		2.64 to 4.40	4	*	2	2
	With	0 to 0.88	*	3	1	*
		0.88 to 1.76	4	*	1	*
		1.76 to 2.64	5	*	2	
		2.64 to 3.52	9	*	2	*
		3.52 to 4.40	10	*	4	*
Mass outside insul.	Without	0 to 0.88	*	2	*	*
		0.88 to 1.76	*	3	*	*
		1.76 to 2.64	*	4	*	*
		2.64 to 4.40	*	5	*	*
	With	0 to 1.76	*	3	*	*
		1.76 to 2.64	*	4	*	*
		2.64 to 3.52	*	5	*	*

*Denotes roof that is not possible with the chosen parameters

Table 11.10 CLTDs (°C) for Sunlit Walls—40°N Latitude, July

CAUTION: Approximate data—Use for preliminary computations only.

Solar time, h

Low Mass, Low R-Value Wall (No. 1)

Wall Facing	6	7	8	9	10	11	12	13	14	15	16	17	18	19	20	21	22
N	−1	4	6	6	7	9	12	14	15	16	16	16	16	15	9	6	4
NE	1	13	23	26	24	19	16	15	16	16	16	15	13	11	8	6	4
E	1	16	28	34	36	33	27	20	17	17	17	16	14	11	8	6	4
SE	0	8	18	26	31	32	31	27	22	18	17	16	14	11	8	6	4
S	−1	0	2	6	12	18	24	28	29	28	24	19	15	11	8	6	4
SW	−1	0	2	4	7	9	14	22	29	36	39	38	34	25	13	7	4
W	−1	1	2	4	7	9	12	15	23	33	41	44	44	34	18	9	5
NW	−1	0	2	4	7	9	12	14	16	21	28	34	36	31	16	8	5

Low Mass, Medium R-Value Wall (No. 4)

Wall Facing	6	7	8	9	10	11	12	13	14	15	16	17	18	19	20	21	22
N	0	0	1	2	3	4	6	7	9	11	12	13	14	15	15	14	12
NE	0	0	2	7	12	16	18	18	18	17	17	17	16	16	14	13	11
E	1	1	3	8	15	21	25	27	26	24	22	21	19	18	16	14	12
SE	1	0	1	4	9	15	20	24	26	26	24	23	21	19	17	14	12
S	1	0	−1	0	1	3	7	11	16	19	23	24	23	22	19	17	13
SW	1	0	0	0	1	3	4	7	10	15	20	26	29	32	32	28	23
W	1	1	0	1	1	3	4	6	8	12	17	22	28	33	36	33	28
NW	1	0	0	0	1	2	4	6	8	11	13	17	21	25	27	27	23

Low Mass, High R-Value Wall (No. 2)

Wall Facing	6	7	8	9	10	11	12	13	14	15	16	17	18	19	20	21	22
N	−1	−1	2	4	5	6	8	10	12	13	14	15	16	16	15	12	9
NE	−1	1	7	14	20	22	21	18	17	16	16	16	16	14	13	10	8
E	−1	1	8	18	26	31	32	29	24	21	19	18	17	15	13	11	8
SE	−1	0	4	11	18	24	28	29	28	25	22	19	17	16	13	11	8
S	−1	−1	−1	1	4	8	13	18	23	26	27	26	22	18	15	12	8
SW	−1	−1	0	1	3	5	7	11	17	23	29	34	36	34	29	22	15
W	−1	−1	0	1	3	5	7	9	13	18	26	33	38	41	37	28	19
NW	−1	−1	−1	1	3	5	7	9	12	14	18	23	28	32	30	23	16

Table 11.10 CLTDs (°C) for Sunlit Walls—40°N Latitude, July *(Continued)*

CAUTION: Approximate data—Use for preliminary computations only.

Wall Facing	High Mass, Low R-Value Wall (No. 5)																
	6	**7**	**8**	**9**	**10**	**11**	**12**	**13**	**14**	**15**	**16**	**17**	**18**	**19**	**20**	**21**	**22**
N	2	1	2	3	3	4	5	7	8	9	11	12	13	13	13	13	12
NE	2	2	4	8	11	14	15	16	16	16	16	16	16	15	14	13	12
E	2	2	4	9	14	18	22	22	22	21	21	19	19	18	16	14	13
SE	2	2	3	6	10	14	18	21	22	22	21	21	19	18	17	15	13
S	2	2	1	2	2	4	7	11	14	17	19	20	20	19	18	16	13
SW	3	3	2	2	3	3	5	7	9	14	18	22	26	27	27	24	21
W	4	3	2	2	3	4	5	6	8	11	16	21	25	29	30	28	24
NW	3	2	2	2	2	3	4	6	8	9	12	15	19	22	23	22	19

Wall Facing	High Mass, Medium R-Value Wall (No. 11)																
	6	**7**	**8**	**9**	**10**	**11**	**12**	**13**	**14**	**15**	**16**	**17**	**18**	**19**	**20**	**21**	**22**
N	5	4	4	4	4	4	5	6	6	7	8	8	9	10	11	11	11
NE	6	5	5	6	8	9	11	12	12	13	13	13	13	14	14	13	13
E	7	6	6	7	9	12	14	16	17	17	17	17	17	17	17	16	15
SE	7	6	6	6	8	9	12	13	15	16	17	17	17	17	17	16	15
S	6	6	5	4	4	4	6	7	9	11	13	14	15	16	16	15	14
SW	9	8	7	6	6	6	6	7	8	9	12	14	17	18	20	20	19
W	10	9	8	7	7	6	7	7	7	8	11	13	16	18	21	22	21
NW	8	7	6	6	5	5	6	6	7	7	8	10	12	14	16	17	17

Wall Facing	High Mass, High R-Value Wall (No. 16)																
	6	**7**	**8**	**9**	**10**	**11**	**12**	**13**	**14**	**15**	**16**	**17**	**18**	**19**	**20**	**21**	**22**
N	6	6	5	4	4	4	4	4	5	6	6	7	8	9	9	10	11
NE	7	6	6	6	6	7	8	9	11	12	12	13	13	13	14	14	13
E	8	7	6	6	7	8	11	12	14	16	17	17	17	18	18	17	17
SE	8	7	6	6	6	7	8	10	12	14	15	16	17	17	17	17	17
S	8	7	6	5	4	4	4	5	6	8	9	11	13	14	15	15	15
SW	11	10	8	7	7	6	6	6	6	7	8	10	12	15	17	18	19
W	12	11	9	8	7	7	6	6	6	7	8	9	11	14	17	19	21
NW	10	9	8	7	6	6	5	5	6	6	7	8	9	11	13	15	16

Table 11.11 Solar Cooling Load (W/m^2) for Sunlit Glass—40° N Latitude, July

CAUTION: Approximate data—Use for preliminary computations only.

Glass	Solar Time, h																		
Facing	5	6	7	8	9	10	11	12	13	14	15	16	17	18	19	20	21	22	23
N	3	79	85	88	101	110	120	126	126	123	113	98	98	113	38	19	9	3	3
NE	6	268	406	422	353	236	173	151	139	126	117	101	82	57	22	9	6	3	0
E	6	293	495	583	576	485	334	211	167	142	123	104	82	57	22	9	6	3	0
SE	3	148	299	413	473	473	413	306	198	154	129	107	85	57	22	9	6	3	0
S	0	28	54	79	129	202	268	306	302	265	198	132	98	63	25	13	6	3	0
SW	0	28	54	76	95	110	123	202	318	419	476	479	419	293	110	54	25	13	6
W	0	28	54	76	95	110	120	126	205	359	498	589	605	491	180	85	41	19	9
NW	0	28	54	76	95	110	120	126	126	158	265	381	450	410	145	69	35	16	9
Hor	0	76	217	378	532	665	759	810	816	772	684	554	394	221	91	44	22	9	6

Apply data directly to standard double strength glass with no inside shade and clear sky on 21st day of month.

Adjustments to table data:
- For design purposes, data will suffice for plus or minus 2 weeks from the 21st day of July.
- For other types of glass and internal shade, multiply values by appropriate shading coefficients.
- For externally shaded glass, use north orientation values.

Table 11.12 Shading Coefficients* for Glass without or with Interior Shading by Venetian Blinds

Type of Glass	Nominal Thickness Each Pane[a], mm	No Interior Shading		Venetian Blinds[b]	
		$h_o = 22.7$	$h_o = 17.0$	Medium	Light
Single Glass					
Single					
Clear	2.4 to 6	1.00	1.00		
Clear	6 to 13	0.94	0.95		
Clear	9.5	0.90	0.92	0.64	0.55
Clear	13	0.87	0.88		
Clear pattern	3 to 7	0.83	0.85		
Heat absorbing pattern	3	0.83	0.85		
Heat absorbing[c]	5.8 to 6.4	0.69	0.73		
Heat absorbing pattern	5.8 to 6.4	0.69	0.73	0.57	0.53
Tinted	3 to 5.6	0.69	0.73		
Heat absorbing[c]	9.5	0.60	0.64	0.54	0.52
Reflective coated glass		0.30		0.25	0.23
		0.40		0.33	0.29
		0.50		0.42	0.38
		0.60		0.50	0.44
Insulating Glass					
Double[d]					
Clear out, clear in	2.4 to 3.2	0.88	0.88	0.57	0.51
Clear out, clear in	6	0.81	0.82		
Heat absorbing out, clear in	6	0.55	0.58	0.39	0.36
Reflective coated glass		0.20		0.19	0.18
		0.30		0.27	0.26
		0.40		0.34	0.33
Triple					
Clear	6	0.71			
Clear	3	0.80			

[a]Refer to manufacturer's literature for values.
[b]For vertical blinds with opaque white and beige louvers in the tightly closed position. SC is 0.25 and 0.29 when used with glass of 0.71 to 0.80 transmittance.
[c]Refers to gray, bronze, and green tinted heat-absorbing glass.
[d]Refers to factory-fabricated units with 5, 6, or 13 mm air space or to prime windows plus storm windows.

* Note: Shading coefficient (SC) has been superseded by solar heat gain coefficient (SHGC) including the effect of incident angle of solar radiation on the glass, and the effect of type of framing. This shading coefficient table is sufficiently accurate for the approximate cooling load calculations of this publication. For the glazing portion of single-pane clear and tinted fenestration, SC = SHGC/0.87. This does not include frame effects.

Table 11.13 Representative Rates at Which Heat and Moisture are Given Off by Human Beings in Different States of Activity [2021F, Ch 18, Tbl 1]

Degree of Activity	Location	Total Heat, W		Sensible Heat, W	Latent Heat, W	% Sensible Heat that is Radiant[b]	
		Adult Male	Adjusted, M/F[a]			Low V	High V
Seated at theater	Theater, matinee	115	95	65	30		
Seated at theater, night	Theater, night	115	105	70	35	60	27
Seated, very light work	Offices, hotels, apartments	130	115	70	45		
Moderately active office work	Offices, hotels, apartments	140	130	75	55		
Standing, light work; walking	Department store; retail store	160	130	75	55	58	38
Walking, standing	Drug store, bank	160	145	75	70		
Sedentary work	Restaurant[c]	145	160	80	80		
Light bench work	Factory	235	220	80	140		
Moderate dancing	Dance hall	265	250	90	160	49	35
Walking 4.8 km/h; light machine work	Factory	295	295	110	185		
Bowling[d]	Bowling alley	440	425	170	255		
Heavy work	Factory	440	425	170	255	54	19
Heavy machine work; lifting	Factory	470	470	185	285		
Athletics	Gymnasium	585	525	210	315		

Notes:
1. Tabulated values are based on 24°C room dry-bulb temperature. For 27°C room dry bulb, total heat remains the same, but sensible heat values should be decreased by approximately 20%, and latent heat values increased accordingly.
2. Also see Table 4 of Chapter 9 of the 2021 *ASHRAE Handbook—Fundamentals* for additional rates of metabolic heat generation.
3. All values are rounded to nearest 5 W.

[a] Adjusted heat gain is based on normal percentage of men, women, and children for the application listed, and assumes that gain from an adult female is 85% of that for an adult male, and gain from a child is 75% of that for an adult male.
[b] Values approximated from data in Table 6 of Chapter 9 of the 2021 *ASHRAE Handbook—Fundamentals*, where V is air velocity.
[c] Adjusted heat gain includes 18 W for food per individual (9 W sensible and 9 W latent).
[d] Figure one person per alley actually bowling, and all others as sitting (117 W) or standing or walking slowly (231 W).

Heat Gain from Lighting

Because lighting is often a major space cooling load component, an accurate estimate of the space heat gain it imposes is needed. Calculation of this load component is not straightforward; the rate of cooling load from lighting at any given moment can be quite different from the heat equivalent of power supplied instantaneously to those lights, because of heat storage.

Instantaneous rate of heat gain from lights, q_{el} kW:

$$q_{el} = WF_{ul}F_{sa} \tag{11.6}$$

where

W = total light wattage installed

F_{ul} = lighting use factor (proportion in use)

F_{sa} = lighting special allowance factor

The **total light wattage** is obtained from the ratings of all lamps installed, both for general illumination and for display use. Ballasts are not included, but are addressed by a separate factor. Wattages of magnetic ballasts are significant; the energy consumption of high-efficiency electronic ballasts might be insignificant compared to that of the lamps.

The **lighting use factor** is the ratio of wattage in use, for the conditions under which the load estimate is being made, to total installed wattage. For commercial applications such as stores, the use factor is generally 1.0.

The **special allowance factor** is the ratio of the lighting fixtures' power consumption, including lamps and ballast, to the nominal power consumption of the lamps. For incandescent lights, this factor is 1. For fluorescent lights, it accounts for power consumed by the ballast as well as the ballast's effect on lamp power consumption. The special allowance factor can be less than 1 for electronic ballasts that lower electricity consumption below the lamp's rated power consumption. Use manufacturers' values for system (lamps + ballast) power, when available.

For high-intensity-discharge lamps (e.g. metal halide, mercury vapor, high- and low-pressure sodium vapor lamps), the actual lighting system power consumption should be available from the manufacturer of the fixture or ballast. Ballasts available for metal halide and high pressure sodium vapor lamps may have special allowance factors from about 1.3 (for low-wattage lamps) down to 1.1 (for high-wattage lamps).

An alternative procedure is to estimate the lighting heat gain on a per square metre basis. Such an approach may be required when final lighting plans are not available. Table 11.14 shows the maximum lighting power density (LPD) (lighting heat gain per square metre) allowed by ANSI/ASHRAE/IES Standard 90.1 for a range of space types.

Table 11.14 Lighting Power Densities Using Space-by-Space Method
[2021F, Ch 18, Tbl 2]

Common Space Types*	LPD, W/m^2
Atrium	
<6.1 m in height	4.6
≥6.1 m and ≤12.2 m in height	5.2
>12.2m in height	6.5
Audience Seating Area	
Auditorium	6.5
Gymnasium	2.5
Motion picture theater	2.9
Penitentiary	7.2
Performing arts theater	12.5
Religious building	7.8
Sports arena	3.5
All other audience seating areas	2.5
Banking Activity Area	6.5
Breakroom (*See* Lounge/Breakroom)	
Classroom/Lecture Hall/Training Room	
Penitentiary	9.5
All other classrooms/lecture halls/ training rooms	7.6
Conference/Meeting/Multipurpose Room	10.5
Confinement Cells	7.5
Copy/Print Room	3.3
Corridor[b]	
Facility for visually impaired (and not used primarily by staff)[c]	7.7
Hospital	7.7
All other corridors	4.4
Courtroom	12.9
Computer Room	10.1
Dining Area	
Penitentiary	4.5
Facility for visually impaired (and not used primarily by staff)[c]	13.7
Bar/lounge or leisure dining	9.3
Cafeteria or fast food dining	4.3
Family dining	6.5
All other dining areas	4.7
Electrical/Mechanical Room[f]	4.6
Emergency Vehicle Garage	5.6
Food Preparation Area	11.7

Common Space Types[a]	LPD, W/m^2
Guest Room	4.4
Laboratory	
In or as classroom	11.9
All other laboratories	14.3
Laundry/Washing Area	5.7
Loading Dock, Interior	9.5
Lobby	
Facility for the visually impaired (and not used primarily by staff)[c]	18.2
Elevator	7.0
Hotel	5.4
Motion picture theater	2.5
Performing arts theater	13.5
All other lobbies	9.0
Locker Room	5.6
Lounge/Breakroom	
Health care facility	4.5
All other lounges/breakrooms	6.3
Office	
Enclosed and ≤23.2 m^2	8.0
Enclosed and >23.2 m^2	7.1
Open plan	6.6
Parking Area, Interior	1.6
Pharmacy Area	17.9
Restroom	
Facility for the visually impaired (and not used primarily by staff)[c]	13.5
All other restrooms	6.8
Sales Area[d]	11.3
Seating Area, General	2.5
Stairway	
Space containing stairway determines LPD and control requirements for stairway.	
Stairwell	5.3
Storage Room	
<4.6 m^2	5.5
>4.6 m^2	4.1
Vehicular Maintenance Area	6.5
Workshop	13.5

Table 11.14 Lighting Power Densities Using Space-by-Space Method [2021F, Ch 18, Tbl 2] *(Continued)*

Building-Specific Space Types*	LPD, W/m²
Facility for Visually Impaired[c]	
Chapel (used primarily by residents)	7.5
Recreation room/common living room (and not used primarily by staff)	19.0
Automotive (*See* **Vehicular Maintenance Area**)	
Convention Center: Exhibit Space	6.6
Dormitory/Living Quarters	5.4
Fire Station: Sleeping Quarters	2.5
Gymnasium/Fitness Center	
Exercise area	9.6
Playing area	9.2
Health Care Facility	
In exam/treatment room	15.1
In imaging room	10.1
Medical supply room	6.7
Nursery	9.9
Nurses' station	12.6
Operating room	24.3
Patient room	7.3
Physical therapy room	9.8
Recovery room	13.5
Library	
Reading area	10.3
Stacks	12.7
Manufacturing Facility	
Detailed manufacturing area	8.6
Equipment room	8.2

Building-Specific Space Types*	LPD, W/m²
Extra-high-bay area (15.2 m floor-to-ceiling height)	15.3
High-bay area (7.6 to 15.2 m floor-to-ceiling height)	13.4
Low-bay area (<7.6 m floor-to-ceiling height)	9.3
Museum	
General exhibition area	3.3
Restoration room	11.9
Performing Arts Theater, Dressing Room	4.4
Post Office, Sorting Area	8.1
Religious Buildings	
Fellowship hall	5.8
Worship/pulpit/choir area	9.2
Retail Facilities	
Dressing/fitting room	5.4
Mall concourse	8.8
Sports Arena, Playing Area[g]	
Class I facility	31.6
Class II facility	21.6
Class III facility	13.9
Class IV facility	9.3
Transportation Facility	
Baggage/carousel area	4.2
Airport concourse	2.7
Ticket counter	5.5
Warehouse—Storage Area	
Medium to bulky, palletized items	3.6
Smaller, hand-carried items[e]	7.4

Source: ANSI/ASHRAE/IES Standard 90.1-2019, Table 9.6.1.

[a]In cases where both a common space type and a building-specific type are listed, the building-specific space type applies.

[b]In corridors, the extra lighting power density allowance is granted when corridor width is <2.4 m and is not based on room/corridor ratio (RCR).

[c]A "Facility for the Visually Impaired" is a facility that can be documented as being designed to comply with light levels in ANSI/IES RP-28 and is licensed or will be licensed by local/state authorities for either senior long-term care, adult daycare, senior support, and/or people with special visual needs.

[d]For accent lighting, see Section 9.6.2(b) of ASHRAE/IES Standard 90.1-2019.

[e]Sometimes called a "Picking Area".

[f]An additional 5.6 W/m² shall be allowed, provided that the additional lighting is controlled separately from the base allowance of 4.6 W/m². The additional 5.6 W/m² allowance shall not be used for any other purpose.

[g]Class of play as defined by IES RP-6.

Table 11.15 provides a range of design data under typical operating conditions: airflow 5 L/s·m², supply air between 15°C and 18°C, room temperature between 22°C and 24°C, and lighting heat input in a range from 10 to 28 W/m². For a fluorescent luminaire without lens, Figures 11.1 gives more precise data. The data should be used with judgment.

Table 11.15 Lighting Heat Gain Parameters for Typical Operating Conditions [2021F, Ch 18, Tbl 3]

Luminaire Category	Space Fraction	Radiative Fraction	Notes
Recessed fluorescent luminaire without lens	0.64 to 0.74	0.48 to 0.68	• Use middle values in most situations • May use higher space fraction, and lower radiative fraction for luminaire with side-slot returns • May use lower values of both fractions for direct/indirect luminaire • May use higher values of both fractions for ducted returns
Recessed fluorescent luminaire with lens	0.40 to 0.50	0.61 to 0.73	• May adjust values in the same way as for recessed fluorescent luminaire without lens
Downlight compact fluorescent luminaire	0.12 to 0.24	0.95 to 1.0	• Use middle or high values if detailed features are unknown • Use low value for space fraction and high value for radiative fraction if there are large holes in luminaire's reflector
Downlight incandescent luminaire	0.70 to 0.80	0.95 to 1.0	• Use middle values if lamp type is unknown • Use low value for space fraction if standard lamp (i.e. A-lamp) is used • Use high value for space fraction if reflector lamp (i.e. BR-lamp) is used
Non-in-ceiling fluorescent luminaire	1.0	0.5 to 0.57	• Use lower value for radiative fraction for surface-mounted luminaire • Use higher value for radiative fraction for pendant luminaire
Recessed LED troffer partial aperture diffuser	0.49 to 0.64	0.37 to 0.47	• Use middle value in most cases • May use higher space fraction for ducted return configuration and lower space fraction for high supply air temperature • May use higher radiant value for ducted return configuration and lower value for large supply airflow rate
Recessed LED troffer uniform diffuser	0.44 to 0.66	0.32 to 0.41	• Use middle value in most cases. • May use higher space fraction for smaller supply airflow rate and lower value for larger supply airflow rate. • May use higher radiant value for ducted return configuration and lower value for larger supply airflow rate.
Recessed high-efficacy LED troffer	0.59	0.51	

Table 11.15 Lighting Heat Gain Parameters for Typical Operating Conditions [2021F, Ch 18, Tbl 3] ***(Continued)***

Luminaire Category	Space Fraction	Radiative Fraction	Notes
Recessed LED downlight	0.40 to 0.56	0.15 to 0.18	• Use middle value in most cases. • May use higher space fraction value for high supply air temperature and lower value for smaller air flowrate. • May use higher radiant value for dimming control and lower value for large supply air flowrate.
Recessed LED retrofit kit 2×4	0.41 to 0.53	0.31 to 0.42	• Use middle value in most cases. • May use higher space fraction value for large supply air flowrate and lower value for ducted return configuration. • May use higher radiant value for ducted return configuration and lower value for larger supply airflow rate
Recessed LED color tuning fixture	0.53 to 0.56	0.40 to 0.42	Use middle value in most cases.
High-bay LED fixture	1.0	0.42 to 0.51	Use middle value in most cases.
Linear pendant LED fixture	1.0	0.55 to 0.60	Use middle value in most cases.

Sources: Fisher and Chantrasrisalai (2006); Zhou et al. (2016).

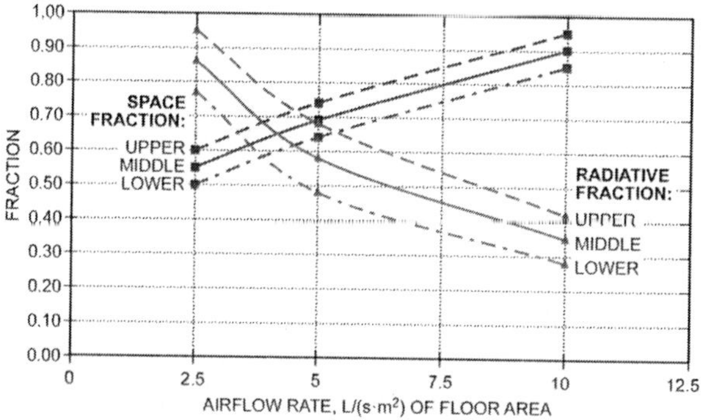

Figure 11.1 Lighting Heat Gain Parameters for Recessed Fluorescent Luminaire Without Lens [2021F, Ch 18, Fig 3]

Heat Gain from Motors and their Loads

Instantaneous rate of heat gain from equipment operated by electric motors within a conditioned space:

$$q_{em} = (P/E_M)\, F_{UM}\, F_{LM} \tag{11.7}$$

where

q_{em}	=	heat equivalent of equipment operation, kW
P	=	motor power rating, kW
E_M	=	motor efficiency, decimal fraction < 1.0
F_{UM}	=	motor use factor, 1.0 or decimal fraction <1.0
F_{LM}	=	motor load factor, 1.0 or decimal fraction <1.0

When motor is outside the conditioned space or airstream but load is inside:

$$q_{em} = P\, F_{UM}\, F_{LM} \tag{11.8}$$

When motor is inside the conditioned space or airstream but load is outside:

$$q_{em} = P\left(\frac{1.0 - E_M}{E_M}\right) F_{UM}\, F_{LM} \tag{11.9}$$

Heat output of a motor is generally proportional to motor load, within rated overload limits. Because of typically high no-load motor current, fixed losses, and other reasons, F_{LM} is generally assumed to be unity, and no adjustment should be made for underloading or overloading unless the situation is fixed and can be accurately established, and reduced-load efficiency data can be obtained from the motor manufacturer.

Unless the manufacturer's technical literature indicates otherwise, motor heat gain normally should be equally divided between radiant and convective components for the subsequent cooling load calculations.

Table 11.16 Minimum Nominal Full-Load Efficiency for NEMA Design A, NEMA Design B, and IEC Design N Motors (Excluding Fire Pump Electric Motors) at 60 Hz[a,b] [2021F, Ch 18, Tbl 4A]

Number of Poles ⇒	2	2	4	4	6	6	8	8
Motor Type ⇒	Enclosed	Open	Enclosed	Open	Enclosed	Open	Enclosed	Open
Motor Kilowatts								
0.75	77.0	77.0	85.5	85.5	82.5	82.5	75.5	75.5
1.1	84.0	84.0	86.5	86.5	87.5	86.5	78.5	77.0
1.5	85.5	85.5	86.5	86.5	88.5	87.5	84.0	86.5
2.2	86.5	85.5	89.5	89.5	89.5	88.5	85.5	87.5
3.7	88.5	86.5	89.5	89.5	89.5	89.5	86.5	88.5
5.5	89.5	88.5	91.7	91.0	91.0	90.2	86.5	89.5
7.5	90.2	89.5	91.7	91.7	91.0	91.7	89.5	90.2
11	91.0	90.2	92.4	93.0	91.7	91.7	89.5	90.2
15	91.0	91.0	93.0	93.0	91.7	92.4	90.2	91.0
18.5	91.7	91.7	93.6	93.6	93.0	93.0	90.2	91.0
22	91.7	91.7	93.6	94.1	93.0	93.6	91.7	91.7
30	92.4	92.4	94.1	94.1	94.1	94.1	91.7	91.7
37	93.0	93.0	94.5	94.5	94.1	94.1	92.4	92.4
45	93.6	93.6	95.0	95.0	94.5	94.5	92.4	93.0
55	93.6	93.6	95.4	95.0	94.5	94.5	93.6	94.1
75	94.1	93.6	95.4	95.4	95.0	95.0	93.6	94.1
90	95.0	94.1	95.4	95.4	95.0	95.0	94.1	94.1
110	95.0	94.1	95.8	95.8	95.8	95.4	94.1	94.1
150	95.4	95.0	96.2	95.8	95.8	95.4	94.5	94.1
186	95.8	95.0	96.2	95.8	95.8	95.8	95.0	95.0
224	95.8	95.4	96.2	95.8	95.8	95.8	NA	NA
261	95.8	95.4	96.2	95.8	95.8	95.8	NA	NA
298	95.8	95.8	96.2	95.8	NA	NA	NA	NA
336	95.8	96.2	96.2	96.2	NA	NA	NA	NA
373	95.8	96.2	96.2	96.2	NA	NA	NA	NA

Source: ANSI/ASHRAE/IES Standard 90.1-2019, Table 10.8-1

[a] Nominal efficiencies shall be established in accordance with 10 CFR 431.

[b] For purposes of determining the required minimum nominal full-load efficiency of an electric motor that has a horsepower or kilowatt rating between two horsepower or two kilowatt ratings listed in this table, each such motor shall be deemed to have a listed horsepower or kilowatt rating, determined as follows:

1. A horsepower at or above the midpoint between the two consecutive horsepowers shall be rounded up to the higher of the two horsepowers.
2. A horsepower below the midpoint between the two consecutive horsepowers shall be rounded down to the lower of the two horsepowers.
3. A kilowatt rating shall be directly converted from kilowatts to horsepower using the formula 1 kilowatt = (1/0.746) horsepower. The conversion should be calculated to three significant decimal places, and the resulting horsepower shall be rounded in accordance with paragraph (1) or (2), whichever applies.

Heat Gain From Restaurant Equipment

Heat gain: $q_s = q_{input} F_U F_R$, where F_U is the usage factor and F_R is the radiation factor.

Table 11.17 Recommended Rates of Radiant and Convective Heat Gain from Unhooded Electric Appliances during Idle (Ready-to-Cook) Conditions [2021F, Ch 18, Tbl 5A]

Appliance	Energy Rate, W		Rate of Heat Gain, W				Usage Factor F_U	Radiation Factor F_R
	Rated	Standby	Sensible Radiant	Sensible Convective	Latent	Total		
Cabinet: hot serving (large), insulated[a]	1993	352	117	234	0	352	0.18	0.33
hot serving (large), uninsulated	1993	1026	205	821	0	1026	0.51	0.20
proofing (large)[a]	5099	410	352	0	59	410	0.08	0.86
proofing (small-15 shelf)	4191	1143	0	264	879	1143	0.27	0.00
Cheesemelter[b]	2400	976	443	533	0	976	0.41	0.45
Coffee brewing urn	3810	352	59	88	205	352	0.08	0.17
Drawer warmers, 2-drawer (moist holding)[a]	1202	147	0	0	59	59	0.12	0.00
Egg cooker[b]	2380	249	65	184	0	249	0.10	0.26
Espresso machine[a]	2403	352	117	234	0	352	0.15	0.33
Food warmer: steam table (2-well-type)	1495	1026	88	176	762	1026	0.69	0.08
Freezer (small)	791	322	147	176	0	322	0.41	0.45
Fryer, countertop, open deep fat[b]	4600	431	202	229	0	431	0.09	0.47
Griddle, countertop[b]	8000	1771	848	923	0	1771	0.22	0.48
Hot dog roller[b]	1600	1240	267	973	0	1240	0.77	0.22
Hot plate: single element, high speed[b]	1100	982	314	668	0	982	0.89	0.32
Hot-food case (dry holding)[a]	9115	733	264	469	0	733	0.08	0.36
Hot-food case (moist holding)[a]	9115	967	264	528	176	967	0.11	0.27
Induction hob, countertop[b]	5000	0	0	0	0	0	0.00	0.00

Table 11.17 Recommended Rates of Radiant and Convective Heat Gain from Unhooded Electric Appliances during Idle (Ready-to-Cook) Conditions [2021F, Ch 18, Tbl 5A] ***(Continued)***

Appliance	Energy Rate, W		Rate of Heat Gain, W				Usage Factor F_U	Radiation Factor F_R
	Rated	Standby	Sensible Radiant	Sensible Convective	Latent	Total		
Microwave oven: commercial[b]	1700	0	0	0	0	0	0	0.00
Oven: countertop conveyorized bake/finishing[b]	5000	3932	718	3214	0	3932	0.79	0.18
Panini[b]	1800	673	195	478	0	673	0.37	0.29
Popcorn popper[b]	850	115	28	87	0	115	0.14	0.24
Rapid-cook oven (quartz-halogen)[a]	12 016	0	0	0	0	0	0	0.00
Rapid-cook oven (microwave/convection)[b]	5700	1141	96	1045	0	1141	0.20	0.08
Reach-in refrigerator[a]	1407	352	88	264	0	352	0.25	0.25
Refrigerated prep table[a]	586	264	176	88	0	264	0.45	0.67
Rice cooker[b]	1550	82	14	68	0	82	0.05	0.17
Soup warmer[b]	800	390	0	53	337	390	0.49	0.00
Steamer (bun)[b]	1500	200	32	168	0	200	0.13	0.16
Steamer, countertop[b]	8300	344	0	248	96	344	0.04	0.00
Toaster: 4-slice pop up (large): cooking	1788	879	59	410	293	762	0.49	0.07
contact (vertical)[b]	2600	759	180	579	0	759	0.29	0.24
conveyor (large)	9613	3019	879	2139	0	3019	0.31	0.29
small conveyor[b]	1745	1702	358	1344	0	1702	0.98	0.21
Tortilla grill[b]	2200	1034	254	780	0	1034	0.47	0.25
Waffle iron[b]	2700	267	60	207	0	267	0.10	0.22

Sources: Swierczyna et al. (2008, 2009), with the following exceptions as noted.
[a]Swierczyna et al. (2009) only.
[b]Additions and updates from ASHRAE research project RP-1631 (Kong and Zhang 2016; Zhang et al. 2016).

Table 11.18 Recommended Rates of Radiant and Convective Heat Gain from Unhooded Electric Appliances during Cooking Conditions [2021F, Ch 18, Tbl 5B]

Appliance	Energy Rate, W		Rate of Heat Gain, W				Usage Factor F_U	Radiation Factor F_R
	Rated	Cooking	Sensible Radiant	Sensible Convective	Latent	Total		
Cheesemelter	2400	2714	443	1094	599	2136	1.13	0.16
Egg cooker	2380	1191	65	369	630	1065	0.50	0.05
Fryer, countertop, open deep fryer	4600	3818	202	492	1629	2323	0.83	0.05
Griddle, countertop	8000	3280	848	631	1277	2757	0.41	0.26
Hot dog roller	1600	1577	267	611	679	1556	0.99	0.17
Hot plate, single burner	1100	985	313	627	44	985	0.90	0.32
Induction hob, countertop	5000	653	0	318	335	653	0.13	0.00
Oven, conveyor	5000	4292	718	2454	193	3365	0.86	0.17
Microwave	1700	2363	0	934	995	1929	1.39	0.00
Rapid cook	5700	2310	96	1234	771	2102	0.41	0.04
Panini grill	1800	1374	195	718	150	1062	0.76	0.14
Popcorn popper	850	576	28	236	192	457	0.68	0.05
Rice cooker	1550	1159	14	95	44	153	0.75	0.01
Soup warmer	800	842	0	85	716	801	1.05	0.00
Steamer (bun)	1500	791	32	240	511	783	0.53	0.04
Steamer, countertop	8300	7731	0	499	6934	7433	0.93	0.00
Toaster, conveyor	1745	1705	358	974	373	1705	0.98	0.21
Vertical	2600	1841	180	715	322	1218	0.71	0.10
Tortilla grill	2200	2194	254	1267	673	2194	1.00	0.12
Waffle maker	2700	1180	60	357	559	975	0.44	0.05

Source: ASHRAE research project RP-1631 (Zhang et al. 2015).

Table 11.19 Recommended Rates of Radiant Heat Gain from Hooded Electric Appliances during Idle (Ready-to-Cook) Conditions [2021F, Ch 18, Tbl 5C]

Appliance	Energy Rate, W		Rate of Heat Gain, W	Usage Factor	Radiation Factor
	Rated	Standby	Sensible Radiant	F_U	F_R
Broiler: underfired 900 mm	10 814	9 056	3165	0.84	0.35
Cheesemelter*	3 605	3 488	1348	0.97	0.39
Fryer, kettle	29 014	528	147	0.02	0.28
Open deep-fat, 1-vat	14 008	821	293	0.06	0.36
Pressure	13 511	791	147	0.06	0.19
Griddle, double-sided 900 mm (clamshell down)*	21 218	2 022	410	0.10	0.20
(Clamshell up)*	21 218	3 370	1055	0.16	0.31
Flat 900 mm	17 115	3 370	1319	0.20	0.39
Small 900 mm*	8 997	1 788	791	0.20	0.44
Induction cooktop*	21 013	0	0	0.00	0.00
Induction wok*	3 488	0	0	0.00	0.00
Oven, combi: combi-mode*	16 411	1 612	234	0.10	0.15
Combi: convection mode	16 412	1 612	410	0.10	0.25
Oven, convection full-size	12 103	1 964	440	0.16	0.22
Convection half-size*	5 510	1 084	147	0.20	0.14
Pasta cooker*	22 010	2 491	0	0.11	0.00
Range top, top off/oven on*	4 865	1 172	293	0.24	0.25
3 elements on/oven off	15 005	4 513	1846	0.30	0.41
6 elements on/oven off	15 005	9 730	4074	0.65	0.42
6 elements on/oven on	19 870	10 668	4250	0.54	0.40
Range, hot-top	15 826	15 035	3458	0.95	0.23
Rotisserie*	11 107	4 044	1319	0.36	0.33
Salamander*	7 004	6 829	2051	0.97	0.30
Steam kettle, large (225 L), simmer lid down*	32 414	762	29	0.02	0.04
small (150 L), simmer lid down*	21 599	528	88	0.02	0.17
Steamer, compartment, atmospheric*	9 789	4 484	59	0.46	0.01
Tilting skillet/braising pan	9 642	1 553	0	0.16	0.00

*Items with an asterisk appear only in Swierczyna et al. (2009); all others appear in both Swierczyna et al. (2008) and (2009).

Table 11.20 Recommended Rates of Radiant Heat Gain from Hooded Gas Appliances during Idle (Ready-to-Cook) Conditions [2021F, Ch 18, Tbl 5D]

Appliance	Standby Energy Rate, W		Rate of Heat Sensible Gain, W	Usage Factor F_U	Radiation Factor F_R
Broiler: batch*	27 842	20 280	2374	0.73	0.12
Chain (conveyor)	38 685	28 340	3869	0.73	0.14
Overfired (upright)*	29 307	25 761	733	0.88	0.03
Underfired 900 mm	28 135	21 658	2638	0.77	0.12
Fryer: doughnut	12 895	3634	850	0.28	0.23
Open deep-fat, 1 vat	23 446	1377	322	0.06	0.23
Pressure	23 446	2638	234	0.11	0.09
Griddle: double sided 900 mm, clamshell down*	31 710	2345	528	0.07	0.23
Clamshell up*	31 710	4308	1436	0.14	0.33
Flat 900 mm	26 376	5979	1084	0.23	0.18
Oven: combi: combi-mode*	22 185	1758	117	0.08	0.07
Convection mode	22 185	1700	293	0.08	0.17
Convection, full-size	12 895	3488	293	0.27	0.08
Conveyor (pizza)	49 822	20 017	2286	0.40	0.11
Deck	30 772	6008	1026	0.20	0.17
Rack mini-rotating*	16 500	1319	322	0.08	0.24
Pasta cooker*	23 446	6946	0	0.30	0.00
Range top: top off/oven on*	7327	2169	586	0.30	0.27
3 burners on/oven off	35 169	17 614	2081	0.50	0.12
6 burners on/oven off	35 169	35 403	3370	1.01	0.10
6 burners on/oven on	42 495	36 018	3986	0.85	0.11

Table 11.20 Recommended Rates of Radiant Heat Gain from Hooded Gas Appliances during Idle (Ready-to-Cook) Conditions [2021F, Ch 18, Tbl 5D] ***(Continued)***

Appliance	Standby Energy Rate, W		Rate of Heat Sensible Gain, W	Usage Factor F_U	Radiation Factor F_R
Range: wok*	29 014	25 614	1524	0.88	0.06
Rethermalizer*	26 376	6829	3370	0.26	0.49
Rice cooker*	10 257	147	88	0.01	0.60
Salamander*	10 257	9759	1553	0.95	0.16
Steam kettle: large (225 L) simmer lid down*	42 495	1583	0	0.04	0.00
Small (38 L) simmer lid down*	15 240	967	88	0.06	0.09
Medium (150 L) simmer lid down	29 307	1260	0	0.04	0.00
Steamer: compartment: atmospheric*	7620	2432	0	0.32	0.00
Tilting skillet/braising pan	30 479	3048	117	0.10	0.04

*Items with an asterisk appear only in Swierczyna et al. (2009); all others appear in both Swierczyna et al. (2008) and (2009).

Table 11.21 Recommended Rates of Radiant Heat Gain from Hooded Solid-Fuel Appliances during Idle (Ready-to-Cook) Conditions [2021F, Ch 18, Tbl 5E]

Appliance	Rated	Standby Energy Rate, W	Rate of Sensible Heat Gain, W	Usage Factor F_U	Radiation Factor F_R
Broiler: solid fuel: charcoal	18 kg	12 309	1817	N/A	0.15
Broiler: solid fuel: wood (mesquite)	18 kg	14 536	2051	N/A	0.14

Source: Swierczyna et al. (2008).

Table 11.22 Recommended Rates of Sensible and Latent Heat Gain to Space from Warewashing Equipment [2021F, Ch 18, Tbl 5F]

Equipment Type	Supply Water Flow Rate, L/s	Operating Water Temp., C	Rate of Heat Gain to Space, W[a]				Usage Factor F_U [c]
			Sensible Radiant[b]	Sensible Convective	Latent	Total	
Pre-Rinse Equipment							
Pre-rinse spray valve	0.04	49	0	59	2403	2462	1
Pre-rinse spray valve	0.08	49	0	88	3429	3517	1
Pre-rinse spray valve	0.25	49	0	322	4044	4367	1
3-Compartment sink, rinsing	NA	49	0	264	1436	1700	1
Idle	NA	NA	0	205	586	791	NA
Power wash sink, rinsing	NA	49	0	586	909	1495	1
Idle	NA	NA	0	440	469	909	NA
Scrapper	1.14[d]	49	0	352	3224	3575	1
Scrapper with trough	4.42[d]	49	0	821	4074	4894	1
Unhooded Dishwashers							
Under-counter dishwasher, low temperature	0.05	60	0	645	1436	2081	1
Under-counter dishwasher, high temperature	0.05	82	0	1172	1348	2520	1
Under-counter dishwasher, high temperature with heat recovery	0.04	82	0	645	322	967	1
Upright door type, low temperature dump and fill	0.06	49	0	938	1026	1964	1
Upright door type, low temperature with tank	0.04	60	0	1143	3869	5012	1
Upright door type, high temperature	0.06	82	0	2345	6272	8616	1
Upright door type, high temperature with heat recovery	0.06	82	0	1407	3810	5217	1
Pot and pan washer	0.15	82	0	1758	6887	8646	1
Pot and pan washer with heat recovery	0.15	82	0	1612	5568	7180	1
1120 mm Conveyor dishwasher, unvented	0.11	82	0	2931	17379	20310	1
1680 mm Conveyor dishwasher, unvented	0.11	82	0	4718	13188	17907	1

Table 11.22 **Recommended Rates of Sensible and Latent Heat Gain to Space from Warewashing Equipment** [2021F, Ch 18, Tbl 5F] ***(Continued)***

Equipment Type	Supply Water Flow Rate, L/s	Operating Water Temp., C	Rate of Heat Gain to Space, W[a]				Usage Factor F_U[c]
			Sensible Radiant[b]	Sensible Convective	Latent	Total	
Hooded or Ducted High-Temperature Dishwashers							
Upright door type, high temperature under a 0.9 × 0.9 m hood at 142 L/s	NA	82	0	1026	3810	4836	1
Upright door type, high temperature under a 1.5 × 1.2 m hood at 236 L/s	NA	82	0	469	2315	2784	1
1120 mm conveyor, high temperature under a 0.9 m hood at 472 L/s	0.11	88	0	293	5861	6154	1
Ducted Conveyor Dishwashers							
Ducted 1680 mm conveyor dishwasher	0.11	82	0	3107	1846	4953	1
Ducted flight type conveyor dishwasher	0.06	88	0	3927	2608	6535	1
Ducted flight type conveyor dishwasher with heat recovery	0.06	88	0	3605	1055	4660	1
Ducted flight type conveyor dishwasher with blow dryer	0.06	88	0	6213	4601	10814	1

Source: Livchak and Swierczyna (2020)

[a] Heat gain rates for pre-rinse equipment and unhooded dishwashers are for unhooded appliances only. For these equipment items the total appliance heat gain affects the space directly. If an appliance is hooded or vented, the heat gain rates in the hooded and ducted sections of the table must be used. Hooded and ducted line items account for the heat gain captured by the hood or duct and therefore only list the portion of heat gain affecting the space.

[b] Average surface temperature of pre-rinse equipment does not exceed 49°C which is the maximum temperature of water flowing through these devices. Average surface temperatures of the dishwashers are less than the wash tank temperature of 75°C. The surfaces are stainless steel and produce negligible radiant heat gain.

[c] Values given are for continuous rinsing or washing and are peak values. Rinse and wash times usually extend 1 to 3 hours after each meal period served by the food service facility. Duration of heat gains will vary based on the operating schedule of the food service.

[d] For scrapper equipment the recirculation flow rate is listed.

Heat Gain From Hospital and Laboratory Equipment

Heat gain varies significantly. In a laboratory, heat gain ranges from 50 to 220 W/m^2. Medical equipment is highly varied in type and application. Table 11.23 is relevant for portable and bench-type equipment. For large equipment, such as MRI, obtain heat gain from the manufacturer.

Table 11.23 Recommended Heat Gain from Typical Medical Equipment [2021F, Ch 18, Tbl 6]

Equipment	Nameplate, W	Peak, W	Average, W
Anesthesia system	250	177	166
Blanket warmer	500	504	221
Blood pressure meter	180	33	29
Blood warmer	360	204	114
ECG/RESP	1440	54	50
Electrosurgery	1000	147	109
Endoscope	1688	605	596
Harmonical scalpel	230	60	59
Hysteroscopic pump	180	35	34
Laser sonics	1200	256	229
Optical microscope	330	65	63
Pulse oximeter	72	21	20
Stress treadmill	N/A	198	173
Ultrasound system	1800	1063	1050
Vacuum suction	621	337	302
X-ray system	968		82
	1725	534	480
	2070		18

Source: Hosni et al. (1999)

Table 11.24 Recommended Heat Gain from Typical Laboratory Equipment [2021F, Ch 18, Tbl 7]

Equipment	Nameplate, W	Peak, W	Average, W
Analytical balance	7	7	7
Centrifuge	138	89	87
	288	136	132
	5500	1176	730
Electrochemical analyzer	50	45	44
	100	85	84
Flame photometer	180	107	105
Fluorescent microscope	150	144	143
	200	205	178
Function generator	58	29	29
Incubator	515	461	451
	600	479	264
	3125	1335	1222
Orbital shaker	100	16	16
Oscilloscope	72	38	38
	345	99	97
Rotary evaporator	75	74	73
	94	29	28
Spectronics	36	31	31
Spectrophotometer	575	106	104
	200	122	121
	N/A	127	125
Spectro fluorometer	340	405	395
Thermocycler	1840	965	641
	N/A	233	198
Tissue culture	475	132	46
	2346	1178	1146

Source: Hosni et al. (1999).

Heat Gain from Office Equipment

Table 11.25 Recommended Heat Gain for Typical Desktop Computers
[2021F, Ch 18, Tbl 8A]

Description	Nameplate Power,[a] W	Peak Heat Gain,[b, d] W
Manufacturer 1		
3.0 GHz processor, 4 GB RAM, *n* = 1	NA	83
3.3 GHz processor, 8 GB RAM, *n* = 8	NA	50
3.5 GHz processor, 8 GB RAM, *n* = 2	NA	42
3.6 GHz processor, 16 GB RAM, *n* = 2	NA	66
3.3 GHz processor, 16 GB RAM, *n* = 2	NA	52
4.0 GHz processor, 16 GB RAM, *n* = 1	NA	83
3.3 GHz processor, 8 GB RAM, *n* = 1	NA	84
3.7 GHz processor, 32 GB RAM, *n* = 1	750	116
	NA	102
3.5 GHz processor, 16 GB RAM, *n* = 3[c]	550	144
	NA	93
Manufacturer 2		
3.6 GHz processor, 32 GB RAM, *n* = 8	NA	80
3.6 GHz processor, 16 GB RAM, *n* = 1	NA	78
3.4 GHz processor, 32 GB RAM, *n* = 1	NA	72
3.4 GHz processor, 24 GB RAM, *n* = 1	NA	86
3.50 GHz processor, 4 GB RAM, *n* = 1	NA	26
3.3 GHz processor, 8 GB RAM, *n* = 1	NA	78
3.20 GHz processor, 8 GB RAM, *n* = 1	NA	61
3.20 GHz processor, 4 GB RAM, *n* = 1	NA	44
2.93 GHz processor, 16 GB RAM, *n* = 1	NA	151
2.67 GHz processor, 8 GB RAM, *n* = 1	NA	137
Average 15-min peak power consumption (range)	82 (26-151)	

Source: Bach and Sarfraz (2018)
n = number of tested equipment of same configuration.
[a]Nameplate for desktop computer is present on its power supply, which is mounted inside desktop, hence not accessible for most computers, where NA = not available.
[b]For equipment peak heat gain value, highest 15-min interval of recorded data is listed in tables.
[c]For tested equipment with same configuration, increasing power supply size does not increase average power consumption.
[d]Approximately 90% convective heat gain and 10% radiative heat gain.

Table 11.26 Recommended Heat Gain for Typical Laptops and Laptop Docking Station [2021F, Ch 18, Tbl 8B]

Equipment	Description	Nameplate Power,[a] W	Peak Heat Gain,[b, c] W
Laptop computer	Manufacturer 1, 2.6 GHz processor, 8 GB RAM, *n* = 1	NA	46
	Manufacturer 2, 2.4 GHz processor, 4 GB RAM, *n* = 1	NA	59
Average 15-min peak power consumption (range)		53 (46-59)	
Laptop with docking station	Manufacturer 1, 2.7 GHz processor, 8 GB RAM, *n* = 1	NA	38
	1.6 GHz processor, 8 GB RAM, *n* = 2	NA	45
	2.0 GHz processor, 8 GB RAM, *n* = 1	NA	50
	2.6 GHz processor, 4 GB RAM, *n* = 1	NA	51
	2.4 GHz processor, 8 GB RAM, *n* = 1	NA	40
	2.6 GHz processor, 8 GB RAM, *n* = 1	NA	35
	2.7 GHz processor, 8 GB RAM, *n* = 1	NA	59
	3.0 GHz processor, 8 GB RAM, *n* = 3	NA	70
	2.9 GHz processor, 32 GB RAM, *n* = 3	NA	58
	3.0 GHz processor, 32 GB RAM, *n* = 1	NA	128
	3.7 GHz processor, 32 GB RAM, *n* = 1	NA	63
	3.1 GHz processor, 32 GB RAM, *n* = 1	NA	89
Average 15-min peak power consumption (range)		61 (26-151)	

Source: Bach and Sarfraz (2018)
n = number of tested equipment of same configuration.
[a]Voltage and amperage information for laptop computer and laptop docking station is available on power supply nameplates; however, nameplate does not provide information on power consumption, where NA = not available.
[b]For equipment peak heat gain value, the highest 15-min interval of recorded data is listed in tables.
[c]Approximately 75% convective heat gain and 25% radiative heat gain.

Table 11.27 Recommended Heat Gain for Typical Tablet PC [2021F, Ch 18, Tbl 8C]

Description	Nameplate Power,[a] W	Peak Heat Gain,[b] W
1.7 GHz processor, 4 GB RAM, *n* = 1	NA	42
2.2 GHz processor, 16 GB RAM, *n* = 1	NA	40
2.3 GHz processor, 8 GB RAM, *n* = 1	NA	30
2.5 GHz processor, 8 GB RAM, *n* = 1	NA	31
Average 15-min peak power consumption (range)	36 (31-42)	

Source: Bach and Sarfraz (2018)
n = number of tested equipment of same configuration.
[a]Voltage and amperage information for tablet PC is available on power supply nameplate; however, nameplate does not provide information on power consumption, where NA = not available.
[b]For equipment peak heat gain value, highest 15-min interval of recorded data is listed in tables.

Table 11.28 Recommended Heat Gain for Typical Monitors [2021F, Ch 18, Tbl 8D]

Description[a]	Nameplate Power, W	Peak Heat Gain,[b, c] W
Manufacturer 1		
1397 mm LED flat screen, $n = 1$ (excluded from average because atypical size)	240	50
686 mm LED flat screen, $n = 2$	40	26
546 mm LED flat screen, $n = 2$	29	25
Manufacturer 2		
1270 mm 3D LED flat screen, $n = 1$ (excluded from average because atypical size)	94	49
Manufacturer 3		
864 mm LCD curved screen, $n = 1$ (excluded from average because atypical size and curved)	130	48
584 mm LED flat screen, $n = 3$	50	17
584 mm LED flat screen, $n = 1$	38	21
584 mm LED flat screen, $n = 1$	38	14
Manufacturer 4		
610 mm LED flat screen, $n = 1$	42	25
Manufacturer 5		
600 mm LED flat screen, $n = 1$	26	17
546 mm LED flat screen, $n = 1$	29	22
Manufacturer 6		
546 mm LED flat screen, $n = 1$	28	24
Average 15-min peak power consumption (range)	21 (14-26)	

Source: Bach and Sarfraz (2018)
n = number of tested equipment of same configuration.
[a]Screens with atypical size and shape are excluded for calculating average 15-min peak power consumption.
[b]For equipment peak heat gain value, highest 15-min interval of recorded data is listed in tables.
[c]Approximately 60% convective heat gain and 40% radiative heat gain.

Table 11.29 Recommended Heat Gain for Typical Printers [2021F, Ch 18, Tbl 9]

Equipment	Description	Max. Printing Speed, Pages per Minute	Nameplate Power, W	Peak Heat Gain,[a] W
Multifunction printer (copy, print, scan)	Large, multiuser, office type	40	1010	540 (Idle 29 W)
		30	1300	303 (Idle 116 W)
		28	1500	433 (Idle 28 W)
Average 15-min peak power consumption (range)			**425 (303-540)**	
	Multiuser, medium-office type	35	900	732 (Idle 18 W)
	Desktop, small-office type	25	470	56 (Idle 3 W)
Monochrome printer	Desktop, medium-office type	55	1000	222
		45	680	61
Average 15-min peak power consumption (range)			**142 (61-222)**	
Color printer	Desktop, medium-office type	40	620	120
Laser printer	Desktop, small-office type	14	310	89
		24	495	67
		26	1090	65
Average 15-min peak power consumption (range)			**74 (65-89)**	
Plotter	Manufacturer 1		1600	571
	Manufacturer 2		270	173
Average 15-min peak power consumption (range)				**372 (173-571)**
Fax machine	Medium		1090	92
	Small		600	46
Average 15-min peak power consumption (range)			**69 (46-92)**	

Source: Bach and Sarfraz (2018)
[a]Approximately 70% convective heat gain and 30% radiative heat gain.

Table 11.30 Recommended Heat Gain for Miscellaneous Equipment
[2021F, Ch 18, Tbl 10]

Equipment	Nameplate Power,[a] W	Peak Heat Gain,[b] W
Vending machine		
Drinks, 280 to 400 items	NA	940
Snacks	NA	54
Food (e.g., for sandwiches)	NA	465
Thermal binding machine, 2 single documents up to 340 pages	350	28.5
Projector, resolution 1024 × 768	340	308
Paper shredder, up to 28 sheets	1415	265
Electric stapler, up to 45 sheets	NA	1.5
Speakers	220	15
Temperature-controlled electronics soldering station	95	16
Cell phone charger	NA	5
Battery charger		
40 V	NA	19
AA	NA	5.5
Microwave oven, 25 to 34 L	1000 to 1550	713 to 822
Coffee maker		
Single cup	1400	385
Up to 12 cups	950	780
With grinder	1350	376
Coffee grinder, up to 12 cups	NA	73
Tea kettle, up to 6 cups	1200	1200
Dorm fridge, 88 L	NA	57
Freezer, 510 L	130	125
Fridge, 510 L	NA	387 to 430
Ice maker and dispenser, 9 kg bin capacity	NA	658
Top mounted bottled water cooler	NA	114 to 350
Cash register	25	9
Touch screen computer, 380 mm standard LCD and 2.2 GHz processor	NA	58
Self-checkout machine	NA	15

Source: Bach and Sarfraz (2018)
[a]For some equipment, nameplate power consumption is not available, where NA = not available.
[b]For equipment peak heat gain value, highest 15-min interval of recorded data is listed in tables.

Table 11.31 Recommended Load Factors for Various Types of Offices [2021F, Ch 18, Tbl 11]

Type of Use	Load Factor*, W/m²	Description
100% laptop, docking station		
light	3.67	15.5 m²/workstation, all laptop docking station use, 1 printer per 10
medium	4.91	11.6 m²/workstation, all laptop docking station use, 1 printer per 10
50% laptop, docking station		
light	4.75	15.5 m²/workstation, 50% laptop docking station/ 50% desktop, 1 printer per 10
medium	6.35	11.6 m²/workstation, 50% laptop docking station/ 50% desktop, 1 printer per 10
100% desktop		
light	5.83	15.5 m²/workstation, all desktop use, 1 printer per 10
medium	7.79	11.6 m²/workstation, all desktop use, 1 printer per 10
100% laptop, docking station		
2 screens	7.44	11.6 m²/workstation, all laptop docking station use, 2 screens, 1 printer per 10
100% desktop		
2 screens	9.06	11.6 m²/workstation, all laptop use, 2 screens, 1 printer per 10
3 screens	10.33	11.6 m²/workstation, all desktop use, 3 screens, 1 printer per 10
100% desktop		
heavy, 2 screens	11.00	7.9 m²/workstation, all desktop use, 2 screens, 1 printer per 8
heavy, 3 screens	12.49	7.9 m²/workstation, all desktop use, 3 screens, 1 printer per 8
100% laptop, docking station		
full on, 2 screens	12.23	7.9 m²/workstation, all laptop docking use, 2 screens, 1 printer per 8, no diversity
100% desktop		
full on, 2 screens	14.35	7.9 m²/workstation, all desktop use, 2 screens, 1 printer per 8, no diversity
full on, 3 screens	16.48	7.9 m²/workstation, all desktop use, 3 screens, 1 printer per 8, no diversity

Source: Bach and Sarfraz (2018)

*Medium office type monochrome printer is used for load factor calculator with 15- min peak power consumption of 142 W.

Table 11.32 Diversity Factor for Different Equipment [2021F, Ch 18, Tbl 12]

Equipment	Diversity Factor, %	Diversity Factor,[a] %
Desktop PC	75	75
Laptop docking station	70	NA
Notebook computer	75[b]	75
Screen	70	60
Printer	45	NA

Source: Bach and Sarfraz (2018)
[a]2013 *ASHRAE Handbook—Fundamentals*
[b]Insufficient data from RP-1742; values based on previous data from 2013 *ASHRAE Handbook—Fundamentals* and judgment of Bach and Sarfraz (2018).

Table 11.33 Refrigerating Effect Produced by Open Refrigerated Display Fixtures

Type of Display Fixture	W/m of Fixture*		
	Latent Heat	Sensible Heat	Total Refrigerating Effect
Low temperature			
Frozen Food			
Single Deck	36	199	236
Single Deck, Double Island	67	384	452
2 Deck	138	554	692
3 Deck	310	1238	1540
4 or 5 Deck	384	1538	1923
Ice Cream			
Single Deck	62	352	413
Single Deck, Double Island	67	384	452
Standard Temperature			
Meats			
Single Deck	50	286	336
Multideck	211	842	1053
Dairy			
Multideck	188	754	942
Produce			
Single Deck	35	196	231
Multideck	184	738	923

*These figures are general magnitudes for fixtures adjusted for average desired product temperatures and apply to store ambients in front of the display cases of 22°C to 23°C with 50% to 55% rh. Raising the dry bulb only 2°C to 3°C and the humidity 5% to 10% can increase heat removal 25% or more. Equally lower temperatures and humidities as in winter, have an equally marked effect on lowering heat removal from the space.

12. VENTILATION

Residential

Adapted Excerpt from ANSI/ASHRAE Standard 62.2-2019, *Ventilation and Acceptable Indoor Air Quality in Residential Buildings*

(*See complete standard for detailed guidance.*)

A mechanical exhaust system, supply system, or combination thereof, shall be installed to operate for each dwelling unit to provide continuous dwelling-unit ventilation with outdoor air at a rate not less than specified below.

The total required ventilation rate (Q_{tot}) shall be as specified in Table 12.2 or alternatively calculated using Equation 12.1:

$$Q_{tot} = 0.15A_{floor} + 3.5(N_{br} + 1) \tag{12.1}$$

where

Q_{tot} = total required ventilation rate, L/s

A_{floor} = dwelling-unit floor area, m^2

N_{br} = number of bedrooms (not to be less than 1)

Exception: Dwelling-unit mechanical ventilation systems are not required if the authority having jurisdiction determines that window operation is a locally permissible method of providing ventilation and provided that at least one of the following conditions is met:

1. The building has no mechanical cooling and is in Climate Zone 1 or 2 of the map shown in Figure 8-1 of Standard 62.2.
2. The building is thermally conditioned for human occupancy for less than 876 h per year.

Ventilation air requirements are shown in Table 12.2. Local mechanical exhaust rates are shown in Tables 12.2 and 12.3.

Ventilation openings: not less than 4% of floor nor less than 0.5 m^2 for habitable rooms; and not less than 4% of floor space nor less than 0.15 m^2 for toilets and utility rooms.

Mechanical systems that supply air to an occupiable space through ductwork exceeding 3 m in length and through a thermal conditioning component, except evaporative coolers, shall be provided with a filter having a designated minimum efficiency of MERV 6 or better when tested in accordance with ANSI/ASHRAE Standard 52.2, or a minimum particle size efficiency of 50% in the 3.0 to 10 μm range in accordance with AHRI Standard 680. The system shall be designed such that all recirculated and mechanically supplied outdoor air is filtered before passing through the thermal conditioning components. The filter shall be located and installed in such a manner as to facilitate access and regular service by the owner.

Airflow quantities refer to the amount of air exhausted by the ventilation system as installed and measured. The prescriptive requirements of Table 12.3 are permitted in place of measurements.

Table 12.1 Ventilation Air Requirements, cfm [Std 62.2-2019, Tbl 4.1b]

Floor Area, m²	Bedrooms				
	1	2	3	4	5
<47	14	18	21	25	28
47–93	21	24	28	31	35
94–139	28	31	35	38	42
140–186	35	38	42	45	49
187–232	42	45	49	52	56
233–279	49	52	56	59	63
280–325	56	59	63	66	70
326–372	63	66	70	73	77
373–418	70	73	77	80	84
419–465	77	80	84	87	91

Table 12.2 Demand-Controlled Local Ventilation Exhaust Airflow Rates [Std 62.2-2019, Tbl 5-1]

Application	Airflow
Enclosed Kitchen	• Vented range hood (including appliance-range hood combinations): 50 L/s • Other kitchen exhaust fans, including downdraft: 150 L/s or a capacity of 5 ach
Nonenclosed Kitchen	• Vented range hood (including appliance-range hood combinations): 50 L/s • Other kitchen exhaust fans, including downdraft: 150 L/s
Bathroom	25 L/s

Table 12.3 Continuous Local Ventilation Exhaust Airflow Rates [Std 62.2-2019, Tbl 5-2]

Application	Airflow
Enclosed Kitchen	5 ach, based on kitchen volume
Bathroom	10 L/s

Table 12.4 Prescriptive Duct Sizing [Std 62.2-2019, Tbl 5-3]

	Fan Airflow Rating, CFM at minimum static pressure of L/s at minimum 62.5 Pa												
	≤25	≤40	≤50	≤60	≤70	≤85	≤95	≤120	≤165	≤190	≤210	≤330	≤380
Duct Type	**Minimum Duct Diameter, mm** [a,b]												
Rigid duct	100 [e]	125	125	150	150	180	180	205	230	255	255	305	305 [d]
Flex duct [c]	100	125	150	150	150	180	205	205	230	255	NP	NP	NP

a. For noncircular ducts, calculate the diameter as four times the cross-sectional area divided by the perimeter.
b. NP = application of the prescriptive table is not permitted for this scenario.
c. Use of this table for verification of flex duct systems requires flex duct to be fully extended and any flex duct elbows to have a minimum bend radius to duct diameter ratio of 1.0.
d. For this scenario, use of elbows is not permitted.
e. For this scenario, 100 mm oval duct shall be permitted, provided the minor axis of the oval is greater than or equal to 75 mm.

Nonresidential

Adapted Excerpt from ANSI/ASHRAE Standard 62.1-2019, *Ventilation for Acceptable Indoor Air Quality*

(See complete standard for detailed guidance.)

General

Use of natural ventilation systems is permitted in lieu of or in conjunction with mechanical ventilation. Naturally ventilated spaces shall be permanently open to operable wall or roof openings to the outdoors; free openable area at least 4% of net occupiable floor area. If interior spaces are ventilated through adjoining rooms, free area between rooms shall be permanently unobstructed and at least 8% of the area of the interior room, nor less than 2.3 m^2. Occupants must have ready access to the openings.

All airstream surfaces shall be designed to resist mold growth and resist erosion. Ductwork construction shall meet SMACNA standards. In accordance with manufacturer instructions, fuel-burning appliances shall have sufficient air for combustion and adequate removal of combustion products, which shall be vented directly outdoors. Filters or air cleaners with minimum MERV 8 by ANSI/ASHRAE Standard 52.2 shall be provided upstream of all cooling coils or other devices with wetted surfaces through which air is supplied to occupiable space. Relative humidity should be 65% or less when system performance is analyzed with outdoor at the design dew point and mean coincident dry bulb, sensible and latent space interior loads at cooling design values, and space solar loads at zero. Drain pans slope minimum 10 mm per metre to outlet at lowest point, and drain line shall have P-trap or other seal when drain pan is at negative static pressure relative to the outlet. Drain pan shall extend from leading edge of the coil to a distance of half the vertical dimension of the coil.

Discharge from noncombustion equipment that captures contaminants generated by the equipment shall be discharged directly outdoors.

Investigate outdoor air quality. Survey and document local and regional outdoor air quality, with description of noticeable air problems and conditions regarding its acceptability. If unacceptable, treat it. Cleaning for ozone is required only if in a high-ozone area (see Informative Appendix F of the standard) and if the minimum design outdoor airflow is 1.5 air changes or more.

Outdoor air intakes shall be located so the shortest distance from intake to any specific contaminant source shall equal or exceed values in Table 12.5 or that result from the calculation method in Normative Appendix B of ANSI/ASHRAE Standard 62.1.

Design intakes to manage rain and snow entrainment and include bird screens.

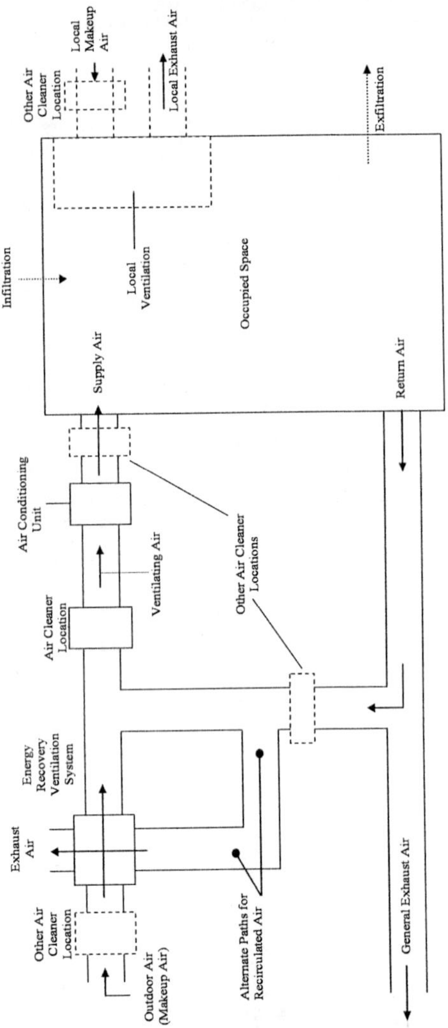

Figure 12.1 Ventilation System [Std 62.1-2019, Fig 3.1]

Table 12.5 Air Intake Minimum Separation Distance [Std 62.1-2019, Tbl 5-1]

Object	Minimum Distance, m
Class 2 air exhaust/relief outlet[a]	3
Class 3 air exhaust/relief outlet[a]	5
Class 4 air exhaust/relief outlet[b]	10
Plumbing vents terminating less than 1 m above the level of the outdoor air intake	3
Plumbing vents terminating at least 1 m above the level of the outdoor air intake	1
Vents, chimneys, and flues from combustion appliances and equipment[c]	5
Garage entry, automobile loading area, or drive-in queue[d]	5
Truck loading area or dock, bus parking/idling area[d]	7.5
Driveway, street, or parking place[d]	1.5
Thoroughfare with high traffic volume	7.5
Roof, landscaped grade, or other surface directly below intake[e,f]	0.30
Garbage storage/pick-up area, dumpsters	5
Cooling tower intake or basin	5
Cooling tower exhaust	7.5

a. This requirement applies to the distance from the outdoor air intakes for one ventilation system to the exhaust outlets and relief outlets for any other ventilation system.

b. Minimum distance listed does not apply to laboratory fume hood exhaust air outlets. Separation criteria for fume hood exhaust shall be in compliance with ANSI/AIHA Z9.5. Informative Appendix J contains sources of additional information on separation criteria. These include the *ACGIH Industrial Ventilation Manual*, *ASHRAE Handbook—HVAC Applications*, *ASHRAE Laboratory Design Guide*, and NSF/ANSI 49.

c. The minimum distances relative to fuel-fired appliances shall be as required by ANSI Z223.1/NFPA 54 for fuel gas burning appliances and equipment, NFPA 31 for oil burning appliances and equipment, and NFPA 211 for other combustion appliances and equipment.

d. Distance measured to closest place that vehicle exhaust is likely to be located.

e. The minimum separation distance shall not apply where outdoor surfaces below the air intake are sloped more than 45 degrees from horizontal or where such surfaces are less than 30 mm in width.

f. Where snow accumulation is expected, the surface of the snow at the expected average snow depth shall be considered to be a surface directly below an intake.

Air classifications:

- Class 1: Air with low contaminant concentration, low sensory-irritation intensity, and inoffensive odor.
- Class 2: Air with moderate contaminant concentration, mild sensory-irritation intensity, or mildly offensive odors. (Class 2 air also includes air that is not necessarily harmful or objectionable but that is inappropriate for transfer or recirculation to spaces used for different purposes.)
- Class 3: Air with significant contaminant concentration, significant sensory-irritation intensity, or offensive odor.
- Class 4: Air with highly objectionable fumes or gases or with potentially dangerous particles, bioaerosols, or gases, at concentrations high enough to be considered harmful.

Procedures from ANSI/ASHRAE Standard 62.1-2019

6.1 General. The Ventilation Rate Procedure, the IAQ Procedure, the Natural Ventilation Procedure, or a combination thereof shall be used to meet the requirements of this section. In addition, the requirements for exhaust ventilation in Section 6.5 shall be met regardless of the method used to determine minimum outdoor airflow rates.

Informative Note: Although the intake airflow determined using each of these approaches may differ significantly because of assumptions about the design, any of these approaches is a valid basis for design.

6.1.1 Ventilation Rate Procedure. The prescriptive design procedure presented in Section 6.2, in which outdoor air intake rates are determined based on space type/application, occupancy level, and floor area, shall be permitted to be used for any zone or system.

6.1.2 Indoor Air Quality (IAQ) Procedure. The performance-based design procedure presented in Section 6.3, in which the building outdoor air intake rates and other system design parameters are based on an analysis of contaminant sources, contaminant concentration limits, and level of perceived indoor air acceptability, shall be permitted to be used for any zone or system.

6.1.3 Natural Ventilation Procedure. The prescriptive or engineered system design procedure presented in Section 6.4, in which outdoor air is provided through openings to the outdoors, shall be permitted to be used for any zone or portion of a zone in conjunction with mechanical ventilation systems in accordance with Section 6.4.

6.1.4 Outdoor Air Treatment. Each ventilation system that provides outdoor air shall comply with Sections 6.1.4.1 through 6.1.4.4.

Exception to 6.1.4: Systems supplying air for enclosed parking garages, warehouses, storage rooms, janitor's closets, trash rooms, recycling areas, shipping/receiving/distribution areas.

Informative Note: Occupied spaces ventilated with outdoor air that is judged to be unacceptable are subject to reduced air quality when outdoor air is not cleaned prior to introduction to the occupied spaces.

6.1.4.1 Particulate Matter Smaller than 10 Micrometers (PM10). In buildings located in an area where the national standard or guideline for PM10 is exceeded, particle filters or air-cleaning devices shall be provided to clean the outdoor air at any location prior to its introduction to occupied spaces. Particulate matter filters or air cleaners shall have either

a. MERV of not less than 8 where rated in accordance with ANSI/ASHRAE Standard 52.2 or

b. the minimum efficiency within ISO ePM10 where rated in accordance with ISO 16890.

Informative Note: See Informative Appendix D for resources regarding selected PM10 national standards and guidelines.

Table 12.6 Airstreams or Sources [Std 62.1-2019, Tbl 6-3]

Description	Air Class
Commercial kitchen grease hoods	4
Commercial kitchen hoods other than grease	3
Diazo printing equipment discharge	4
Hydraulic elevator machine room	2
Laboratory hoods	4
Paint spray booths	4
Refrigerating machinery rooms	3
Residential kitchen hoods in transient occupancy	3

Ventilation

6.1.4.2 Particulate Matter Smaller than 2.5 Micrometers (PM2.5). In buildings located in an area where the national standard or guideline for PM2.5 is exceeded, particle filters or air-cleaning devices shall be provided to clean the outdoor air at any location prior to its introduction to occupied spaces. Particulate matter filters or air cleaners shall have either

a. MERV of not less than 11 where rated in accordance with ANSI/ASHRAE Standard 52.2 or
b. the minimum efficiency within ISO ePM2.5 where rated in accordance with ISO 16890.

Informative Note: See Informative Appendix D for resources regarding selected PM2.5 national standards and guidelines.

6.1.4.3 Ozone. Air-cleaning devices for ozone shall be provided when the most recent three-year average annual fourth-highest daily maximum eight-hour average ozone concentration exceeds 195 $\mu g/m^3$.

Such air-cleaning devices shall have a volumetric ozone removal efficiency of not less than 40% where installed, operated, and maintained in accordance with manufacturer recommendations and shall be approved by the authority having jurisdiction (AHJ). Such devices shall be operated where the outdoor ozone levels are expected to exceed 195 $\mu g/m^3$.

Exception to 6.1.4.3: Air cleaning for ozone shall not be required where

1. the system design outdoor air intake flow is 1.5 ach or less,
2. controls are provided that sense outdoor ozone level and reduce intake airflow to 1.5 ach or less while complying with the outdoor airflow requirements of Section 6, or
3. outdoor air is brought into the building and heated by direct-fired makeup air units.

Informative Note: In the U.S., a most recent three-year average annual fourth-highest daily maximum eight-hour average ozone concentration exceeding 195 $\mu g/m^3$ equates to a USEPA eight-hour ozone classification of "Serious" or higher (Severe 15, Severe 17, or Extreme).

6.1.4.4 Other Outdoor Contaminants. In buildings located in an area where the national standard for one or more contaminants not addressed in Section 6.1.4 is exceeded, any design assumptions and calculations related to the impact on IAQ shall be included in the design documents.

Table 12.7 Minimum Ventilation Rates in Breathing Zone
[Adapted from Std 62.1-2019, Tbl 6-1]

Occupancy Category	People Outdoor Air Rate R_p L/s·person	Area Outdoor Air Rate R_a L/s·m^2	Default Values Occupant Density #/100 m^2	Air Class	OS (6.2.6.1.4)
Animal Facilities					
Animal exam room (veterinary office)	5	0.6	20	2	
Animal imaging (MRI/CT/PET)	5	0.9	20	3	
Animal operating rooms	5	0.9	20	3	
Animal postoperative recovery room	5	0.9	20	3	
Animal preparation rooms	5	0.9	20	3	
Animal procedure room	5	0.9	20	3	
Animal surgery scrub	5	0.9	20	3	
Large-animal holding room	5	0.9	20	3	
Necropsy	5	0.9	20	3	
Small-animal-cage room (static cages)	5	0.9	20	3	
Small-animal-cage room (ventilated cages)	5	0.9	20	3	
Correctional Facilities					
Booking/waiting	3.8	0.3	50	2	
Cell	2.5	0.6	25	2	
Dayroom	2.5	0.3	30	1	
Guard stations	2.5	0.3	15	1	
Educational Facilities					
Art classroom	5	0.9	20	2	
Classrooms (ages 5 to 8)	5	0.6	25	1	
Classrooms (age 9 plus)	5	0.6	35	1	
Computer lab	5	0.6	25	1	
Daycare sickroom	5	0.9	25	3	
Daycare (through age 4)	5	0.9	25	2	
Lecture classroom	3.8	0.3	65	1	✓
Lecture hall (fixed seats)	3.8	0.3	150	1	✓
Libraries	2.5	0.6	10		
Media center	5	0.6	25	1	
Multiuse assembly	3.8	0.3	100	1	✓
Music/theater/dance	5	0.3	35	1	✓
Science laboratories	5	0.9	25	2	
University/college laboratories	5	0.9	25	2	
Wood/metal shop	5	0.9	20	2	

Table 12.7 Minimum Ventilation Rates in Breathing Zone
[Adapted from Std 62.1-2019, Tbl 6-1] ***(Continued)***

Occupancy Category	People Outdoor Air Rate R_p	Area Outdoor Air Rate R_a	Default Values Occupant Density	Air Class	OS (6.2.6.1.4)
	L/s·person	L/s·m^2	#/100 m^2		
Food and Beverage Service					
Bars, cocktail lounges	3.8	0.9	100	2	
Cafeteria/fast-food dining	3.8	0.9	100	2	
Kitchen (cooking)	3.8	0.6	20	2	
Restaurant dining rooms	3.8	0.9	70	2	
General					
Break rooms	2.5	0.3	25	1	✓
Coffee stations	2.5	0.3	20	1	✓
Conference/meeting	2.5	0.3	50	1	✓
Corridors	—	0.3	—	1	✓
Occupiable storage rooms for liquids or gels	2.5	0.6	2	2	
Hotels, Motels, Resorts, Dormitories					
Barracks sleeping areas	2.5	0.3	20	1	✓
Bedroom/living room	2.5	0.3	10	1	✓
Laundry rooms, central	2.5	0.6	10	2	
Laundry rooms within dwelling units	2.5	0.6	10	1	
Lobbies/prefunction	3.8	0.3	30	1	✓
Multipurpose assembly	2.5	0.3	120	1	✓
Miscellaneous Spaces					
Banks or bank lobbies	3.8	0.3	15	1	✓
Bank vaults/safe deposit	2.5	0.3	5	2	✓
Computer (not printing)	2.5	0.3	4	1	✓
Freezer and refrigerated spaces (<10°C)	5	0	0	2	
Manufacturing where hazardous materials are not used	5.0	0.9	7	2	
Manufacturing where hazardous materials are used (excludes heavy industrial and chemical processes)	5.0	0.9	7	3	
Pharmacy (prep. area)	2.5	0.9	10	2	
Photo studios	2.5	0.6	10	1	
Shipping/receiving	5	0.6	2	2	
Sorting, packing, light assembly	3.8	0.6	7	2	
Telephone closets	—	0.0	—	1	
Transportation waiting	3.8	0.3	100	1	✓
Warehouses	5	0.3	—	2	

Table 12.7 Minimum Ventilation Rates in Breathing Zone
[Adapted from Std 62.1-2019, Tbl 6-1] ***(Continued)***

Occupancy Category	People Outdoor Air Rate R_p	Area Outdoor Air Rate R_a	Default Values Occupant Density	Air Class	OS (6.2.6.1.4)
	L/s·person	L/s·m²	#/100 m²		
Office Buildings					
Breakrooms	2.5	0.6	50	1	
Main entry lobbies	2.5	0.3	10	1	✓
Occupiable storage rooms for dry materials	2.5	0.3	2	1	
Office space	2.5	0.3	5	1	✓
Reception areas	2.5	0.3	30	1	✓
Telephone/data entry	2.5	0.3	60	1	✓
Outpatient Health Care Facilities [a,b]					
Birthing room	5	0.9	15	2	
Class 1 imaging rooms	2.5	0.6	5	1	
Dental operatory	5	0.9	20	1	
General examination room	3.8	0.6	20	1	
Other dental treatment areas	2.5	0.3	5	1	
Physical therapy exercise area	10	0.9	7	2	
Physical therapy individual room	5	0.3	20	1	
Physical therapeutic pool area	—	2.4	—	2	
Prosthetics and orthotics room	5	0.9	20	1	
Psychiatric consultation room	2.5	0.3	20	1	
Psychiatric examination room	2.5	0.3	20	1	
Psychiatric group room	2.5	0.3	50	1	
Psychiatric seclusion room	5	0.3	5	1	
Speech therapy room	2.5	0.3	20	1	
Urgent care examination room	3.8	0.6	20	1	
Urgent care observation room	2.5	0.3	20	1	
Urgent care treatment room	3.8	0.9	20	1	
Urgent care triage room	5	0.9	20	1	
Public Assembly Spaces					
Auditorium seating area	2.5	0.3	150	1	✓
Courtrooms	2.5	0.3	70	1	✓
Legislative chambers	2.5	0.3	50	1	✓
Libraries	2.5	0.6	10	1	
Lobbies	2.5	0.3	150	1	✓
Museums (children's)	3.8	0.6	40	1	
Museums/galleries	3.8	0.3	40	1	✓
Places of religious worship	2.5	0.3	120	1	✓

Table 12.7 Minimum Ventilation Rates in Breathing Zone
[Adapted from Std 62.1-2019, Tbl 6-1] ***(Continued)***

Occupancy Category	People Outdoor Air Rate R_p	Area Outdoor Air Rate R_a	Default Values: Occupant Density	Air Class	OS (6.2.6.1.4)
	L/s·person	L/s·m²	#/100 m²		
Retail					
Sales (except as below)	3.8	0.6	15	2	
Barbershop	3.8	0.3	25	2	✓
Beauty and nail salons	10	0.6	25	2	
Coin-operated laundries	3.8	0.6	20	2	
Mall common areas	3.8	0.3	40	1	✓
Pet shops (animal areas)	3.8	0.9	10	2	
Supermarket	3.8	0.3	8	1	✓
Sports and Entertainment					
Bowling alley (seating)	5	0.6	40	1	
Disco/dance floors	10	0.3	100	2	✓
Gambling casinos	3.8	0.9	120	1	
Game arcades	3.8	0.9	20	1	
Gym, sports arena (play area)	10	0.9	7	2	
Health club/aerobics room	10	0.3	40	2	
Health club/weight rooms	10	0.3	10	2	
Spectator areas	3.8	0.3	150	1	✓
Stages, studios	5	0.3	70	1	✓
Swimming (pool and deck)	—	2.4	—	2	
Transient Residential					
Common corridors	—	0.3		1	✓
Dwelling unit	2.5	0.3	F	1	✓

a. Outpatient facilities to which the rates apply are freestanding birth centers, urgent care centers, neighborhood clinics and physicians offices, Class 1 imaging facilities, outpatient psychiatric facilities, outpatient rehabilitation facilities, and outpatient dental facilities.

b. The requirements of this table provide for acceptable IAQ. The requirements of this table do not address the airborne transmission of airborne viruses, bacteria, and other infectious contagions.

Informative Note: These rates are intended only for outpatient dental clinics where the amount of nitrous oxide is limited. They are not intended for dental operatories in institutional buildings where nitrous oxide is piped.

OTHER NOTES

1 The rates in this table are based on all other applicable requirements of ANSI/ASHRAE Standard 62.1 being met.

2 For DCV (demand control ventilation) zones in the occupied standby mode, breathing zone outdoor airflow shall be permitted to be reduced to zero for the occupancy categories indicated "OS," provided that airflow is restored to design value whenever occupancy is detected.

Table 12.8 Minimum Exhaust Rates [Std 62.1-2019, Tbl 6-2]

Occupancy Category	Notes	Exhaust Rate, L/s·unit	Exhaust Rate, L/s·m²	Air Class
Animal Facilities				
Animal imaging (MRI/CT/PET)		—	4.5	3
Animal operating rooms		—	15	3
Animal postoperative recovery room		—	7.5	3
Animal preparation rooms		—	7.5	3
Animal procedure room		—	11.3	3
Animal surgery scrub		—	7.5	3
Large-animal holding room		—	11.3	3
Necropsy		—	11.3	3
Small-animal-cage room (static cages)		—	11.3	3
Small-animal-cage room (ventilated cages)		—	7.5	3
Arenas	B	—	—	1
Art classrooms		—	3.5	2
Auto repair rooms	A	—	7.5	2
Barber shops		—	2.5	2
Beauty and nail salons		—	3.0	2
Cells with toilet		—	5.0	2
Copy, printing rooms		—	2.5	2
Darkrooms		—	5.0	2
Educational science laboratories		—	5.0	2
Janitor closets, trash rooms, recycling		—	5.0	3
Kitchenettes		—	1.5	2
Kitchens—commercial		—	3.5	2
Locker rooms for athletic, industrial, and health care facilities		—	2.5	2
All other locker rooms	—	—	1.25	2
Shower rooms	G,I	10/25		2
Paint spray booths	F	—	—	4
Parking garages	C	—	3.7	2
Pet shops (animal areas)	—	—	4.5	2
Refrigerating machinery rooms	F	—	—	3
Residential kitchens	G	25/50	—	2
Soiled laundry storage rooms	F	—	5.0	3
Storage rooms, chemical	F	—	7.5	4
Toilets—private	E, H	12.5/25	—	2
Toilets—public	D, H	25/35	—	2
Woodwork shop/classrooms	—	—	2.5	2

NOTES:

A Stands where engines are run shall have exhaust systems that directly connect to the engine exhaust and prevent escape of fumes.

B Where combustion equipment is intended to be used on the playing surface, additional dilution ventilation, source control, or both shall be provided.

C Exhaust shall not be required where two or more sides compose walls that are at least 50% open to the outside.

D Rate is per water closet, urinal, or both. Provide the higher rate where periods of heavy use are expected to occur. The lower rate shall be permitted to be used otherwise.

E Rate is for a toilet room intended to be occupied by one person at a time. For continuous system operation during hours of use, the lower rate shall be permitted to be used. Otherwise the higher rate shall be used.

F See other applicable standards for exhaust rate.

G For continuous system operation, the lower rate shall be permitted to be used. Otherwise the higher rate shall be used.

H Exhaust air that has been cleaned to meet Class 1 criteria from Section 5.18.1 shall be permitted to be recirculated.

I Rate is per showerhead.

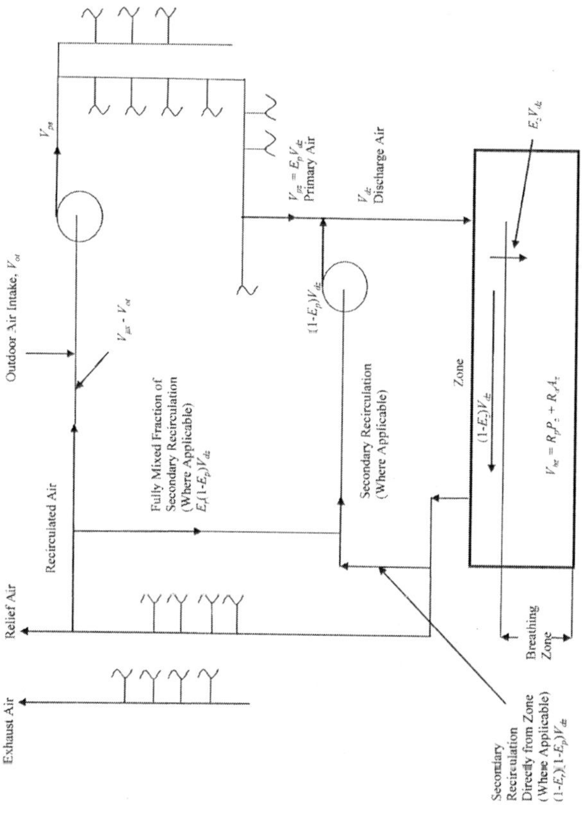

Figure 12.2 Ventilation System Schematic [Std 62.1-2019, Fig A-1]

Operation and Maintenance

An O&M manual, either written or electronic, must be developed and maintained on site or in a centrally accessible location for the working life of the applicable ventilation system equipment or components. This manual must be updated as necessary. The manual must include the O&M procedures, ventilation system operating schedules and any changes made thereto, final design drawings, maintenance schedules based on manufacturer instructions, and the maintenance requirements and frequencies provided in Table 12.9.

Table 12.9 Minimum Maintenance Activity and Frequency for Ventilation System Equipment and Associated Components
[Std 62.1-2019, Tbl 8-1]

Inspection/Maintenance Task	Frequency*
a. Investigate system for water intrusion or accumulation. Rectify as necessary.	As necessary
b. Verify that the space provided for routine maintenance and inspection of open cooling tower water systems, closed cooling tower water systems, and evaporative condensers is unobstructed.	Monthly
c. Open cooling tower water systems, closed cooling tower water systems, and evaporative condensers shall be treated to limit the growth of microbiological contaminants, including *legionella sp.*	Monthly
d. Verify that the space provided for routine maintenance and inspection of equipment and components is unobstructed.	Quarterly
e. Check pressure drop and scheduled replacement date of filters and air-cleaning devices. Clean or replace as necessary to ensure proper operation.	Quarterly
f. Check ultraviolet lamp. Clean or replace as needed to ensure proper operation.	Quarterly
g. Visually inspect dehumidification and humidification devices. Clean and maintain to limit fouling and microbial growth. Measure relative humidity and adjust system controls as necessary.	Quarterly
h. Maintain floor drains and trap primer located in air plenums or rooms that serve as air plenums to prevent transport of contaminants from the floor drain to the plenum.	Semiannually
i. Check ventilation and indoor air quality related control systems and devices for proper operation. Clean, lubricate, repair, adjust, or replace as needed to ensure proper operation.	Semiannually
j. Check P-traps in floor drains located in plenums or rooms that serve as air plenums. Prime as needed to ensure proper operation.	Semiannually
k. Check fan belt tension. Check for belt wear and replace if necessary to ensure proper operation. Check sheaves for evidence of improper alignment or evidence of wear and correct as needed.	Semiannually
l. Check variable-frequency drive for proper operation. Correct as needed.	Semiannually
m.Check for proper operation of cooling or heating coil for damage or evidence of leaks. Clean, restore, or replace as required.	Semiannually
n. Visually inspect outdoor air intake louvers, bird screens, mist eliminators, and adjacent areas for cleanliness and integrity; clean as needed; remove all visible debris or visible biological material observed and repair physical damage to louvers, screens, or mist eliminators if such damage impairs the item from providing the required outdoor air entry.	Semiannually
o. Visually inspect natural ventilation openings and adjacent areas for cleanliness and integrity; clean as needed. Remove all visible debris or visible biological material observed and repair physical damage to louvers, and screens if such damage impairs the item from providing the required outdoor air entry. Manual and/or automatic opening apparatus shall be physically tested for proper operation and repaired or replaced as necessary.	Semiannually

Table 12.9 Minimum Maintenance Activity and Frequency for Ventilation System Equipment and Associated Components
[Std 62.1-2019, Tbl 8-1] ***(Continued)***

Inspection/Maintenance Task	Frequency*
p. Verify the operation of the outdoor air ventilation system and any dynamic minimum outdoor air controls.	Annually
q. Check air filter fit and housing seal integrity. Correct as needed.	Annually
r. Check control box for dirt, debris, and/or loose terminations. Clean and tighten as needed.	Annually
s. Check motor contactor for pitting or other signs of damage. Repair or replace as needed.	Annually
t. Check fan blades and fan housing. Clean, repair, or replace as needed to ensure proper operation.	Annually
u. Check integrity of all panels on equipment. Replace fasteners as needed to ensure proper integrity and fit/finish of equipment.	Annually
v. Assess field serviceable bearings. Lubricate if necessary.	Annually
w. Check drain pans, drain lines, and coils for biological growth. Check adjacent areas for evidence of unintended wetting. Repair and clean as needed.	Annually
x. Check for evidence of buildup or fouling on heat exchange surfaces. Restore as needed to ensure proper operation.	Annually
y. Inspect unit for evidence of moisture carryover from cooling coils beyond the drain pan. Make corrections or repairs as necessary.	Annually
z. Check for proper damper operation. Clean, lubricate, repair, replace, or adjust as needed to ensure proper operation.	Annually
aa. Visually inspect areas of moisture accumulation for biological growth. If present, clean or disinfect as needed.	Annually
ab. Check condensate pump. Clean or replace as needed.	Annually
ac. Visually inspect exposed ductwork and external piping for insulation and vapor barrier for integrity. Correct as needed.	Annually
ad. Verify the accuracy of permanently mounted sensors whose primary function is outdoor air delivery monitoring, outdoor air delivery verification, or dynamic minimum outdoor air control, such as flow stations at an air handler and those used for demand-control ventilation. A sensor failing to meet the accuracy specified in the O&M manual shall be recalibrated or replaced. Performance verification shall include output comparison to a measurement reference standard consistent with those specified for similar devices in ANSI/ASHRAE Standard 41.2 or ANSI/ASHRAE Standard 111.	5 years
ae. Verify the total quantity of outdoor air delivered by air handlers set to minimum outdoor air mode. If measured minimum airflow rates are less than the design minimum rate documented in the O&M manual, ± a 10% balancing tolerance, (1) confirm the measured rate does not conform with the provisions of this standard and (2) adjust or modify the air-handler components to correct the airflow deficiency. Ventilation systems shall be balanced in accordance with ASHRAE Standard 111 or its equivalent, at least to the extent necessary to verify conformance with the total outdoor airflow and space supply airflow requirements of this standard. **Exception:** Units under 1000 L/s of supply air are exempt from this requirement.	5 years

* Minimum frequencies may be increased or decreased if indicated in the O&M manual.
** National Institute of Standards and Technology, U.S. Department of Commerce, Gaithersburg, MD.

Healthcare Facilities

ANSI/ASHRAE/ASHE Standard 170 provides detailed listings of design parameters for inpatient and outpatient space types based on function. These parameters include the following:

- Pressure relationship to adjacent areas
- Minimum outdoor air changes per hour
- Minimum total air changes per hour
- Requirement that all room air be exhausted directly to outdoors
- Whether air can be recirculated by room units
- Whether unoccupied turndown is permitted
- Minimum filter efficiency
- Design relative humidity
- Design temperature

13. ENVIRONMENTALLY RESPONSIBLE DESIGN

Climate Change

The ASHRAE Position Document on Climate Change states unequivocally that climate change "is the most formidable environmental challenge ever faced by society." Historically, designers and planners assumed a stable climate, and heating and cooling system designs were based on statistics calculated from past weather data, typically for a recent 30-year period. However, we are now experiencing major changes in climate both globally and locally, at rates 10 times greater than those seen since the end of the last ice age 20,000 years ago—over decades instead of centuries or millennia. This presents a challenge for designers of buildings and their systems, as well as for building operators, since buildings are expected to operate robustly for decades.

While there are currently no standardized methods endorsed by ASHRAE to integrate changing climatic information into building and HVAC&R system design, this does not preclude action by practicing engineers and architects. Repeatable and defensible methods can inform and enhance existing ASHRAE standards and guidelines, like those of other professional organizations such as The Chartered Institution of Building Services Engineers (CIBSE), American Society of Civil Engineers (ASCE), American Society of Mechanical Engineers (ASME), and American Association of State Highway and Transportation Officials (AASHTO). Resources for assessing the impact of a changing climate on building and system performance exist, both within and outside ASHRAE. It is important to recognize that, while climate change will have impacts of varying severity and duration around the world, many designers and operators may find that the buildings and systems they are designing or refurbishing currently will end up in a different climate during their operational life.

From the findings of peer-reviewed published research and assessments applying the basic principles of conservation of energy and thermal radiation governed by Planck's law, it is virtually certain that the observed increase in greenhouse gas (GHG) concentrations has had, and is having, a global warming effect, with resulting implications on severe weather, sea level rise, and other aspects of the Earth's climate system.

Temperatures are increasing in the atmosphere (troposphere), on land, and in the oceans. Global decadal averaged temperature (as calculated from instrumental records over both land and oceans) has increased by about 1.1°C since the lowest ten-year average temperature in the decade of 1900 to 1909. This change is by far the largest increase in the globally averaged temperature over at least the last 1700 years and probably much longer. Additionally, this increase is occurring at a much faster rate than the climate tends to change historically. The most recent six years are the six warmest in the instrumental record: 2014 (6th), 2015 (3rd), 2016 (1st), 2017 (4th), 2018 (5th), and 2019 (2nd).

While it has long been recognized that the Earth's climate can change because of natural influences, over the last century it has become increasingly clear that human-related forcings (i.e., drivers for changing the energy balance in the Earth system that lead to climate change) can also have, and are having, a large impact upon the climate system. Though climate drivers of significance over the industrial era include both types of forcings, the most significant factor in the large changes in climate over the last century has been the increasing concentrations in GHGs, especially carbon dioxide (CO_2), methane (CH_4), and nitrous oxide (N_2O). That is, it is very likely that more than half of the observed warming is from anthropogenic forces.

Certain types of extreme weather conditions have changed and have been attributed to global warming. The frequency and intensity of extreme heat has increased, and extreme cold has become less intense in most areas of the globe. Extreme precipitation has become more frequent at most long-term reporting stations, likely the result of larger amounts of water vapor over the warming oceans. Heating degree-days (HDDs) are projected to decrease in most places while cooling degree-days (CDDs) are expected to increase. This change is expected to be neither uniform across the globe nor monotonic for a given weather station. A monotonic change would imply that the given value always either increases or decreases for consecutive time steps but not both. Under a high-emissions (RCP8.5) scenario, climate zones, which are calculated from degree-days, are projected to change all around the world. Recent changes in zone classifications for North America are evident when comparing area classifications based on 2000–2019 data vs data from 1980–1999. These historical changes in climate zones are almost uniformly toward warmer zones (i.e., a lower number in the

ASHRAE system). The changes are clustered at the boundaries between climate zones, and the pattern is particularly clear in central North America. To accommodate warmer climates, a new zone 0 was added in 2016 to extend the climate zone scale.

Both heating and cooling design temperatures are increasing for large numbers of stations around the world and are projected to continue increasing.

Creating projected weather "input files" from global climate models (GCMs) suitable for use in building design and analysis is a complex process. For simulating estimated future building performance, reliable sets of future weather conditions are needed. Similarly, sizing calculations require estimates of future design conditions. Currently, there are no standardized methods to integrate information about a changing climate into building and system design and analysis, but this does not preclude action and the need for professional judgment. The output from GCMs can be useful for stress-testing buildings and systems using multiple plausible futures. This allows a designer to test the impact of multiple plausible acute or chronic climatic events such as heat waves or gradually increasing temperature over decades.

Limiting climate change requires a substantial and sustained global effort to significantly reduce anthropogenic GHG emission rates to avoid further warming and long-lasting changes in all components of the climate system. The consequences of a lack of timely action threaten the health and well-being of both current and future generations. Climate change mitigation (simply called *mitigation*) is an intentional human intervention to reduce the sources or enhance the sinks of GHGs in order to limit the magnitude of future warming. The ultimate goal of mitigation is to stabilize GHG concentrations in the atmosphere at a level that avoids dangerous changes in the climate system. Any changes in climate that are not avoided through mitigation must be considered by engineers as factors to adjust to through climate change adaptation. Mitigation can be understood as "avoiding the unmanageable," whereas adaptation can be understood as "managing the unavoidable."

ASHRAE GreenGuide: Design, Construction, and Operation of Sustainable Buildings, Fifth Edition, provides extensive guidance for designers and building operators to mitigate climate change by modifying their methods, designs, and operations for new buildings; using passive designs; appropriately retrofitting existing buildings by improving energy efficiency; constructing buildings using low-embodied energy materials; and eventually operating buildings using only renewable energy systems (RESs). RESs can be building integrated, on site, or off site. The pursuit of building electrification for both new and existing buildings using electric-powered systems and equipment instead of combustion-based systems will assist in easing the transition from GHG-emitting fossil-fuel sources to RESs. It is also important to reduce the use of fluorinated gases (F-gases) having high global warming potential (GWP) ratings while maintaining system efficiency.

Reduce Carbon Emissions by Design and Construction

Net zero or near zero buildings demonstrate the maximum level of climate change mitigation that can be achieved through building design and operation. These buildings consume minimal energy as limited by energy-efficiency opportunities and are then powered using carbon-free sources of energy. Environmentally friendly building design has progressed beyond simply reducing energy use. Architects, engineers, and other building professionals now aspire to achieving net zero buildings or even negative zero buildings where the building can supply more energy to the grid than it consumes overall. While there are many names, variations, and building certifications for it, net zero building design generally accounts for, and eliminates as much as possible, the carbon emissions from a building over its service life. At a minimum, net zero buildings are designed to use as little energy as possible and source the remaining energy needs from renewable, carbon-free sources. To be considered net zero, direct carbon emissions are usually eliminated completely. This is achieved by forbidding onsite fossil fuel combustion (with the possible exception of emergency generators at least for the near future) and by selecting only electrically powered components and systems for all end uses, since future energy will be provided either as grid or mini-grid electricity or through photovoltaic (PV) solar- or wind-generated on-site electricity. Net zero carbon buildings additionally track the carbon emissions associated with building construction (including embodied energy) and operation and, until energy is produced only by renewable energy, offset the carbon emissions by generating a surplus of renewable energy on site or procuring carbon offsets through a certified carbon offset program.

Table 13.1 Refrigerant Environmental Properties [2021F, Ch 36, Tbl 4]

Refrigerant	Atmospheric Lifetime, years	ODP	GWP100
CFC-11	52	1	5,160
CFC-113	93	0.81	6,080
CFC-114	85	0.5	8,580
CFC-115	540	0.26	7,310
CFC-12	102	0.73	10,300
CFC-13	640	1	13,900
HC-1270	0.001	0	<1
HC-290	0.034	0	<1
HC-600	12 ± 3[a]	0[a]	4[a]
HC-600a	0.016	0	<1
HC-601a	0.009[a]	0[a]	–20[a]
HCFC-123	1.3	0.01	80
HCFC-124	5.9	0.02	530
HCFC-142b	18	0.057	2,070
HCFC-22	12	0.034	1,780
HCFO1224yd(Z)	0.058[a]	0.00012[a]	<1[a]
HCFO1233zd(E)	0.071	0.00034	4
HE-E170	0.015[a]	0[a]	1[a]
HFC/HFC410A		0	1,920[b]
HFC-125	31	0	3,450
HFC-134a	14	0	1,360
HFC-143a	51	0	5,080
HFC-152a	1.6	0	148
HFC-227ea	36	0	3,140
HFC-23	228	0	12,690
HFC-236fa	222	0	7,680
HFC-245fa	7.9	0	880
HFC-32	5.4	0	705
HFO-1234yf	0.029	0	<1
HFO-1234ze(E)	0.045	0	<1
HFO1336mzz(E)	0.045–0.088	0	16
HFO1336mzz(Z)	0.06	0	2
PFC-116	10,000	0	11,100
PFC-218	2,600	0	8,900
PFC-c318	3,200	0	9,540
R-717	(few days)	—	<1
R-744		0	1

Atmospheric lifetime, ODP, and GWP100 from Table A-1 of Fahey et al. (2018) except where indicated.

[a] Chapter 29, 2017 *ASHRAE Handbook—Fundamentals*.

[b] IPCC (2013).

Net zero design fiscal prioritizations include:

- **Step 1:** Assess and prioritize the site and community resources such as optimizing zoning for renewable energy and building proximity to local or community waste heat resources.
- **Step 2:** Maximize architectural passive design strategies to meet property and occupant needs. Passive design strategies can include site planning, building orientation and massing, envelope tightness to reduce infiltration, continuous insulation, window and daylighting optimization, and designing a building to take advantage of natural ventilation and solar thermal opportunities.
- **Step 3:** Anticipate decarbonization of electric grids. Emphasize electric systems or provide dual fuel systems.
- **Step 4:** Maximize mechanical, electrical, and plumbing system efficiencies. Incorporate heat recovery systems, advanced controls, and
- **Step 5:** Offset remaining energy needs with on-site or purchased off-site renewable energy to reduce the facility's carbon footprint.

Climate Change Metrics

Carbon Dioxide Equivalent (MMTCO2Eq): A metric measure used to compare the emissions from various greenhouse gases (GHGs) based on their global warming potential (GWP). Carbon dioxide (CO_2) equivalents are commonly expressed as "million metric tons of carbon dioxide equivalents (MMTCO2Eq)." The CO_2 equivalent for a gas is derived by multiplying the tons of the gas by the associated GWP. MMTCO2Eq = (million metric tons of a gas) × (GWP of the gas).

Carbon Footprint: The total amount of GHGs that are emitted into the atmosphere each year by a person, family, building, organization, or company. A person's carbon footprint includes GHG emissions from fuel that an individual burns directly, such as by heating a home or riding in a car. It also includes GHGs that come from producing the goods or services that the individual uses, including emissions from power plants that make electricity, factories that make products, and landfills where trash gets sent.

CO_2-Equivalent Emission: The amount of CO_2 emission that would cause the same integrated radiative forcing, over a given time horizon, as an emitted amount of a GHG or a mixture of GHGs.

Global Warming Potential (GWP): An index developed to provide a simplified means of describing the relative ability of a chemical compound to affect radiative forcing, if emitted to the atmosphere, over its lifetime in the atmosphere, and thereby to affect the global climate. Radiative forcing reflects the factors that affect the balance between the energy absorbed by the earth and the energy emitted by it in the form of longwave infrared radiation. GWP is defined on a mass basis relative to CO_2. The GWP for a compound must be calculated up to a particular integrated time horizon, for example, 20, 100, or 500 years. The time horizon most widely accepted is 100 years (GWP100).

Ozone Depletion Potential (ODP): A numerical quantity describing the extent of ozone depletion calculated to arise from the release to the atmosphere of 1 kg of a compound relative to the ozone depletion calculated to arise from a similar release of the refrigerant R-11. The calculation is an integration of all known potential effects on ozone over the whole time that any portion of the compound could remain in the atmosphere.

Energy Efficiency

Most energy used in buildings is from nonrenewable resources, the cost of which historically has not considered replenishment or environmental impact. Thus, consideration of energy use in design has been based primarily on economic advantages, which are weighted to encourage more rather than less use.

As resources become less readily available and more exotic, and as replenishable sources are investigated, the need to operate buildings effectively using less energy becomes paramount. Extensive studies since the mid-1970s have shown that building energy use can be significantly reduced by applying the fundamental principles discussed in the following paragraphs.

Energy Ethic-Resource Conservation Design Principles. The basic approach to energy-efficient design is reducing loads (power), improving transport systems, and providing efficient components and "intelligent" controls. Important design concepts include understanding the relationship between energy and power, maintaining simplicity, using self-imposed budgets, and applying energy-smart design practices.

Energy and Power. From an economic standpoint, more energy-efficient systems need not be more expensive than less-efficient systems. Quite the opposite is true because of the simple relationship between energy and power, in which power is simply the time rate of energy use (or, conversely, energy is power times time). The power term *kilowatts* is used in expressing the size of a motor, chiller, boiler, or transformer. Generally, the smaller the equipment, the less it costs. Other things being equal, as smaller equipment operates over time, it consumes less energy. Thus, in designing for energy efficiency, the first objective is always to reduce the power required to the bare minimum necessary to provide the desired performance, starting with the building's heating and cooling loads (a power term, in *kilowatts*) and continuing with the various systems and subsystems.

Simplicity. Complex designs to save energy seldom function in the manner intended unless the systems are continually managed and operated by technically skilled individuals. Experience has shown that long-term, energy-efficient performance with a complex system is seldom achievable. Further, when complex systems are operated by minimally skilled individuals, both energy efficiency and performance suffer. Most techniques discussed in this chapter can be implemented with great simplicity.

Self-Imposed Budgets. Just as an engineer must work to a cost budget with most designs, self-imposed power budgets can be similarly helpful in achieving energy-efficient design. The Advanced Energy Design Guide series, published by ASHRAE, provides guidance on achievable design budgets. For example, the following are possible categories of power (or power-affecting) design budgets for a mid-rise office building:

- Installed lighting (overall), W/m^2
- Space sensible cooling, W/m^2
- Space heating load, W/m^2
- Electric power (overall), W/m^2
- Thermal power (overall), W/m^2
- Hydronic system head, kPa
- Water chiller (water-cooled), kW/kW (COP)
- Chilled-water system auxiliaries, kW/kW
- Unitary air-conditioning systems, kW/kW (COP)
- Annual electric energy, kWh/(m^2·yr)
- Annual thermal energy, kJ/(m^2·yr·K·day)

As the building and systems are designed, all decisions become interactive as each subsystem's power or energy performance is continually compared to the budget.

Design Process for Energy-Efficient Projects. Consider energy efficiency at the beginning of the building design process, because energy-efficient features are most easily and effectively incorporated at that time. Seek the active participation of all members of the design team, including the owner, architect, engineer, and often the contractor, early in the design process. Consider building attributes such as function, form, orientation, window-to-wall ratio, and HVAC system types early in the process, because each has major energy implications. Identify meaningful energy performance benchmarks suited to the project and set project-specific goals. Energy benchmarks for a sample project are shown in Table 13.2. Consider energy resources, on-site energy sources, and use of renewable energy, credits, utility rebates, or carbon offsets to mitigate environmental impacts of energy use.

Table 13.2 Example Benchmark and Energy Targets for University Research Laboratory [2021F, Ch 35, Tbl 1]

	Gross	Lit/ Conditioned						
Building area, m^2	15 793	10 266						
Electric	**Electricity for Lighting**	**Electricity for Ventilation (Fans)**	**Electricity for In-Building Pumps**	**Electricity for Plug Loads**	**Electricity for Unidentified Loads**	**Total Electricity**	**Cogenerated Electricity**	**Grid Electricity**
Design load, W/m^2	5.60	5.38	6.46	10.4	—	28.0	—	
Peak demand, W/m^2	4.52	5.38	4.52	7.86	0.0017	22.3	—	
Peak demand, kW (Projected submetered peak)	71	85	72	124	20	372	—	
Annual consumption, kWh/yr (Projected submetered reading)	218 154	346 598	191 245	891 503	175 200	1 823 000	966 000	857 000
Annual use index goal, kWh/yr	1.28	2.04	1.12	5.24	1.03	10.72		
Annual use index goal, site MJ/m^2 gross·yr	4378	6956	3838	17 893	3516	36 583		
Annual use index, kWh/m^2 gross·yr*	27.0 to 35.7	48.2 to 74.0	included elsewhere	47.3 to 61.0	NA	158.7 to 192.8		
Annual use index, site MJ/m^2 gross·yr*	97.3	173.6	—	170.1	—	571.1 to 694.0		

*From Labs21 program of U.S. Environmental Protection Agency (EPA) and U.S. Department of Energy (DOE). See labs21benchmarking.lbl.gov.

Address a building's energy requirements in the following sequence:

1. Minimize the impact of the building's functional requirements by analyzing how the building relates to its external environment. Advocate changes in building form, aspect ratio, and other attributes that reduce, redistribute, or delay (shift) loads. The load calculation should be interactive so that the effect of those factors can be seen immediately.
2. Minimize loads by analyzing external and internal loads imposed on the building's energy-using subsystems, both for peak- and part-load conditions. Design for efficient and effective operation off-peak, where the majority of operating hours and energy use typically occurs.
3. Maximize subsystem efficiency by analyzing the diversified energy and power requirements of each energy-using subsystem serving the building's functional requirements. Consider static and dynamic efficiencies of energy conversion and energy transport subsystems, and consider opportunities to reclaim, redistribute, and store energy for later use.
4. Study alternative ways to integrate subsystems into the building by considering both the power and time components of energy use. Identify, evaluate, and design each of these components to control overall design energy consumption. Consider the following when integrating major building subsystems:
   - Address more than one problem at a time when developing design solutions, and make maximum use of the building's advantageous features (e.g., windows, structural mass).
   - Examine design solutions that consider time (i.e., when energy use occurs), because sufficient energy may already be present from the environment (e.g., solar heat, night cooling) or from internal equipment (e.g., lights, computers) but available at times different from when needed. Storage techniques that are active (e.g., heat pumps with water tanks) or passive (e.g., building mass) may need to be considered.
   - Examine design solutions that consider the anticipated use of space. For example, in large but relatively unoccupied spaces, consider task or zone lighting. Consider transporting excess energy (light and heat) from locations of production and availability to locations of need instead of purchasing additional energy.
   - Evaluate the economic benefit of energy recovery and heat recovery systems whenever waste energy is available at temperatures usable for space conditioning or other practical purposes.
   - Consider or advocate design solutions that provide more comfortable surface temperatures or increase the availability of controlled daylight in buildings where human occupancy is a primary function.
   - Use easily understood design solutions, because they have a greater probability of use by building operators and occupants.
   - Where the functional requirements of a building are likely to change over time, design the installed environmental system to adapt to meet anticipated changes and to provide flexibility in meeting future changes in use, occupancy, or other functions.
   - Develop energy performance benchmarks, metrics, and targets that allow building owners and operators to better achieve the design intent.
   - Differentiate between peak loads for system design and selection and lower operating loads that determine actual energy use.

Building Energy Use Elements

Envelope

- Control thermal conductivity by using insulation (including movable insulation), thermal mass, and/or phase-change thermal storage at levels that minimize net heating and cooling loads on a time-integrated (annual) basis.
- Minimize unintentional or uncontrolled thermal bridges and include them in energy-related calculations because they can radically alter building envelope conductivity. Examples include wall studs, balconies, ledges, and extensions of building slabs.

- Minimize infiltration so that it approaches zero. (An exception is when infiltration provides the sole means of ventilation, such as in small residential units.) This minimizes fan energy consumption in pressurized buildings during occupied periods and minimizes heat loss (or unwanted heat gain, in warm climates) during unoccupied periods. In warm, humid climates, a tight envelope also improves IAQ. Reduce infiltration through design details that enhance the fit and integrity of building envelope joints in ways that may be readily achieved during construction (e.g., caulking, weatherstripping, vestibule doors, revolving doors), with construction meeting accepted specifications. Building envelope commissioning or testing can help verify these design and construction targets.
- Consider operable windows to allow occupant-controlled ventilation. This requires careful design of the building's mechanical system to minimize unnecessary HVAC energy consumption, and building operators and occupants should be cautioned about improper use of operable windows. CIBSE also provides comprehensive design considerations for natural ventilation. See *ASHRAE Design Guide for Natural Ventilation* for more detailed design information.
- Strive to maintain occupant radiant comfort regardless of whether the building envelope is designed to be a static or a dynamic membrane. Design opaque surfaces so that average inside surface temperatures remain within 3°C of room temperature in the coldest anticipated weather (i.e., winter design conditions) and so that the coldest inside surface remains within 14°C of room temperature (but always above the indoor dew point). In a building with time-varying internal heat generation, consider thermal mass for controlling radiant comfort. In the perimeter zone, thermal mass is more effective when it is positioned inside the envelope's insulation.
- Effective control of solar radiation is critical to energy-efficient design because of the high level of internal heat production in most commercial buildings. In some climates, lighting energy consumption savings from daylighting techniques can be greater than the heating and cooling energy penalties that result from the additional glazed surface area required, if the building envelope is properly designed for daylighting and lighting controls are installed and used. In other climates and in buildings with light emitting diode (LED) lighting, there may not be net savings. Daylighting designs are most effective if direct solar beam radiation is not allowed to cause glare in building spaces.
- Design transparent parts of the building envelope to prevent solar radiant gain above that necessary for effective daylighting and solar heating. On south-facing facades (in the northern hemisphere), using low shading coefficients is generally not as effective as using external physical shading devices in achieving this balance. Consider low-emissivity, high-visible-transmittance glazings for effective control of radiant heat gains and losses. For shading control, judicious use of vegetation may block excess gain year round or seasonally, depending on the plant species chosen.
- Use the envelope and roof for photovoltaic or solar thermal panels. The solar energy panels can be part of the envelope (building integrated photovoltaic [BIPV]) or added to roofs and shading devices.

Lighting

Lighting is both a major energy end use in commercial buildings (especially office buildings) and a major contributor to internal loads by increasing cooling loads and decreasing heating loads. Design should both meet the lighting functional criteria of the space and minimize energy use. The Illuminating Engineering Society (IES) recommends illuminance levels for visual tasks and surrounding lighted areas. Principles of energy-conserving design within that context include the following:

- Energy use is determined by the lighting load (demand power) and its duration of use (time). Minimize actual demand load rather than just apparent connected load. Control the load rather than just area switching, if switching may adversely affect the quality of the luminous environment.
- Consider daylighting with proper controls to reduce costs of electric lighting. Design should be sensitive to window glare, sudden changes in luminance, and general user acceptance of daylighting controls. Carefully select window treatment (blinds,

drapes, and shades) and glazing to control direct solar penetration and luminance extremes while maintaining the view and daylight penetration.

- Design the lighting system so that illumination required for tasks is primarily limited to the location of the task and comes from a direction that minimizes direct glare and veiling reflections on the task. When the design is based on nonuniform illuminance, walls should be a light to medium color or illuminated to provide visual comfort. In densely occupied work spaces, uniform distribution of general lighting may be most appropriate. Where necessary, provide supplementary task illumination. General ambient illumination should not be lower than one-third of the luminance required for the task, to help maintain visually comfortable luminance ratios.
- Use local task lighting to accommodate needs for higher lighting levels because of task visual difficulty, glare, intermittently changing requirements, or individual visual differences (poor or aging eyesight).
- Group similar activities so that high illuminance or special lighting for particular tasks can be localized in certain rooms or areas and so that less-efficient fixtures required for critical glare control do not have to be installed uniformly when they are only required sparsely.
- Use lighting controls throughout so lighting is available when and where it is needed but not wasted when tasks are less critical or spaces are not fully occupied. Also consider user acceptance of control strategies to maximize energy savings.
- Use LED lighting. Only use incandescent and fluorescent lamps in applications where their characteristics cannot be duplicated by other sources, because manufacturing of most incandescent and fluorescent lamps will be discontinued during the life of the building.
- Carry lighting design through the rest of the building's interior design. Reduced light absorption may be achieved by using lighter finishes, particularly on ceilings, walls, and partitions.

Other Loads

- Minimize the thermal impact of equipment and appliances on HVAC systems by using hoods, radiation shields, or other confining techniques, and by using controls to turn off equipment when not needed. Where practical, locate major heat-generating equipment where it can balance other heat losses. Computer centers or kitchen areas usually have separate, dedicated HVAC equipment. In addition, consider heat recovery for this equipment.
- Use storage techniques to level or distribute loads that vary on a time or spatial basis to allow operation of a device at maximum (often full-load) efficiency.

HVAC System Design

- Consider separate HVAC systems to serve areas expected to operate on widely differing operating schedules or design conditions. For instance, systems serving office areas should generally be separate from those serving retail areas.
- Arrange systems so that spaces with relatively constant, weather-independent loads are served by systems separate from those serving perimeter spaces. Areas with special temperature or humidity requirements (e.g., computer rooms) should be served by systems separate from those serving areas that require comfort heating and cooling only. Alternatively, provide these areas with supplementary or auxiliary systems.
- Sequence the supply of zone cooling and heating to prevent simultaneous operation of heating and cooling systems for the same space, to the extent possible. Where this is not possible because of ventilation, humidity control, or air circulation requirements, reduce air quantities as much as possible before incorporating reheating, recooling, or mixing hot and cold airstreams. For example, if reheat is needed to dehumidify and prevent overcooling, only ventilation air needs to be treated, not the entire recirculated air quantity. Finally, reset supply air temperature to the extent possible to reduce reheating, recooling, or mixing losses.
- Provide local and central controls to allow operation in occupied and unoccupied modes. In unoccupied mode, ventilation and exhaust systems should be shut off centrally or locally when possible, and comfort heating and cooling systems should be

shut off except to maintain space conditions ready for the next occupancy cycle. Use standby mode with no ventilation for spaces that are unoccupied during the occupied mode but maintain occupied space temperature set points.

- In geographical areas where diurnal temperature swings and humidity levels permit, consider judicious coupling of air distribution and building structural mass to allow nighttime cooling to reduce the requirement for daytime mechanical cooling.
- Use heat recovery and dedicated outdoor air systems (DOASs) to reduce the cooling, dehumidification, humidification, and heating load to reduce the size of components such as coils, pipes, pumps, chillers, boilers, transformers, and other electrical equipment and to reduce the energy consumption. See Chapter 51 of the 2020 *ASHRAE Handbook—HVAC Systems and Equipment* and/or *ASHRAE Design Guide for Dedicated Outdoor Air Systems* for more details on DOASs.

HVAC Equipment Selection

- To allow HVAC equipment operation at the highest efficiencies, match conversion devices to load increments, and sequence the operation of modules. Oversized or large-scale systems should never serve small seasonal loads (e.g., a large heating boiler serving a summer-service water-heated load). Include specific low-load units and auxiliaries where prolonged use at minimal capacities is expected.
- Select the most efficient (or highest-COP) equipment practical at both design and reduced capacity (part-load) operating conditions.
- When selecting large-power devices such as chillers (including their auxiliary energy burdens), perform an economic analysis of the complete life-cycle costs. See Chapter 37 of the 2019 *ASHRAE Handbook—HVAC Applications* for more information on detailed economic analysis.
- Keep fluid temperatures for heating equipment devices as low as practical and for cooling equipment as high as practical, while still meeting loads and minimizing flow quantities.

Energy Transport Systems

Energy should be transported as efficiently as possible. The following options are listed in order of theoretical efficiency, from the lowest energy transport burden (most efficient) to the highest (least efficient):

1. Electric wire or fuel pipe
2. Two-phase fluid pipe (steam or refrigerant)
3. Single-phase liquid/fluid pipe (water, glycol, etc.)
4. Air duct

Select a distribution system that complements other parameters such as control strategies, storage capabilities, conversion efficiency, and utilization efficiency. The following specific design techniques may be applied to thermal energy transport systems:

Steam Systems

- Include provisions for seasonal or non-use shutdown.
- Require that all steam pipe and components are well insulated.
- Minimize venting of steam and ingestion of air, with design directed toward full-vapor performance.
- Avoid subcooling, if practical.
- Return condensate to boilers or source devices at the highest possible temperature.

Hydronic Systems

- Minimize flow quantity by designing for the maximum practical temperature range.
- Vary flow quantity with load where possible.
- Design for the lowest practical pressure loss in the critical pressure path. Require that the differential pressure set point is reset when at part flow.
- Provide operating and idle control modes.
- When locating equipment, identify the critical pressure path and size runs for the minimum reasonable pressure drop.

Air Systems

- Minimize airflow by careful load analysis and an effective distribution system. If the application allows, supply air quantity should vary with sensible load (i.e., variable-air-volume [VAV] systems). Hold the fan pressure requirement to the lowest practical value and avoid using fan pressure as a source for control power. Require that the static pressure set point is reset when at part flow.
- Provide normal and idle control modes for fan and psychrometric systems.
- Keep duct runs as short as possible, and keep runs on the critical pressure path sized for minimum practical pressure drop.

Power Distribution

- Size transformers and generating units as closely as possible to the actual anticipated load (i.e., avoid oversizing to minimize fixed thermal losses).
- Consider distribution of electric power at the highest practical voltage and load selection at the maximum power factor consistent with safety.
- Consider tenant submetering in commercial and multifamily buildings as a cost-effective energy conservation measure. A large portion of energy use in tenant facilities occurs simply because there is no economic incentive to conserve.

Domestic Hot-Water Systems

- Choose showerheads that provide and maintain user comfort and energy savings. They should not have removable flow-restricting inserts to meet flow limitation requirements.
- Consider point-of-use water heaters where their use will reduce energy consumption and annual energy cost.
- Consider using storage to facilitate heat recovery when the heat to be recovered is out of phase with the demand for hot water or when energy use for water heating can be shifted to take advantage of off-peak rates.

Controls

Well-designed digital control provides information to managers and operators as well as to the data processor that serves as the intelligent controller. Include the energy-saving concepts discussed previously throughout the operating sequences and control logic. Energy conservation should not be sought at the expense of adequate performance, however; in a well-designed system, these two parameters are compatible. See Chapter 7 of the 2021 *ASHRAE Handbook—Fundamentals*, Chapter 48 of the 2019 *ASHRAE Handbook—HVAC Applications*, and ASHRAE Guideline 36 for more information on controls.

Commissioning

Building systems are typically complicated and often do not work as intended. This causes reduced comfort levels, reduced IAQ, and/or increased energy consumption. A commissioning process of documentation, adjusting, testing, verification, and training to ensure that the finished facility operates in accordance with the Owner's Project Requirements and the construction documents will improve the performance of the building. As a result of the commissioning process, the facility will be more comfortable, have better IAQ, and/or use less energy. ANSI/ASHRAE/IES Standard 90.1 recognizes this and requires commissioning of the mechanical systems.

ASHRAE has developed an extensive suite of commissioning standards and guidelines that deal with new and existing projects, including ANSI/ASHRAE/IES Standard 202, *Commissioning Process for Buildings and Systems*; ASHRAE Guideline 0, *The Commissioning Process*; and ASHRAE Guideline 0.2, *Commissioning Process for Existing Systems and Assemblies*.

Energy Design Resources

ASHRAE, in conjunction with partner organizations, has developed The Advanced Energy Design Guide series (www.ashrae.org/aedgs), which includes design recommendations to achieve three levels of energy efficiency for a range of building types.

Green Design and Sustainability

Sustainability is defined in *ASHRAE GreenGuide*, Fifth Edition, in general terms, as "providing for the needs of the present without detracting from the ability to fulfill the needs of the future"—a definition very similar to that developed in 1987 by the United Nations' Brundtland Commission. Others have defined sustainability as "the concept of maximizing the effectiveness of resource use while minimizing the impact of that use on the environment" and an environment in which "an equilibrium… exists between human society and stable ecosystems." Sustaining (i.e., keeping up or prolonging) those elements on which humankind's existence and that of the planet depend, such as energy, the environment, and health, are worthy goals.

Sustainability Addresses the Future. Sustainability is focused on the future. Actions taken under the name of sustainability must address the impact of present actions on conditions likely to prevail in that future time frame. As stressed in Chapter 36 of the 2021 *ASHRAE Handbook—Fundamentals*, climates are changing at a faster rate than previously thought. In designing the built environment, the emphasis has often been on the present or the near future, usually in the form of capital (or first-cost) impact. As is apparent when life-cycle cost analysis is applied, capital cost assumes less importance the longer the future period under consideration.

Sustainability Has Many Contributors. Sustainability is not just about energy, carbon emissions, pollution, waste disposal, or population growth. Although these are central ideas in thinking about sustainability, it is an oversimplification to think that addressing one factor, or even any one set of factors, can result in a sustainable future for the planet.

It is likewise a mistake to think that HVAC&R design practitioners, by themselves and just through activities within their purview, can create a sustainable result. To be sure, their activities can contribute to sustainability by creating a sustainable building, development, or other related project. But they cannot by themselves create global sustainability. Such an endeavor depends on many outside factors that cannot be controlled by HVAC&R engineers; however, these engineers should make their fair-share contribution to sustainability in all their endeavors and encourage other individuals and entities to do the same.

Sustainability is Comprehensive. Sustainability has no borders or limits. A good-faith effort to make a project sustainable does not mean that sustainability will be achieved globally. A superb design job on a building with sustainability as a goal will probably not contribute much to the global situation if a significant number of other buildings are not so designed, or if the transportation sector makes an inadequate contribution, or if only a few regions of the world do their fair share toward making the planet sustainable. A truly sustainable outcome depends on comprehensive efforts in all sectors the world around.

Sustainable versus Green. An emphasis on the distant future can differentiate sustainable design from green design. Whereas green design addresses many of the same characteristics as sustainable design, it may also emphasize near-term impacts such as indoor environmental quality, operation and maintenance features, and meeting current client needs. Thus, green design may focus more on the immediate future (i.e., starting when the building is first constructed and then occupied). Sustainable design is of paramount importance to the global environment in the long term while still incorporating features of green design that focus on the present and near future.

The design process is the first crucial element in producing a green building. For design efficiency, it is necessary to define the owner's objectives and criteria, including sustainable/green goals, before beginning the design so as to minimize potential increased design costs. Once designed, the building must be constructed, its performance verified, and it must be operated in a way that supports the green concept. If it is not designed with the intent to make it green, the desired results will never be achieved.

Figure 13.1 conceptually shows the impact of providing design input at succeeding stages of a project, relative to the cost and effort required. The solid curve shows it is much easier to have a major impact on the performance (potential energy savings, water efficiency, maintenance costs, etc.) of a building if one starts at the very earliest stages of the design process. The available impacts diminish thereafter as one proceeds through the subsequent design and construction phases. A corollary to this is the cost of implementing changes to improve build-

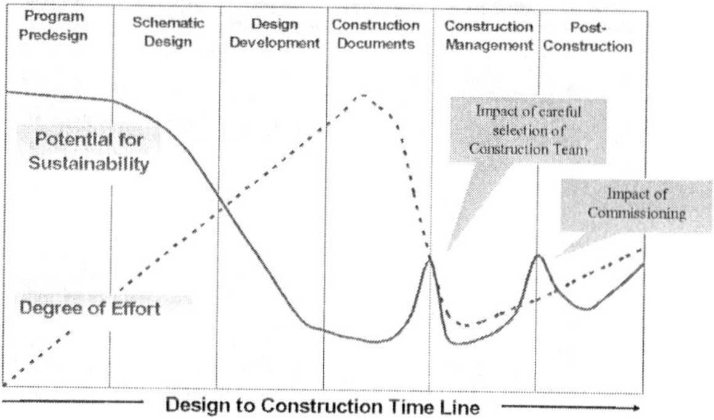

(Image courtesy of Malcolm Lewis, CTG Energetics, Inc.)

Figure 13.1 The Changing Potential for Sustainability over a Project Time Line and the Impact of Commissioning [ASHRAE GreenGuide 5th, Fig 3-2]

ing performance rises at each successive stage of the project (cost is shown as the dotted curve of the graph).

HVAC&R designers know how to design buildings that are much more energy efficient than they have been in the past, but such buildings are still relatively rare, especially in the general commercial market (as opposed to those owned by high-profile entities). ASHRAE's long-standing guidance in designing energy-efficient (now green and/or sustainable) buildings, and the motivation provided by its own and other entities' programs, have pointed the way technologically for the built environment and related industries to make their fair-share contribution to sustainability. Such programs include 1) ASHRAE's net-zero energy buildings (NZEB) thrust, 2) the U.S. Green Building Council (USGBC) Leadership in Energy and Environmental Design (LEED®) Green Building Rating System™, 3) the American Institute of Architects (AIA) 2030 Challenge, 4) the Green Building Institute (GBI) Green Globes program, and 5) the U.S. Environmental Protection Agency (EPA) ENERGY STAR® program.

Embodied Energy and Carbon Emissions

As buildings become more energy efficient and their operational energy is reduced, emphasis will shift toward reducing their embodied energy. Materials used directly in the construction and the components, equipment, and systems for building operation embody the energy and carbon used during their manufacturing, transportation, and installation. Material selection should also consider the environmental impact of demolition and disposal after the service life of the products. Building life-cycle assessment (LCA) focuses on the environmental impact of a product system (from materials acquisition to manufacturing, use, and final disposition) and plays a major role in promoting sustainability. This is a cradle-to-grave approach that evaluates all stages of individual materials and the product's life to determine their cumulative environmental impact. Designers should give preference to resource-efficient materials and reduce waste by recycling and reusing whenever possible.

LCA is an internationally standardized methodology (ISO Standard 14040). Consecutive parts of an LCA include a life-cycle inventory (e.g., collection and analysis of air and water emissions, waste generation, and resource consumption over the product's life) and life-cycle impact assessment (LCIA), which is an estimation of indicators of the environmental pressures in terms of climate change, summer smog, resource depletion, acidification, human health effects, etc., attributable to the product's life cycle. More information on the LCA approach is available online from the EPA (www.epa.gov/saferchoice/design-environment-life-cycle-assessments), the European Commission (ec.europa.eu/environment/ipp/lca.htm), and the Carbon Leadership Forum (http://carbonleadershipforum.org). Various LCA databases and tools available from commercial as well as governmental or public-domain sources can be used to calculate and compare the embodied energy (e.g., total energy per unit mass), related embodied emissions (e.g., total mass of CO_2 per unit mass of material or product), or other embodied environmental impacts of common building materials and products.

A challenge for modeling life-cycle energy use in buildings is using consistent system boundaries and data collection. Available software to address these concerns include Building for Environmental and Economic Sustainability (BEES; www.nist.gov/services-resources/software/bees), the Athena Sustainable Materials Institute's Environmental Impact Estimator (www.athenasmi.org/what-we-do/lca-data-software/), OneClick LCA (www.oneclicklca.com), and the Embodied Carbon in Construction Calculator (EC3) (www.buildingtransparency.org/en/). Regional data are important because conversion factors to primary energy and GHG emissions can differ by country, depending on energy sources used (e.g., coal- or oil-fired power plants versus solar or natural-gas-based generation).

LCA-based information may be in the form of environmental product declarations (EPD) based on CEN Standard 15804 or the product environmental footprints (PEFs) inspired by ISO Standards 14040 and 14044 and voluntary environmental declarations (ISO Standard 14025). They have been gaining in popularity beyond building construction materials, given the growing criteria of green building certifications such as LEED v4 credit, which now rewards selection of HVAC products with EPDs, based on the updated LEED credit interpretation in early 2015.

LCA can also assess specific refurbishments intended to improve energy performance of systems in existing buildings. For example, replacing electric or gas water heaters with solar hot-water systems can provide net emissions savings compared with the conventional systems after 0.6 month to 2.5 years, depending on the auxiliary fuel. For solar domestic hot-water systems and solar central space heating, the energy consumed by producing and installing the solar systems is recovered in about 1.2 years, and the payback time for the systems' embodied energy emissions varies from a few months (for solar domestic water heating) to 9.5 years (for solar central space heating), again depending on the energy carrier for the conventional system and the specific environmental emission indicators considered. The energy payback times of photovoltaic systems have been shown to be between 0.75 and 2 years for a roof-mounted system in a location with 1700 kWh per m^2 irradiation per year and will vary with climate, type of photovoltaic technology, mounting, and other system parameters.

Table 14.1 Characteristics of AC Motors (Nonhermetic) [2020S, Ch 45, Tbl 4]

	Split-Phase	Permanent Split-Capacitor	Capacitor-Start Induction-Run	Capacitor-Start Capacitor-Run	Shaded-Pole	Polyphase, 60-Hz
Connection Diagram		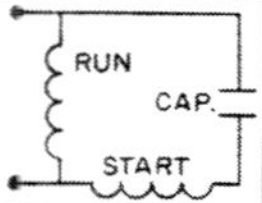	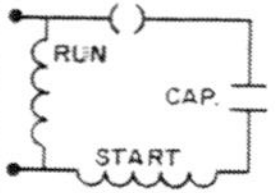	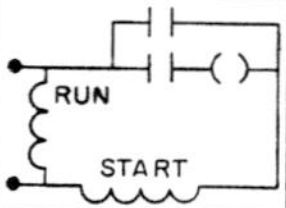		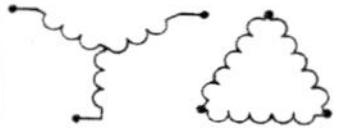
Speed Torque Curves	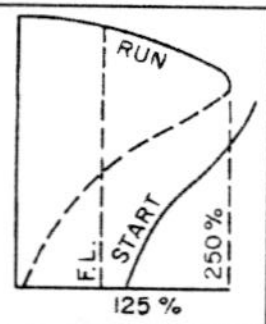	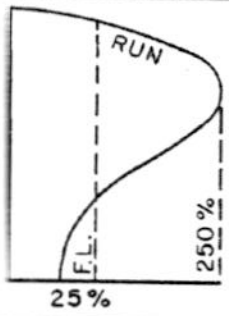	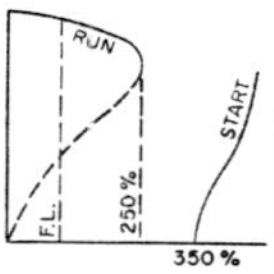	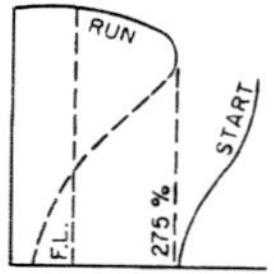	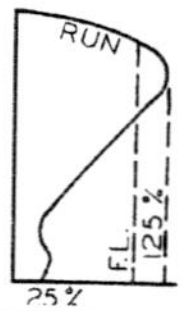	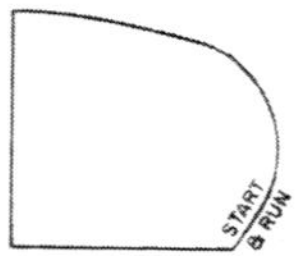
Starting Method	Centrifugal switch	None	Centrifugal switch	Centrifugal switch	None	Motor controller
Ratings, kW	0.04 to 0.4	0.04 to 3.7	0.04 to 3.7	0.04 to 3.7	0.04 to 0.19	0.4 and up
Full-Load Speeds at 60-Hz (Two-Pole, Four-Pole)	3450 to 1725	3450 to 1725	3450 to 1725	3500 to 1750	3100 to 1550	3500 to 1750
Torque[a] **Locked Rotor Breakdown**	125% to 150% 250% to 300%	25% 250% to 300%	250% to 350% 250% to 300%	250% 250%	25% 125%	150% to 350% 250% to 350%
Speed Classification	Constant	Constant	Constant	Constant	Constant or adjustable	Constant
Full-Load Power Factor	60%	95%	65%	95%	60%	80%
Efficiency	Medium	High	Medium	High	Low	High-Medium

[a] Expressed as percent of rated horsepower torque.

Table 14.2 Motor and Motor Control Equipment Voltages (Alternating Current) [2020S, Ch 45, Tbl 1]

System Nominal Voltage	U.S. Domestic Equipment Nameplate Voltage Ratings (60 Hz)			
	>1 kW Motors		<1 kW Motors	
	Three-Phase	Single-Phase	Three-Phase	Single-Phase
120	—	115	—	115
208	208/230 or 200/230	208/230 or 200/230	208/230 or 200/230	208/230 or 200/230
240	208/230 or 200/230	208/230 or 200/230	208/230 or 200/230	208/230 or 200/230
277	—	265	—	265
480	460	—	460	—
600*	575	—	575	—
2 400	2 300	—	—	—
4 160	4 000	—	—	—
4 800	4 600	—	—	—
6 900	6 600	—	—	—
13 800	13 200	—	—	—

*Some control and protective equipment has maximum voltage limit of 600 V. Consult manufacturer, power supplier, or both to ensure proper application.

System Nominal Voltage	International Equipment Nameplate Voltage Ratings			
	50 Hz		60 Hz	
	Three-Phase	Single-Phase	Three-Phase	Single-Phase
127	—	127	—	127
200	220/200	200	230/208 or 230/200	—
220	220/240	220/240 or 230/208	230/208 or 230/200	230/208
230	230/208	220/240 or 230/208	230/208 or 230/200	230/208
240	230/208	220/240	230/208	230/208
250	—	250	—	—
380	380/415	—	460/380	—
400	380/415	—	—	—
415	380/415	—	—	—
440	440	—	460	—
480	500	—	—	—

Note: Primary operating voltage for a dual-voltage rating is usually listed first (e.g., 220 is primary for a 220/240 volt rating).

Table 14.3 Motor Full-Load Amperes

Nominal Power, kW	Recommended Starter Size Three Phase		Three-Phase AC Squirrel-Cage and Wound-Rotor (Induction Type)			Recommended Starter Size Single Phase	Single-Phase AC			Nominal Power, kW
	230 V	460 V	200 V	230 V	460 V	230 V	115 V	200 V	230 V	
0.12						00	4.4	2.5	2.2	0.12
0.19						00	5.8	3.3	2.9	0.19
0.37	00	00	2.3	2	1	00	9.8	5.6	4.9	0.37
0.56	00	00	3.2	2.8	1.4	00	13.8	7.9	6.9	0.56
0.75	00	00	4.1	3.6	1.8	00	16	9.2	8	0.75
1.12	00	00	6.0	5.2	2.6	0	20	11.5	10	1.12
1.5	0	00	7.8	6.8	3.4	0	24	13.8	12	1.5
2.2	0	0	11.0	9.6	4.8	1	34	19.6	17	2.2
3.7	1	0	17.5	15.2	7.6	1	56	32.2	28	3.7
5.6	1	1	25.3	22	11	2	80	46	40	5.6
7.5	2	1	32.2	28	14	2	100	57.5	50	7.5
11	2	2	48.3	42	21	3				11
15	3	2	62.1	54	27					15
19	3	2	78.2	68	34					19
22	3	3	92	80	40					22
30	4	3	119.6	104	52					30
37	4	3	149.5	130	65					37
45	5	4	177.1	154	77					45
56	5	4	220.8	192	96					56
75	5	4	285.2	248	124					75
93	6	5	358.8	312	156					93
112	6	5	414	360	180					112
150	6	5	552	480	240					150

Values are for motors with normal torque characteristics running at usual belted speeds.

Table 14.4 Useful Electrical Formulas

To Find	Direct Current	Single Phase	Three Phase
Amperes when power known	$1000P\eta$	$\frac{1000P}{(\eta F)}$	$\frac{1000P}{(1.73\eta F)}$
Amperes when kVA known	—	1000 kVA/η	1000 kVA/(1.73η)
kVA	—	$IE/1000$	1.73 $IE/1000$
Power out, kW	$\eta IE/1000$	$\eta IEF/1000$	1.73 $\eta IEF/1000$

I = amperes; E = volts; η = efficiency expressed as decimal; F = power factor; P = kilowatts; kVA = kilovolt-amperes.

Motor Protection and Control

In general, four functions are accomplished by motor protection and control. Separate or integral control components are provided to (1) disconnect the motor and controller from the power supply and protect the operator; (2) start and stop the motor and, in some applications, control the speed or direction of rotation; (3) protect motor branch circuit conductors and control apparatus against short-circuiting; and (4) protect the motor itself from overloading and overheating.

Variable-Frequency Drives (VFDs)

By far the most energy-efficient means of varying flow of fans and pumps driven by electric motors are VFDs. Their application involves careful consideration of their effects (here VFD is considered synonymous with variable-speed drive [VSD], pulse-width modulated drive [PWM drive], adjustable-speed drive [ASD], and adjustable-frequency drive [AFD].) An alternating-current VFD consists of a diode bridge AC-to-DC converter and a pulse-width-modulation (PWM) controller with fast-rise power transistors—usually insulated-gate bipolar transistors (IGBTs)—and an induction motor. The IGBT changes the characteristics of waveforms applied to a motor due to the speed at which the IGBT cycles on and off. At switching speed up to 20 k Hz, the impedance in the connecting cable is far less than the motor impedance, particularly for small motors, causing pulse reflectance at the motor terminals to form damaging high voltage. NEMA motor standard MG1 states PWM drive limits and establishes a peak of 1600 V and a minimum rise time of 0.1 μs for motors rated less than 600 V. Typical manufacturer maximum voltage withstand levels range from 1000 V to 1800 V. When specifying motors for operation on VSDs, the voltage withstand level based on the dV/dt of the drive and the known cable distance should be specified.

Harmonics caused by the portion of a VSD converting line power LDC affect input lines and are termed *line-side harmonics*. Output line harmonics are caused solely by the inverter section of the VSD and are known as load side or motor harmonics. Generally, PWM drives containing internal bus reaction or three-phase AC line reactors do not cause interference with other electrical equipment. There may be problems when a VSD is switched onto a standby generator, or when power factor correction capacitors are used.

Table 15.1 Maximum Capacity of Gas Pipe in Litres per Second [2021F, Ch 22, Tbl 40]
(At gas pressures of 3.5 kPa above atm. or less and a pressure drop of 75 Pa. Density = 0.735 kg/m^3)

Nominal Iron Pipe Size, mm	Internal Diameter, mm	Length of Pipe, m													
		10	20	30	40	50	60	70	80	90	100	125	150	175	200
8	9.25	0.19	0.13	0.11	0.09	0.08	0.07	0.07	0.06	0.06	0.06	0.05	0.05	0.19	0.13
10	12.52	0.43	0.29	0.24	0.20	0.18	0.16	0.15	0.14	0.13	0.12	0.12	0.11	0.43	0.29
15	15.80	0.79	0.54	0.44	0.37	0.33	0.30	0.28	0.26	0.24	0.23	0.22	0.21	0.79	0.54
20	20.93	1.65	1.13	0.91	0.78	0.69	0.63	0.58	0.54	0.50	0.47	0.45	0.43	1.65	1.13
25	26.14	2.95	2.03	1.63	1.40	1.24	1.12	1.03	0.96	0.90	0.85	0.81	0.77	2.95	2.03
32	35.05	6.4	4.4	3.5	3.0	2.7	2.4	2.2	2.1	1.9	1.8	1.7	1.7	6.4	4.4
40	40.89	9.6	6.6	5.3	4.5	4.0	3.6	3.3	3.1	2.9	2.8	2.6	2.5	9.6	6.6
50	52.50	18.4	12.7	10.2	8.7	7.7	7.0	6.4	6.0	5.6	5.3	5.0	4.8	18.4	12.7
65	62.71	29.3	20.2	16.2	13.9	12.3	11.1	10.2	9.5	8.9	8.4	8.0	7.7	29.3	20.2
80	77.93	51.9	35.7	28.6	24.5	21.7	19.7	18.1	16.8	15.8	14.9	14.2	13.5	51.9	35.7
100	102.26	105.8	72.7	58.4	50.0	44.3	40.1	36.9	34.4	32.2	30.4	28.9	27.6	105.8	72.7

Note: Capacity is in litres per second at gas pressures of 3.5 kPa (gage) or less and a pressure drop of 75 kPa; density = 0.735 kg/m^3.

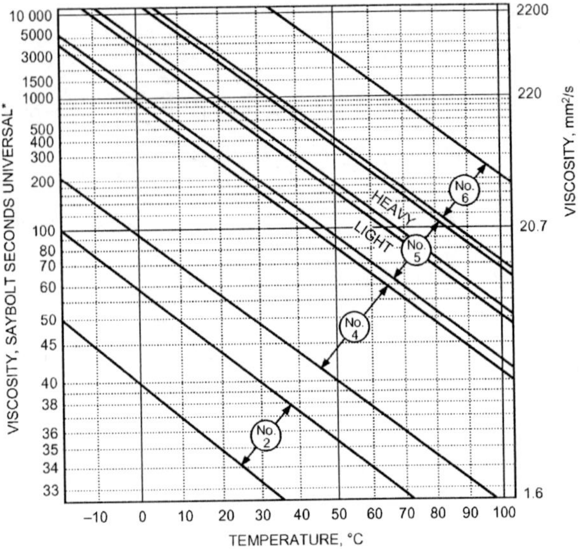

* 1 Saybolt Second (SSU, or SUS) = time required for 60 mL to gravity-flow through Saybolt universal viscometer.

Figure 15.1 Approximate Viscosity of Fuel Oils [2021F, Ch 28, Fig 2]

Table 15.2 Typical API Gravity, Density, and Heating Value of Standard Grades of Fuel Oil [2021F, Ch 28, Tbl 10]

Grade No.	Density, kg/m³	Higher Heating Value, GJ/m³
1	833 to 800	38.2 to 37.0
2	874 to 834	39.5 to 38.2
4	933 to 886	41.3 to 39.9
5L	951 to 921	41.8 to 40.9
5H	968 to 945	42.4 to 41.6
6	1012 to 965	43.5 to 42.2

Types of Fuel Oils

Fuel oils for heating are broadly classified as distillate fuel oils (lighter oils) or residual fuel oils (heavier oils). ASTM has established specifications for fuel oil properties which subdivide the oils into various grades. Grades No. 1 and 2 are distillate fuel oils. Grades 4, 5 (Light), 5 (Heavy), and 6 are residual fuel oils. Specifications for the grades are based on required characteristics of fuel oils for use in different types of burners.

Grade No. 1 is a light distillate intended for vaporizing-type burners. High volatility is essential to continued evaporation of the fuel oil with minimum residue.

Grade No. 2 is a heavier (API Gravity) distillate than No. 1. It is used primarily with pressure-atomizing (gun) burners that spray the oil into a combustion chamber. The atomized oil vapor mixes with air and burns. This grade is used in most domestic burners and many medium capacity commercial-industrial burners.

Grade No. 4 is an intermediate fuel that is considered either a light residual or a heavy distillate. Intended for burners that atomize oils of higher viscosity than domestic burners can handle, its permissible viscosity range allows it to be pumped and atomized at relatively low storage temperatures.

Grade No. 5 (Light) is a residual fuel of intermediate viscosity for burners that handle fuel more viscous than No. 4 without preheating. Preheating may be necessary in some equipment for burning and, in colder climates, for handling.

Grade No. 5 (Heavy) is a residual fuel more viscous than No. 5 (Light), but intended for similar purposes. Preheating is usually necessary for burning and, in colder climates, for handling.

Grade No. 6, sometimes referred to as Bunker C, is a high-viscosity oil used mostly in commercial and industrial heating. It requires preheating in the storage tank to permit pumping, and additional preheating at the burner to allow atomizing.

Types and Properties of Liquid Fuels for Engines

The primary stationary engine fuels are diesel and gas turbine oils, natural gases, and liquefied petroleum gases. Other fuels include sewage gas, manufactured gas, and other commercial gas mixtures. Gasoline and the JP series of gas turbine fuels are rarely used for stationary engines.

Properties of the three grades of diesel fuel oils (1-D, 2-D, and 4-D) are listed in ASTM Standard D975.

Grade No. 1-D includes the class of volatile fuel oils from kerosene to intermediate distillates. They are used in high-speed engines with frequent and relatively wide variations in loads and speeds and where abnormally low fuel temperatures are encountered.

Grade No. 2-D includes the class of lower volatility distillate gas oils. These fuels are used in high-speed engines with relatively high loads and uniform speeds, or in engines not requiring fuels with the higher volatility or other properties specified for Grade No. 1-D.

Grade No. 4-D covers the class of more viscous distillates and blends of these distillates with residual fuel oils. These fuels are used in low- and medium-speed engines involving sustained loads at essentially constant speed.

Property specifications and test methods for Grade No. 1-D, 2-D, and 4-D diesel fuel oils are essentially identical to specifications of Grade No. 1, 2, and 4 fuel oils, respectively. However, diesel fuel oils have an additional specification for **cetane number**, which measures ignition quality and influences combustion roughness. Cetane number requirements depend on engine design, size, speed and load variations, and starting and atmospheric conditions. An increase in cetane number over values actually required does not improve engine performance. Thus, the cetane number should be as low as possible to assure maximum fuel availability. ASTM Standard D975 provides several methods for estimating cetane number from other fuel oil properties.

ASTM Standard D2880 for gas turbine fuel oils relates gas turbine fuel oil grades to fuel and diesel fuel oil grades.

Table 15.3 Approximate Air Requirements for Stoichiometric Combustion of Various Fuels [2021F, Ch 28, Tbl 13]

Type of Fuel	Theoretical Air Required for Combustion
Solid fuels	kg/kg fuel
Anthracite	9.6
Semibituminous	11.2
Bituminous	10.3
Lignite	6.2
Coke	11.2
Liquid fuels	Mg/m^3 fuel
No. 1 fuel oil	12.34
No. 2 fuel oil	12.70
No. 5 fuel oil	13.42
No. 6 fuel oil	13.66
Gaseous fuels	m^3/m^3 fuel
Natural gas	9.6
Butane	31.1
Propane	24.0

Table 15.4 Approximate Maximum Theoretical (Stoichiometric) CO_2 Values and CO_2 Values of Various Fuels with Different Percentages of Excess Air [2021F, Ch 28, Tbl 15]

Type of Fuel	Theoretical or Maximum CO_2, %	Percent CO_2 at Given Excess Air Values 20%	40%	60%
Gaseous Fuels				
Natural gas	12.1	9.9	8.4	7.3
Propane gas (commercial)	13.9	11.4	9.6	8.4
Butane gas (commercial)	14.1	11.6	9.8	8.5
Mixed gas (natural and carbureted water gas)	11.2	12.5	10.5	9.1
Carbureted water gas	17.2	14.2	12.1	10.6
Coke oven gas	11.2	9.2	7.8	6.8
Liquid Fuels				
No. 1 and 2 fuel oil	15.0	12.3	10.5	9.1
No. 6 fuel oil	16.5	13.6	11.6	10.1
Solid Fuels				
Bituminous coal	18.2	15.1	12.9	11.3
Anthracite	20.2	16.8	14.4	12.6
Coke	21.0	17.5	15.0	13.0

Table 15.5 Recommended Nominal Size for Fuel Oil Suction Lines from Tank to Pump (Distillate Grades No. 1 and No. 2) [2021F, Ch 22, Tbl 42]

Pumping Rate, L/h	Length of Run in Metres at Maximum Suction Lift of 9.0 kPa 10	20	30	40	50	60	70	80	90	100
50	15	15	15	15	15	20	20	20	25	25
100	15	15	15	15	20	20	20	20	25	25
200	15	20	20	20	20	20	25	25	25	25
300	15	20	20	20	20	25	25	25	25	32
400	20	20	20	20	25	25	25	25	32	32
500	20	25	25	25	25	25	32	32	32	32
600	20	25	25	25	25	32	32	32	32	50
700	20	25	25	25	25	32	32	32	50	50
800	20	25	25	25	32	32	32	32	50	50

Note: Sizes (in millimetres) are nominal.

Table 15.6 Recommended Nominal Size for Fuel Oil Suction Lines from Tank to Pump (Residual Grades No. 5 and No. 6) [2021F, Ch 22, Tbl 41]

Pumping Rate, L/h	Length of Run in Metres at Maximum Suction Lift of 4.5 kPa 10	20	30	40	50	60	70	80	90	100
50	40	40	40	50	50	50	65	65	65	80
100	40	40	50	50	65	65	65	65	80	80
200	40	50	50	50	65	65	65	80	80	80
300	50	50	65	65	65	80	80	80	80	80
400	50	50	65	65	80	80	80	80	80	100
500	50	65	65	65	80	80	80	80	100	100
600	65	65	65	80	80	80	100	100	100	100
700	65	65	65	80	80	100	100	100	100	100
800	65	65	80	80	100	100	100	100	100	100

Notes:
1. Sizes (in millimetres) are nominal.
2. Pipe sizes smaller than 25 mm ISO are not recommended for use with residual grade fuel oils.
3. Lines conveying fuel oil from pump discharge port to burners and tank return may be reduced by one or two sizes, depending on piping length and pressure losses.

16. OWNING AND OPERATING

Maintenance Costs

The maintenance cost of mechanical systems varies widely depending on configuration, equipment locations, accessibility, system complexity, service duty, geography, and system reliability requirements.

Dohrmann and Alereza (1986) obtained maintenance costs and HVAC system information from 342 buildings located in 35 states in the United States. In 1983 U.S. dollars, data collected showed a mean HVAC system maintenance cost of $3.40/m^2 per year, with a median cost of $2.60/m^2 per year. Building age has a statistically significant but minor effect on HVAC maintenance costs. Analysis also indicated that building size is not statistically significant in explaining cost variation. The type of maintenance program or service agency that building management chooses can also have a significant effect on total HVAC maintenance costs. Although extensive or thorough routine and preventive maintenance programs cost more to administer, they usually extend equipment life; improve reliability; and reduce system downtime, energy costs, and overall life-cycle costs.

Some maintenance cost data are available, both in the public domain and from proprietary sources used by various commercial service providers. These sources may include equipment manufacturers, independent service providers, insurers, government agencies (e.g., the U.S. General Services Administration), and industry-related organizations [e.g., the Building Owners and Managers Association (BOMA)] and service industry publications. More traditional, widely used products and components are likely to have statistically reliable records. However, design changes or modifications necessitated by industry changes, such as alternative refrigerants, may make historical data less relevant.

Newer HVAC products, components, system configurations, control systems and protocols, and upgraded or revised system applications present an additional challenge. Care is required when using data not drawn from broad experience or field reports. In many cases, maintenance information is proprietary or was sponsored by a particular entity or group. Particular care should be taken when using such data. It is the user's responsibility to obtain these data and to determine their appropriateness and suitability for the application being considered.

ASHRAE research project TRP-1237 (Abramson et al. 2005) developed a standardized Internet-based data collection tool and database on HVAC equipment service life and maintenance costs. The database was seeded with data on 163 buildings from around the country. Maintenance cost data were gathered for total HVAC system maintenance costs from 100 facilities. In 2004 dollars, the mean HVAC maintenance cost from these data was $5.06/m^2, and the median cost was $4.74/m^2. Table 16.1 compares these figures with estimates reported by Dohrmann and Alereza (1986), both in terms of contemporary dollars, and in 2004 dollars, and shows that the cost per square metre varies widely between studies.

Estimating Maintenance Costs

Total HVAC maintenance cost for new and existing buildings with various types of equipment may be estimated several ways, using several resources. Equipment maintenance requirements can be obtained from the equipment manufacturers for large or custom pieces of equipment. Estimating in-house labor requirements can be difficult; BOMA provides guidance on this topic. Many independent mechanical service companies provide preventative maintenance contracts. These firms typically have proprietary estimating programs developed through their experience, and often provide generalized maintenance costs to engineers and owners upon request, without obligation.

Table 16.1 Comparison of Maintenance Costs Between Studies
[2019A, Ch 38, Tbl 6]

Survey	Cost per m^2, as Reported		Consumer Price Index	Cost per m^2, 2004 Dollars	
	Mean	Median		Mean	Median
Dohrmann and Alereza (1986)	$3.44	$2.58	99.6	$6.57	$4.95
Abramson et al. (2005)	$5.06	$4.74	188.9	$5.06	$4.74

When evaluating various HVAC systems during design or retrofit, the absolute magnitude of maintenance costs may not be as important as the relative costs. Whichever estimating method or resource is selected, it should be used consistently throughout any evaluation. Mixing information from different resources in an evaluation may provide erroneous results.

Applying simple costs per unit of building floor area for maintenance is highly discouraged. Maintenance costs can be generalized by system types. When projecting maintenance costs for different HVAC systems, the major system components need to be identified with a required level of maintenance. The potential long-term costs of environmental issues on maintenance costs should also be considered.

Table 16.2 Owning and Operating Cost Data and Summary
[2019A, Ch 38, Tbl 1]

OWNING COSTS			
I. Initial Cost of System		______	
II. Periodic Costs			
A. Income taxes	______		
B. Property taxes	______		
C. Insurance	______		
D. Rent	______		
E. Other periodic costs	______		
Total Periodic Costs		______	
III. Replacement Cost		______	
IV. Salvage Value		______	
Total Owning Costs			______
OPERATING COSTS			
V. Annual Utility, Fuel, Water, etc., Costs			
A. Utilities			
1. Electricity	______		
2. Natural gas	______		
3. Water/sewer	______		
4. Purchased steam	______		
5. Purchased hot/chilled water	______		
B. Fuels			
1. Propane	______		
2. Fuel oil	______		
3. Diesel	______		
4. Coal	______		
C. On-site generation of electricity	______		
D. Other utility, fuel, water, etc., costs	______		
Total		______	
VI. Annual Maintenance Allowances/Costs			
A. In-house labor	______		
B. Contracted maintenance service	______		
C. In-house materials	______		
D. Other maintenance allowances/costs (e.g., water treatment)	______		
Total		______	
VII. Annual Administration Costs		______	
Total Annual Operating Costs			______
TOTAL ANNUAL OWNING AND OPERATING COSTS			______

Economic Analysis Techniques

Simple Payback

In the simple payback technique, a projection of the revenue stream, cost savings, and other factors is estimated and compared to the initial capital outlay. This simple technique ignores the cost of borrowing money (interest) and lost opportunity costs. It also ignores inflation and the time value of money.

Present-Value (Present Worth) Analysis

All sophisticated economic analysis methods use the basic principles of present value analysis to account for the time value of money. The total present value (present worth) for any analysis is determined by summing the present worths of all individual items under consideration, both future single-payment items and series of equal future payments. The scenario with the highest present value is the preferred alternative.

Single-Payment Present-Value Analysis. The cost or value of money is a function of the available interest rate and inflation rate. The future value F of a present sum of money P over n periods with compound interest rate i per period is

$$F = P(1 + i)^n \tag{16.1}$$

Conversely, the present value or present worth P of a future sum of money F is given by

$$P = F/(1 + i)^n \tag{16.2}$$

or

$$P = F \times \mathrm{PWF}(i,n)_{sgl} \tag{16.3}$$

where the single-payment present-worth factor $\mathrm{PWF}(i,n)_{sgl}$ is defined as

$$\mathrm{PWF}(i,n)_{sgl} = 1/(1 + i)^n \tag{16.4}$$

Series of Equal Payments. The present-worth factor for a series of future equal payments (e.g., operating costs) is given by

$$\mathrm{PWF}(i,n)_{ser} = \frac{(1 + I)^n - 1}{i(1 + i)^n} \tag{16.5}$$

The present value P of those future equal payments (PMT) is then the product of the present-worth factor and the payment [i.e., $P = \mathrm{PWF}(i,n)_{ser} \times \mathrm{PMT}$].

The number of future equal payments to repay a present value of money is determined by the capital recovery factor (CRF), which is the reciprocal of the present-worth factor for a series of equal payments:

$$\mathrm{CRF} = \mathrm{PMT}/P \tag{16.6}$$

$$\mathrm{CRF}(i,n)_r = \frac{i(1 + i)^n}{(1 + I)^n - 1} = \frac{i}{1 - (1 + i)^n} \tag{16.7}$$

The CRF is often used to describe periodic uniform mortgage or loan payments.

Note that when payment periods other than annual are to be studied, the interest rate must be expressed per appropriate period. For example, if monthly payments or return on investment are being analyzed, then interest must be expressed per month, not per year, and n must be expressed in months.

Improved Payback Analysis. This somewhat more sophisticated payback approach is similar to the simple payback method except that the cost of money (interest rate, discount rate, etc.) is considered. Solving Equation 16.7 for n yields the following:

$$n = \frac{\ln[\mathrm{CRF}/(\mathrm{CRF} - i)]}{\ln(1 + i)} \tag{16.8}$$

Given known investment amounts and earnings, CRFs can be calculated for the alternative investments. Subsequently, the number of periods until payback has been achieved can be calculated using Equation 16.8.

Accounting for Inflation. Different economic goods may inflate at different rates. Inflation reflects the rise in the real cost of a commodity over time and is separate from the time value of money. Inflation must often be accounted for in an economic evaluation. One way to account for inflation is to substitute effective interest rates that account for inflation into the equations given above.

The effective interest rate i', sometimes called the real rate, accounts for inflation rate j and interest rate i or discount rate i_d; it can be expressed as follows:

$$i' = \frac{1+i}{1+j} - 1 = \frac{i-j}{1+j} \quad (16.9)$$

Different effective interest rates can be applied to individual components of cost. Projections for future fuel and energy prices are available in the *Annual Supplement to NIST Handbook 135*.

The following are three common methods of present-value analysis that include life-cycle cost factors (life of equipment, analysis period, discount rate, energy escalation rates, maintenance cost, etc., as shown in Table 16.2):

- Savings-to-Investment Ratio (SIR)
- Internal Rate of Return (IRR)
- Adjusted Internal Rate of Return (AIRR)

These comparison techniques rely on the same assumptions and economic analysis theories but display the results in different forms. They also use the same definition of each term. All can be displayed as a single calculation or as a cash flow table using a series of calculations for each year of the analysis period.

Table 16.3 Two Alternative LCC Examples[2019A, Ch 38, Tbl 8]

Alternative 1: Purchase Chilled Water from Utility

	Year										
	0	1	2	3	4	5	6	7	8	9	10
First costs		—	—	—	—	—	—	—	—	—	—
Chilled-water costs		$65 250	$66 881	$68 553	$70 267	$72 024	$73 824	$75 670	$77 562	$79 501	$81 488
Replacement costs		—	—	—	—	—	—	—	—	—	—
Maintenance costs		—	—	—	—	—	—	—	—	—	—
Net annual cash flow		65 250	66 881	68 553	70 267	72 024	73 824	75 670	77 501	79 501	81 488
Present value of cash flow		60 417	57 340	54 420	51 648	49 018	46 522	44 153	41 904	39 770	37 745

	Year										
		11	12	13	14	15	16	17	18	19	20
Financing annual payments		—	—	—	—	—	—	—	—	—	—
Chilled-water costs		$83 526	$85 614	$87 754	$89 948	$92 197	$94 501	$96 864	$99 286	$101 768	$104 312
Replacement costs		—	—	—	—	—	—	—	—	—	—
Maintenance costs		—	—	—	—	—	—	—	—	—	—
Net annual cash flow		83 526	85 614	87 754	89 948	92 197	94 501	96 864	99 286	101 768	104 312
Present value of cash flow		35 823	33 998	32 267	30 624	29 064	27 584	26 179	24 846	23 581	22 380
20-year life-cycle cost	$769 823										

Table 16.3 Two Alternative LCC Examples[2019A, Ch 38, Tbl 8] (Continued)

Alternative 2: Install Chiller and Tower

	Year										
	0	1	2	3	4	5	6	7	8	9	10
First costs	$220 000	—	—	—	—	—	—	—	—	—	—
Energy costs		$18 750	$19 688	$20 672	$21 705	$22 791	$23 930	$25 127	$26 383	$27 702	$29 087
Replacement costs		—	—	—	—	—	—	—	—	—	90 000
Maintenance costs		15 200	15 656	16 126	16 609	17 108	17 621	18 150	18 694	19 255	19 833
Net annual cash flow	220 000	33 950	35 344	36 798	38 315	39 898	41 551	43 276	45 077	46 957	138 920
Present value of cash flow	220 000	31 435	30 301	29 211	28 163	27 154	26 184	25 251	24 354	23 490	64 347

		Year									
		11	12	13	14	15	16	17	18	19	20
Financing annual payments		—	—	—	—	—	—	—	—	—	—
Energy costs		$30 542	$32 069	$33 672	$35 356	$37 124	$38 980	$40 929	$42 975	$45 124	$47 380
Replacement costs		—	—	—	—	—	—	—	—	—	—
Maintenance costs		20 428	21 040	21 672	22 322	22 991	23 681	24 392	25 123	25 877	26 653
Net annual cash flow		50 969	53 109	55 344	57 678	60 115	62 661	65 320	68 099	71 001	74 034
Present value of cash flow		21 860	21 090	20 350	19 637	18 951	18 290	17 654	17 042	16 452	15 884
20-year life-cycle cost	$717 100										

17. SOUND

A primary objective in the design of HVAC systems and equipment is to evaluate noise and vibration to ensure that the acoustical environment in a given space is acceptable for various occupant activities. Sound and vibration are created by a source, are transmitted along one or more paths, and reach a receiver. Treatments and modifications can be applied to any or all of these elements to reduce unwanted noise and vibration, although it is usually most effective and least expensive to reduce noise at the source.

Levels

The magnitude of sound and vibration physical properties are almost always expressed in levels. The level L is based on the common (base 10) logarithm of a ratio of the magnitude of a physical property of power, intensity, or energy to a reference magnitude of the same type of property:

$$L = 10 \log \left(\frac{A}{A_{ref}} \right) \tag{17.1}$$

where A is the magnitude of the physical property of interest and A_{ref} is the reference value. Note that the ratio is dimensionless. In this equation, a factor of 10 is included to convert bels to decibels (dB).

Sound Pressure and Sound Pressure Level

Sound waves in air are variations in pressure above and below atmospheric pressure. **Sound pressure** is measured in pascals (Pa). The human ear responds across a broad range of sound pressures; the threshold of hearing to the threshold of pain covers a range of approximately 10^{14}:1. Table 17.1 gives approximate values of sound pressure by various sources at specified distances from the source.

The range of sound pressure in Table 17.1 is so large that it is more convenient to use a scale proportional to the logarithm of this quantity. Therefore, the **decibel** (dB) scale is the preferred method of presenting quantities in acoustics, not only because it collapses a large range of pressures to a more manageable range but also because its levels correlate better with human responses to the magnitude of sound than do sound pressures. Equation 17.2 describes levels of power, intensity, and energy, which are proportional to the square of other physical properties, such as sound pressure and vibration acceleration. Thus, the **sound pressure level** L_p corresponding to a sound pressure is given by

$$L_p = 10 \log \left(\frac{p}{p_{ref}} \right)^2 = 20 \log \left(\frac{p}{p_{ref}} \right) \tag{17.2}$$

where p is the root mean square (RMS) value of acoustic pressure in pascals. The root mean square is the square root of the time average of the square of the acoustic pressure ratio. The ratio p/p_{ref} is squared to give quantities proportional to intensity or energy. A reference quantity is needed so the term in parentheses is nondimensional. For sound pressure levels in air, the reference pressure p_{ref} is 20 μPa, which corresponds to the approximate threshold of hearing for a young person with good hearing exposed to a pure tone with a frequency of 1000 Hz.

The decibel scale is used for many different descriptors relating to sound: source strength, sound level at a specified location, and attenuation along propagation paths; each has a different reference quantity. For this reason, it is important to be aware of the context in which the term *decibel* or *level* is used. For most acoustical quantities, there is an internationally accepted reference value. A reference quantity is always implied even if it does not appear.

Sound pressure level is relatively easy to measure and thus is used by most noise codes and criteria. (The human ear and microphones are pressure sensitive.) Sound pressure levels for the corresponding sound pressures are also given in Table 17.1.

A number of AHRI, AMCA, CTI, and ANSI sound standards are used by equipment manufacturers to provide accurate sound data. Manufacturer-supplied data in accordance with the appropriate standard should be used in preference to empirical information in evaluating the noise resulting from a particular equipment item.

Table 17.1 Typical Sound Pressures and Sound Pressure Levels [2021F, Ch 8, Tbl 1]

Source	Sound Pressure, Pa	Sound Pressure Level, dB re 20 μPa	Subjective Reaction
Military jet takeoff at 30 m	200	140	Extreme danger
Artillery fire at 3 m	63.2	130	
Passenger jet takeoff at 15 m	20	120	Threshold of pain
Loud rock band	6.3	110	Threshold of discomfort
Automobile horn at 3 m	2	100	
Unmuffled large diesel engine at 40 m	0.6	90	Very loud
Accelerating diesel truck at 15 m	0.2	80	
Freight train at 30 m	0.06	70	
Conversational speech at 1 m	0.02	60	
Window air conditioner at 3 m	0.006	50	Moderate
Quiet residential area	0.002	40	
Whispered conversation at 2 m	0.0006	30	
Buzzing insect at 1 m	0.0002	20	
Threshold of good hearing	0.00006	10	Faint
Threshold of excellent youthful hearing	0.00002	0	Threshold of hearing

SOUND

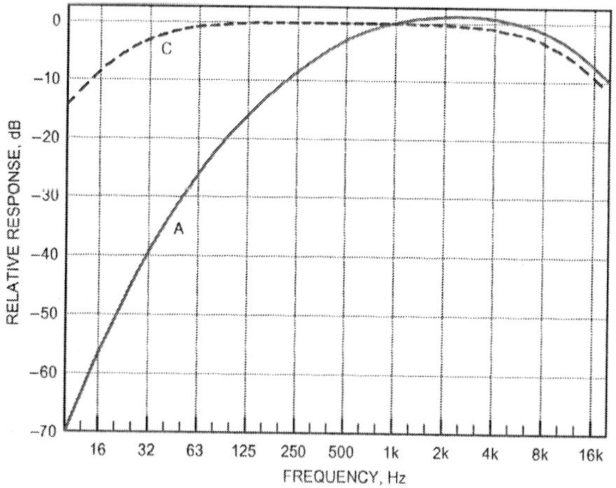

Figure 17.1 Curves Showing A- and C-Weighting Responses for Sound Level Meters [2021F, Ch 8, Fig 1]

Combining Sound Levels

To estimate the levels from multiple sources from the levels from each source, the intensities (not the levels) must be added. Thus, the levels must first be converted to find intensities, the intensities summed, and then converted to a level again, so the combination of multiple levels L_1, L_2, etc., produces a level L_{sum} given by

$$L_{sum} = 10 \log\left(\sum_i 10^{L_i/10}\right) \qquad (17.3)$$

where for sound pressure level (L_p), $10^{L_i/10}$ is p_i^2/p_{ref}^2, and L_i is the sound pressure level for the ith source.

A simpler and slightly less accurate method is outlined in Table 17.2. This method, although not exact, results in errors of 1 dB or less. The process with a series of levels may be shortened by combining the largest with the next largest, then combining this sum with the third largest, then the fourth largest, and so on until the combination of the remaining levels is 10 dB lower than the combined level. The process may then be stopped.

Sound Power and Sound Power Level

The **sound power** of a source is its rate of emission of acoustical energy and is expressed in watts. Sound power depends on operating conditions but not distance of observation location from the source or surrounding environment. Approximate sound power outputs for common sources are shown in Table 17.2 with corresponding sound power levels. For **sound power level L_w**, the power reference is 10^{-12} W or 1 picowatt. The definition of sound power level is therefore

$$L_w = 10 \log(w/10^{-12}) \qquad (17.4)$$

where w is the sound power emitted by the source in watts.

Table 17.2 Combining Two Sound Levels [2021F, Ch 8, Tbl 3]

Difference between levels to be combined, dB	0 to 1	2 to 4	5 to 9	10 and More
Number of decibels to add to highest level to obtain combined level	3	2	1	0

Table 17.3 Midband and Approximate Upper and Lower Cutoff Frequencies for Octave and 1/3 Octave Band Filters [2021F, Ch 8, Tbl 4]

Octave Bands, Hz			1/3 Octave Bands, Hz		
Lower	Midband	Upper	Lower	Midband	Upper
			11.2	12.5	14
11.2	16	22.4	14	16	18
			18	20	22.4
			22.4	25	28
22.4	31.5	45	28	31.5	35.5
			35.5	40	45
			45	50	56
45	63	90	56	63	71
			71	80	90
			90	100	112
90	125	180	112	125	140
			140	160	180
			180	200	224
180	250	355	224	250	280
			280	315	355
			355	400	450
355	500	710	450	500	560
			560	630	710
			710	800	900
710	1 000	1 400	900	1 000	1 120
			1 120	1 250	1 400
			1 400	1 600	1 800
1 400	2 000	2 800	1 800	2 000	2 240
			2 240	2 500	2 800
			2 800	3 150	3 550
2 800	4 000	5 600	3 550	4 000	4 500
			4 500	5 000	5 600
			5 600	6 300	7 100
5 600	8 000	11 200	7 100	8 000	9 000
			9 000	10 000	11 200
			11 200	12 500	14 000
11 200	16 000	22 400	14 000	16 000	18 000
			18 000	20 000	22 400

Table 17.4 Design Guidelines for HVAC-Related Background Sound in Rooms [2019A, Ch 49, Tbl 1]

Room Types		Octave Band Analysis[a]	Approximate Overall Sound Pressure Level[a]	
		NC/RC[b]	dBA[c]	dBC[c]
Rooms with Intrusion from Outdoor Noise Sources[d]	Traffic noise	N/A	45	70
	Aircraft flyovers	N/A	45	70
Residences, Apartments, Condominiums	Living areas	30	35	60
	Bathrooms, kitchens, utility rooms	35	40	60
Hotels/Motels	Individual rooms or suites	30	35	60
	Meeting/banquet rooms	30	35	60
	Corridors and lobbies	40	45	65
	Service/support areas	40	45	65
Office Buildings	Executive and private offices	30	35	60
	Conference rooms	30	35	60
	Teleconference rooms	25	30	55
	Open-plan offices	40	45	65
	Corridors and lobbies	40	45	65
Courtrooms	Unamplified speech	30	35	60
	Amplified speech	35	40	60
Performing Arts Spaces	Drama theaters, concert and recital halls	20	25	50
	Music teaching studios	25	30	55
	Music practice rooms	30	35	60
Hospitals and Clinics	Patient rooms	30	35	60
	Wards	35	40	60
	Operating and procedure rooms	35	40	60
	Corridors and lobbies	40	45	65

Table 17.4 Design Guidelines for HVAC-Related Background Sound in Rooms [2019A, Ch 49, Tbl 1] ***(Continued)***

Room Types		Octave Band Analysis[a]	Approximate Overall Sound Pressure Level[a]	
		NC/RC[b]	dBA[c]	dBC[c]
Laboratories	Testing/research w/minimal speech communication	50	55	75
	Extensive phone use and speech communication	45	50	70
Churches, Mosques, Synagogues	Group teaching	35	40	60
	General assembly with critical music programs[e]	25	30	55
Schools[f]	Classrooms	30	35	60
	Large lecture rooms with speech amplification	30	35	60
	Large lecture rooms without speech amplification	25	30	55
Libraries		30	35	60
Indoor Stadiums, Gymnasiums	Gymnasiums and natatoriums[g]	45	50	70
	Large-seating-capacity spaces with speech amplification[g]	50	55	75

N/A = Not applicable

[a]Values and ranges are based on judgment and experience, and represent general limits of acceptability for typical building occupancies.

[b]NC: this metric plots octave band sound levels against a family of reference curves, with the number rating equal to the highest tangent line value.

RC: when sound quality in the space is important, the RC metric provides a diagnostic tool to quantify both the speech interference level and spectral imbalance.

[c]dBA and dBC: these are overall sound pressure level measurements with A- and C-weighting, and serve as good references for a fast, single-number measurement. They are also appropriate for specification in cases where no octave band sound data are available for design.

[d]Intrusive noise is addressed here for use in evaluating possible non-HVAC noise that is likely to contribute to background noise levels.

[e]An experienced acoustical consultant should be retained for guidance on acoustically critical spaces (below RC 30) and for all performing arts spaces.

[f]Some educators and others believe that HVAC-related sound criteria for schools, as listed in previous editions of this table, are too high and impede learning for affected groups of all ages. See ANSI/ASA Standard S12.60 for classroom acoustics and a justification for lower sound criteria in schools. The HVAC component of total noise meets the background noise requirement of that standard if HVAC-related background sound is approximately NC/RC 25. Within this category, designs for K-8 schools should be quieter than those for high schools and colleges.

[g]RC or NC criteria for these spaces need only be selected for the desired speech and hearing conditions.

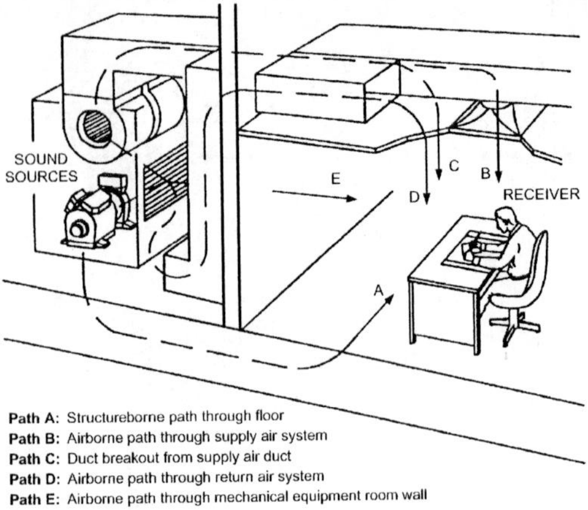

Figure 17.2 Typical Paths of Noise and Vibration Propagation in HVAC Systems [2019A, Ch 49, Fig 1]

Table 17.5 Comparison of Sound Rating Methods [2019A, Ch 49, Tbl 4]

Method	Overview	Considers Speech Interference Effects	Evaluates Sound Quality	Components Presently Rated by Each Method
dBA	No quality assessment Frequently used for outdoor noise ordinances	Yes	No	Cooling towers Water chillers Condensing units
NC	Can rate components Limited quality assessment Does not evaluate low-frequency rumble	Yes	Somewhat	Air terminals Diffusers
RC Mark II	Used to evaluate systems Should not be used to evaluate components Evaluates sound quality Provides improved diagnostics capability	Yes	Yes	Not used for component rating
NCB	Can rate components Some quality assessment	Yes	Somewhat	See NC
RNC	Some quality assessment Attempts to quantify fluctuations	Yes	Somewhat	Not used for component rating

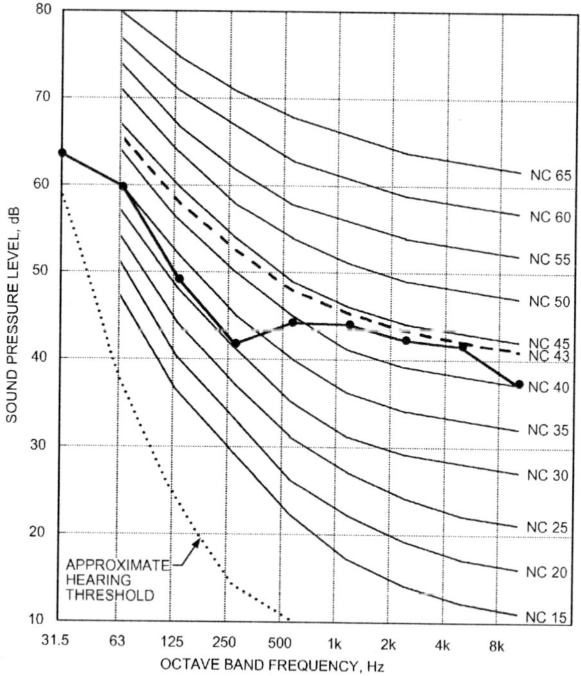

Figure 17.3 NC (Noise Criteria) Curves and Sample Spectrum (Curve with Symbols) [2021F, Ch 8, Fig 7]

Sound

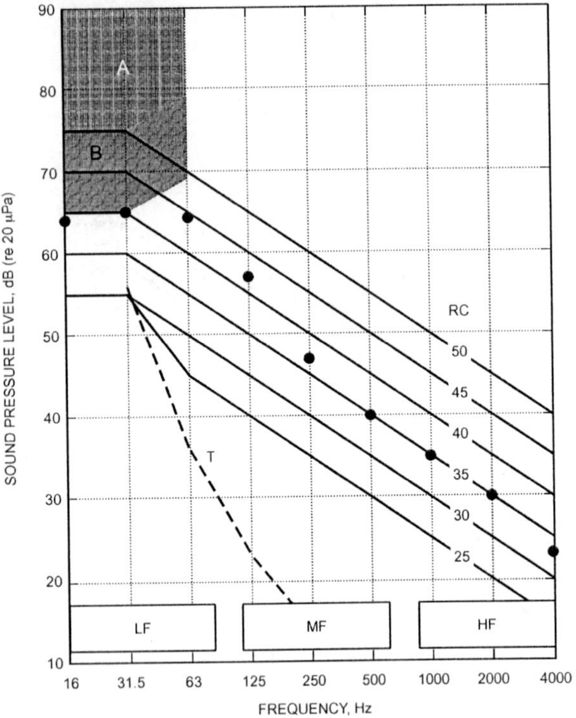

Note:

- Noise levels for lightweight wall and ceiling constructions:
  - In shaded region B are likely to generate vibration that may be perceptible. There is a slight possibility of rattles in light fixtures, doors, windows, etc.
  - In shaded region A have a high probability of generating easily perceptible noise-induced vibration. Audible rattling in light fixtures, doors, windows, etc. may be anticipated.
- LF, MF, and HF refer to low-frequency rumble, mid-frequency roar, and high-frequency hiss, respectively.
- Solid dots are sound pressure levels for the example discussed in Chapter 49 of the 2021 *ASHRAE Handbook—Fundamentals*.

Figure 17.4 Room Criteria Curves, Mark II [2019A, Ch 49, Fig 6]

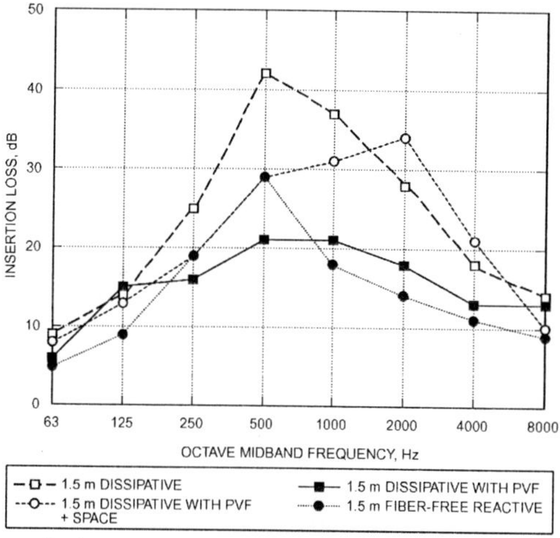

Note: Dissipative silencers use sound-absorptive media such as fiberglass as the primary means of attenuating sound. Reactive silencers are constructed only of metal, both solid and perforated.

Figure 17.5 Comparison of 1.5 m Dissipative and Reactive Silencer Performance [2019A, Ch 49, Fig 23]

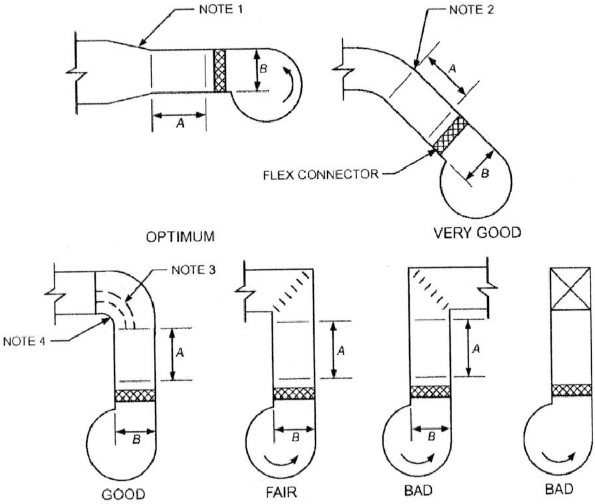

Notes:

1. Slopes of 1 in 7 preferred. Slopes of 1 in 4 permitted below 10 m/s.
2. Dimension *A* should be at least 1.5 times *B*, where *B* is largest discharge duct dimension.
3. Rugged turning vanes should extend full radius of elbow.
4. Minimum 150 mm radius required.

Figure 17.6 Various Outlet Configurations for Centrifugal Fans and Their Possible Rumble Conditions [2019A, Ch 49, Fig 25]

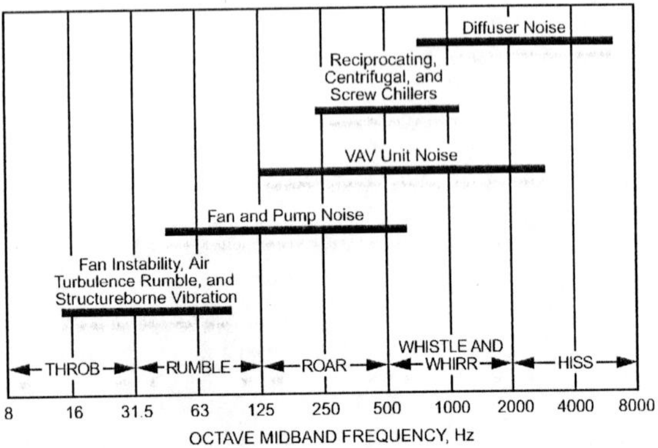

Figure 17.7 Frequencies at Which Different Types of Mechanical Equipment Generally Control Sound Spectra [2019A, Ch 49, Fig 4]

Table 17.6 Sound Transmission Class (STC) and Transmission Loss Values of Typical Mechanical Equipment Room Wall, Floor, and Ceiling Types, dB
[2019A, Ch 49, Tbl 40]

		Octave Midband Frequency, Hz						
Room Construction Type	**STC**	**63**	**125**	**250**	**500**	**1000**	**2000**	**4000**
200 mm CMU*	50	35	35	41	44	50	57	64
200 mm CMU with 16 mm GWB* on furring strips	53	33	32	44	50	56	59	65
16 mm GWB on both sides of 92 mm metal studs	38	18	16	33	47	55	43	47
16 mm GWB on both sides of 92 mm metal studs with fiberglass insulation in cavity	49	16	23	44	58	64	52	53
2 layers of 16 mm GWB on both sides of 92 mm metal studs with fiberglass insulation in cavity	56	19	32	50	62	67	58	63
Double row of 92 mm metal studs, 25 mm apart, each with 2 layers of 16 mm GWB and fiberglass insulation in cavity	64	23	40	54	62	71	69	74
150 mm solid concrete floor/ceiling	53	40	40	40	49	58	67	76
150 mm solid concrete floor with 100 mm isolated concrete slab and fiberglass insulation in cavity	72	44	52	58	73	87	97	100
150 mm solid concrete floor with two layers of 16 mm GWB hung on spring isolators with fiberglass insulation in cavity	84	53	63	70	84	93	104	105

Note: Actual material composition (e.g., density, porosity, stiffness) affects transmission loss and STC values.
*CMU = concrete masonry unit; GWB = gypsum wallboard.

Table 17.7 Sound Sources, Transmission Paths, and Recommended Noise Reduction Methods [2019A, Ch 49, Tbl 6]

Sound Source	Path No.
Circulating fans; grilles; registers; diffusers; unitary equipment in room	1
Induction coil and fan-powered VAV mixing units	1, 2
Unitary equipment located outside of room served; remotely located air-handling equipment, such as fans, blowers, dampers, duct fittings, and air washers	2, 3
Compressors, pumps, and other reciprocating and rotating equipment (excluding air-handling equipment)	4, 5, 6
Cooling towers; air-cooled condensers	4, 5, 6, 7
Exhaust fans; window air conditioners	7, 8
Sound transmission between rooms	9, 10

No.	Transmission Paths	Noise Reduction Methods
1	Direct sound radiated from sound source to ear	Direct sound can be controlled only by selecting quiet equipment.
	Reflected sound from walls, ceiling, and floor	Reflected sound is controlled by adding sound absorption to the room and to equipment location.
2	Air- and structureborne sound radiated from casings and through walls of ducts and plenums is transmitted through walls and ceiling into room	Design duct and fittings for low turbulence; locate high-velocity ducts in noncritical areas; isolate ducts and sound plenums from structure with neoprene or spring hangers.
3	Airborne sound radiated through supply and return air ducts to diffusers in room and then to listener by Path 1	Select fans for minimum sound power; use ducts lined with sound-absorbing material; use duct silencers or sound plenums in supply and return air ducts.
4	Noise transmitted through equipment room walls and floors to adjacent rooms	Locate equipment rooms away from critical areas; use masonry blocks or concrete for equipment room walls and floor.
5	Vibration transmitted via building structure to adjacent walls and ceilings, from which it radiates as noise into room by Path 1	Mount all machines on properly designed vibration isolators; design mechanical equipment room for dynamic loads; balance rotating and reciprocating equipment.
6	Vibration transmission along pipes and duct walls	Isolate pipe and ducts from structure with neoprene or spring hangers; install flexible connectors between pipes, ducts, and vibrating machines.
7	Noise radiated to outside enters room windows	Locate equipment away from critical areas; use barriers and covers to interrupt noise paths; select quiet equipment.
8	Inside noise follows Path 1	Select quiet equipment.
9	Noise transmitted to an air diffuser in a room, into a duct, and out through an air diffuser in another room	Design and install duct attenuation to match transmission loss of wall between rooms.
10	Sound transmission through, over, and around room partition	Extend partition to ceiling slab and tightly seal all around; seal all pipe, conduit, duct, and other partition penetrations.

18. VIBRATION

Mechanical vibration and vibration-induced noise are common sources of occupant complaints in modern buildings. Low-mass construction in buildings provides conditions that can result in vibration-related problems. Mandates for energy conservation have resulted in many buildings being designed with variable air volume systems with variable-speed equipment. As rotating equipment spins slower, its forcing frequency approaches the structure's resonant frequency, which can amplify vibration-induced noise and generate vibration at the structure's resonant frequency. Mechanical equipment is often located in penthouses or on the roof, where structures are typically the most susceptible to inducing vibration-related problems. Mechanical equipment rooms are typically located on intermediate level floors, close to the occupied areas they serve.

Occupant complaints associated with building vibration typically take one or more of three forms:

- The level of vibration perceived by building occupants is of sufficient magnitude to cause annoyance, concern, or alarm
- Vibration energy from mechanical equipment, which is transmitted to the building structure, is transmitted to various parts of the building and then is radiated as structureborne noise
- Vibration in a building may interfere with proper operation of sensitive equipment or instrumentation

At the natural frequency (in Hz) of the system, the mass's vibration response in to the applied excitation is a maximum:

$$\text{Natural frequency, } f_n = \frac{1}{2\pi}\sqrt{\frac{k}{M}} \tag{18.1}$$

where k is the stiffness of vibration isolator (N/m) and M is the mass of equipment supported by the isolator.

$$f_n = \frac{3.13}{\sqrt{\delta_{st}}} \tag{18.2}$$

where δ_{st} is the static deflection of the isolator in inches.

Transmissibility is the ratio of the amplitudes of the force transmitted to the building structure to the exciting force produced by the vibrating equipment. Transmissibility is inversely proportional to the square of the disturbing frequency f_d to the natural frequency f_n.

$$T = \left|\frac{1}{1-(f_d/f_n)^2}\right| \tag{18.3}$$

At $f_d = f_n$, resonance occurs. Vibration isolation is effective only at a f_d/f_n ratio > 3.5.

When supporting structure stiffness is not large with respect to stiffness of isolator, it becomes a two-degree of freedom system. In this case, choose an isolator that will provide static deflection eight to ten times that of the estimated floor static deflection due to the added weight of the equipment. Seismic snubbers must be included in or with isolators to limit equipment movement.

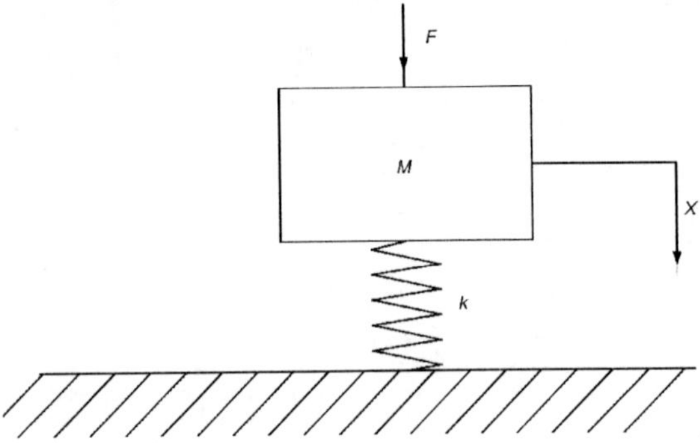

Figure 18.1 Single-Degree-of-Freedom System [2021F, Ch 8, Fig 8]

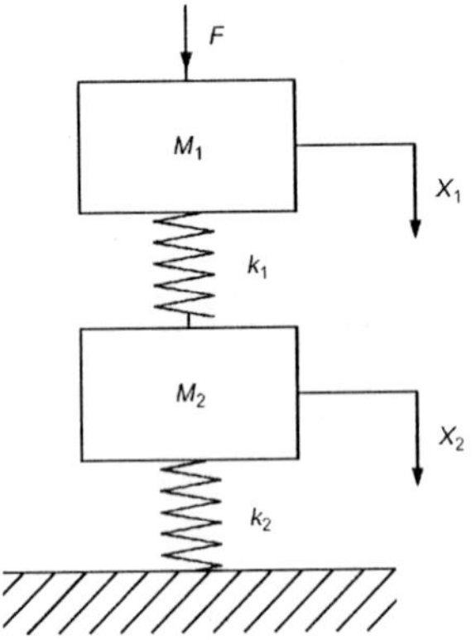

Figure 18.2 Two-Degrees-of-Freedom System [2021F, Ch 8, Fig 11]

Table 18.1 Selection Guide for Vibration Isolation [2019A, Ch 49 Tbl 47]

Equipment Type	Shaft Power kW and Other	RPM	Equipment Location (see Notes for Table 18.1, Item 1)												Reference Notes
			Slab on Grade			Floor Span									
						Up to 6 m			6 to 9 m			9 to 12 m			
			Base Type	Isolator Type	Min. Defl., mm	Base Type	Isolator Type	Min. Defl., mm	Base Type	Isolator Type	Min. Defl., mm	Base Type	Isolator Type	Min. Defl., mm	
Refrigeration Machines and Chillers															
Water-cooled reciprocating	All	All	A	2	6.4	A	4	19	A	4	38	A	4	64	2,3,12
Water-cooled centrifugal, scroll	All	All	A	1	6.4	A	4	19	A	4	38	A	4	38	2,3,4,8,12
Water-cooled screw	All	All	A	4	25	A	4	38	A	4	64	A	4	64	2,3,4,12
Absorption	All	All	A	1	6.4	A	4	19	A	4	38	A	4	38	
Air-cooled recip., scroll	All	All	A	1	6.4	A	4	38	A	4	38	A	4	64	2,4,5,12
Air-cooled screw	All	All	A	4	25	A	4	38	B	4	64	B	4	64	2,4,5,8,12
Air Compressors and Vacuum Pumps															
Tank-mounted horiz.	≤7.5	All	A	3	19	A	3	19	A	3	38	A	3	38	3,15
	≥7.5	All	C	3	19	C	3	19	C	3	38	C	3	38	3,15
Tank-mounted vert.	All	All	C	3	19	C	3	19	C	3	38	C	3	38	3,15
Base-mounted	All	All	C	3	19	C	3	19	C	3	38	C	3	38	3,14,15
Large reciprocating	All	All	C	3	19	C	3	19	C	3	38	C	3	38	3,14,15
Pumps															
Close-coupled	≤5.6	All	B	2	6.4	C	3	19	C	3	19	C	3	19	16
	≥5.6	All	C	3	19	C	3	19	C	3	38	C	3	38	16
Large inline	3.7 to 19	All	A	3	19	A	3	38	A	3	38	A	3	38	
	≥19	All	A	3	38	A	3	38	A	3	38	A	3	64	
End suction and split case	≤30	All	C	3	19	C	3	19	C	3	38	C	3	38	16
	30 to 93	All	C	3	19	C	3	19	C	3	38	C	3	64	10,16
	≥93	All	C	3	19	C	3	38	C	3	64	C	3	89	10,16
Packaged pump systems	All	All	A	3	19	A	3	19	A	3	38	C	3	64	

Table 18.1 Selection Guide for Vibration Isolation [2019A, Ch 49, Tbl 47] ***(Continued)***

Equipment Type	Shaft Power kW and Other	RPM	Equipment Location (see Notes for Table 18.1, Item 1)												Reference Notes
			Slab on Grade			Floor Span									
						Up to 6 m			6 to 9 m			9 to 12 m			
			Base Type	Isolator Type	Min. Defl., mm	Base Type	Isolator Type	Min. Defl., mm	Base Type	Isolator Type	Min. Defl., mm	Base Type	Isolator Type	Min. Defl., mm	
Cooling Towers		Up to 300	A	1	6.4	A	4	89	A	4	89	A	4	89	5,8,18
	All	301 to 500	A	1	6.4	A	4	64	A	4	64	A	4	64	5,18
		501 and up	A	1	6.4	A	4	19	A	4	19	A	4	38	5,18
Boilers															
Fire-tube	All	All	A	1	6.4	B	4	19	B	4	38	B	4	64	4
Water-tube, copper fin	All	All	A	1	3	A	1	3	A	1	3	B	4	6.4	
Axial Fans, Plenum Fans, Cabinet Fans, Fan Sections, Centrifugal Inline Fans															
Up to 560 mm diameter	All	All	A	2	6.4	A	3	19	A	3	19	C	3	19	4,9
610 mm diameter and up	≤500 Pa SP	Up to 300	B	3	64	C	3	89	C	3	89	C	3	89	9,8
		300 to 500	B	3	19	B	3	38	C	3	64	C	3	64	9,8
		501 and up	B	3	19	B	3	38	B	3	38	B	3	38	9,8
	≥501 Pa SP	Up to 300	C	3	64	C	3	89	C	3	89	C	3	89	3,8,9
		300 to 500	C	3	38	C	3	38	C	3	64	C	3	64	3,8,9
		501 and up	C	3	19	C	3	38	C	3	38	C	3	64	3,8,9
Centrifugal Fans															
Up to 560 mm diameter	All	All	B	2	6.4	B	3	19	B	3	19	B	3	38	9,19
610 mm diameter and up		Up to 300	B	3	64	B	3	89	B	3	89	B	3	89	8,19
	≤30	300 to 500	B	3	38	B	3	38	B	3	64	B	3	64	8,19
		501 and up	B	3	19	B	3	19	B	3	19	B	3	38	8,19
	≥37	Up to 300	C	3	64	C	3	89	C	3	89	C	3	89	2,3,8,9,19
		300 to 500	C	3	38	C	3	38	C	3	64	C	3	64	2,3,8,9,19
		501 and up	C	3	25.4	C	3	38	C	3	38	C	3	64	2,3,8,9,19
Propeller Fans															
Wall-mounted	All	All	A	1	6.4	A	1	6.4	A	1	6.4	A	1	6.4	
Roof-mounted	All	All	A	1	6.4	A	1	6.4	B	4	38	D	4	38	

Table 18.1 Selection Guide for Vibration Isolation [2019A, Ch 49, Tbl 47] *(Continued)*

Equipment Type	Shaft Power kW and Other	RPM	Equipment Location (see Notes for Table 18.1, Item 1)												Reference Notes
			Slab on Grade			Floor Span									
						Up to 6 m			6 to 9 m			9 to 12 m			
			Base Type	Isolator Type	Min. Defl., mm	Base Type	Isolator Type	Min. Defl., mm	Base Type	Isolator Type	Min. Defl., mm	Base Type	Isolator Type	Min. Defl., mm	
Heat Pumps, Fan-Coils, Computer Room Units	All	All	A	3	19	A	3	19	A	3	19	A/D	3	38	
Condensing Units	All	All	A	1	6.4	A	4	19	A	4	38	A/D	4	38	
Packaged AH, AC, H and V Units															
All	7.5	All	A	3	19	A	3	19	A	3	19	A	3	19	19
	7.5 to 11	Up to 300	A	3	19	A	3	89	A	3	89	C	3	89	2,4,8,19
	≤ 1kPa SP	301 to 500	A	3	19	A	3	64	A	3	64	A	3	64	4,19
		501 and up	A	3	19	A	3	38	A	3	38	A	3	38	4,19
	>11,	Up to 300	B	3	19	C	3	89	C	3	89	C	3	89	2,3,4,8,9
	> 1kPa SP	301 to 500	B	3	19	C	3	38	C	3	64	C	3	64	2,3,4,9
		501 and up	B	3	19	C	3	38	C	3	38	C	3	64	2,3,4,9
Packaged Rooftop Equipment	All	All	A/D	1	6.4	D	3	19	See Reference Note 17						5,6,8,17
Ducted Rotating Equipment															
Small fans, fan-powered boxes	≤300 L/s	All	A	3	12.7	A	3	12.7	A	3	12.7	A	3	12.7	7
	≥301 L/s	All	A	3	19	A	3	19	A	3	19	A	3	19	7
Engine-Driven Generators	All	All	A	3	19	C	3	38	C	3	64	C	3	89	2,3,4
Piping and Ducts (See sections on Isolating Vibration and Noise in Piping Systems and Isolating Duct Vibration for isolator selection.)															

Base Types:
A. No base, isolators attached directly to equipment (Note 28)
B. Structural steel rails or base (Notes 29 and 30)
C. Concrete inertia base (Note 30)
D. Curb-mounted base (Note 31)

Isolator Types:
1. Pad, rubber, or glass fiber (Notes 20 and 21)
2. Rubber floor isolator or hanger (Notes 20 and 25)
3. Spring floor isolator or hanger (Notes 22, 23, and 26)
4. Restrained spring isolator (Notes 22 and 24)
5. Thrust restraint (Note 27)
6. Air spring (Note 25)

Notes for Table 18.1:

These notes are keyed to the column titled Reference Notes in Table 47 of Chapter 49 of the 2019 *ASHRAE Handbook—Applications* and to other reference numbers throughout the table. Although the guide is conservative, cases may arise where vibration transmission to the building is still excessive. If the problem persists after all short circuits have been eliminated, it can almost always be corrected by altering the support path (e.g., from ceiling to floor), increasing isolator deflection, using low-frequency air springs, changing operating speed, improving rotating component balancing, or, as a last resort, changing floor frequency by stiffening or adding more mass. Assistance from a qualified vibration consultant can be very useful in resolving these problems.

Note 1. Isolator deflections shown are based on a reasonably expected floor stiffness according to floor span and class of equipment. Certain spaces may dictate higher levels of isolation. For example, bar joist roofs may require a static deflection of 38 mm over factories, but 64 mm over commercial office buildings.

Note 2. For large equipment capable of generating substantial vibratory forces and structureborne noise, increase isolator deflection, if necessary, so isolator stiffness is less than one-tenth the stiffness of the supporting structure, as defined by the deflection due to load at the equipment support.

Note 3. For noisy equipment adjoining or near noise-sensitive areas, see the section on Mechanical Equipment Room Sound Isolation.

Note 4. Certain designs cannot be installed directly on individual isolators (type A), and the equipment manufacturer or a vibration specialist should be consulted on the need for supplemental support (base type).

Note 5. Wind load conditions must be considered. Restraint can be achieved with restrained spring isolators (type 4), supplemental bracing, snubbers, or limit stops. Also see Chapter 56 of the 2019 *ASHRAE Handbook—HVAC Applications*.

Note 6. Certain types of equipment require a curb-mounted base (type D). Airborne noise must be considered.

Note 7. See section on Resilient Pipe Hangers and Supports for hanger locations adjoining equipment and in equipment rooms.

Note 8. To avoid isolator resonance problems, select isolator deflection so that resonance frequency is 40% or less of the lowest normal operating speed of equipment (see Chapter 8 in the 2021 *ASHRAE Handbook—Fundamentals*). Some equipment, such as variable-frequency drives, and high-speed equipment, such as screw chillers and vaneaxial fans, contain very-high-frequency vibration. This equipment creates new technical challenges in the isolation of high-frequency noise and vibration from a building's structure. Structural resonances both internal and external to the isolators can significantly degrade their performance at high frequencies. Unfortunately, at present no test standard exists for measuring the high-frequency dynamic properties of isolators, and commercially available products are not tested to determine their effectiveness for high frequencies. To reduce the chance of high-frequency vibration transmission, add a minimum 20 mm thick elastomeric pad (type 1, Note 20) to the base plate of spring isolators (type 3, Note 22, 23, 24). For some sensitive locations, air springs (Note 25) may be required. If equipment is located near extremely noise-sensitive areas, follow the recommendations of an acoustical consultant.

Note 9. To limit undesirable movement, thrust restraints (type 5) are required for all ceiling-suspended and floor-mounted units operating at 500 Pa or more total static pressure.

Note 10. Pumps over 55 kW may need extra mass and restraints.

Note 11. See text for full discussion.

Isolation for Specific Equipment

Note 12. Refrigeration Machines: Large centrifugal, screw, and reciprocating refrigeration machines may generate very high noise levels; special attention is required when such equipment is installed in upper-story locations or near noise-sensitive areas. If equipment is located near extremely noise-sensitive areas, follow the recommendations of an acoustical consultant.

Note 13. Compressors: The two basic reciprocating compressors are (1) single- and double-cylinder vertical, horizontal or L-head, which are usually air compressors; and (2) Y, W, and multihead or multicylinder air and refrigeration compressors. Single- and double-cylinder compressors generate high vibratory forces requiring large inertia bases (type C) and are generally not suitable for upper-story locations. If this equipment must be installed in an upper-story location or at-grade location near noise-sensitive areas, the expected maximum unbalanced force data must be obtained from the equipment manufacturer and a vibration specialist consulted for design of the isolation system.

Note 14. Compressors: When using Y, W, and multihead and multicylinder compressors, obtain the magnitude of unbalanced forces from the equipment manufacturer so the need for an inertia base can be evaluated.

Note 15. Compressors: Base-mounted compressors through 4 kW and horizontal tank-type air compressors through 8 kW can be installed directly on spring isolators (type 3) with structural bases (type B) if required, and compressors 10 to 75 kW on spring isolators (type 3) with inertia bases (type C) with a mass 1 to 2 times the compressor mass.

Note 16. Pumps: Concrete inertia bases (type C) are preferred for all flexible-coupled pumps and are desirable for most close-coupled pumps, although steel bases (type B) can be used. Close-coupled pumps should not be installed directly on individual isolators (type A) because the impeller usually overhangs the motor support base, causing the rear mounting to be in tension. The primary requirements for type C bases are strength and shape to accommodate base elbow supports. Mass is not usually a factor, except for pumps over 55 kW, where extra mass helps limit excess movement due to starting torque and forces. Concrete bases (type C) should be designed for a thickness of one-tenth the longest dimension with minimum thickness as follows: (1) for up to 20 kW, 150 mm; (2) for 30 to 55 kW, 200 mm; and (3) for 75 kW and up, 300 mm.

Pumps over 55 kW and multistage pumps may exhibit excessive motion at start-up ("heaving"); supplemental restraining devices can be installed if necessary. Pumps over 90 kW may generate high starting forces; consult a vibration specialist.

Note 17. Packaged Rooftop Air-Conditioning Equipment: This equipment is usually installed on low-mass structures that are susceptible to sound and vibration transmission problems. The noise problems are compounded further by curb-mounted equipment, which requires large roof openings for supply and return air.

The table shows type D vibration isolator selections for all spans up to 6 m, but extreme care must be taken for equipment located on spans of over 6 m, especially if construction is open web joists or thin, low-mass slabs. The recommended procedure is to determine the additional deflection caused by equipment in the roof. If additional roof deflection is 6 mm or less, the isolator should be selected for up to 10 times the additional roof deflection. If additional roof deflection is over 6 mm, supplemental roof stiffening should be installed to bring the roof deflection down below 6 mm, or the unit should be relocated to a stiffer roof position.

For mechanical units capable of generating high noise levels, mount the unit on a platform above the roof deck to provide an air gap (buffer zone) and locate the unit away from the associated roof penetration to allow acoustical treatment of ducts before they enter the building.

Some rooftop equipment has compressors, fans, and other equipment isolated internally. This isolation is not always reliable because of internal short-circuiting, inadequate static deflection, or panel resonances. It is recommended that rooftop equipment over 135 kg be isolated externally, as if internal isolation was not used.

Note 18. Cooling Towers: These are normally isolated with restrained spring isolators (type 4) directly under the tower or tower dunnage. High-deflection isolators proposed for use directly under the motor-fan assembly must be used with extreme caution to ensure stability and safety under all weather conditions. See Note 5.

Note 19. Fans and Air-Handling Equipment: Consider the following in selecting isolation systems for fans and air-handling equipment:

1. Fans with wheel diameters of 560 mm and less and all fans operating at speeds up to 300 rpm do not generate large vibratory forces. For fans operating under 300 rpm, select isolator deflection so the isolator natural frequency is 40% or less than the fan speed. For example, for a fan operating at 275 rpm, $0.4 \times 275 = 110$ rpm. Therefore, an isolator natural frequency of 110 rpm or lower is required. This can be accomplished with a 75 mm deflection isolator (type 3).
2. Flexible duct connectors should be installed at the intake and discharge of all fans and air-handling equipment to reduce vibration transmission to air duct structures.
3. Inertia bases (type C) are recommended for all class 2 and 3 fans and air-handling equipment because extra mass allows the use of stiffer springs, which limit heaving movements.
4. Thrust restraints (type 5) that incorporate the same deflection as isolators should be used for all fan heads, all suspended fans, and all base-mounted and suspended air-handling equipment operating at 500 Pa or more total static pressure. Restraint movement adjustment must be made under normal operational static pressures.

Vibration Isolators: Materials, Types, and Configurations

Notes 20 through 32 include figures to assist in evaluating commercially available isolators for HVAC equipment. The isolator selected for a particular application depends on the required deflection, life, cost, and compatibility with associated structures.

RUBBER PADS (Type 1)

RUBBER MOUNTS (Type 2)

Note 20. Rubber isolators are available in pad (type 1) and molded (type 2) configurations. Pads are used in single or multiple layers. Molded isolators come in a range of 30 to 70 durometer (a measure of stiffness). Material in excess of 70 durometer is usually ineffective because durometers are not a measure of stiffness of an isolator. Isolators are designed for up to 13 mm deflection, but are used where 8 mm or less deflection is required. Solid rubber and composite fabric and rubber pads are also available. They provide high load capacities with small deflection and are used as noise barriers under columns and for pipe supports. These pad types work well only when they are properly loaded and the mass load is evenly distributed over the entire pad surface. Metal loading plates can be used for this purpose.

GLASS FIBER PADS (Type 1)

Note 21. Glass fiber with elastic coating (type 1). This type of isolation pad is precompressed molded fiberglass pads individually coated with a flexible, moisture-impervious elastomeric membrane. Natural frequency of fiberglass vibration isolators should be essentially constant for the operating load range of the supported equipment. Mass load is evenly distributed over the entire pad surface. Metal loading plates can be used for this purpose.

SPRING ISOLATOR (Type 3) 	**Note 22.** Steel springs are the most popular and versatile isolators for HVAC applications because they are available for almost any deflection and have a virtually unlimited life. Spring isolators may have a rubber acoustical barrier to reduce transmission of high-frequency vibration and noise that can migrate down the steel spring coil. They should be corrosion protected if installed outdoors or in a corrosive environment. The basic types include the following: **Note 23.** *Open spring isolators* (type 3) consist of top and bottom load plates with adjustment bolts for leveling equipment. Springs should be designed with a horizontal stiffness of at least 80% of the vertical stiffness (k_x/k_y) to ensure stability. Similarly, the springs should have a minimum ratio of 0.8 for the diameter divided by the deflected spring height.
RESTRAINED SPRING ISOLATOR (Type 4) 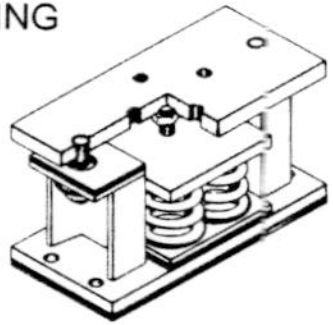	**Note 24.** *Restrained spring isolators* (type 4) have hold-down bolts to limit vertical as well as horizontal movement. They are used with (a) equipment with large variations in mass (e.g., boilers, chillers, cooling towers) to restrict movement and prevent strain on piping when water is removed, (b) outdoor equipment, such as condensing units and cooling towers, to prevent excessive movement due to wind loads, and (c) with any equipment subject to seismic forces. Spring criteria should be the same as open spring isolators, and snubbers should have adequate clearance so that they are activated only when a temporary restraint is needed. See Chapter 56 of the 2019 *ASHRAE Handbook—HVAC Applications* for typical snubber types. *Closed mounts* or *housed spring isolators* consist of two telescoping housings separated by a resilient material. These provide lateral snubbing and some vertical damping of equipment movement, but do not limit the vertical movement. Additional vertical snubbers must be used where vertical travel must be limited (see Chapter 56 of the 2019 *ASHRAE Handbook—HVAC Applications*). Care should be taken in selection and installation to minimize binding and short circuiting.
AIR SPRINGS (Type 6)  ROLLING LOBE  BELLOWS	**Note 25.** Air springs (type 6) can be designed for any frequency, but are economical only in applications with natural frequencies of 1.33 Hz or less (150 mm or greater deflection). They do not transmit high-frequency noise and are often used to replace high-deflection springs on problem jobs (e.g., large transformers on upper-floor installations). A constant air supply (an air compressor with an air dryer) and leveling valves are typically required.

RUBBER HANGER
(Type 2)

SPRING HANGER
(Type 3)

Note 26. Isolation hangers (types 2 and 3) are used for suspended pipe and equipment and have rubber, springs, or a combination of spring and rubber elements. Criteria should be similar to open spring isolators, though lateral stability is less important. Where support rod angular misalignment is a concern, use hangers that have sufficient clearance and/or incorporate rubber bushings to prevent the rod from touching the housing. Swivel or traveler arrangements may be necessary for connections to piping systems subject to large thermal movements.

Precompressed spring hangers incorporate some means of precompression or preloading of the isolator spring to minimize movement of the isolated equipment or system. These are typically used on piping systems that can change mass substantially between installation and operation.

THRUST RESTRAINT
(Type 5)

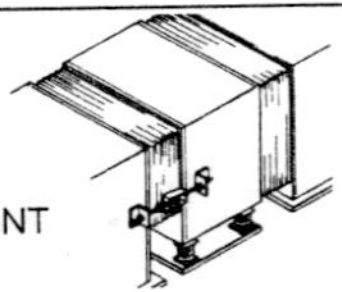

Note 27. Thrust restraints (type 5) are similar to spring hangers or isolators and are installed in pairs to resist the thrust caused by air pressure. These are typically sized to limit lateral movement to 6.4 mm or less.

DIRECT ISOLATION (Type A)

Note 28. Direct isolation (type A) is used when equipment is unitary and rigid and does not require additional support. Direct isolation can be used with large chillers, some fans, packaged air-handling units, and air-cooled condensers. If there is any doubt that the equipment can be supported directly on isolators, use structural bases (type B) or inertia bases (type C), or consult the equipment manufacturer.

STRUCTURAL
BASES (Type B)

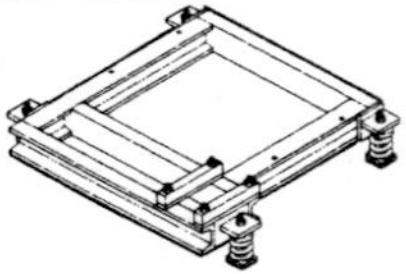

Note 29. Structural bases (type B) are used where equipment cannot be supported at individual locations and/or where some means is necessary to maintain alignment of component parts in equipment. These bases can be used with spring or rubber isolators (types 2 and 3) and should have enough rigidity to resist all starting and operating forces without supplemental hold-down devices. Bases are made in rectangular configurations using structural members with a depth equal to one-tenth the longest span between isolators. Typical base depth is between 100 and 300 mm, except where structural or alignment considerations dictate otherwise.

STRUCTURAL RAILS (Type B)

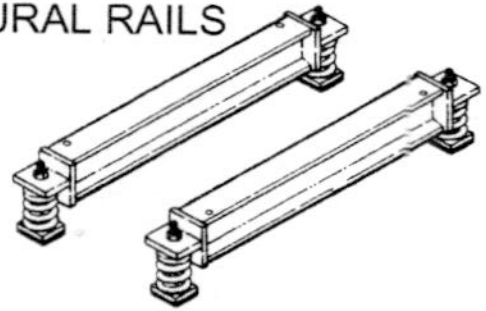

Note 30. Structural rails (type B) are used to support equipment that does not require a unitary base or where the isolators are outside the equipment and the rails act as a cradle. Structural rails can be used with spring or rubber isolators and should be rigid enough to support the equipment without flexing. Usual practice is to use structural members with a depth one-tenth of the longest span between isolators, typically between 100 and 300 mm, except where structural considerations dictate otherwise.

CONCRETE BASES (Type C)

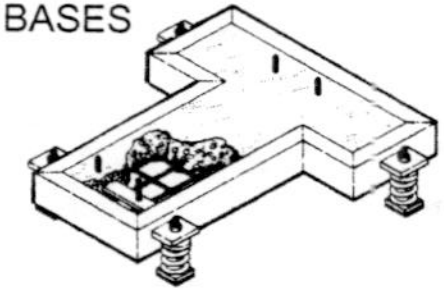

Note 31. Concrete bases (type C) are used where the supported equipment requires a rigid support (e.g., flexible-coupled pumps) or excess heaving motion may occur with spring isolators. They consist of a steel pouring form usually with welded-in reinforcing bars, provision for equipment hold-down, and isolator brackets. Like structural bases, concrete bases should be sized to support piping elbow supports, rectangular or T-shaped, and for rigidity, have a depth equal to one-tenth the longest span between isolators. Base depth is typically between 150 and 300 mm unless additional depth is specifically required for mass, rigidity, or component alignment.

CURB ISOLATION (Type D)

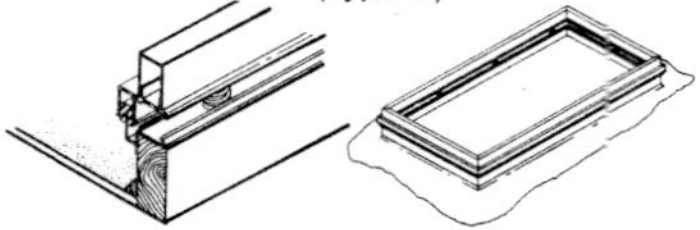

Note 32. Curb isolation systems (type D) are specifically designed for curb-supported rooftop equipment and have spring isolation with a watertight, and sometimes airtight, assembly. *Rooftop rails* consist of upper and lower frames separated by nonadjustable springs and rest on top of architectural roof curbs. *Isolation curbs* incorporate the roof curb into their design as well. Both kinds are designed with springs that have static deflections in the 25 to 75 mm range to meet the design criteria described in type 3. Flexible elastomeric seals are typically most effective for weatherproofing between the upper and lower frames. A continuous sponge gasket around the perimeter of the top frame is typically applied to further weatherproof the installation.

19. HVAC SYSTEMS AND EQUIPMENT

For information on boilers, compressors, chillers, and cooling towers, see the 2020 *ASHRAE Handbook—HVAC Systems and Equipment.*

Furnaces

Furnaces are self-enclosed, permanently installed major appliances that provide heated air through ductwork to the space being heated. In addition, a furnace may provide the indoor fan necessary for circulating heated or cooled air from a split or single-package air conditioner or heat pump. Furnaces may be used in either residential or commercial applications, and may be grouped according to the following characteristics:

- Heat source: electricity, natural gas/propane (fan assisted, condensing or noncondensing), or oil (forced draft with power atomizing burner)
- Installation location: within conditioned space (indoors), or outside conditioned space (either outdoors, or inside the structure but not within the conditioned space)
- Combustion air source: direct vent (outdoor air) or indoor air
- Mounting arrangement and airflow: horizontal, vertical upflow, vertical downflow, or multiposition

Furnaces that use electricity as a heat source include one or more resistance-type heating elements that heats the circulating air either directly or through a metal sheath that encloses the resistance element. In gas- or oil-fired furnaces, combustion occurs in the heat exchanger sections or in a combustion chamber, with direct-spark, hot-surface, or electric ignition. Circulating air passes over the outer surfaces of a heat exchanger so that it does not contact the fuel or the products of combustion, which are passed to the outdoor atmosphere through a vent.

In North America, natural gas is the most common fuel supplied for residential heating, and the central-system forced-air furnace (Figure 19.1) is the most common way of heating with natural gas.

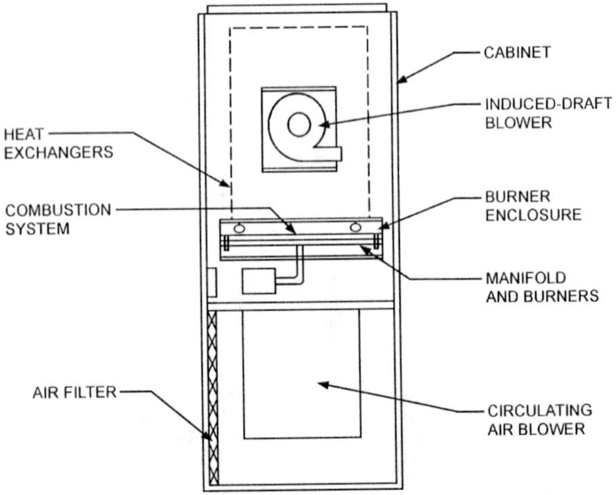

Figure 19.1 Induced-Draft Gas Furnace [2020S, Ch 33, Fig 1]

Furnaces with gas-fired burners have heat exchangers that are typically made either of left/right sets of formed parts that are joined together to form a clamshell, finless tubes bent into a compact form, or finned-tube (condensing) heat exchangers. Standard indoor furnace heat exchangers are generally made of alloy steel. Common corrosion-resistant materials include aluminized steel and stainless steel. Furnaces certified for use downstream of a cooling coil must have corrosion-resistant heat exchangers.

Heat exchangers of oil-fired furnaces are normally heavy-gage steel formed into a welded assembly. Hot flue products flow through the inside of the heat exchanger into the chimney, and conditioned air flows over the outside of the heat exchanger and into the air supply plenum.

Fan-assisted combustion furnaces use a small blower to induce flue products through the furnace. Induced-draft furnaces may or may not have a relief air opening, but they meet the same safety requirements regardless. Residential furnaces built since 1987 are equipped with a blocked-vent shutoff switch to shut down the furnace in case the vent becomes blocked.

Direct-vent furnaces use outdoor air for combustion. Outdoor air is supplied to the furnace combustion chamber by direct connections between the furnace and the outdoor air. If the vent or the combustion air supply becomes blocked, the furnace control system will shut down the furnace.

ANSI Standard Z21.47/CSA 2.3 classifies venting systems. Central furnaces are categorized by temperature and pressure attained in the vent and by the steady-state efficiency attained by the furnace. Although ANSI Standard Z21.47/CSA 2.3 uses 83% as the steady-state efficiency dividing central furnace categories, a general rule of thumb is as follows:

Category I: nonpositive vent pressure and flue loss of 17% or more
Category II: nonpositive vent pressure and flue loss less than 17%
Category III: positive vent pressure and flue loss of 17% or more
Category IV: positive vent pressure and flue loss less than 17%

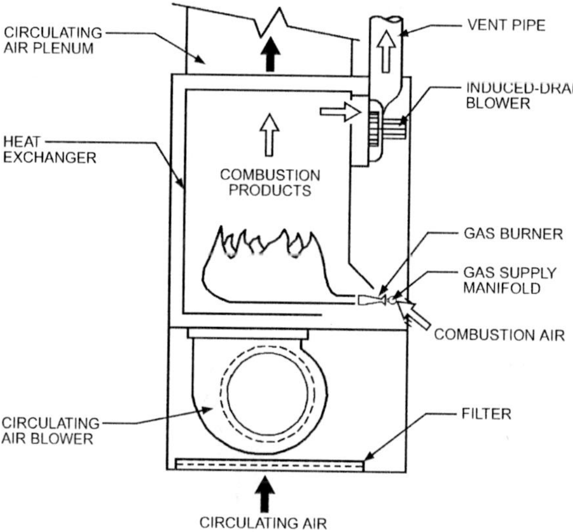

Figure 19.2 Upflow Category I Furnace with Induced-Draft Blower
[2020S, Ch 33, Fig 2]

Furnaces rated in accordance with ANSI Standard Z21.47/CSA 2.3 that are not direct vent are marked to show that they are in one of these four venting categories.

Ducted-system, oil-fired, forced-air furnaces are usually forced draft.

Furnaces with capacities above 150,000 Btu/h are classified as commercial furnaces. The other basic differences between residential and commercial furnaces are available options such as economizers, outdoor air dampers, and the type of electrical service required (three-phase).

Commercial heating equipment comes in almost as many flow arrangements and design variations as residential equipment. Some are identical to residential equipment, whereas others are unique to commercial applications. Some commercial units function as a part of a ducted system, and others operate as unducted space heaters.

Externally, the furnace is controlled by a low-voltage room thermostat.

Several types of gas valves perform various functions within the furnace. The type of valve available relates closely to the type of ignition device used. **Two-stage valves**, available on some furnaces, operate at full gas input or at a reduced rate, and are controlled by either a two-stage thermostat or a software algorithm programmed in the furnace control system.

The **fan control switch** controls the circulating air blower. This switch may be temperature-sensitive and exposed to the circulating airstream in the furnace cabinet, or it may be an electronically operated relay. Blower start-up is typically delayed about 1 min after burner start-up. This delay gives the heat exchangers time to warm up and reduces the flow of cold air when the blower comes on. Blower shutdown is also delayed several minutes after burner shutdown to remove residual heat from the heat exchangers and to improve the annual efficiency of the furnace.

The **limit switch** prevents overheating in the event of severe reduction in circulating airflow. This temperature-sensitive switch is exposed to the circulating airstream and shuts off

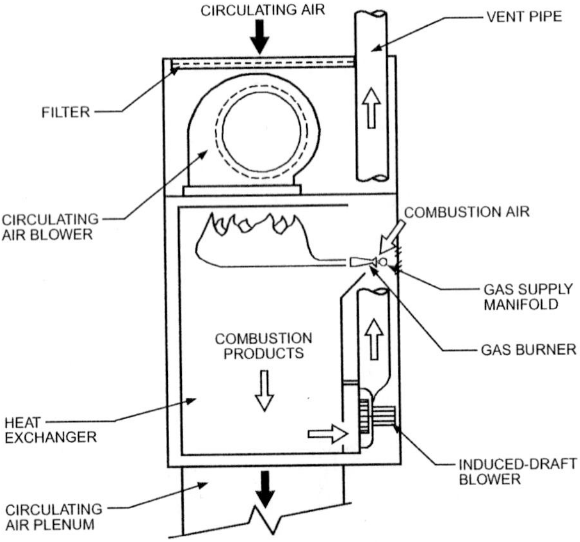

Figure 19.3 Downflow (Counterflow) Category I Furnace with Induced-Draft Blower [2020S, Ch 33, Fig 3]

the heat source (e.g., gas valve or electric element) if the temperature of air leaving the furnace is excessive. The fan control and limit switches are sometimes incorporated in the same housing and may be operated by the same thermostatic element. In the United States, the **blocked-vent shutoff switch** and **flame rollout switch** shut off the gas valve if the vent is blocked or when insufficient combustion air is present.

Furnaces using fan-assisted combustion feature a **pressure switch** to verify the flow of combustion air before opening the gas valve.

Electronic control systems are available in furnaces to provide sequencing of the inducer prepurge, ignition, circulating air blower operation, and inducer postpurge functions according to an algorithm provided by the manufacturer.

Furnaces can be installed inside or outside a building. For ideal air distribution, locate the unit in the center of the structure being heated. Furnaces are typically located in a closet, mechanical room, basement, attic, crawlspace, garage, or outdoors.

The type of fuel selected for heating is based on relative fuel cost, number of heating degree-days, and availability of utilities in the area. The most common fuel is natural gas because of its clean burning characteristics, and because of the continuous supply of this fuel through underground distribution networks to most urban settings. Propane and oil fuels are

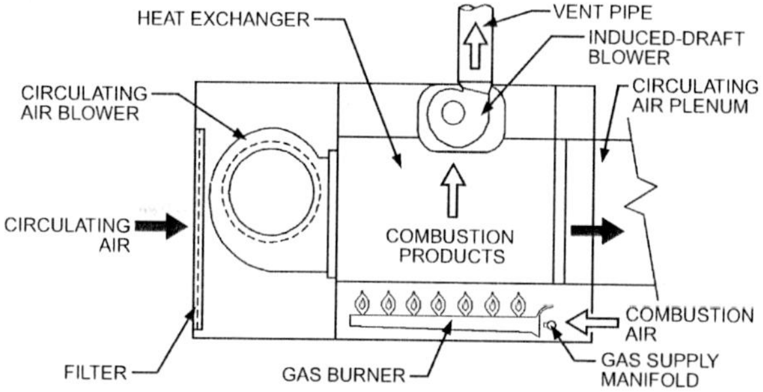

Figure 19.4 Horizontal Category I Furnace with Induced-Draft Blower [2020S, Ch 33, Fig 4]

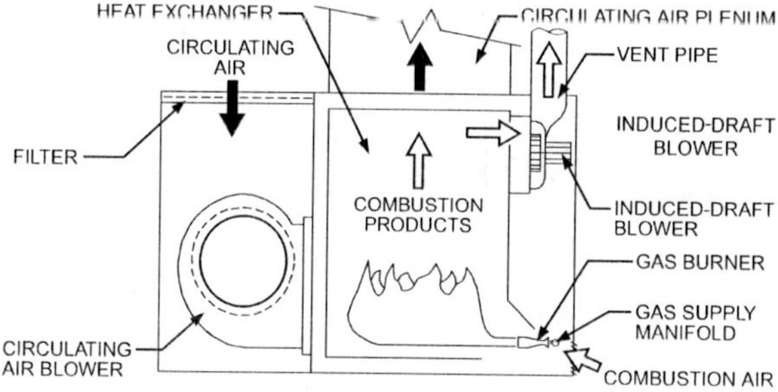

Figure 19.5 Basement (Lowboy) Category I Furnace with Induced-Draft Blower [2020S, Ch 33, Fig 5]

also commonly used. These fuels require on-site storage and periodic fuel deliveries. Electric heat is also continuously available through electrical power grids and is common especially where natural gas is not provided, or where the heating demand is small relative to the cooling demand.

Furnaces are clearly marked for the type of fuel to be used. In some cases, a manufacturer-approved conversion kit may be necessary to convert a furnace from one fuel type to another. If the fuel type is changed after the original installation, the conversion must be done by a qualified service person per the manufacturer's instructions and using the manufacturer's specified conversion kit. After conversion, the unit must be properly inspected by the local code authority.

All fuel-burning furnaces must be properly vented to the outdoors. Metal vents, masonry chimneys, and plastic vents are commonly used for furnaces. Manufacturers provide installation instructions for venting their furnaces, and Chapter 35 of the 2020 *ASHRAE Handbook—HVAC Systems and Equipment* has a detailed discussion on venting.

Air for combustion enters the combustion zone through louvers or pipes. Outdoor air usually has lower levels of pollutants than are typically found in air from indoors, garages, utility rooms, and basements.

The furnace's heating capacity (i.e., the maximum heating rate the furnace can provide) is provided on the appliance rating plate; it is also available through the manufacturers' product literature.

Other factors should be considered when determining furnace capacity. Thermostat setback recovery may require additional heating capacity. Increasing furnace capacity may increase space temperature swing, and thus reduce comfort. Two-stage or step-modulating equipment could help by using the unit's maximum capacity to meet the setback recovery needs, and providing a lower stage of heating capacity at other times.

Fuel-burning furnaces are typically subdivided into two primary categories:

- **Condensing** furnaces typically have high efficiencies, ranging from 89% to 98%, because they have a specially designed secondary heat exchanger that extracts the heat of vaporization of water vapor in the exhaust. The dew-point temperatures of flue gases of condensing furnaces are significantly above the vent temperature, so

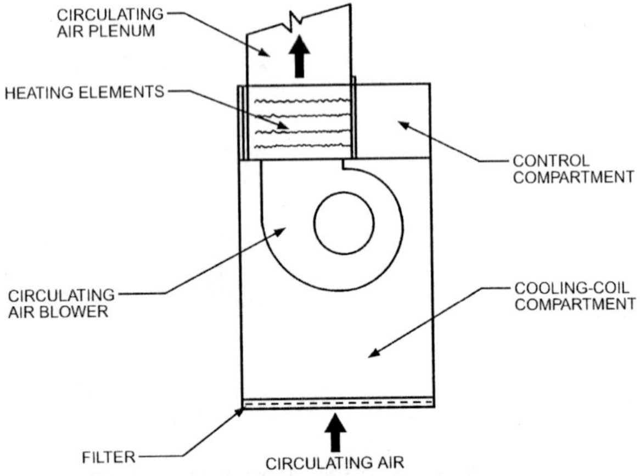

Figure 19.6 Electric Forced-Air Furnace [2020S, Ch 33, Fig 7]

plastic or other corrosion-resistant venting material is required. Condensing furnaces must be plumbed for condensate disposal.

- **Noncondensing** furnaces have generally less than 82% steady-state efficiency. This type of furnace has higher flue gas temperatures and requires either metal, masonry, or a combination of the two for venting materials.

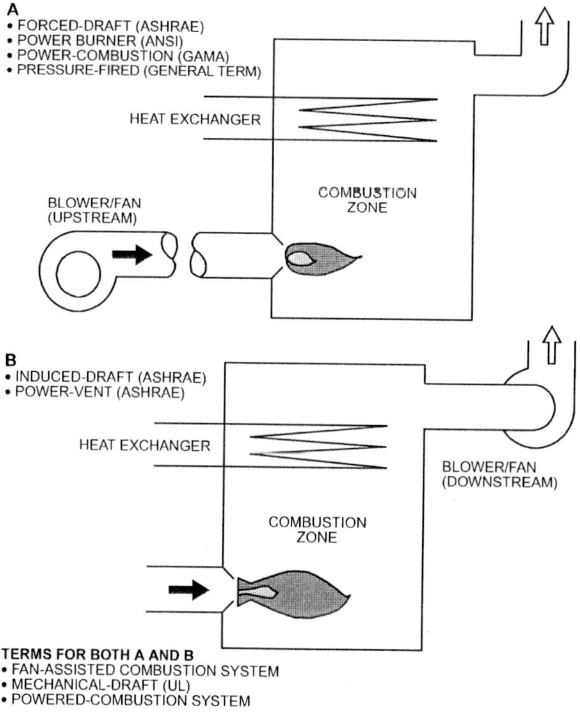

Figure 19.7 Terminology Used to Describe Fan-Assisted Combustion
[2020S, Ch 33, Fig 6]

Hydronic Heating Units and Radiators

Radiators, convectors, and baseboard and finned-tube units are heat-distributing devices used in hot-water and steam heating systems. They supply heat by a **combination of radiation and convection** and maintain the desired air temperature and/or mean radiant temperature in a space without fans. Figure 19.8 shows sections of typical heat-distributing units. In heating systems, radiant panels are also used. Units are inherently self-adjusting in the sense that heat output is based on temperature differentials; cold spaces receive more heat and warmer spaces receive less heat.

The following are the most common types of radiators:

- **Sectional radiators** are fabricated from welded sheet metal sections (generally two, three, or four tubes wide), and resemble freestanding cast-iron radiators.
- **Panel radiators** consist of fabricated flat panels (generally one, two, or three deep), with or without an exposed extended fin surface attached to the rear for increased output. These radiators are most common in Europe.
- **Tubular steel radiators** consist of supply and return headers with interconnecting parallel steel tubes in a wide variety of lengths and heights. They may be specially shaped to coincide with the building structure. Some are used to heat bathroom towel racks.
- **Specialty radiators** are fabricated of welded steel or extruded aluminum and are designed for installation in ceiling grids or floor-mounting. Various unconventional shapes are available.

Pipe coils have largely been replaced by finned tubes. See Table 5 in Chapter 28 of the 1988 *ASHRAE Handbook—Equipment* for the heat emission of such pipe coils.

A **convector** is a heat-distributing unit that operates with gravity-circulated air (natural convection). It has a heating element with a large amount of secondary surface and contains two or more tubes with headers at both ends. The heating element is surrounded by an enclosure with an air inlet below and an air outlet above the heating element.

Baseboard (or baseboard radiation) units are designed for installation along the bottom of walls in place of the conventional baseboard. They may be made of cast iron, with a substantial portion of the front face directly exposed to the room, or with a finned-tube element in a sheet metal enclosure. They use gravity-circulated room air.

Baseboard heat-distributing units are divided into three types: radiant, radiant convector, and finned tube.

- The **radiant** unit, which is made of aluminum, has no openings for air to pass over the wall side of the unit. Most of this unit's heat output is by radiation.
- The **radiant-convector** baseboard is made of cast iron or steel. The units have air openings at the top and bottom to allow circulation of room air over the wall side of the unit, which has extended surface to provide increased heat output. A large portion of the heat emitted is transferred by convection.
- The **finned-tube** baseboard has a finned-tube heating element concealed by a long, low sheet metal enclosure or cover. A major portion of the heat is transferred to the room by convection. The output varies over a wide range, depending on the physical dimensions and the materials used. A unit with a high relative output per unit length compared to overall heat loss (which would result in a concentration of the heating element over a relatively small area) should be avoided. Optimum comfort for room occupants is obtained when units are installed along as much of the exposed wall as possible.

Finned-tube (or fin-tube) units are fabricated from metallic tubing, with metallic fins bonded to the tube. They operate with gravity-circulated room air. Finned-tube elements are available in several tube sizes, in either steel or copper (1 to 2 in. nominal steel or 3/4 to 1 1/4 in. nominal copper) with various fin sizes, spacings, and materials. Resistance to steam or water flow is the same as that through standard distribution piping of equal size and type.

Finned-tube elements installed in occupied spaces generally have covers or enclosures in a variety of designs. When human contact is unlikely, they are sometimes installed bare or provided with an expanded metal grille for minimum protection.

The heat output ratings of heat-distributing units are expressed in Btu/h or in square feet equivalent direct radiation (EDR). By definition, 240 Btu/h = 1 ft^2 EDR with 1 psig steam.

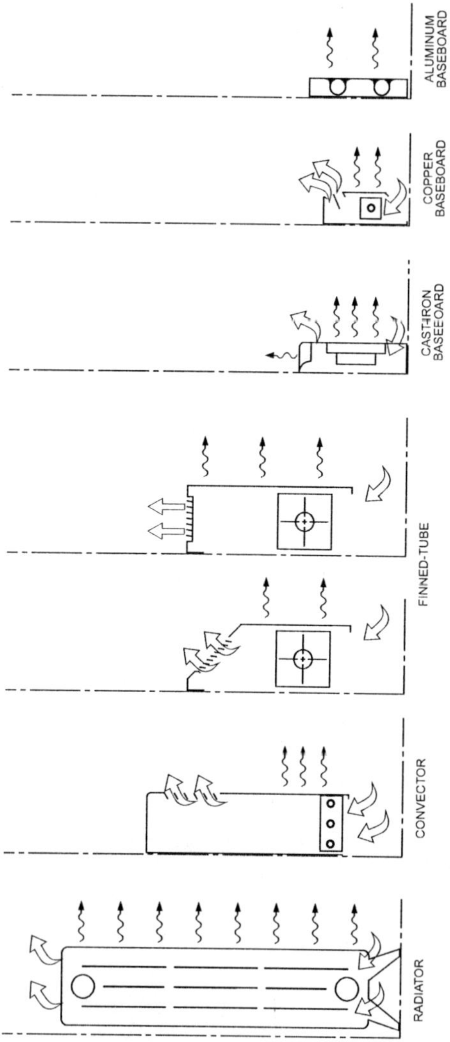

Figure 19.8 Terminal Units [2020S, Ch 36, Fig 1]

Table 19.1 Small-Tube Cast-Iron Radiators [2020S, Ch 36, Tbl 1]

Number of Tubes per Section	Catalog Rating per Section,[a]		Section Dimensions				
			A Height, in.[b]	B Width, in.		C Spacing, in.[c]	D Leg Height, in.[b]
	ft^2	Btu/h		Min.	Max.		
3	1.6	384	25	3.25	3.50	1.75	2.50
4	1.6	384	19	4.44	4.81	1.75	2.50
	1.8	432	22	4.44	4.81	1.75	2.50
	2.0	480	25	4.44	4.81	1.75	2.50
5	2.1	504	22	5.63	6.31	1.75	2.50
	2.4	576	25	5.63	6.31	1.75	2.50
6	2.3	552	19	6.81	8	1.75	2.50
	3.0	720	25	6.81	8	1.75	2.50
	3.7	888	32	6.81	8	1.75	2.50

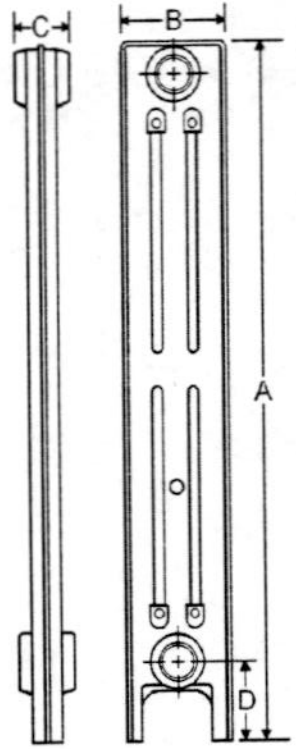

[a] Ratings based on steam at 215°F and air at 70°F. They apply only to installed radiators exposed in a normal manner, not to radiators installed behind enclosures, behind grilles, or under shelves. For Btu/h ratings at other temperatures, multiply table values by factors found in Table 2 of Chapter 36 of the 2020 *ASHRAE Handbook—HVAC Systems and Equipment*.

[b] Overall height and leg height, as produced by some manufacturers, are 1 in. greater than shown in columns A and D. Radiators may be furnished without legs. Where greater than standard leg heights are required, leg height should be 4.5 in.

[c] Length equals number of sections multiplied by 1.75 in.

Table 19.2 Correction Factors *c* for Various Types of Heating Units [2020S, Ch 36, Tbl 2]

Steam Pressure (Approx.)		Steam or Water Temp., °F	Cast-Iron Radiator Room Temp., °F					Convector Inlet Air Temp., °F					Finned-Tube Inlet Air Temp., °F					Baseboard Inlet Air Temp., °F				
			80	75	70	65	60	75	70	65	60	55	75	70	65	60	55	75	70	65	60	55
		100											0.10	0.12	0.15	0.17	0.20	0.08	0.10	0.13	0.15	0.18
		110											0.15	0.17	0.20	0.23	0.26	0.13	0.15	0.18	0.21	0.25
		120											0.20	0.23	0.26	0.29	0.33	0.18	0.21	0.25	0.28	0.31
		130											0.26	0.29	0.33	0.36	0.40	0.25	0.28	0.31	0.34	0.38
		140											0.33	0.36	0.40	0.42	0.45	0.31	0.34	0.38	0.42	0.45
in. Hg Vac.	**psia**																					
22.4	3.7	150	0.39	0.42	0.46	0.50	0.54	0.35	0.39	0.43	0.46	0.50	0.40	0.42	0.45	0.49	0.53	0.38	0.42	0.45	0.49	0.53
20.3	4.7	160	0.46	0.50	0.54	0.58	0.62	0.43	0.47	0.51	0.54	0.58	0.45	0.49	0.53	0.57	0.61	0.45	0.49	0.53	0.57	0.61
17.7	6.0	170	0.54	0.58	0.62	0.66	0.69	0.51	0.54	0.58	0.63	0.67	0.53	0.57	0.61	0.65	0.69	0.53	0.57	0.61	0.65	0.69
14.6	7.5	180	0.62	0.66	0.69	0.74	0.78	0.58	0.63	0.67	0.71	0.76	0.61	0.65	0.69	0.73	0.78	0.61	0.65	0.69	0.72	0.78
10.9	9.3	190	0.69	0.74	0.78	0.83	0.87	0.67	0.71	0.76	0.81	0.85	0.69	0.73	0.78	0.81	0.86	0.69	0.73	0.78	0.82	0.86
6.5	11.5	200	0.78	0.83	0.87	0.91	0.95	0.76	0.81	0.85	0.90	0.95	0.77	0.81	0.86	0.90	0.95	0.81	0.86	0.92	0.95	1.00
psig	**psia**																					
1	15.6	215	0.91	0.95	1.00	1.04	1.09	0.90	0.95	1.00	1.05	1.10	0.91	0.94	1.00	1.06	1.11	0.91	0.95	1.00	1.05	1.09
6	21	230	1.04	1.09	1.14	1.18	1.23	1.05	1.10	1.15	1.20	1.26	1.03	1.08	1.14	1.19	1.24	1.04	1.09	1.14	1.19	1.25
15	30	250	1.23	1.28	1.32	1.37	1.43	1.27	1.32	1.37	1.43	1.47	1.20	1.26	1.31	1.37	1.43	1.22	1.27	1.32	1.37	1.43
27	42	270	1.43	1.47	1.52	1.56	1.61	1.47	1.54	1.59	1.67	1.72	1.38	1.44	1.50	1.56	1.62	1.43	1.47	1.52	1.59	1.64
52	67	300	1.72	1.75	1.82	1.89	1.92	1.85	1.89	1.96	2.04	2.08	1.67	1.73	1.79	1.86	1.92	1.75	1.82	1.89	1.92	1.96

Note: Use these correction factors to determine output ratings for radiators, convectors, and finned-tube and baseboard units at operating conditions other than standard. Standard conditions in the United States for a radiator are 215°F heating medium temperature and 70°F room temperature (at center of space and at 5 ft level).
Standard conditions for convectors and finned-tube and baseboard units are 215°F heating medium temperature and 65°F inlet air temperature at 29.92 in. Hg atmospheric pressure. Water flow is 3 fps for finned-tube units. Inlet air at 65°F for convectors and finned-tube or baseboard units represents the same room comfort conditions as 70°F room air temperature for a radiator. Standard conditions for radiant panels are 122°F heating medium temperature and 68°F for room air temperature; *c* depends on panel construction.
To determine output of a heating unit under nonstandard conditions, multiply standard heating capacity by appropriate factor for actual operating heating medium and room or inlet air temperatures.

Corrections for Nonstandard Conditions

The heating capacity of a radiator, convector, baseboard, finned-tube heat-distributing unit, or radiant panel is a power function of the temperature difference between the air in the room and the heating medium in the unit, shown as

$$q = c(t_s - t_a)^n \tag{19.1}$$

where

q = heating capacity, Btu/h

c = constant determined by test

t_s = average temperature of heating medium, °F. For hot water, the arithmetic average of the entering and leaving water temperatures is used.

t_a = room air temperature, °F. Air temperature 60 in. above the floor is generally used for radiators, whereas entering air temperature is used for convectors, baseboard units, and finned-tube units.

n = exponent that equals 1.3 for cast-iron radiators, 1.4 for baseboard radiation, 1.5 for convectors, 1.0 for ceiling heating and floor cooling panels, and 1.1 for floor heating and ceiling cooling panels. For finned-tube units, n varies with air and heating medium temperatures. Correction factors to convert heating capacities at standard rating conditions to heating capacities at other conditions are given in Table 19.2.

Equation 19.1 may also be used to calculate heating capacity at nonstandard conditions.

Designing for high temperature drops through the system (as much as 60°F to 80°F in low-temperature water (LTW) systems and as much as 200°F in high-temperature systems) can result in low water velocities in the finned-tube or baseboard element. Applying very short runs designed for conventional temperature drops (i.e., 20°F) can also result in low velocities.

Figure 19.9 shows the effect of water velocity on the heat output of typical sizes of finned-tube elements. The figure is based on work done by Harris (1957) and Pierce (1963) and tests at the Hydronics Institute. The velocity correction factor F_v is

$$F_v = (V/3.0)^{0.04} \tag{19.2}$$

where V = water velocity, fps.

Heat output varies little over the range from 0.5 to 3 fps, where F_v ranges from 0.93 to 1.00. The factor drops rapidly below 0.5 fps because flow changes from turbulent to laminar at around 0.1 fps. Avoid such a low velocity because the output is difficult to predict accurately when designing a system. In addition, the curve is so steep in this region that small changes in actual flow have a significant effect on output. Not only does the heat transfer rate change, but the temperature drop and, therefore, the average water temperature change (assuming a constant inlet temperature).

The designer should check water velocity throughout the system and select finned-tube or baseboard elements on the basis of velocity as well as average temperature. Manufacturers of finned-tube and baseboard elements offer a variety of tube sizes, ranging from 0.5 in. copper tubes for small baseboard elements to 2 in. for large finned-tube units, to aid in maintenance of turbulent flow conditions over a wide range of flow.

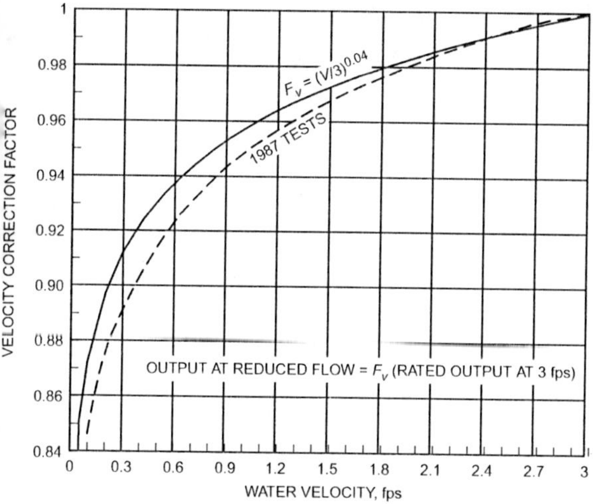

Figure 19.9 Water Velocity Correction Factor for Baseboard and Finned-Tube Radiators [2020S, Ch 36, Fig 3]

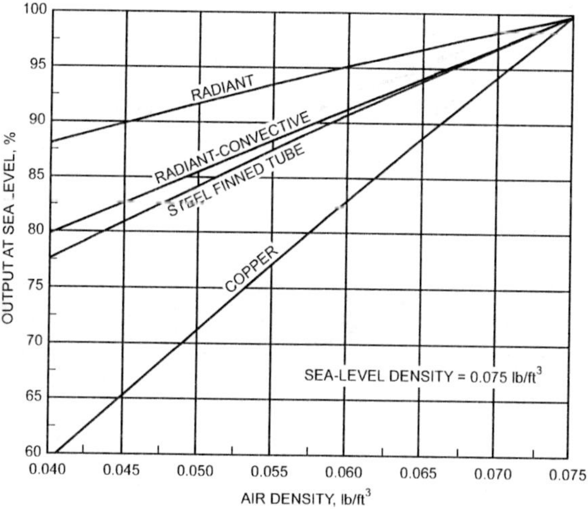

Figure 19.10 Effect of Air Density on Radiator Output [2020S, Ch 36, Fig 4]

HVAC Systems and Equipment

Unit Ventilators, Unit Heaters, and Makeup Air Units

A **heating unit ventilator** is an assembly whose principal functions are to heat, ventilate, and cool a space by introducing outdoor air in quantities up to 100% of its rated capacity. The heating medium may be steam, hot water, gas, or electricity. The essential components of a heating unit ventilator are the fan, motor, heating element, damper, filter, automatic controls, and outlet grille, all of which are encased in a housing.

An **air-conditioning unit ventilator** is similar to a heating unit ventilator; however, in addition to the normal winter function of heating, ventilating, and cooling with outdoor air, it is also equipped to cool and dehumidify during the summer. It is usually arranged and controlled to introduce a fixed quantity of outdoor air for ventilation during cooling in mild weather. The air-conditioning unit ventilator may be provided with a various of combinations of heating and air-conditioning elements. Some of the more common arrangements include

- Combination hot- and chilled-water coil (two-pipe)
- Separate hot- and chilled-water coils (four-pipe)
- Hot-water or steam coil and direct-expansion coil
- Electric heating coil and chilled-water or direct-expansion coil
- Gas-fired furnace with direct-expansion coil

The typical unit ventilator has controls that allow heating, ventilating, and cooling to be varied while the fans operate continuously. In normal operation, the discharge air temperature from a unit is varied in accordance with the room requirements. The heating unit ventilator can provide **ventilation cooling** by bringing in outdoor air whenever the room temperature is above the room set point. Air-conditioning unit ventilators can provide refrigerated cooling when the outdoor air temperature is too high to be used effectively for ventilation cooling.

Unit ventilators are available for floor mounting, ceiling mounting, and recessed applications. They are available with various airflow and capacity ratings, and the fan can be arranged so that air is either blown through or drawn through the unit. With direct-expansion refrigerant cooling, the condensing unit can either be furnished as an integral part of the unit ventilator assembly or be remotely located.

Unit ventilators are used primarily in schools, meeting rooms, offices, and other areas where the density of occupancy requires controlled ventilation to meet local codes.

Floor-model unit ventilators are normally installed on an outer wall near the centerline of the room. Ceiling models are mounted against either the outer wall or one of the inside walls. Ceiling models discharge air horizontally. Best results are obtained if the unit can be placed so that the airflow is not interrupted by ceiling beams or surface-mounted lighting fixtures.

Table 19.3 Typical Unit Ventilator Capacities [2020S, Ch 28, Tbl 1]

Airflow, cfm	Heating Unit Ventilator Total Heating Capacity, Btu/h	A/C Unit Ventilator Total Cooling Capacity, Btu/h
500	38,000	19,000
750	50,000	28,000
1000	72,000	38,000
1250	85,000	47,000
1500	100,000	56,000

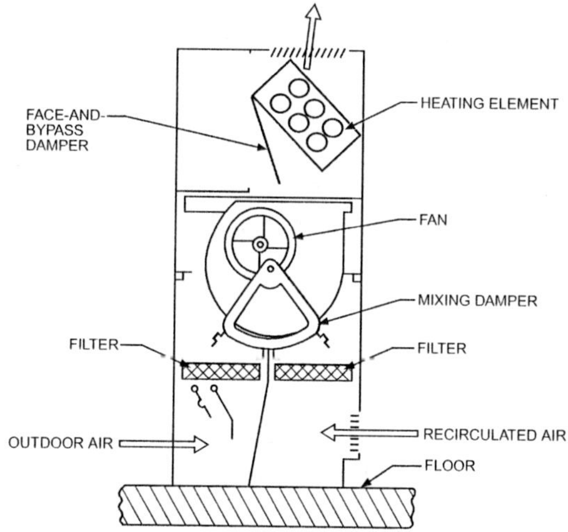

A. HEATING UNIT VENTILATOR

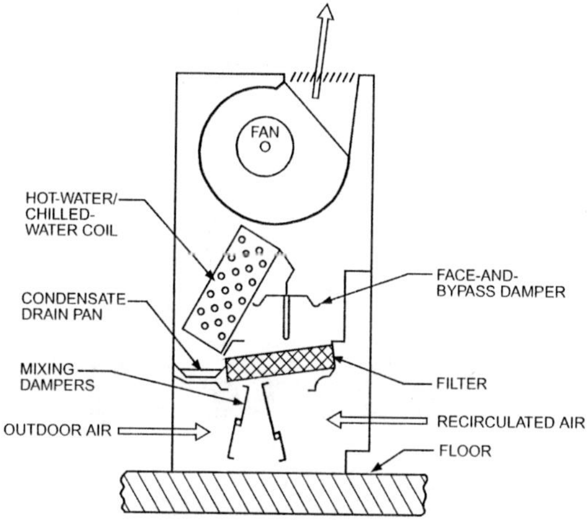

B. AIR-CONDITIONING UNIT VENTILATOR WITH
COMBINED HOT- AND CHILLED-WATER COIL

Figure 19.11 Typical Unit Ventilators [2020S, Ch 28, Fig 1]

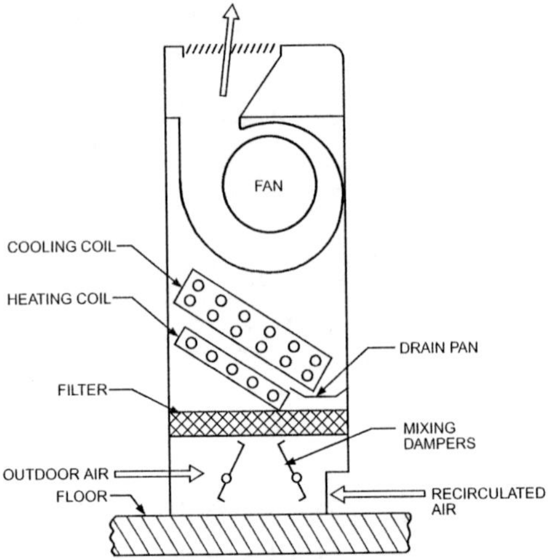

C. AIR-CONDITIONING UNIT VENTILATOR WITH SEPARATE COILS

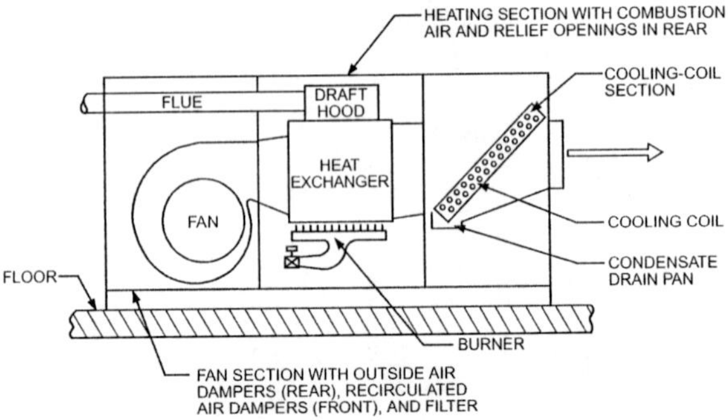

D. GAS-FIRED AIR-CONDITIONING UNIT VENTILATOR

Figure 19.11 Typical Unit Ventilators [2020S, Ch 28, Fig 1]
(Continued)

Unit Heaters

A unit heater is an assembly of elements with the main function of heating a space. The essential elements are a fan and motor, a heating element, and an enclosure. Filters, dampers, directional outlets, duct collars, combustion chambers, and flues may also be included. Some types of unit heaters are shown in Figure 19.12.

Unit heaters have the following principal characteristics:

- Relatively large heating capacities in compact casings
- Ability to project heated air in a controlled manner over a considerable distance
- Relatively low installed cost per unit of heat output
- Application where an elevated sound level is permissible

They are, therefore, usually placed in applications where the heating capacity requirements, physical volume of the heated space, or both, are too large to be handled adequately or economically by other means. By eliminating extensive duct installations, the space is freed for other use.

Unit heaters are mostly used for heating commercial and industrial structures such as garages, factories, warehouses, showrooms, stores, and laboratories, as well as corridors, lobbies, vestibules, and similar auxiliary spaces in all types of buildings. Unit heaters may often be used to advantage in specialized applications requiring spot or intermittent heating, such as at outer doors in industrial plants or in corridors and vestibules. Cabinet unit heaters may be used where heated air must be filtered.

The following factors should be considered when selecting a unit heater:

Heating Medium. The proper heating medium is usually determined by economics and requires examining initial cost, operating cost, and conditions of use.

Steam or **hot-water unit heaters** are relatively inexpensive but require a boiler and piping system. The unit cost of such a system generally decreases as the number of units increases. Therefore, steam or hot-water heating is most frequently used (1) in new installations involving a relatively large number of units, and (2) in existing systems that have sufficient capacity to handle the additional load. High-pressure steam or high-temperature hot-water units are normally used only in very large installations or when a high-temperature medium is required for process work. Low-pressure steam and conventional hot-water units are usually selected for smaller installations and for those concerned primarily with comfort heating.

Gas and **oil indirect-fired unit heaters** are frequently preferred in small installations where the number of units does not justify the expense and space requirements of a new boiler system or where individual metering of the fuel supply is required, as in a shopping center. Gas indirect-fired units usually have either horizontal propeller fans or industrial centrifugal fans. Oil indirect-fired units largely have industrial centrifugal fans. Some codes limit the use of indirect-fired unit heaters in some applications. Indirect-fired oil and gas units are of blow-through design to mitigate the possibility of combustion products entering the occupied space.

Electric unit heaters are used when the cost of available electric power is lower than that of alternative fuel sources and for isolated locations, intermittent use, supplementary heating, or temporary service. Typical applications are ticket booths, security offices, factory offices, locker rooms, and other isolated rooms scattered over large areas. Electric units are particularly useful in isolated and untended pumping stations or pits, where they may be thermostatically controlled to prevent freezing.

Propeller fan units are generally used in non-ducted applications where the heating capacity and distribution requirements can best be met by units of moderate output and where heated air does not need to be filtered. Horizontal-blow units are usually installed in buildings with low to moderate ceiling heights. Downblow units are used in spaces with high ceilings and where floor and wall space limitations dictate that heating equipment be kept out of the way. Downblow units may have an adjustable diffuser to vary the discharge pattern from a high-velocity vertical jet (to achieve the maximum distance of downward throw) to a horizontal discharge of lower velocity (to prevent excessive air motion in the zone of occupancy). Revolving diffusers are also available.

Cabinet unit heaters are used when a more attractive appearance is desired. They are suitable for free-air delivery or low static pressure duct applications. They may be equipped with filters, and they can be arranged to discharge either horizontally or vertically up or down.

Industrial centrifugal fan units are applied where heating capacities and space volumes are large or where filtration of the heated air or operation against static resistance is required. Downblow or horizontal-blow units may be used, depending on the requirements.

Duct unit heaters are used where the air handler is remote from the heater. These heaters sometimes provide an economical means of adding heating to existing cooling or ventilating systems with ductwork. They require flow and temperature limit controls.

Location for Proper Heat Distribution. Units must be selected, located, and arranged to provide complete heat coverage while maintaining acceptable air motion and temperature at an acceptable sound level in the working or occupied zone. Proper application depends on size, number, and type of units; direction of airflow and type of directional outlet used; mounting height; outlet velocity and temperature; and air volumetric flow. Many of these factors are interrelated.

The mounting height may be governed by space limitations or by the presence of equipment such as display cases or machinery. The higher a downblow heater is mounted, the lower the temperature of air leaving the heater must be to force the heated air into the occupied zone. Also, the distance that air leaving the heater travels depends largely on the air temperature and initial velocity. A high discharge temperature reduces the area of effective heat coverage because of its buoyancy.

For area heating, place horizontal-blow unit heaters in exterior zones such that they blow either along the exposure or toward it at a slight angle. When possible, arrange multiple units so that the discharge airstreams support each other and create a general circulatory motion in the space. Interior zones under exposed roofs or skylights should be completely blanketed. Arrange downblow units so that the heated areas from adjacent units overlap slightly to provide complete coverage.

For spot heating of individual spaces in larger unheated areas, single unit heaters may be used, but allowance must be made for the inflow of unheated air from adjacent spaces and the consequent reduction in heat coverage. Such spaces should be isolated by partitions or enclosures, if possible.

Horizontal unit heaters should have discharge outlets located well above head level. Both horizontal and vertical units should be placed so that the heated airstream is delivered to the occupied zone at acceptable temperature and velocity. Outlet air temperature of free-air delivery unit heaters used for comfort heating should be 50°F to 60°F higher than the design room temperature. When possible, locate units so that they discharge into open spaces, such as aisles, and not directly on the occupants.

Manufacturers' catalogs usually include suggestions for the best arrangements of various unit heaters, recommended mounting heights, heat coverage for various outlet velocities, final temperatures, directional outlets, and sound level ratings.

The following factors should be considered when evaluating capacity ratings:

Steam or Hot Water. Heating capacity must be determined at a standard condition. Variations in entering steam or water temperature, entering air temperature, and steam or water flow affect capacity. Typical standard conditions for rating steam unit heaters are dry saturated steam at 2 psig pressure at the heater coil, air at 60°F (29.92 in. Hg barometric pressure) entering the heater, and the heater operating free of external resistance to airflow. Standard conditions for rating hot-water unit heaters are entering water at 200°F, water temperature drop of 20°F, entering air at 60°F and 29.92 in. Hg barometric pressure, and the heater operating free of external resistance to airflow.

Gas-Fired. Gas-fired unit heaters are rated in terms of both input and output, in accordance with the approval requirements of the American Gas Association.

Oil-Fired. Ratings of oil-fired unit heaters are based on heat delivered at the heater outlet.

Electric. Electric unit heaters are rated based on the energy input to the heating element.

Effect of Airflow Resistance on Capacity. Unit heaters are customarily rated at free-air delivery. Airflow and heating capacity decrease if outdoor air intakes, air filters, or ducts on the inlet or discharge are used. The manufacturer should have information on the heat output to be expected at other than free-air delivery.

Effect of Inlet Temperature. Changes in entering air temperature influence the total heating capacity in most unit heaters and the final temperature in all units. Because many unit heaters are located some distance from the occupied zone, possible differences between the temperature of the air actually entering the unit and that of air being maintained in the heated area should be considered, particularly with downblow unit heaters.

Filters. Air from propeller unit heaters cannot be filtered because the heaters are designed to operate with heater friction loss only. If dust in the building must be filtered, centrifugal fan units or cabinet units should be used.

Controls. The controls for a steam or hot water unit heater can provide either (1) on/off operation of the unit fan, or (2) continuous fan operation with modulation of heat output. For on/off operation, a room thermostat is used to start and stop the fan motor or group of fan motors. A limit thermostat, often strapped to the supply or return pipe, prevents fan operation in the event that heat is not being supplied to the unit. An auxiliary switch that energizes the fan only when power is applied to open the motorized supply valve may also be used to prevent undesirable cool air from being discharged by the unit.

Continuous fan operation eliminates both the intermittent blasts of hot air resulting from on/off operation and the stratification of temperature from floor to ceiling that often occurs during off periods. In this arrangement, a proportional room thermostat controls a valve modulating the heat supply to the coil or a bypass around the heating element. A limit thermostat or auxiliary switch stops the fan when heat is no longer available.

One type of control used with downblow unit heaters is designed to automatically return the warm air, which would normally stratify at the higher level, down to the zone of occupancy. Two thermostats and an auxiliary switch are required. The lower thermostat is placed in the zone of occupancy and is used to control a two-position supply valve to the heater. An auxiliary switch is used to stop the fan when the supply valve is closed. The higher thermostat is placed near the unit heater at the ceiling or roof level where the warm air tends to stratify. The lower thermostat automatically closes the steam valve when its setting is satisfied, but the higher thermostat overrides the auxiliary switch so that the fan continues to run until the temperature at the higher level falls below a point sufficiently high to produce a heating effect.

Indirect-fired and electric units are usually controlled by intermittent operation of the heat source under control of the room thermostat, with a separate fan switch to run the fan when heat is being supplied.

Unit heaters can be used to circulate air in summer. In such cases, the heat is shut off and the thermostat has a bypass switch, which allows the fan to run independently of the controls.

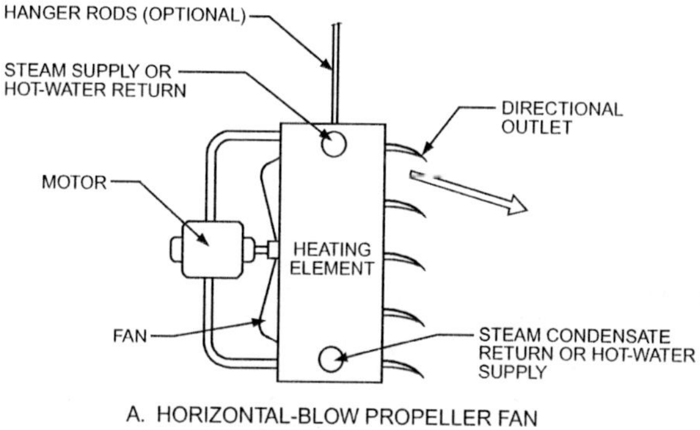

Figure 19.12 Typical Unit Heaters [2020S, Ch 28, Fig 2]

HVAC Systems and Equipment

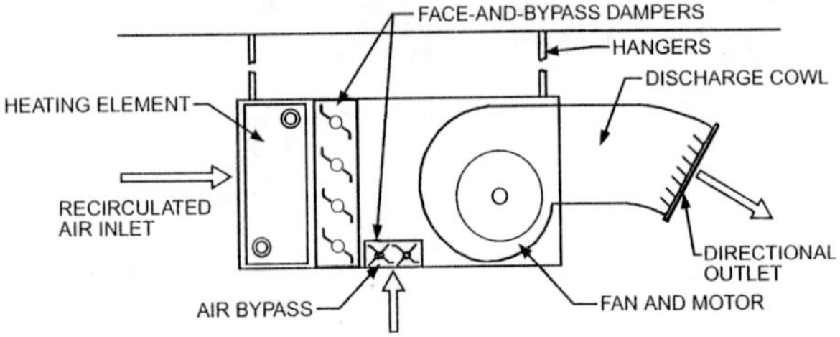

B. SUSPENDED INDUSTRIAL-TYPE WITH CENTRIFUGAL FAN AND BYPASS CONTROL

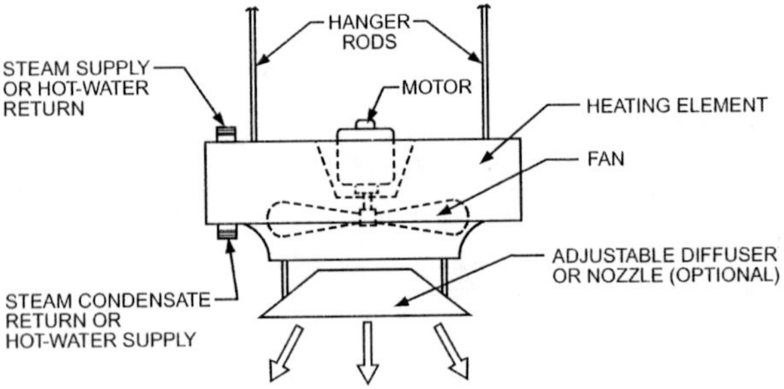

C. DOWNBLOW PROPELLER FAN

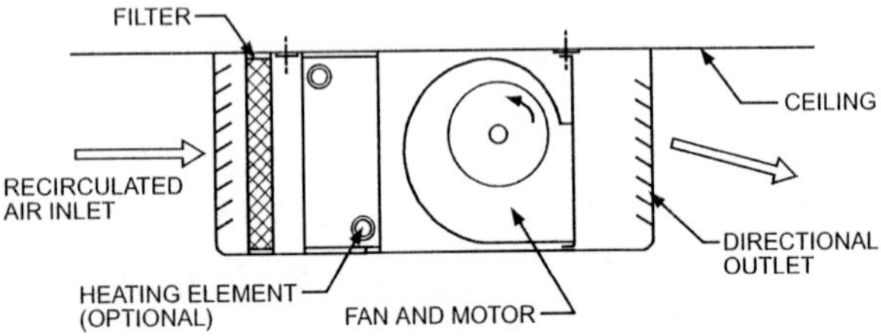

D. SUSPENDED CABINET WITH CENTRIFUGAL FAN

Figure 19.12 Typical Unit Heaters [2020S, Ch 28, Fig 2]
(Continued)

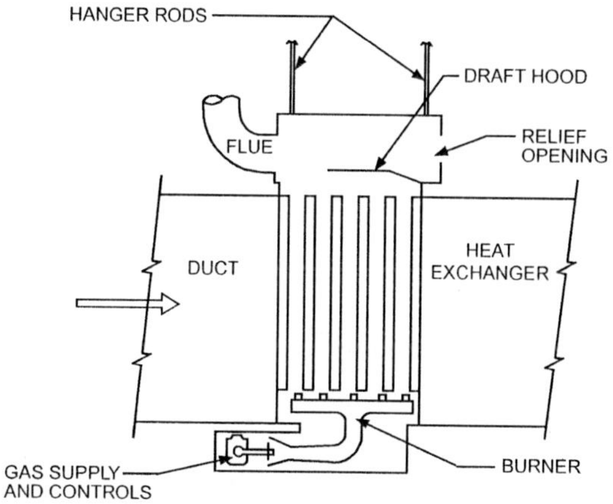

E. GAS INDIRECT-FIRED DUCT MOUNTED

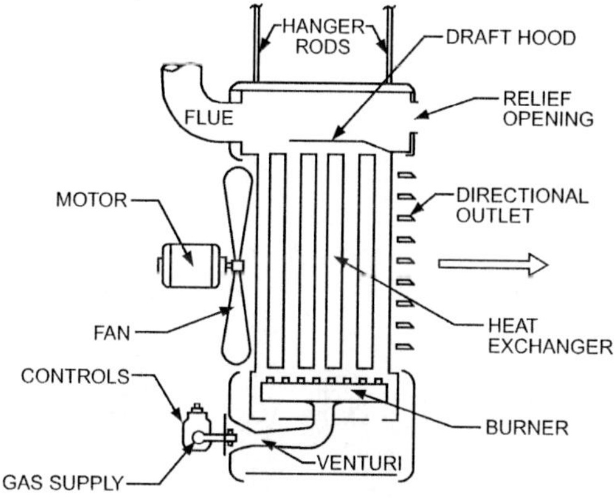

F. GAS INDIRECT-FIRED WITH PROPELLER FAN

Figure 19.12 Typical Unit Heaters [2020S, Ch 28, Fig 2]
(Continued)

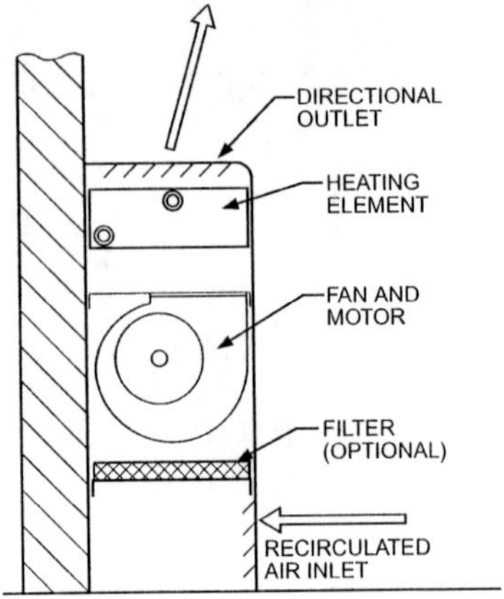

G. FLOOR-MOUNTED CABINET WITH CENTRIFUGAL FAN

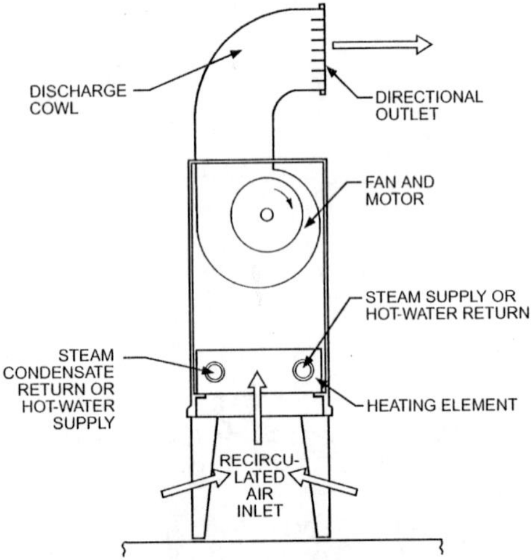

H. FLOOR-MOUNTED INDUSTRIAL-TYPE WITH CENTRIFUGAL FAN

Figure 19.12 Typical Unit Heaters [2020S, Ch 28, Fig 2]
(Continued)

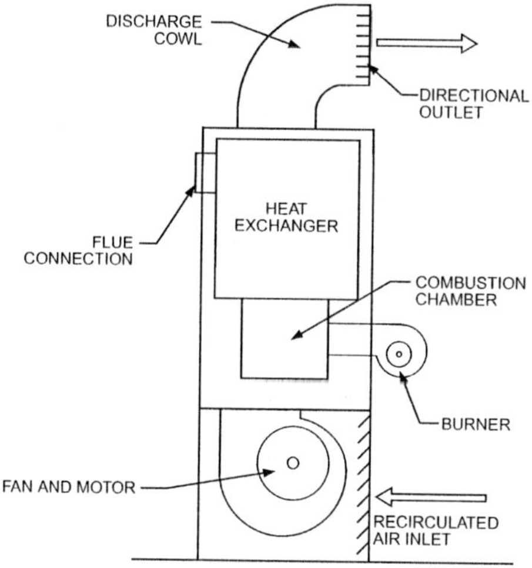

I. FLOOR-MOUNTED, INDUSTRIAL-TYPE, OIL OR GAS INDIRECT-FIRED WITH CENTRIFUGAL FAN

Figure 19.12 Typical Unit Heaters [2020S, Ch 28, Fig 2] ***(Continued)***

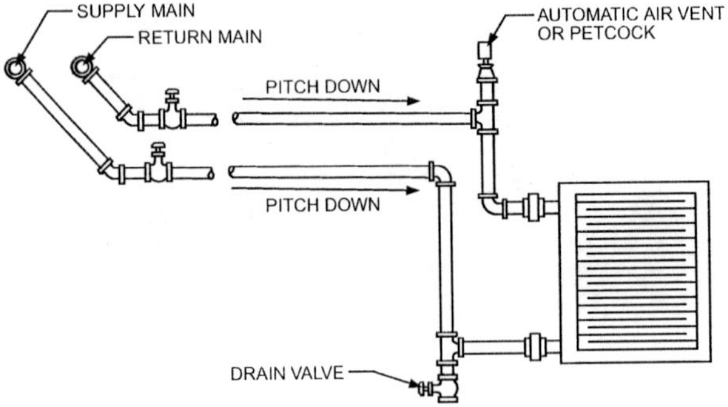

A. OVERHEAD HOT-WATER MAINS

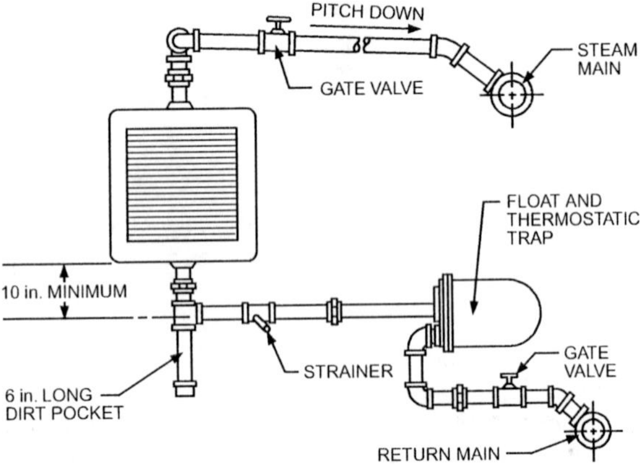

B. LOW-PRESSURE STEAM, OPEN GRAVITY, OR VACUUM RETURN

Figure 19.13 Hot Water and Steam Connections for Unit Heaters
[2020S, Ch 28, Fig 4]

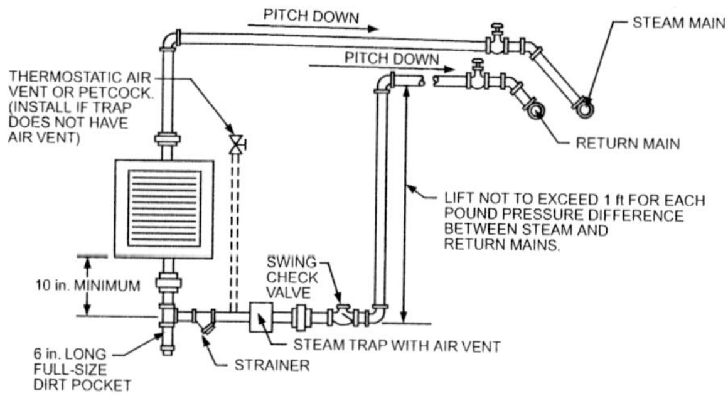

NOTE: This piping arrangement is only for two-position control. Modulating steam control may not provide sufficient pressure to lift condensate to return main in throttled position.

C. OVERHEAD STEAM AND RETURN MAINS

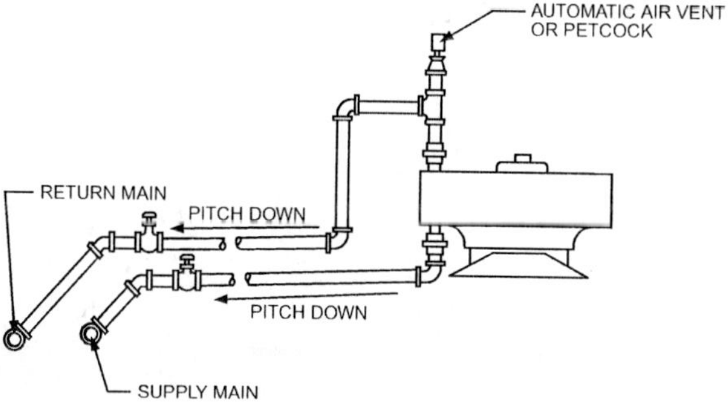

D. LOWER HOT-WATER MAINS

Figure 19.13 Hot Water and Steam Connections for Unit Heaters
[2020S, Ch 28, Fig 4] ***(Continued)***

Makeup Air Units

Makeup air units are designed to condition ventilation air introduced into a space or to replace air exhausted from a building. The air exhausted may be from a process or general area exhaust, through either powered exhaust fans or gravity ventilators. The units may be used to prevent negative pressure within buildings or to reduce airborne contaminants in a space. The units may heat, cool, humidify, dehumidify, and/or filter incoming air. They may be used to replace air in the conditioned space or to supplement or accomplish all or part of the airflow needed to satisfy the heating, ventilating, or cooling airflow requirements.

Makeup air systems used for ventilation may be (1) sized to balance air exhaust volumes or (2) sized in excess of the exhaust volume to dilute contaminants. In applications where contaminant levels vary, variable-flow units should be considered so that the supply air varies for contaminant control and the exhaust volume varies to track supply volume. In critical spaces, the exhaust volume may be based on requirements to control pressure in the space.

Location. Makeup air units are defined by their location or the use of a key component. Examples are rooftop makeup air units, truss- or floor-mounted units, and sidewall units. Some manufacturers differentiate their units by heating mode, such as steam or direct gas-fired makeup air units.

Rooftop units are commonly used for large single-story industrial buildings to simplify air distribution. Access (via roof walks) is more convenient than access to equipment mounted in the truss; truss units are only accessible by installing a catwalk adjacent to the air units. Disadvantages of rooftop units are (1) they increase foot traffic on the roof, thus reducing its life and increasing the likelihood of leaks; (2) inclement weather reduces equipment accessibility; and (3) units are exposed to weather.

Makeup air units can also be placed around the perimeter of a building with air ducted through the sidewall. This approach limits future building expansion, and the effectiveness of ventilating internal spaces decreases as the building gets larger. However, access to the units is good, and minimum support is required because the units are mounted on the ground.

Heating and Cooling Media. Heaters in makeup air systems may be direct gas-fired burners, electric resistance heating coils, indirect gas-fired heaters, steam coils, or hot-water heating coils. Air distribution systems are often required to direct heat to spaces requiring it.

Mechanical refrigeration with direct-expansion or chilled-water cooling coils, direct or indirect evaporative cooling sections, or well water coils may be used. Air distribution systems are often required to direct cooling to specific spaces that experience or create heat gain.

If direct-expansion coils are used in conjunction with direct-fired gas coils, the cooling coils' headers must be isolated from the airstream and directly vented outdoors.

Filters. High-efficiency filters (approximately MERV 16 for near-HEPA performance) are not normally used in a makeup air unit because of their relatively high cost. HVAC prefilters are generally in the MERV 6 to 13 range, depending on particulate removal needs.

Fans. Follow AMCA Standard 205-12 for fan selection. Fans should have variable-speed drives for possible energy savings or for use in variable-airflow systems.

Fans should have variable-speed drives for possible energy savings or for use in variable-airflow systems.

Controls. Controls for a makeup air unit fall into the following categories: (1) local temperature controls, (2) airflow controls, (3) plant-wide controls for proper equipment operation and efficient performance, (4) safety controls for burner gas, and (5) building smoke control systems.

Safety controls for gas-fired units include components to properly light the burner and to provide a safeguard against flame failure. The heater and all attached inlet ducting must be purged with at least four air changes before initiating an ignition sequence and before reignition after a malfunction. A flame monitor and control system must be used to automatically shut off gas to the burner upon burner ignition or flame failure. Critical malfunctions include flame failure, supply fan failure, combustion air depletion, power failure, control signal failure, excessive or inadequate inlet gas supply pressure, excess air temperature, and gas leaks in motorized valves or inlet gas supply piping.

Makeup air units should be interlocked with exhaust units to avoid overpressurization, and should include shutoff dampers with limit switches for when not in use. Damper leakage rates should be within limits set in ANSI/ASHRAE/IES Standard 90.1. These units should also be interlocked to the building's fire alarm system to shut down in the case of a fire, where required by applicable codes.

Consider using automatic safety shutoff valves on interconnecting piping systems where there are risks of overtemperature, overpressure, or gas leaks.

Small Forced-Air Heating and Cooling Systems

Forced-air systems are heating and/or cooling systems that use motor-driven blowers to distribute heated, cooled, and otherwise treated air to multiple outlets for the comfort of individuals in confined spaces. A typical residential or small commercial system includes (1) a heating and/or cooling unit, (2) supply and return ductwork (including registers and grilles), (3) accessory equipment, and (4) controls (see Figure 19.14).

Three types of forced-air heating and cooling devices are (1) furnaces, (2) air conditioners, and (3) heat pumps.

Furnaces are the basic component of most forced-air heating systems. They are manufactured to use specific fuels such as oil, natural gas, or liquefied petroleum gas, and are augmented with an air-conditioning coil when cooling is included. The fuel used dictates installation requirements and safety considerations.

Common **air-conditioning** systems use a split configuration with an air-handling unit, such as a furnace. The air-conditioning evaporator coil (indoor unit) is installed on the discharge air side of the air handler. The compressor and condensing coil (outdoor unit) are located outside the structure, and refrigerant lines connect the outdoor and indoor units.

Self-contained air conditioners contain all necessary air-conditioning components, including circulating air blowers, and may or may not include fuel-fired heat exchangers or electric heating elements.

Heat pumps cool and heat using the refrigeration cycle. They are available in split and packaged (self-contained) configurations. Generally, air-source heat pumps require supplemental heating; therefore, electric heating elements are usually included with the heat pump as part of the forced-air system. Heat pumps offer high efficiency at mild temperatures, but may be combined with fossil-fuel furnaces to minimize heating cost. Heat pump supplemental heating also may be provided by thermostat-controlled gas heating appliances (e.g., fireplaces, free-standing stoves).

Ground-source heat pumps (GSHPs) are becoming more common in residential housing, especially in colder climates. Because underground temperatures are mild year-round, GSHPs typically do not use supplemental heating except in emergency mode (i.e., when the heat pump does not provide enough heat).

Forced-air systems may be equipped with **accessories** that further condition the air. They may modify humidity, remove contaminants, mix outdoor air with the recirculating air, or transfer energy in other ways. Disposable air filters on the return side of forced-air systems are so common that they are not considered accessories.

A simple thermostat controlling on/off cycling of central equipment may be all that is used or needed for temperature **control**. Such thermostats typically have a switch for automatic or continuous fan operation, and another to choose heating, cooling, or neither.

More complex systems may provide control features for timed variations (from simple night setback to a weekly schedule of temperatures); multiple independent zones, power stages, or fan speeds; influence of outdoor sensors; humidity; automatic switching between heating and cooling modes, etc.

The overall system design should proceed as follows:

1. Estimate heating and cooling loads, including target values for duct losses.
2. Determine preliminary ductwork location and materials of ductwork and outlets.
3. Determine heating and cooling unit location.
4. Select accessory equipment. Accessory equipment is not generally provided with initial construction; however, the system may be designed for later addition of these components.
5. Select control components.
6. Select heating/cooling equipment.
7. Determine maximum airflow (cooling or heating) for each supply and return location.
8. Determine airflow at reduced heating and cooling loads (two-speed and variable-speed fans).
9. Finalize heating/cooling equipment.
10. Finalize control system.
11. Finalize duct design and size.
12. Select supply and return grilles.
13. When the duct system is in place, measure duct leakage and compare results with target values used in step 1.

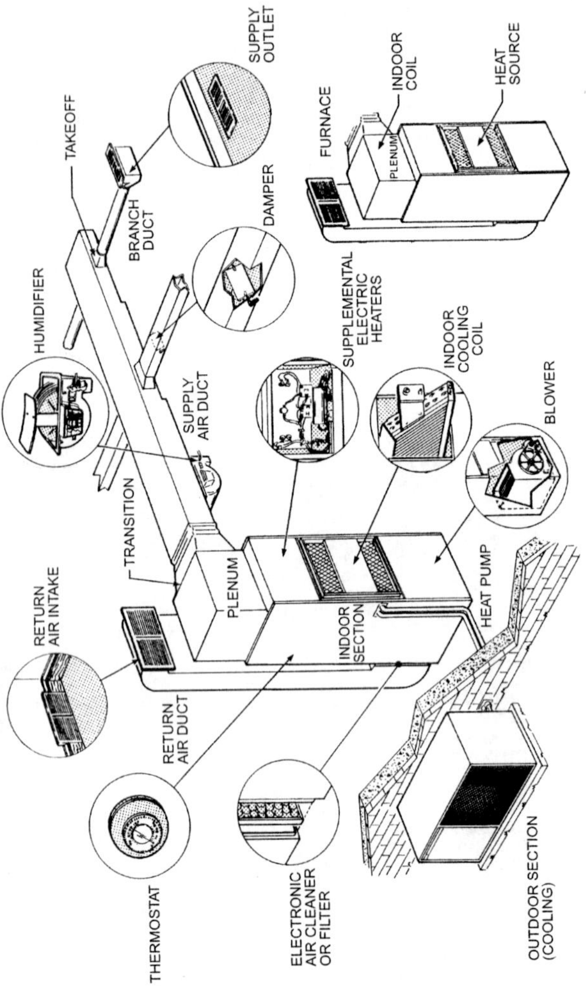

Figure 19.14 Heating and Cooling Components [2020S, Ch 10, Fig 1]

For maximum energy efficiency, ductwork and equipment should be installed in the conditioned space. ANSI/ASHRAE/IES Standard 90.2 gives a credit for installation in this location. The next best location is in a full basement. If a structure has an insulated, unvented, and sealed crawlspace, the ductwork and equipment can be located there (with appropriate provision for combustion air, if applicable), or the equipment can be placed in a closet or utility room. Vented attics and vented crawlspaces are the least preferred locations for ductwork and HVAC equipment.

Furnace heating output should match or slightly exceed the estimated design load. The Air Conditioning Contractors of America (ACCA *Manual* S) recommends a 40% limit on oversizing for fossil fuel furnaces. This limit minimizes venting problems associated with oversized equipment and improves part-load performance. Note that the calculated load must include duct loss, humidification load, and night setback recovery load, as well as building conduction and infiltration heat losses.

To help conserve energy, manufacturers have added features to improve furnace efficiency. Electric ignition has replaced the standing pilot; vent dampers and more efficient motors are also available. Furnaces with fan-assisted combustion systems (FACSs) and condensing furnaces also improve efficiency. Two-stage heating and cooling, variable-speed heat pumps, and two-speed and variable-speed blowers are also available.

Research on the effect of blower performance on residential forced-air heating system performance suggested reductions of 180 to 250 kWh/yr for automatic furnace fan operation and 2600 kWh/yr for continuous fan operation by changing from permanent split capacitor (PSC) blower motors to brushless permanent electronically commutated magnet motors (ECMs) (Phillips 1998).

A system designed to both heat and cool and that cycles cooling equipment on and off by sensing dry-bulb temperature alone should be sized to match the design heat gain as closely as possible. Oversizing under this control strategy could lead to higher-than-desired indoor humidity levels. Chapter 17 of the 2021 *ASHRAE Handbook—Fundamentals* recommends that cooling units not be oversized. Other sources suggest limiting oversizing to 15% of the sensible load. A heat pump should be sized for the cooling load with supplemental heat provided to meet heating requirements. Size air-source heat pumps in accordance with the equipment manufacturer recommendations. ACCA *Manual* S can also be used to assist in the selection and sizing of equipment.

The required airflow and the blower's static pressure limitation are the parameters around which the duct system is designed. The heat loss or gain for each space determines the proportion of the total airflow supplied to each space. Static pressure drop in supply registers should be limited to about 0.03 in. of water. The required pressure drop must be deducted from the static pressure available for duct design.

The flow delivered by a single supply outlet should be determined by considering the (1) space limitations on the number of registers that can be installed, (2) pressure drop for the register at the flow rate selected, (3) adequacy of air delivery patterns for offsetting heat loss or gain, and (4) space use pattern.

Manufacturers' specifications include blower airflow for each blower speed and external static pressure combination. Determining static pressure available for duct design should include the possibility of adding accessories in the future (e.g., electronic air cleaners or humidifiers). Therefore, the highest available fan speed should not be used for design.

For systems that heat only, the blower rate may be determined from the manufacturer's data. The temperature rise of air passing through the heat exchanger of a fossil-fuel furnace must be within the manufacturer's recommended range (usually 40°F to 80°F). The possible later addition of cooling should also be considered by selecting a blower that operates in the midrange of the fan speed and settings.

For cooling only, or for heating and cooling, the design flow can be estimated by Equation 19.3:

$$Q = \frac{q_s U}{\rho c_p \Delta t} \tag{19.3}$$

where

Q	=	flow rate, cfm
q_s	=	sensible load, Btu/h
ρ	=	air density assumed to equal 0.075 lb/ft^3
c_p	=	specific heat of air = 0.24 Btu/lb·°F
Δt	=	dry-bulb temperature difference between air entering and leaving equipment, °F
U	=	unit conversion factor, 1 h/60 min

Replacing all constant values gives the simplified equation in the given units.

$$Q = \frac{1}{1.08} \times \frac{q_s}{\Delta t} = \frac{q_s}{1.08\Delta t} \tag{19.4}$$

For preliminary design, an approximate Δt is as follows:

Sensible Heat Ratio (SHR)	Δt, °F
0.75 to 0.79	21
0.80 to 0.85	19
0.85 to 0.90	17

SHR = Calculated sensible load/Calculated total load

For example, if calculation indicates the sensible load is 23,000 Btu/h and the latent load is 4900 Btu/h, the SHR is calculated as follows:

$$\text{SHR} = \frac{23{,}000}{23{,}000 + 4900} = 0.82 \tag{19.5}$$

and

$$Q = \frac{23{,}000}{1.08 + 19} = 1121 \text{ cfm} \tag{19.6}$$

This value is the estimated design flow. The exact design flow can only be determined after the cooling unit is selected. The unit that is ultimately selected should supply an airflow in the range of the estimated flow, and must also have adequate sensible and latent cooling capacity when operating at design conditions.

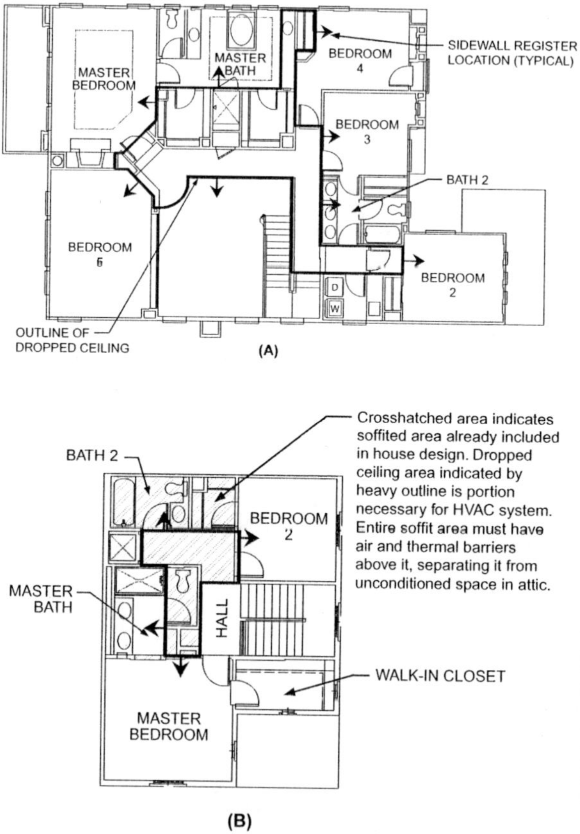

(Hedrick 2002)

Figure 19.15 Sample Floor Plans for Locating Ductwork in Second Floor of (A) Two-Story House and (B) Townhouse [2020S, Ch 10, Fig 2]

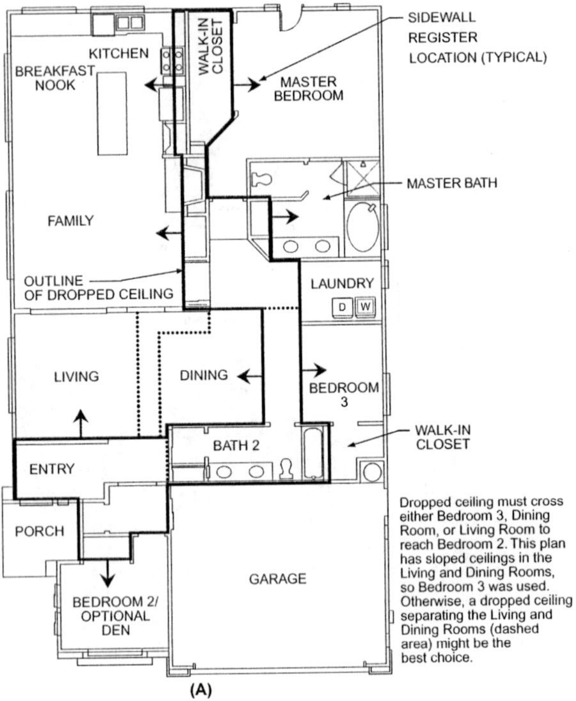

(EPA 2000)

Figure 19.16 Sample Floor Plans for One-Story House with (A) Dropped Ceilings, (B) Ducts in Conditioned Spaces, and (C) Right-Sized Air Distribution in Conditioned Spaces [2020S, Ch 10, Fig 3]

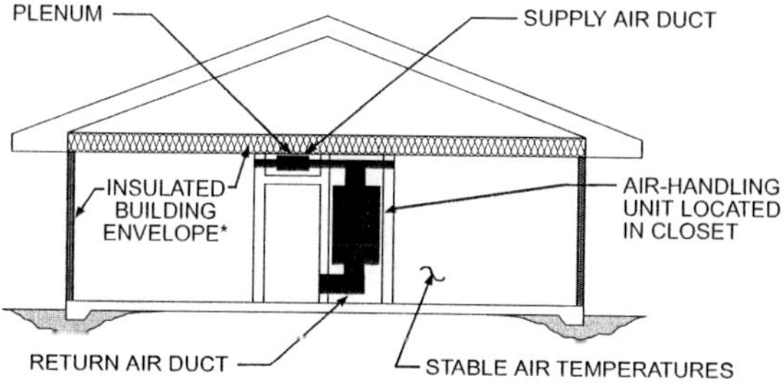

*To minimize duct leakage to attic, continuous drywall or other air barrier is needed at ceiling below insulation in all plenum areas.

(B)

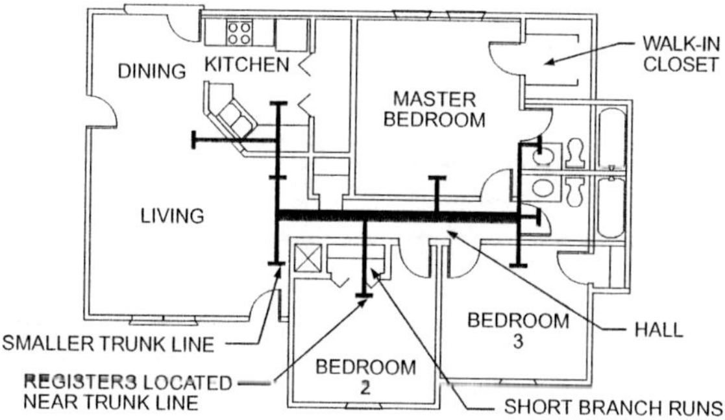

(C)

(EPA 2000)

Figure 19.16 Sample Floor Plans for One-Story House with (A) Dropped Ceilings, (B) Ducts in Conditioned Spaces, and (C) Right-Sized Air Distribution in Conditioned Spaces
[2020S, Ch 10, Fig 3] ***(Continued)***

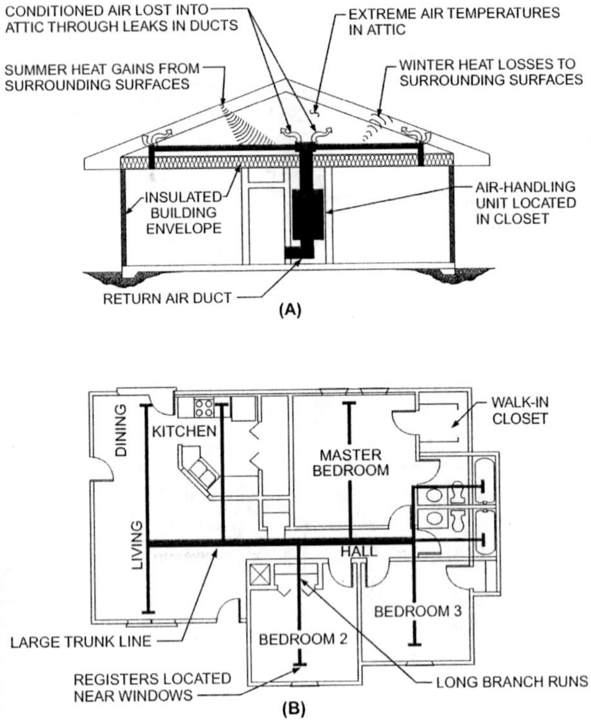

(EPA 2000)

Figure 19.17 (A) Ducts in Unconditioned Spaces and (B) Standard Air Distribution System in Unconditioned Spaces [2020S, Ch 10, Fig 4]

Unitary Air Conditioners and Heat Pumps

Unitary air conditioners are factory-made assemblies that normally include an evaporator or cooling coil and a compressor/ condenser combination, and possibly provide heating as well. An **air-source unitary heat pump** normally includes an indoor conditioning coil, compressor(s), and an outdoor coil. It must provide heating and possibly cooling as well. A **water-source heat pump** rejects or extracts heat to and from a water loop instead of from ambient air. A unitary air conditioner or heat pump with more than one factory-made assembly (e.g., indoor and outdoor units) is commonly called a **split system**.

Unitary equipment is divided into three general categories: residential, light commercial, and commercial. Residential equipment is single-phase unitary equipment with a cooling capacity of 65,000 Btu/h or less and is designed specifically for residential application. Light commercial equipment is generally three-phase, with cooling capacity up to 135,000 Btu/h, and is designed for small businesses and commercial properties. Commercial unitary equipment has cooling capacity higher than 135,000 Btu/h and is designed for large commercial buildings.

Unitary equipment is available in many configurations, such as

- **Single-zone, constant-volume**, which consists of one controlled space with one thermostat that controls to maintain a set point. This equipment may be single stage, multistage, or variable capacity.
- **Multizone, constant-volume**, which has several controlled spaces served by one unit that supplies air of different temperatures to different zones as demanded (Figure 19.18).
- **Single-package, variable-volume**, which consists of several controlled spaces served by one unit. Supply air from the unit is at a constant temperature, with air volume to each space varied to satisfy space demands (Figure 19.19).
- **Multisplit**, which consists of several controlled spaces, each served by a separate indoor unit. All indoor units are connected to an outdoor condensing unit (Figure 19.20). When each indoor unit varies its refrigerant flow in response to heating or cooling load demand, the system is called **variable refrigerant flow (VRF)**.

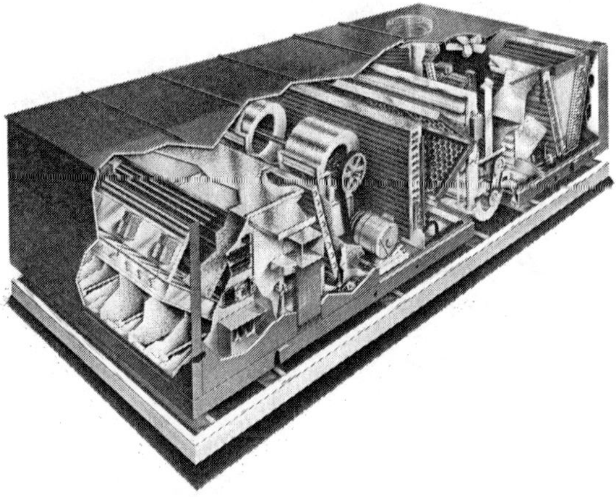

Figure 19.18 Typical Rooftop Air-Cooled Single-Package Air Conditioner
[2020S, Ch 48, Fig 1]

In general, roof-mounted single-package unitary equipment is limited to five or six stories because duct space and available blower power become excessive in taller buildings. Split units are limited by the maximum distance allowed between the indoor and outdoor sections because of losses in refrigerant piping, compressor capability, and refrigerant oil management. Indoor, single-zone equipment is generally less expensive to maintain and service than multizone or multisplit units.

Manufacturers' literature has detailed information about geometry, performance, electrical characteristics, application, and operating limits. The system designer selects suitable equipment with the capacity for the application.

Unitary equipment is designed to keep installation costs low. Adequate planning is important for installing large, roof-mounted equipment because special rigging equipment is frequently required. ACCA Standard 5 describes minimum criteria for the proper installation of HVAC systems in residential and commercial installations.

An advantage of packaged unitary equipment is that proper installation minimizes the risk of field contamination of the circuit. Care must be taken to properly install split-system interconnecting tubing (e.g., proper cleanliness, brazing, and evacuation to remove moisture and other noncondensables). Split systems should be charged according to the manufacturer's instructions. Filter-driers are necessary; if they are not installed at the factory, they should be field installed. When installing split, multisplit, and VRF systems, lines must be properly routed and sized to ensure proper oil return to the compressor.

Unitary equipment must be located to avoid noise and vibration problems. Single-package equipment of over 20 ton capacity should be mounted on concrete pads if vibration control is a concern. Large-capacity equipment should be roof mounted only after the roof's structural adequacy has been evaluated. If they are located over occupied space, roof-mounted units with return fans that use ceiling space for the return plenum should have a lined return plenum according to the manufacturer's recommendations. Use duct silencers where low sound levels are desired. Weight and sound data are available from many manufacturers. Additional installation guidelines include the following:

- In general, install products containing compressors on solid, level surfaces.
- Avoid mounting products containing compressors (such as remote units) on or touching the foundation of a house or building, or outside bedroom windows. A separate pad that does not touch the foundation is recommended to reduce any noise and vibration transmission through the slab.
- Do not box in outdoor air-cooled units with fences, walls, overhangs, or bushes. Doing so reduces the air-moving capability of the unit, reducing efficiency. Manufacturers include minimum clearances in literature.

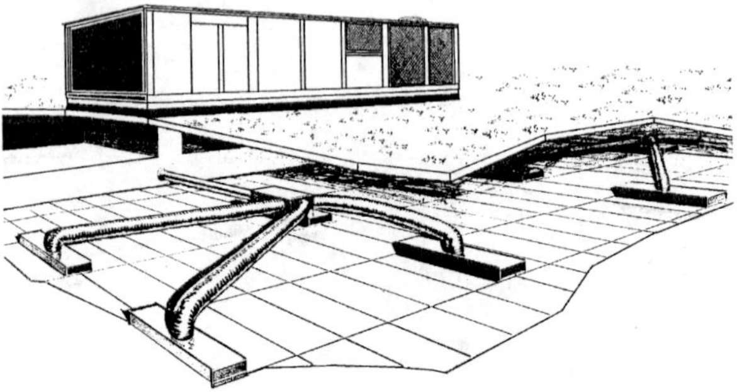

Figure 19.19 Single-Package Air Equipment with Variable Air Volume
[2020S, Ch 48, Fig 2]

- For a split-system remote unit, choose an installation site that is close to the indoor part of the system to minimize pressure drop in the connecting refrigerant tubing. Comply with manufacturers' refrigerant line length limits and required accessories listed in their literature.
- For VRF units, locate the refrigerant pipes' headers so that the length of refrigerant pipes is minimized.
- Contact the manufacturer or consult installation instructions for further information on installation procedures.

Unitary equipment should be listed or certified by nationally recognized testing laboratories to ensure safe operation and compliance with government and utility regulations. Equipment should also be installed to comply with agency standards' rating and application requirements to ensure that it performs according to industry criteria. Larger and more specialized equipment often does not carry agency labeling. However, power and control wiring practices should comply with the *National Electrical Code*® (NFPA Standard 70). Consult local codes before the installation is designed, and consult local inspectors before installation.

Unitary air conditioners have factory-matched refrigerant circuit components that are applied in the field to fulfill the user's requirements. The manufacturer often incorporates a heating function compatible with the cooling system and a control system that requires minimal field wiring.

Products are available to meet the objectives of nearly any system. Many different heating sections (gas- or oil-fired, electric, or condenser reheat), air filters, and heat pumps, which are a specialized form of unitary product, are available. Such matched equipment, selected with compatible accessory items, requires little field design or field installation work.

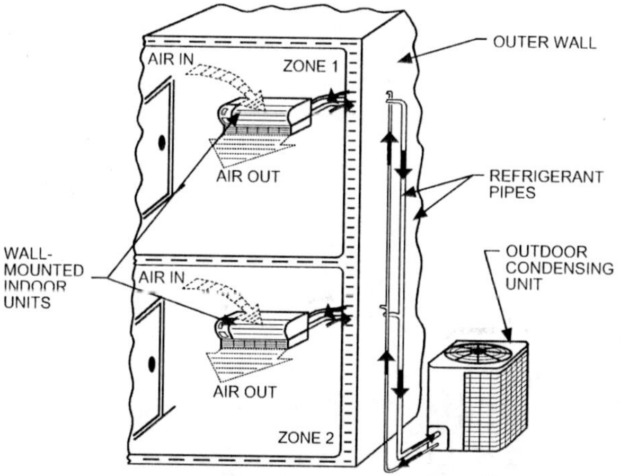

Figure 19.20 Example of Two-Zone Ductless Multisplit System in Typical Residential Installation [2020S, Ch 48, Fig 3]

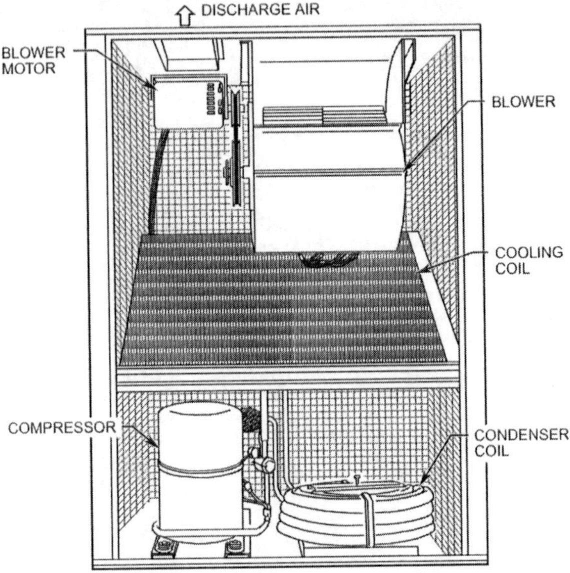

Figure 19.21 Water-Cooled Single-Package Air Conditioner
[2020S, Ch 48, Fig 4]

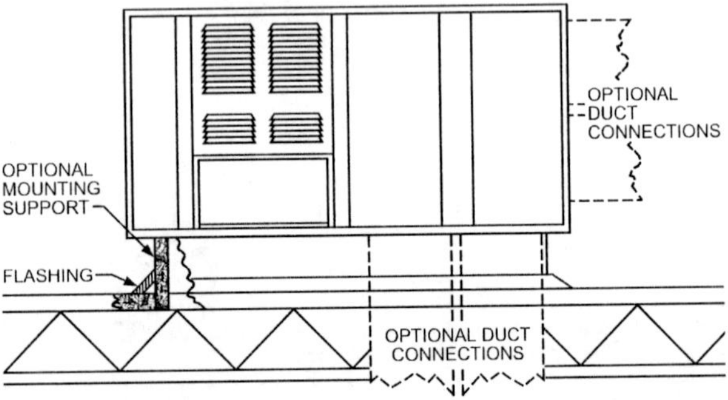

Figure 19.22 Rooftop Installation of Air-Cooled Single-Package Unit
[2020S, Ch 48, Fig 5]

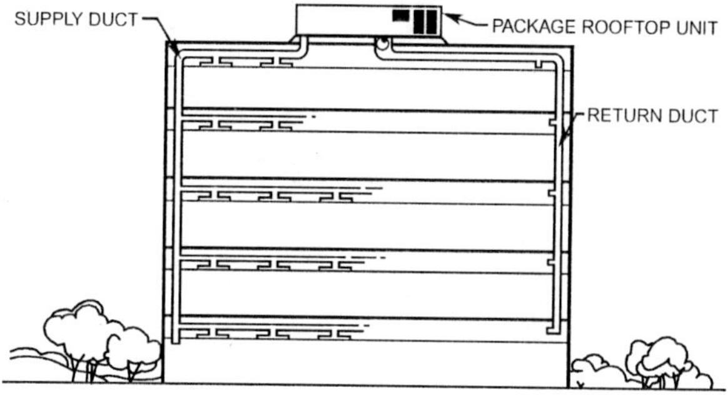

Figure 19.23 Multistory Rooftop Installation of Single-Package Unit [2020S, Ch 48, Fig 6]

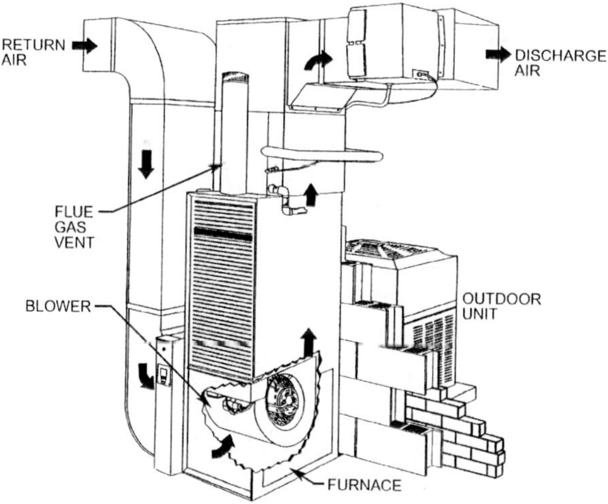

Figure 19.24 Residential Installation of Split-System Air-Cooled Condensing Unit with Coil and Upflow Furnace [2020S, Ch 48, Fig 8]

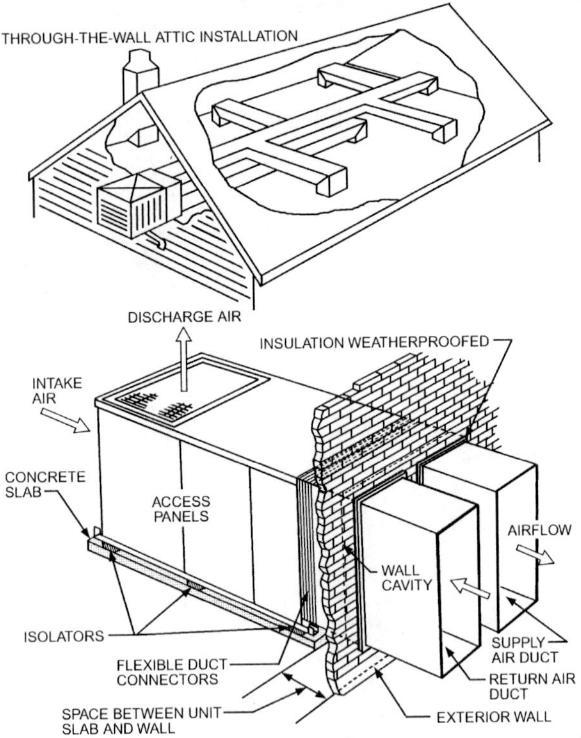

Figure 19.25 Through-the-Wall Installation of Air-Cooled Single-Package Unit
[2020S, Ch 48, Fig 7]

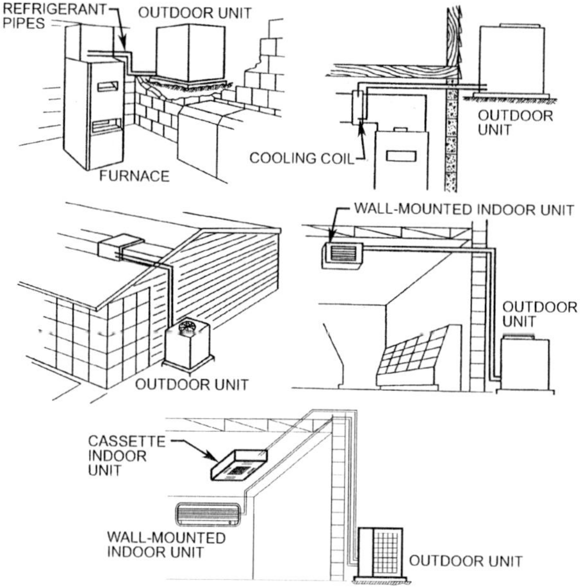

Figure 19.26 Outdoor Installations of Split-System Air-Cooled Condensing Units with Coil and Upflow Furnace or with Indoor Blower-Coils [2020S, Ch 48, Fig 9]

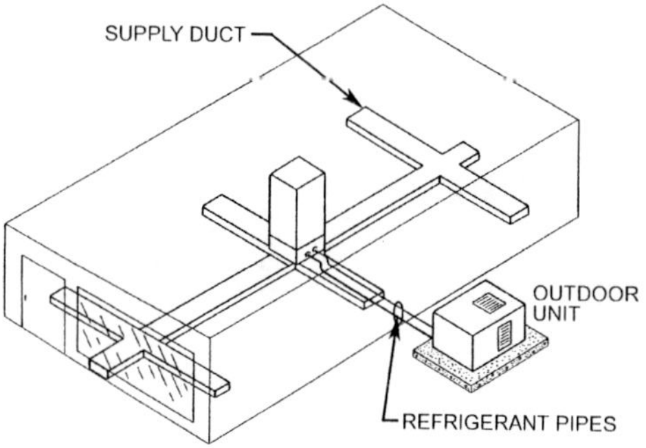

Figure 19.27 Outdoor Installation of Split-System Air-Cooled Condensing Unit with Indoor Coil and Downflow Furnace [2020S, Ch 48, Fig 10]

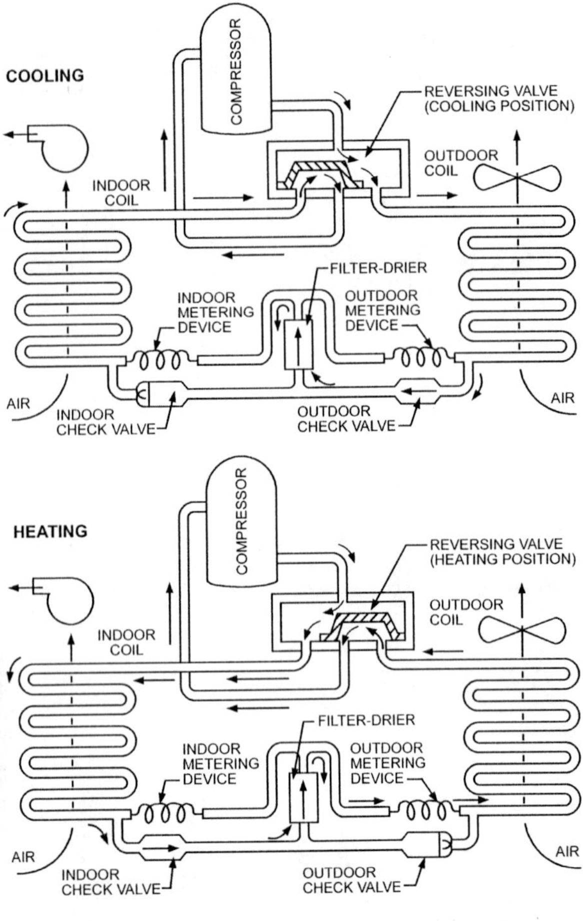

Figure 19.28 Schematic Typical of Air-to-Air Heat Pump System
[2020S, Ch 48, Fig 12]

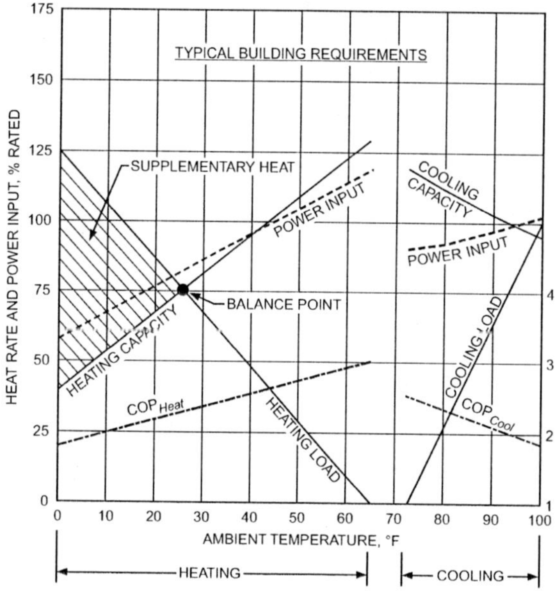

Figure 19.29 Operating Characteristics of Single-Stage Unmodulated Heat Pump [2020S, Ch 48, Fig 13]

HVAC Systems and Equipment

Water-Source Heat Pumps

A water-source heat pump (WSHP) is a single-package reverse-cycle heat pump that uses water as the heat source for heating and as the heat sink for cooling. The water supply may be a recirculating closed loop, a well, a lake, or a stream. Water for closed-loop heat pumps is usually circulated at 2 to 3 gpm per ton of cooling capacity.

WSHPs are used in a variety of systems, such as the following:

A **water-loop heat pump (WLHP)** uses a circulating water loop as the heat source and heat sink. When loop water temperature exceeds a certain level during cooling, a cooling tower dissipates heat from the water loop into the atmosphere. When loop water temperature drops below a prescribed level during heating, heat is added to the circulating loop water, usually with a boiler. In multiple-unit installations, some heat pumps may operate in cooling mode while others operate in heating, and controls are needed to keep loop water temperature within the prescribed limits. Chapter 9 of the 2020 *ASHRAE Handbook—HVAC Systems and Equipment* has more information on water-loop heat pumps.

A **groundwater heat pump (GWHP)** passes groundwater from a nearby well through the heat pump's water-to-refrigerant heat exchanger, where it is warmed or cooled, depending on the operating mode. It is then discharged to a drain, stream, or lake, or is returned to the ground through a reinjection well.

Many state and local jurisdictions have ordinances about use and discharge of groundwater. Because aquifers, the water table, and groundwater availability vary from region to region, these regulations cover a wide spectrum.

A **surface-water heat pump (SWHP)** uses water from a nearby lake, stream, or canal. After passing through the heat pump heat exchanger, it is returned to the source or a drain several degrees warmer or cooler, depending on the operating mode of the heat pump. **Closed-loop** surface water heat pumps use a closed water or brine loop that includes pipes or tubing submerged in the surface water (river, lake, or large pond) that serves as the heat exchanger. The adequacy of the total thermal capacity of the body of water must be considered.

A **ground-coupled heat pump (GCHP), ground-source heat pump (GSHP),** or **geothermal heat pump (GHP)** system uses the earth as a heat source and sink. Usually, plastic

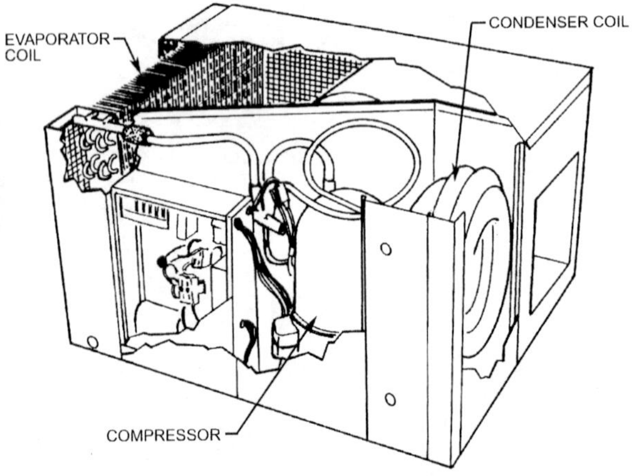

Figure 19.30 Typical Horizontal Water-Source Heat Pump
[2020S, Ch 48, Fig 15]

piping is installed in either a shallow horizontal or deep vertical array to form the heat exchanger. The massive thermal capacity of the earth provides a temperature-stabilizing effect on the circulating loop water or brine. Installing this type of system requires detailed knowledge of the climate; site; soil temperature, moisture content, and thermal characteristics; and performance, design, and installation of water-to-earth heat exchangers. Additional information on GCHP systems is presented in Chapter 35 of the 2019 *ASHRAE Handbook—HVAC Applications*.

Entering Water Temperatures. These various water sources provide a wide range of entering water temperatures to WSHPs. Entering water temperatures vary not only by water source, but also by climate and time of year. Because of the wide range of entering water or brine temperatures encountered, it is not feasible to design a universal packaged product that can handle the full range of possibilities effectively. Therefore, WSHPs are rated for performance at a number of standard rating conditions.

Compressors. WSHPs usually have single-speed compressors, although some high-efficiency models use multispeed compressors. Higher-capacity equipment may use multiple compressors. Compressors may be reciprocating, rotary, or scroll. Single-phase units are available at voltages of 115, 208, 230, and 265. All larger equipment is for three-phase power supplies with voltages of 208, 230, 460, or 575. Compressors usually have electromechanical protective devices.

Indoor Air System. Console WSHP models are designed for free delivery of conditioned air. Other models have ducting capability. Smaller WSHPs have multispeed, direct-drive centrifugal blower wheel fan systems. Large-capacity equipment has belt-drive systems. All units have provisions for fiberglass, metal, or plastic foam air filters.

Indoor Air Heat Exchanger. The indoor air heat exchanger of WSHP units is usually a conventional plate-fin coil of copper tubes and aluminum fins. Microchannel evaporators are also used in some products. The indoor air heat exchanger must be circuited so that it can function effectively as an evaporator with refrigerant flow in one direction and as a condenser when refrigerant flow is reversed.

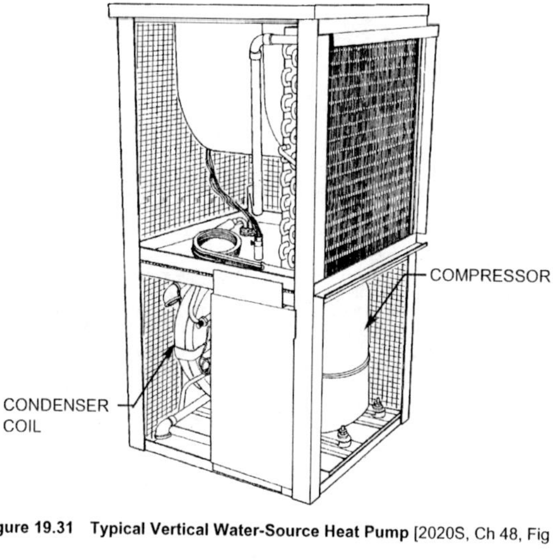

Figure 19.31 Typical Vertical Water-Source Heat Pump [2020S, Ch 48, Fig 16]

Table 19.4 Space Requirements for Typical Packaged Water-Source Heat Pumps [2020S, Ch 48, Tbl 4]

Water-to-Air Heat Pump	Length × Width × Height, ft	Weight, lb
1.5 ton vertical unit	2.0 × 2.0 × 3.0	180
3 ton vertical unit	2.5 × 2.5 × 4.0	250
3 ton horizontal unit	3.5 × 2.0 × 2.0	250
5 ton vertical unit	3.0 × 2.5 × 4.0	330
11 ton vertical unit	3.5 × 3.0 × 6.0	720
26 ton vertical unit	3.5 × 5.0 × 6.0	1550

Note: See manufacturers' specification sheets for actual values.

Refrigerant-to-Water Heat Exchanger. The heat exchanger, which couples the heat pump to source/sink water, is tube-in-tube, tube-in-shell, or brazed-plate. It must function in either condensing or evaporating mode, so special attention is given to refrigerant-side circuitry. Heat exchanger construction is usually of copper and steel, and the source/sink water is exposed only to the copper portions. Cupronickel options to replace the copper are usually available for use with brackish or corrosive water. Brazed-plate heat exchangers are usually constructed of stainless steel, which reduces the need for special materials.

Refrigerant Expansion Devices. These WSHPs operate over a narrow range of entering water temperatures and typically use simple capillaries as expansion devices. However, units may also use thermostatic expansion valves for improved performance over a broader range of inlet fluid temperatures.

Refrigerant-Reversing Valve. The refrigerant-reversing valves in WSHPs are identical to those used in air-source heat pumps.

Condensate Disposal. Condensate, which forms on the indoor coil when cooling, is collected and conveyed to a drain system.

Controls. Console WSHP units have built-in operating mode selector and thermostatic controls. Ducted units use low-voltage remote heat/cool thermostats.

Size. Typical space requirements and weights of WSHPs are presented in Table 19.4.

Special Features. Some WSHPs include the following:

- *Desuperheater*. Uses discharge gas in a special water/refrigerant heat exchanger to heat water for a building.
- *Capacity modulation*. May use multiple compressors, multispeed compressors, or hot-gas bypass.
- *Variable air volume (VAV)*. Reduces fan energy usage and requires some form of capacity modulation.
- *Automatic water valve*. Closes off water flow through the unit when the compressor is off and allows variable water volume in the loop, which reduces pumping energy.
- *Outdoor air economizer*. Cools directly with outdoor air to reduce or eliminate the need for mechanical refrigeration during mild or cold weather when outdoor humidity levels and air quality are appropriate.
- *Water-side economizer*. Cools with loop water to reduce or eliminate the need for mechanical refrigeration during cold weather; requires a hydronic coil in the indoor air circuit that is valved into the circulating loop when loop temperatures are relatively low and cooling is required.
- *Electric heaters*. Used in WLHP systems that do not have a boiler as a source for loop heating.

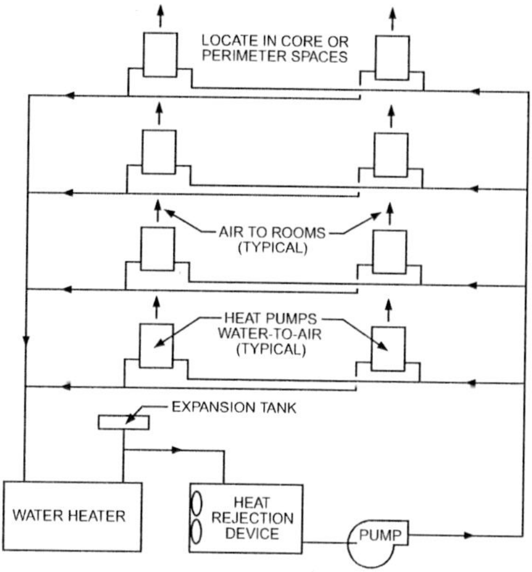

A. WATER-LOOP HEAT PUMP SYSTEM

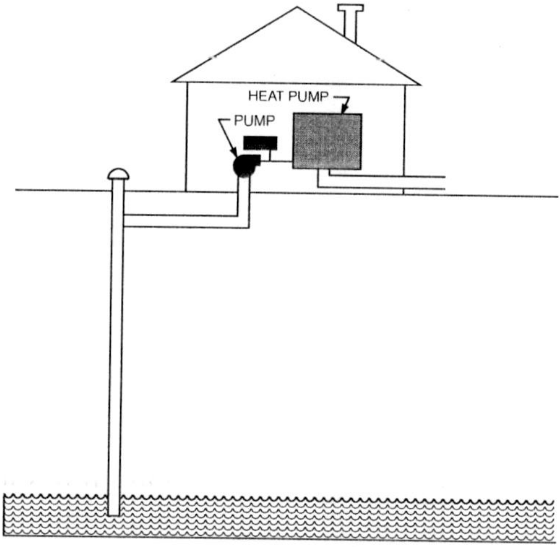

B. GROUNDWATER HEAT PUMP SYSTEM

Figure 19.32 Water-Source Heat Pump Systems [2020S, Ch 48, Fig 17]

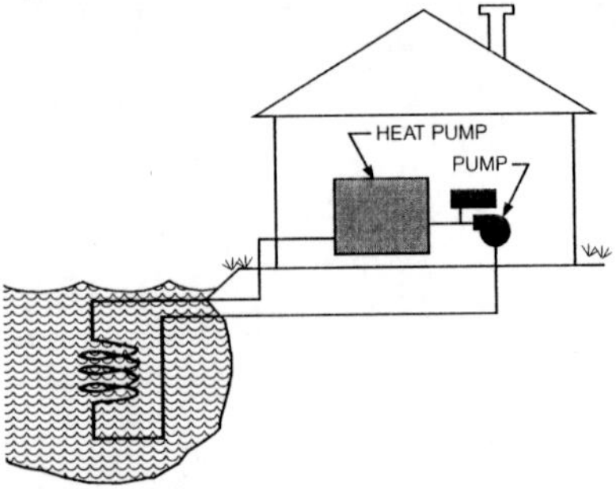

C. CLOSED-LOOP SURFACE-WATER HEAT PUMP SYSTEM

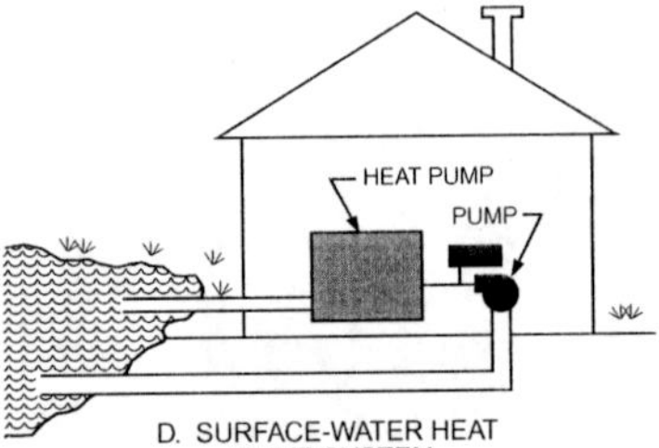

D. SURFACE-WATER HEAT PUMP SYSTEM

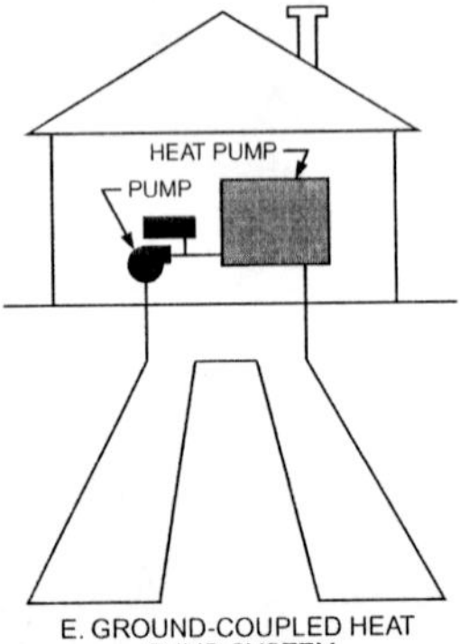

E. GROUND-COUPLED HEAT PUMP SYSTEM

Figure 19.32 Water-Source Heat Pump Systems [2020S, Ch 48, Fig 17]
(Continued)

Variable-Refrigerant-Flow Heat Pumps

A variable-refrigerant-flow (VRF) system typically consists of a condensing section housing compressor(s) and condenser heat exchanger interconnected by a single set of refrigerant piping to multiple indoor direct-expansion (DX) evaporator fan-coil units. Thirty or more DX fan-coil units can be connected to a single condensing section, depending on system design, and with capacity ranging from 0.5 to 8 tons.

The DX fan-coils are constant air volume, but use variable refrigerant flow through an electronic expansion valve. The electronic expansion valve reacts to several temperature-sensing devices such as return air, inlet and outlet refrigerant temperatures, or suction pressure. The electronic expansion valve modulates to maintain the desired set point.

Application. VRF systems are most commonly air-to-air, but are also available in a water-source (water-to-refrigerant) configuration. They can be configured for simultaneous heating and cooling operation (some indoor fan-coil units operating in heating and some in cooling, depending on requirements of each building zone).

Indoor units are typically direct-expansion evaporators using individual electronic expansion devices and dedicated microprocessor controls for individual control. Each indoor unit can be controlled by an individual thermostat. The outdoor unit may connect several indoor evaporator units with capacities 130% or more than the outdoor condensing unit capacity.

Categories. VRF equipment is divided into three general categories: residential, light commercial, and applied. Residential equipment is single-phase unitary equipment with a cooling capacity of 65,000 Btu/h or less. Light commercial equipment is generally three-phase, with cooling capacity greater than 65,000 Btu/h, and is designed for small businesses and commercial properties. Applied equipment has cooling capacity higher than 135,000 Btu/h and is designed for large commercial buildings.

Refrigerant Circuit and Components. VRF heat pump systems use a two-pipe (liquid and suction gas) system; simultaneous heat and cool systems use the same system, as well as a gas flow device that determines the proper routing of refrigerant gas to a particular indoor unit.

VRF systems use a sophisticated refrigerant circuit that monitors mass flow, oil flow, and balance to ensure optimum performance. This is accomplished in unison with variable-speed compressors and condenser fan motors. Both of these components adjust their frequency in reaction to changing mass flow conditions and refrigerant operating pressures and temperatures. A dedicated microprocessor continuously monitors and controls these key components to ensure proper refrigerant is delivered to each indoor unit in cooling or heating.

Heating and Defrost Operation. In heating mode, VRF systems typically must defrost like any mechanical heat pump, using reverse-cycle valves to temporarily operate the outdoor coil in cooling mode. Oil return and balance with the refrigerant circuit is managed by the microprocessor to ensure that any oil entrained in the low side of the system is brought back to the high side by increasing the refrigerant velocity using a high-frequency operation performed automatically based on hours of operation.

More information on VRF heat pumps can be found in Chapter 18 of the 2020 *ASHRAE Handbook—HVAC Systems and Equipment*, and information on performance-rating these systems can be found in AHRI Standard 1230.

Room Air Conditioners and Packaged Terminal Air Conditioners

Room Air Conditioners

Room air conditioners are encased assemblies designed primarily for mounting in a window or through a wall. They are designed to deliver cool or warm conditioned air to the room, either without ducts or with very short ducts (up to a maximum of about 48 in.). Each unit includes a prime source of refrigeration and dehumidification and a means for circulating and filtering air; it may also include a means for ventilating and/or exhausting and heating.

The basic function of a room air conditioner is to provide comfort by cooling, dehumidifying, filtering or cleaning, and circulating the room air. It may also provide ventilation by introducing outdoor air into the room and/or exhausting room air to the outdoors. Room temperature may be controlled by an integral thermostat. The conditioner may provide heating by heat pump operation, electric resistance elements, or a combination of the two.

Figure 19.33 shows a typical room air conditioner in cooling mode. Warm room air passes over the cooling coil and transfers sensible and latent heat. The conditioned air is then recirculated in the room by a fan or blower.

Heat from the warm room air vaporizes the cold (low-pressure) liquid refrigerant flowing through the evaporator. The vapor then carries the heat to the compressor, which compresses the vapor and increases its temperature above that of the outdoor air. In the condenser, the hot (high-pressure) refrigerant vapor liquefies, transferring the heat from the room air to outdoor air. Next, the high-pressure liquid refrigerant passes through a restrictor, which reduces its pressure and temperature. The cold (low-pressure) liquid refrigerant then enters the evaporator to repeat the refrigeration cycle.

Room air conditioners have line cords, which may be plugged into standard or special electric circuits. Most units in the United States are designed to operate at 115, 208, or 230 V; single-phase; 60 Hz power. Some units are rated at 265 V or 277 V, for which the chassis or chassis assembly must provide permanent electrical connection. The maximum amperage of 115 V units is generally 12 A, which is the maximum current permitted by NFPA Standard 70 [the *National Electrical Code*® (NEC)] for a single-outlet, 15 A circuit. Models designed for

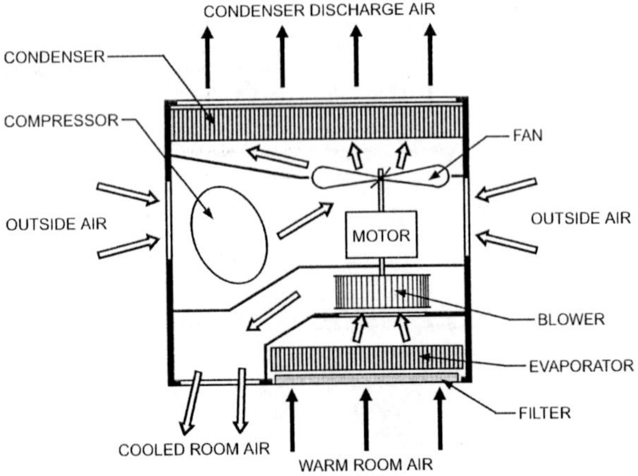

Figure 19.33 Schematic View of Typical Room Air Conditioner
[2020S, Ch 49, Fig 1]

countries other than the United States are generally for 50 or 60 Hz systems, with typical design voltage ranges of 100 to 120 and 200 to 240 V, single-phase.

Popular 115 V models have capacities in the range of 5000 to 8000 Btu/h, and are typically used in single-room applications. Larger-capacity 115 V units are in the 12,000 to 15,000 Btu/h range. Capacities for 230, 208, or 230/208 V units range from 8000 to 36,000 Btu/h. These higher-voltage units are typically used in multiple-room installations.

Heat pump models are also available, usually for 208 or 230 V applications. These units are generally designed for reversed-refrigerant-cycle operation as the normal means of supplying heat, but may incorporate electrical-resistance heat either to supplement heat pump capacity or to provide the total heating capacity when outdoor temperatures drop below a set value.

Another type of heating model incorporates electrical heating elements in regular cooling units so that heating is provided entirely by electrical resistance heat.

Installation procedures vary because units can be mounted in various ways. It is important to select the mounting for each installation that best satisfies the user and complies with applicable building codes. Common mounting methods include the following:

- **Indoor flush mounting.** Interior face of conditioner is approximately flush with inside wall.
- **Balance mounting.** Unit is approximately half inside and half outside window.
- **Outdoor flush mounting.** Outer face of unit is flush with or slightly beyond outside wall.
- **Special mounting.** Examples include casement windows, horizontal sliding windows, and office windows with swinging units (or swinging windows) to allow window washing, and transoms over doorways.
- **Through-the-wall mounts or sleeves.** This mounting is used for installing window-type chassis, complete units, or consoles in walls of apartment buildings, hotels, motels, and residences. Although very similar to window-mounted units, through-the-wall models do not have side louvers for condenser air; air comes from the outdoor end of the unit.

Room air conditioners have become more compact to minimize both loss of window light and projection inside and outside the structure. Several types of expandable mounts are now available for fast, dependable installation in single- and double-hung windows, as well as in horizontal sliding windows. Window air conditioners should fit in the opening without too much space around the sides; most units come with accordion-style flaps attached to the sides, used to fill the open space around the unit. Installation kits include all parts needed for structural mounting, such as gaskets, panels, and seals for weathertight assembly. Sealing the space around the unit (e.g., with caulk or foam seals) is important to prevent air from escaping and limit seeping of hot outdoor air into the conditioned space, and to achieve energy savings during operation.

Adequate wiring and proper breakers or fuses must be provided for the service outlet. Necessary information is usually given on instruction sheets or stamped on the air conditioner near the service cord or on the serial plate. It is important to follow the manufacturer's recommendation for size and type of breaker or fuse. All units are equipped by the manufacturer with grounding plug caps on the service cord. Receptacles with grounding contacts correctly designed to fit these plug caps should be used when units are installed.

Units rated 265 or 277 V must provide for permanent electrical connection with armored cable or conduit to the chassis or chassis assembly. Manufacturers usually provide an adequate cord and plug cap in the chassis assembly to facilitate installation and service.

One type of room air conditioner is the integral chassis design, with the outer cabinet fastened permanently to the chassis. Most electrical components can be serviced by partially dismantling the control area without removing the unit from the installation. Another type is the **slide-out chassis** design, which allows the outer cabinet to remain in place while the chassis is removed for service.

Effective condensate management should be considered during the installation of the air conditioner. Many window air conditioners have connection points for condensate drains. Follow plumbing and building code requirements for handling discharge of the condensate produced when the air conditioner is operating.

Packaged Terminal Air Conditioners

A packaged terminal air conditioner (PTAC) includes a wall sleeve and a separate unencased combination of heating and cooling assemblies intended for mounting through the wall. A PTAC includes refrigeration components, separable outdoor louvers, forced ventilation, and heating by hot water, steam, or electric resistance. PTAC units with indirect-fired gas heaters are also available from some manufacturers. A packaged terminal heat pump (PTHP) is a heat pump version of a PTAC that provides heat with a reverse-cycle operating mode. A PTHP should provide a supplementary heat source, which can be hot water, steam, electric resistance, or another source.

PTACs are designed primarily for commercial installations to provide the total heating and cooling functions for a room or zone and are specifically for through-the-wall installation. The units are mostly used in relatively small zones on the perimeter of buildings such as hotels and motels, apartments, hospitals, nursing homes, and office buildings. In larger buildings, they may be combined with nearly any system selected for environmental control of the building core.

PTACs and PTHPs are similar in design and construction. The most apparent difference is the addition of a refrigerant-reversing valve in the PTHP. Optional components that control the heating functions of the heat pump include an outdoor thermostat to signal the need for changes in heating operating modes, and, in more complex designs, frost sensors, defrost termination devices, and base pan heaters.

PTACs/PTHPs are available in a wide range of rated cooling capacities, typically 6000 to 18,000 Btu/h, with comparable levels of heating output. Units are available as sectional types or integrated types. A sectional-type unit (Figure 19.34) has a separate cooling chassis; an integrated-type unit (Figure 19.35) has an electric or a gas heating option added to the chassis. Hot-water or steam heating options are usually part of the cabinet or wall box. Both types include the following:

- Heating elements available in hot water, steam, electric, or gas heat
- Integral or remote temperature and operating controls
- Wall sleeve or box
- Removable (or separable) outdoor louvers
- Room cabinet
- Means for controlled forced ventilation
- Means for filtering air delivered to the room

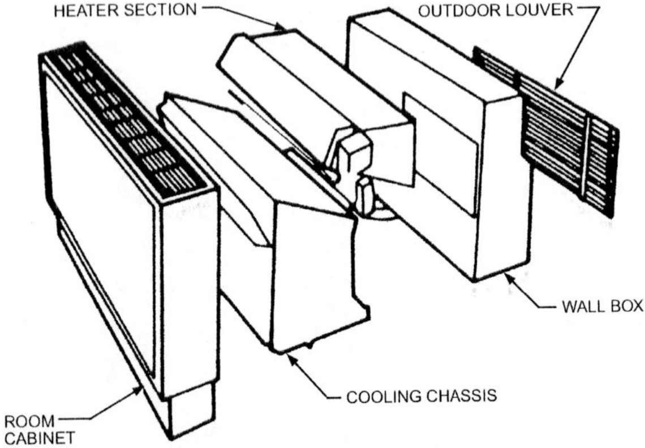

Figure 19.34 Sectional Packaged Terminal Air Conditioner
[2020S, Ch 49, Fig 2]

PTAC assemblies are intended for use in free conditioned-air distribution, but a particular application may require minimal ductwork with a total external static resistance up to 0.1 in. of water.

Packaged terminal air conditioners and packaged terminal heat pumps allow the HVAC designer to integrate the exposed outdoor louver or grille with the building design. Various grilles are available to blend with or accent most construction materials. Because the product becomes part of the building's facade, the architect must consider the product during the conception of the building. Wall sleeve installation is usually done by ironworkers, masons, or carpenters. All-electric units dominate the market.

All the energy of all-electric versions is dispersed through the building via electrical wiring, so the electric designer and electrical contractor play a major role. Final installation is reduced to sliding in the chassis and plugging the unit into an adjacent receptacle. For these all-electric units, the traditional HVAC contractor's work involving ducting, piping, and refrigeration systems is bypassed. This results in a low-cost installation and allows installation of the PTAC/PTHP chassis to be deferred until just before occupancy.

When comparing a gas-fired PTAC to a PTAC with electric resistance heat or a PTHP, evaluate both operating and installation costs. Generally, a gas-fired PTAC is more expensive to install but less expensive to operate in heating mode. A life-cycle cost comparison is recommended (see Chapter 38 of the 2019 *ASHRAE Handbook—HVAC Applications*).

One main advantage of the PTAC/PTHP concept is that it provides excellent zoning capability. Units can be shut down or operated in a holding condition during unoccupied periods. Present equipment efficiency-rating criteria are based on full-load operation, so an efficiency comparison to other approaches may suffer.

The designer must also consider that total capacity is the sum of the peak loads of each zone rather than the peak load of the building. Therefore, total cooling capacity of the zonal system will exceed that of a central system.

Because PTAC units are located in the conditioned space, both appearance and sound level of the equipment are important considerations. Sound attenuation in ducting is not available with the free-discharge PTAC units.

The designer must also consider the added infiltration and thermal leakage load resulting from perimeter wall penetrations. These losses are accounted for during the *on* cycle in equipment cooling ratings and PTHP heating ratings, but during the *off* cycle or with other forms of heating, they could be significant.

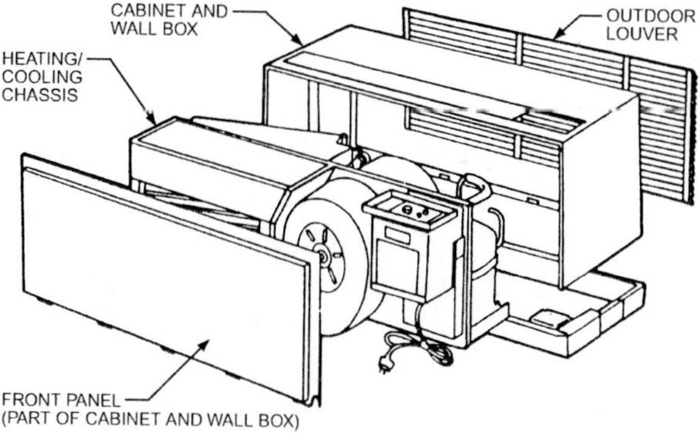

Figure 19.35 Integrated Packaged Terminal Air Conditioner
[2020S, Ch 49, Fig 3]

Most packaged terminal equipment is designed to fit into a wall aperture approximately 42 in. wide and 16 in. high. Although unitary products can increase in size with increasing cooling capacity, PTAC/PTHP units, regardless of cooling capacity, are usually constrained to a few cabinet sizes. The exterior of the equipment must be essentially flush with the exterior wall to meet most building codes. In addition, cabinet structural requirements and the slide-in chassis reduce the available area for outdoor air inlet and relief to less than a total of 3.5 ft^2. Manufacturers' specification sheets should be consulted for more accurate and detailed information.

Basic PTHP units can operate in heat pump mode until outdoor temperature is just above the point at which the outdoor heat exchanger would frost. At that point, heat pump mode is locked out, and other forms of heating are required. Some units include two-stage indoor thermostats and automatically switch from heat pump mode to an alternative heat source if space temperature drops too far below the first-stage set point. Some PTHPs use control schemes that extend heat pump operation to lower temperatures. One approach allows heat pump operation down to outdoor temperatures just above freezing. If the outdoor coil frosts, it is defrosted by shutting down the compressor and allowing the outdoor fan to continue circulating outdoor air over the coil. Another approach allows heat pump operation to even lower outdoor temperatures by using a reverse-cycle defrost sequence. In those cases, the heat pump mode is usually locked out for outdoor temperatures below 10°F.

EVAPORATIVE COOLING

Direct Evaporative Air Coolers

Air is drawn through porous wetted pads or a spray, or rigid media; and its sensible heat energy evaporates some water. The heat and mass transfer between the air and water lowers the air dry-bulb temperature and increases the humidity at a constant wet-bulb temperature. The dry-bulb temperature of the nearly saturated air approaches the ambient air's wet-bulb temperature. The process is adiabatic, so no sensible cooling occurs.

The extent to which the leaving air temperature from a direct evaporative cooler approaches the thermodynamic wet-bulb temperature of the entering air or the extent to which complete saturation is approached is expressed as the **direct saturation efficiency**, defined as

$$\varepsilon_e = 100\frac{t_1 - t_2}{t_1 - t_s'} \tag{19.7}$$

where

ε_e	=	direct evaporative cooling or saturation efficiency, %
t_1	=	dry-bulb temperature of entering air, °F
t_2	=	dry-bulb temperature of leaving air, °F
t_s'	=	thermodynamic wet-bulb temperature of entering air, °F

An efficient wetted pad (with high saturation efficiency) can reduce the air dry-bulb temperature by as much as 95% of the wet-bulb depression (ambient dry-bulb temperature less wet-bulb temperature), while an inefficient and poorly designed pad may only reduce this by 50% or less.

Direct evaporative cooling, though simple and inexpensive, has the disadvantage that if the ambient wet-bulb temperature is higher than about 70°F, the cooling effect is not sufficient for indoor comfort but still may be sufficient for relief cooling applications. Direct evaporative coolers should not recirculate indoor air.

Direct evaporative coolers for residences in low-wet-bulb regions typically require 70% less energy than direct-expansion air conditioners.

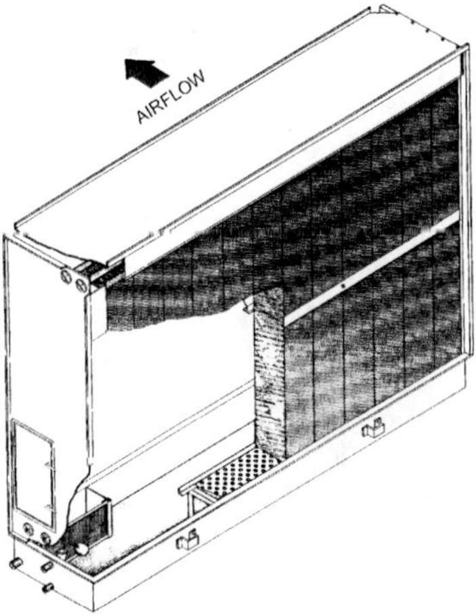

Figure 19.36 Typical Rigid-Media Air Cooler [2020S, Ch 41, Fig 2]

Indirect Evaporative Air Coolers

Either outdoor or recirculated air may be cooled using indirect evaporative coolers (IECs); commercial IECs commonly cool outdoor air, often as part of a dedicated outdoor air system (DOAS) or variable-air-volume (VAV) air-handler design, whereas in industrial applications and data center cooling designs, process recirculation air is more commonly cooled. Air delivered to the room or process flows through the inside of elliptical-shaped polymer tubes, where it is sensibly cooled. A sump pump recirculates water over the outside of the polymer tubes of this cross-flow air-to-air heat exchanger. Scavenger outdoor air or building return/exhaust air (typical for commercial applications) is drawn upward over the exterior of the wetted polymer tubes.

Polymer tube heat exchangers can provide a 60% to 80% approach of the dry-side entering dry-bulb temperature to the wet-side entering wet-bulb temperature. The heat exchanger efficiency calculation is called *wet-bulb depression efficiency* (WBDE) and is defined as

$$\text{WBDE} = 100\frac{(t_1 - t_2)}{(t_1 - t_s')} \tag{19.8}$$

where

WBDE	=	indirect evaporative cooling efficiency,%
t_1	=	dry-bulb temperature of entering primary air, °F
t_2	=	dry-bulb temperature of leaving primary air, °F
t_s'	=	wet-bulb temperature of entering secondary air, °F

Supply air-side static pressure losses for these heat exchangers range from 0.25 to 0.75 in. of water. Wet-side airflow pressure drop penalties range from 0.4 to 0.9 in. of water. Secondary airflow ratios are in the range of 1.5 to 1 down to a low of 1 cfm of outdoor air to 0.7 cfm of secondary airflow. The higher the ratios of wet-side air to dry air, the greater the WBDE, with all other factors remaining constant. Cooling energy efficiency ratios (EERs) for this type of heat exchanger range from 40 to 80.

In a two-stage indirect/direct evaporative cooler, a first-stage indirect evaporative cooler lowers both the dry- and wet-bulb temperature of the incoming air. After leaving the indirect stage, the supply air passes through a second-stage direct evaporative cooler.

This method can lower the supply air dry-bulb temperature by 10°F or more below the secondary air wet-bulb temperature.

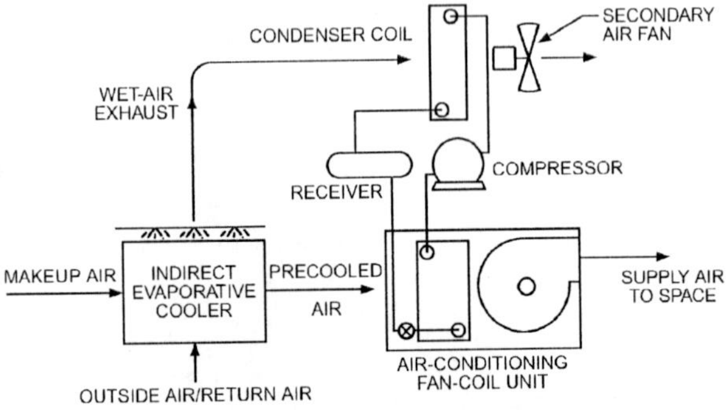

Figure 19.37 Indirect Evaporative Cooler Used as Precooler
[2020S, Ch 41, Fig 4]

In areas with a higher wet-bulb design temperature or where the design requires a supply air temperature lower than that attainable using indirect/direct evaporative cooling, a third cooling stage may be required. This stage may be a direct-expansion refrigeration unit or a chilled-water coil located either upstream or downstream from the direct evaporative cooling stage, but always downstream from the indirect evaporative stage.

Heat Recovery. Indirect evaporative cooling has been used in a number of heat recovery systems, including plate heat exchangers, heat pipe heat exchangers, rotary regenerative heat exchangers, and two-phase thermosiphon loop heat exchangers. Indirect evaporative cooling/heat recovery can be retrofitted on existing systems, lowering operational cost and peak demand. For new installations, equipment can be downsized, lowering overall project and operational costs.

Cooling Tower/Coil Systems. Combining a cooling tower or other evaporative water cooler with a water-to-air heat exchanger coil and water-circulating pump is another type of indirect evaporative cooling. Water flows from the cooling tower reservoir to the coil and returns to the tower's upper distribution header. Both open-water and closed-loop systems are used. Coils in open systems should be cleanable.

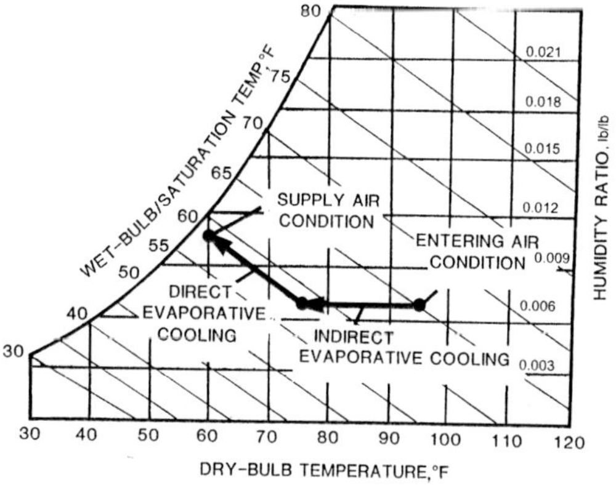

Figure 19.38 Two-Stage Indirect/Direct Evaporative Cooling Process
[2020S, Ch 41, Fig 6]

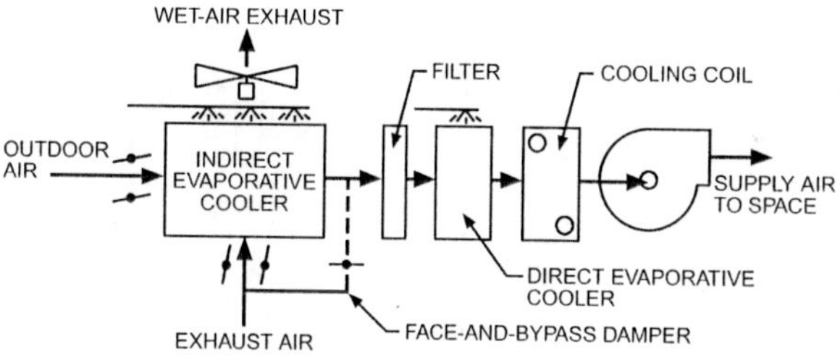

Figure 19.39 Three-Stage Indirect/Direct Evaporative Cooler
[2020S, Ch 41, Fig 8]

HVAC Systems and Equipment

Table 19.5 Indirect Evaporative Cooling Systems Comparison [2019A, Ch 53, Tbl 2]

System Type[a]	WBDE,[b] %	Heat Recovery Efficiency, %	Wet-Side Air ΔP, in. of water	Dry-Side Air ΔP, in. of water	Pump hp per 10,000 cfm	Parasitic Loss Range,[e] kW/ton of Cooling	Equipment Cost Range,[f] $/Supply cfm	Notes
Cooling tower to coil	40 to 60	NA	NA	0.4 to 0.7	Varies	Varies	0.50 to 1.00	Best for serving multiple AHUs from a single cooling tower. No winter heat recovery.
Cross-flow plate	60 to 85	40 to 50	0.7 to 1.0	0.4 to 0.7	0.1 to 0.2	0.12 to 0.20	1.20 to 1.70	Most cost-effective for lower airflows. Some cross contamination possible. Low winter heat recovery.
Heat pipe[c]	65 to 75	50 to 60	0.7 to 1.0	0.5 to 0.7	0.2 to 0.4	0.15 to 0.25	1.50 to 2.50	Most cost-effective for large airflows. Some cross contamination possible. Medium winter heat recovery.
Heat wheel[d]	60 to 70	70 to 80	0.6 to 0.9	0.4 to 0.65	0.1 to 0.2	0.2 to 0.3	1.50 to 2.50	Best for high airflows. Some cross contamination. Highest winter heat recovery rates.
Runaround coil[d]	35 to 50	40 to 60	0.6 to 0.8	0.4 to 0.6	Varies	> 0.35	1.00 to 2.00	Best for applications where supply and return air ducts are separated. Lowest summer WBDE.

WBDE = wet-bulb depression efficiency

Notes:

[a]All air-to-air heat exchangers have equal mass flow on supply and exhaust sides.

[b]Plate and heat pipe are direct spray on exhaust side. Heat wheel and runaround coil systems use 90% WBDE direct evaporative cooling media on exhaust air side.

[c]Assumes six-row heat pipe, 11 fpi, with 500 fpm face velocity on both sides.

[d]Assumes 500 fpm face velocity. Parasitic loss includes wheel rotational power.

[e]Includes air-side static pressure and pumping penalty.

[f]Excludes cooling tower cost and assumes less than 200 ft piping between components.

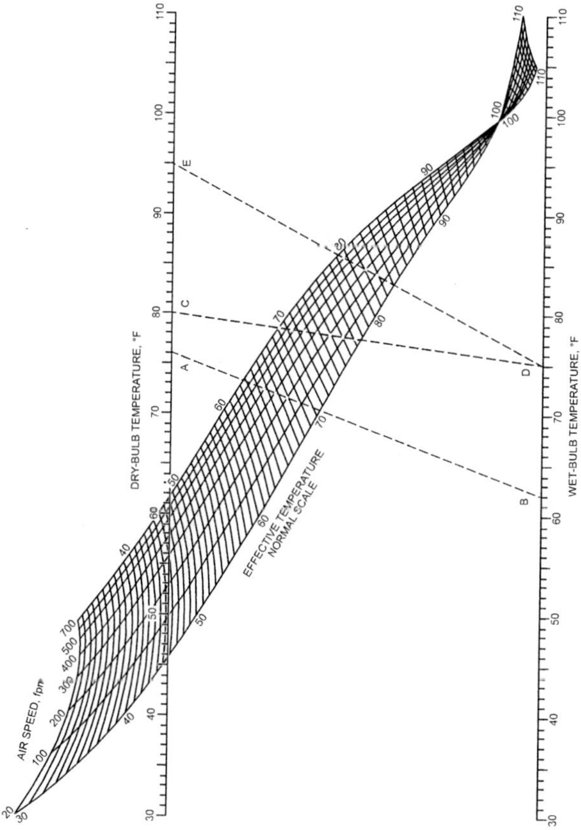

Figure 19.40 Effective Temperature Chart [2019A, Ch 53, Fig 15]

20. AUTOMATIC CONTROLS

Automatic control of HVAC systems and equipment usually includes control of temperature, humidity, pressure, and flow rates of air and water. Automatic controls can sequence equipment operation to meet load requirements and to provide safe equipment operation using direct digital control (DDC), electronic, electrical, mechanical, and/or pneumatic devices. Automatic controls are only fully effective when applied to well-designed mechanical systems; they cannot compensate for misapplied systems, excessive undersizing or oversizing, or highly nonlinear processes.

This chapter illustrates typical control diagrams for common HVAC systems and components.

A building automation system (BAS) with direct digital, electronic, or pneumatic controls has several physical control loops, with each loop including a controlled variable (e.g., temperature), controlled device (e.g., actuator), and the process to be controlled (e.g., heating system). A BAS with DDC controllers can share sensor values with several control loops or have multiple control loops selectively activate an actuator.

A BAS with DDC controllers allows information such as system status or alarms to be shared through a common communication protocol, between HVAC systems or within building systems and services, thereby enabling advanced, energy-saving, system-level applications. ASHRAE Guideline 13 and ANSI/ASHRAE Standard 135 have more detailed discussions of networking and interoperability. See also ASHRAE Guideline 36.

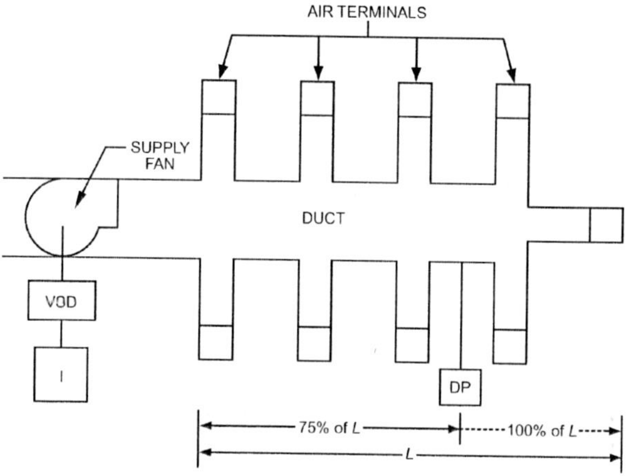

Figure 20.1 Duct Static Pressure Control [2019A, Ch 48, Fig 15]

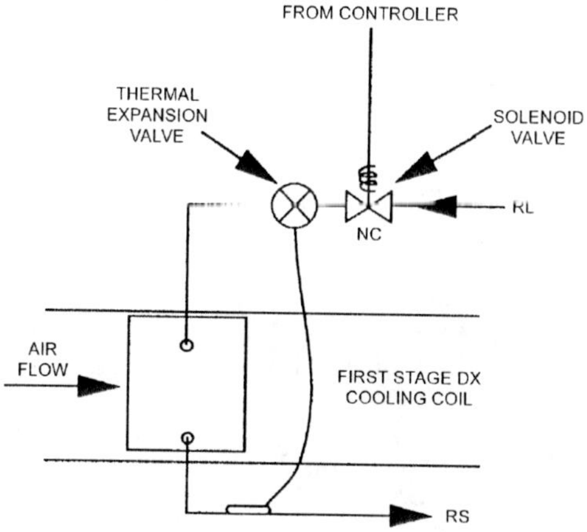

Figure 20.2 Direct Expansion—Two-Position Control

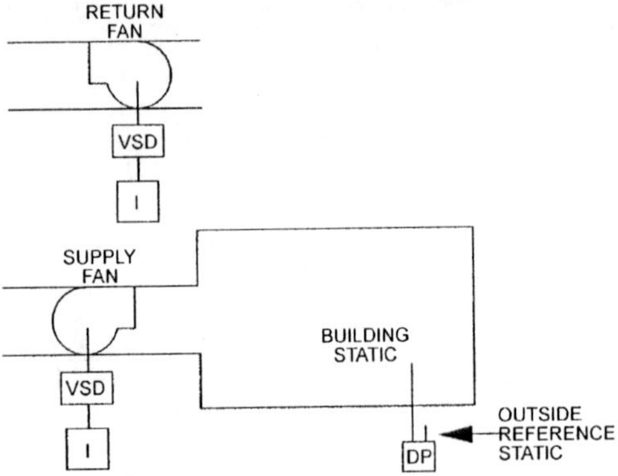

Figure 20.3 Duct Static Control of Return Fan

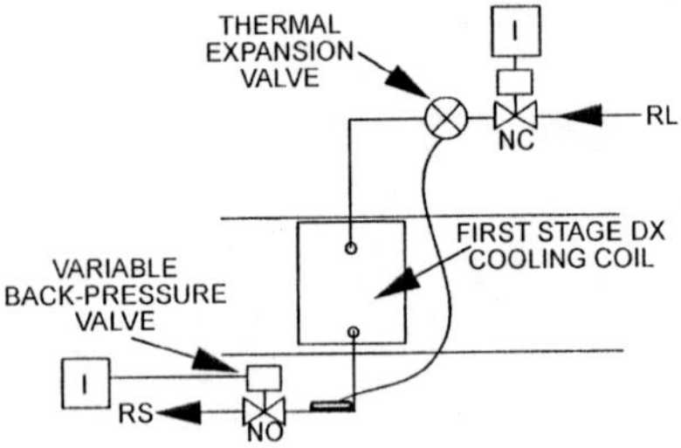

Figure 20.4 Modulating Direct-Expansion Cooling

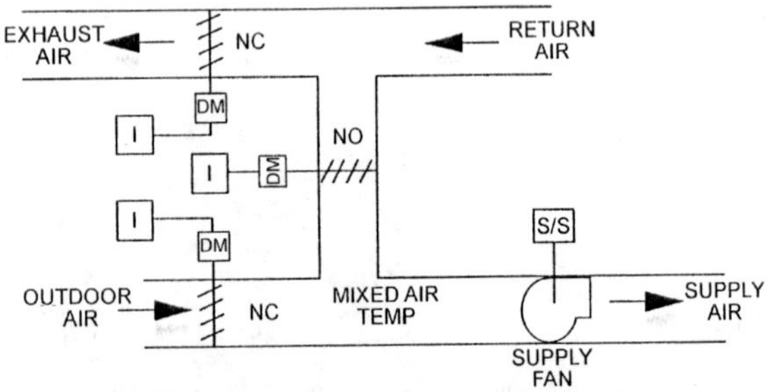

Figure 20.5 Economizer Cycle Control

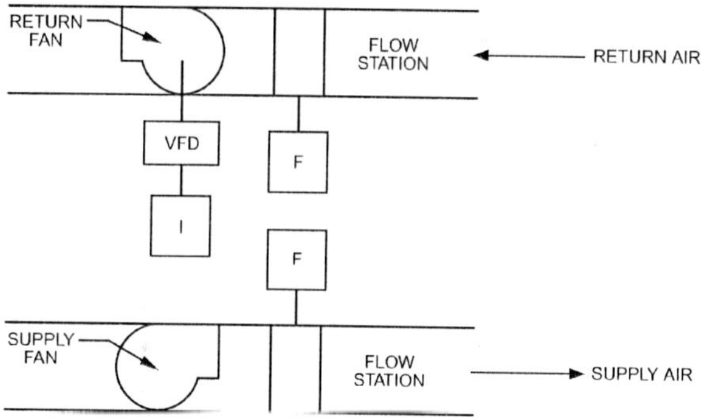

Figure 20.6 Airflow Tracking Control [2019A, Ch 48, Fig 17]

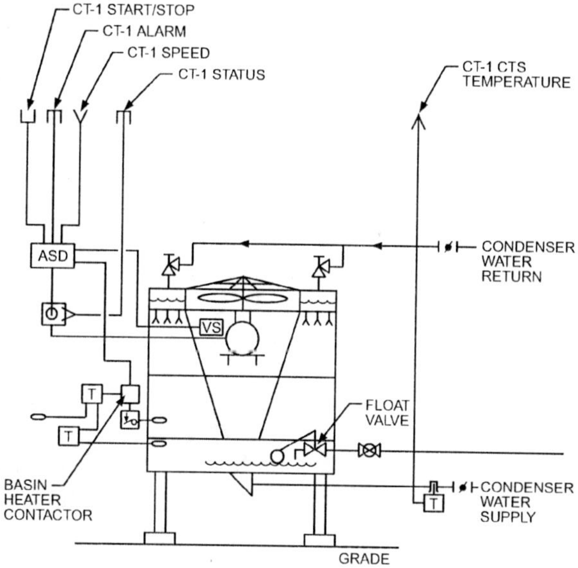

Figure 20.7 Cooling Tower [2019A, Ch 48, Fig 13]

Automatic Controls

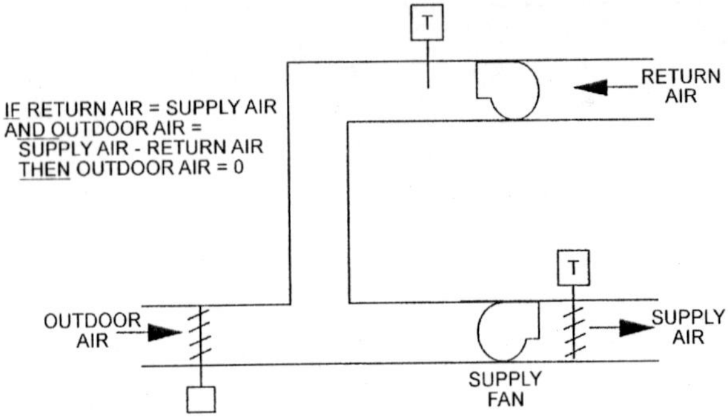

Figure 20.8 Warm-Up Control

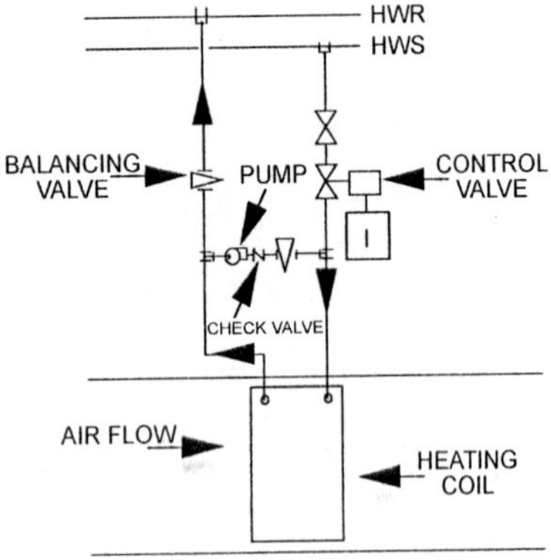

Figure 20.9 Preheat with Secondary Pump and Two-Way Valve

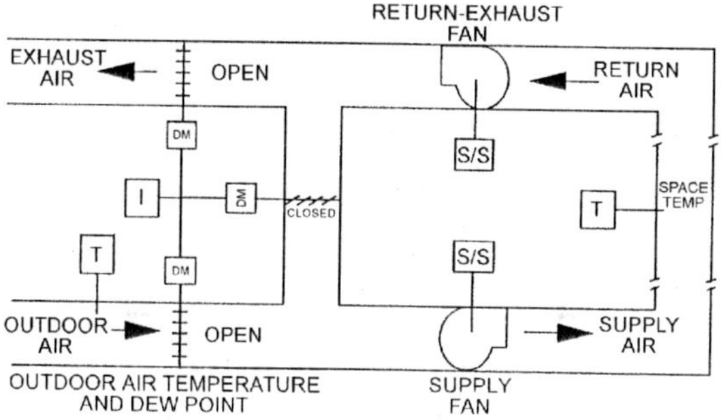

Figuro 20.10 Night Cooldown Control

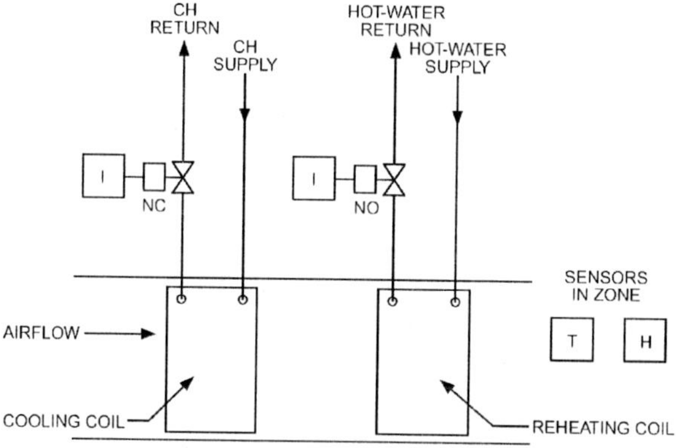

Figure 20.11 Cooling and Dehumidifying with Reheat [2019A, Ch 48, Fig 31]

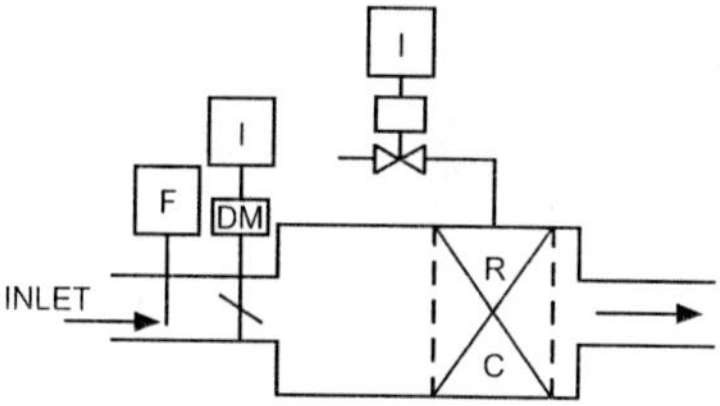

Figure 20.12 Throttling VAV Terminal Unit [2019A, Ch 48, Fig 24]

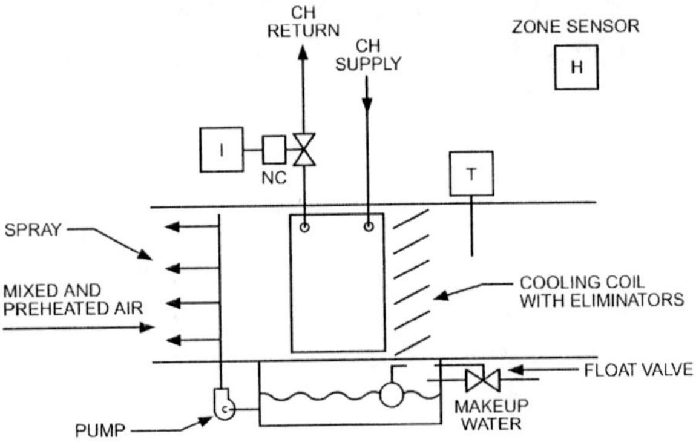

Figure 20.13 Sprayed Coil Dehumidifier [2019A, Ch 48, Fig 32]

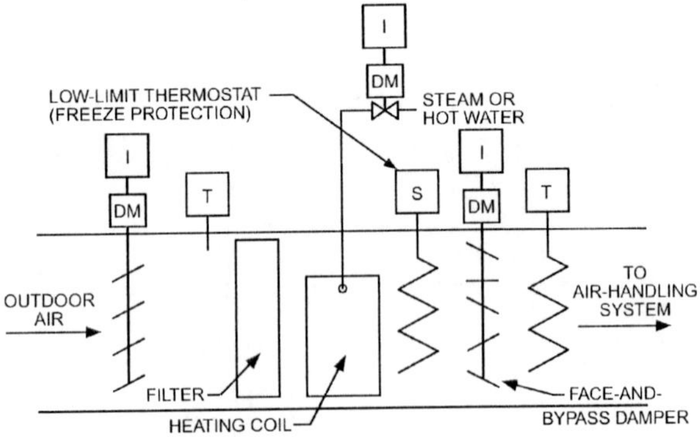

Figure 20.14 Preheat with Face and Bypass Dampers [2019A, Ch 48, Fig 5]

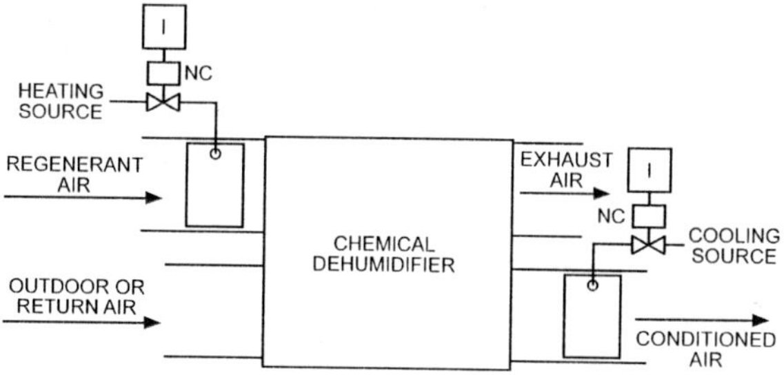

Figure 20.15 Desiccant Dehumidifier [2019A, Ch 48, Fig 34]

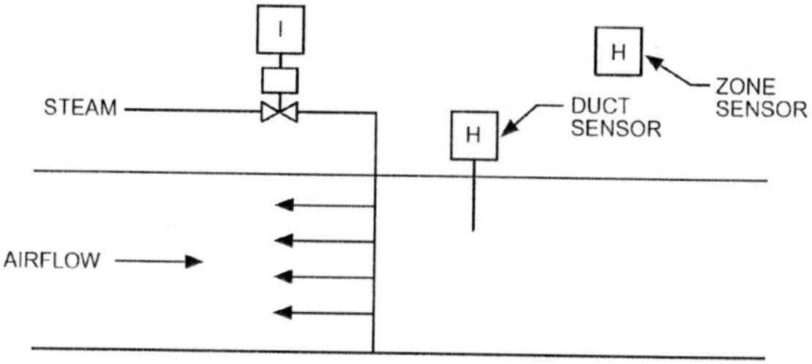

Figure 20.16 Steam Injection Humidifier [2019A, Ch 48, Fig 35]

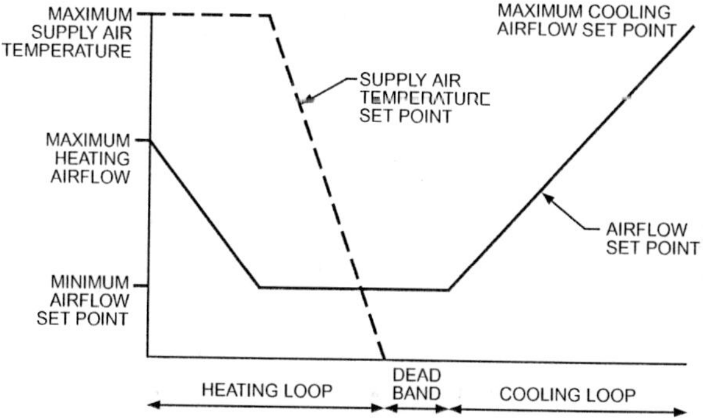

Figure 20.17 Throttling VAV Terminal Unit: Dual Maximum Control Sequence [2019A, Ch 48, Fig 25]

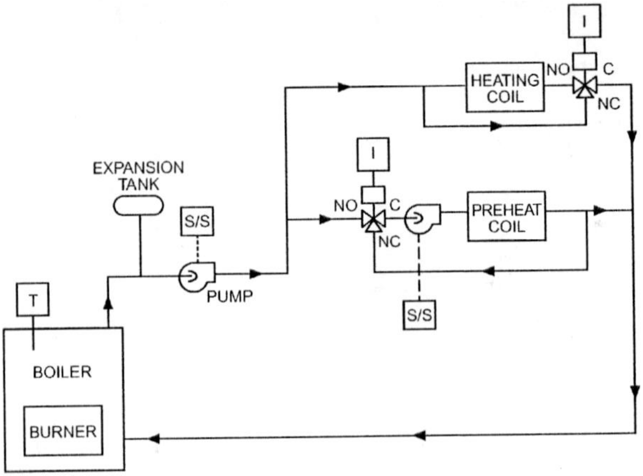

Figure 20.18 Load and Zone Control in Constant-Flow System
[2019A, Ch 48, Fig 3]

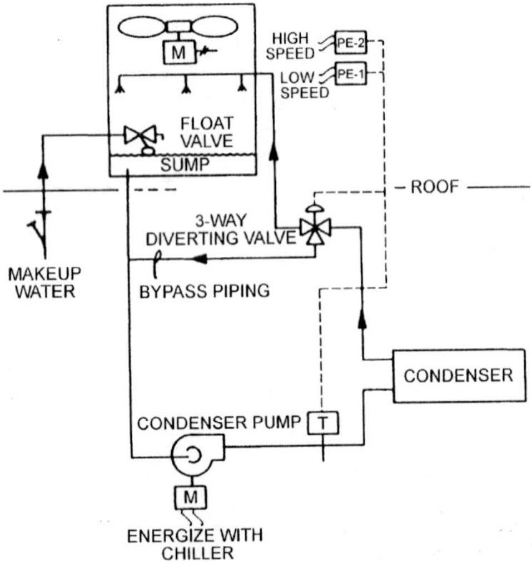

Figure 20.19 Condenser Water Temperature Control

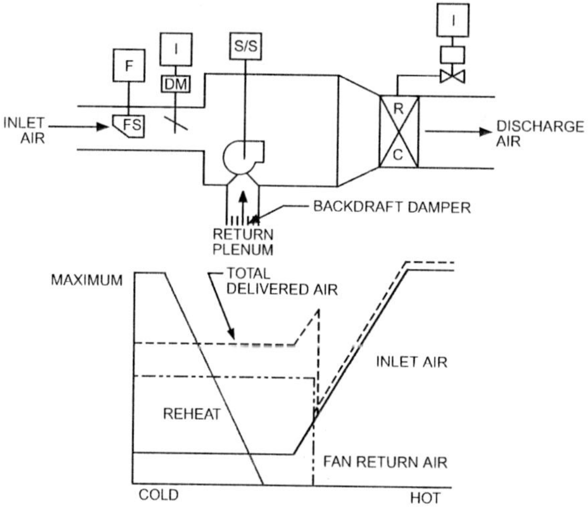

Figure 20.20 Parallel Fan Terminal Unit [2019A, Ch 48, Fig 28]

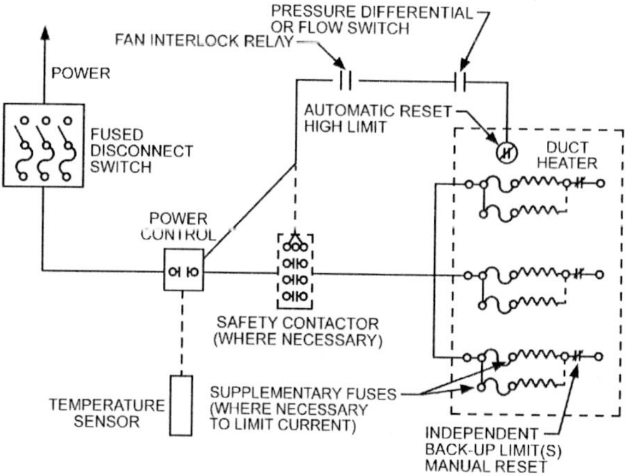

Figure 20.21 Duct Heater Control [2019A, Ch 48, Fig 9]

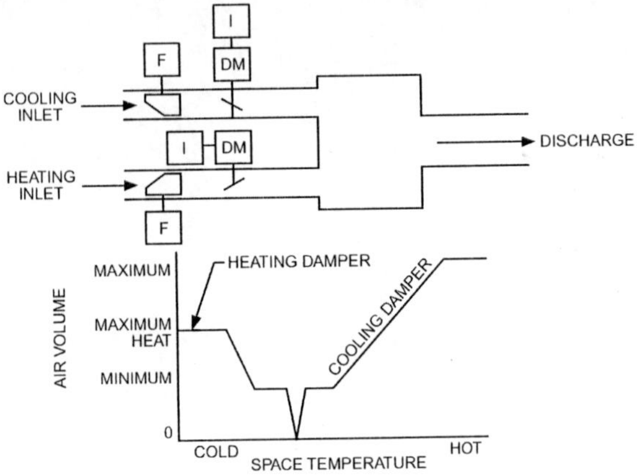

Figure 20.22 Variable-Volume Dual-Duct Terminal Unit [2019A, Ch 48, Fig 29]

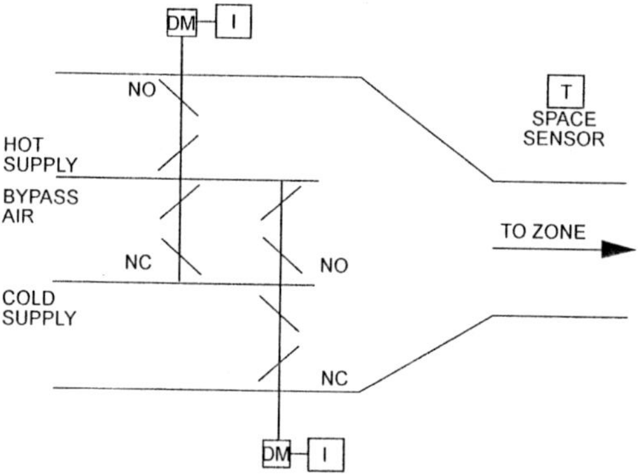

Figure 20.23 Zone Mixing Dampers—Three-Deck Multizone System

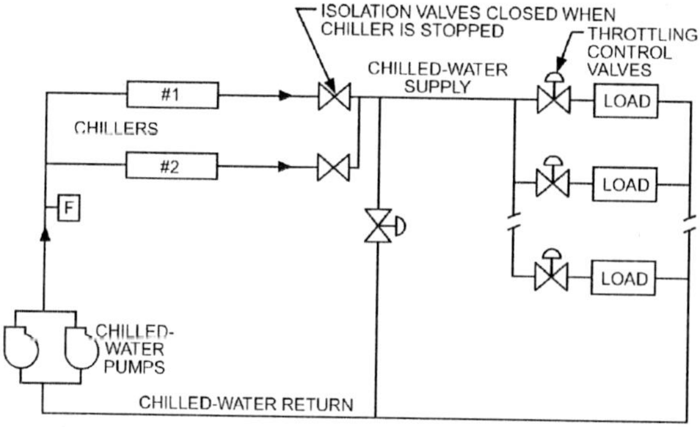

Figure 20.24 Variable-Flow Chilled-Water System (Primary Only)
[2019A, Ch 48, Fig 10]

Automatic Controls

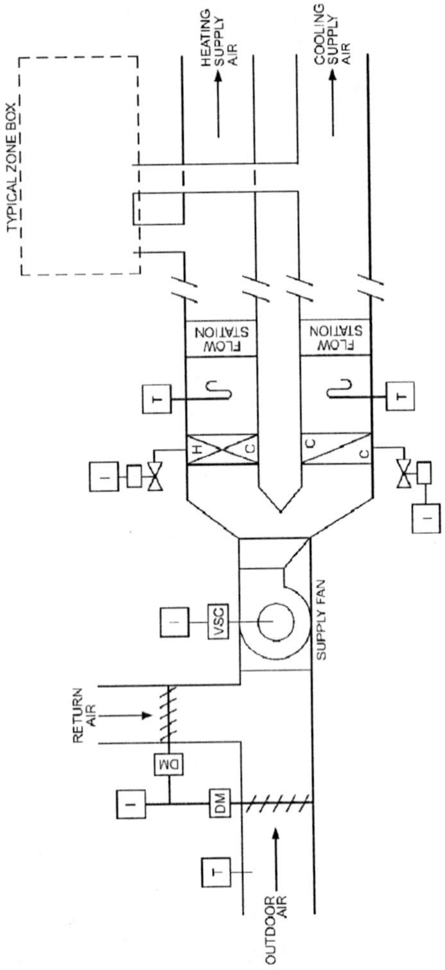

Figure 20.25 Single Fan, Dual-Duct System [2019A, Ch 48, Fig 41]

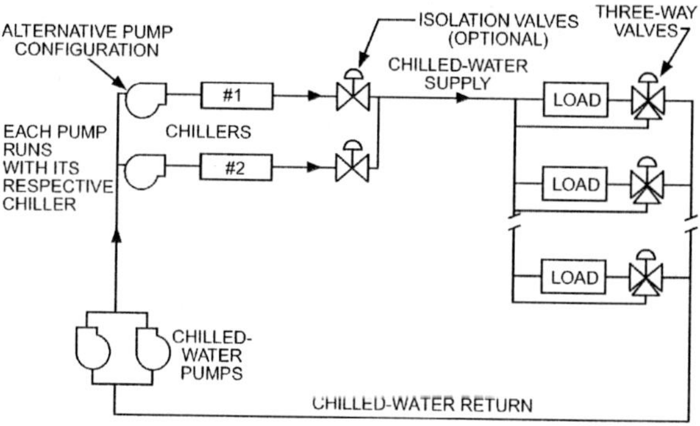

Figure 20.26 Constant-Flow Chilled-Water System (Primary/Only)
[2019A, Ch 48, Fig 12]

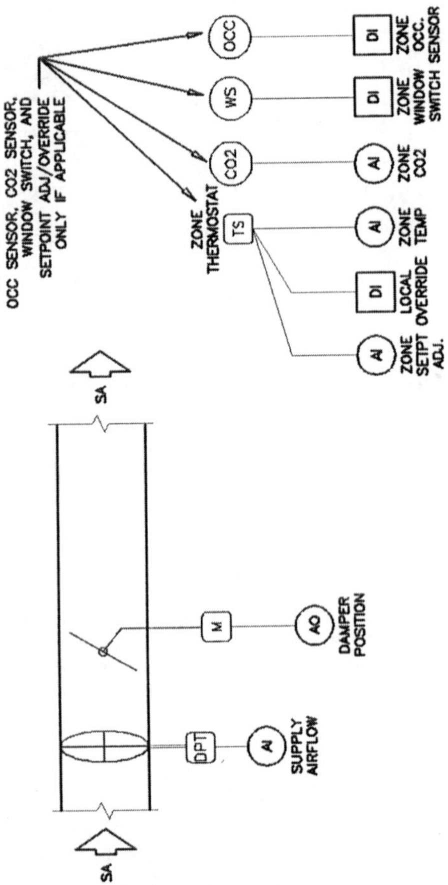

Figure 20.27 VAV Terminal Unit, Cooling Only [Gdln 36-2021, Fig A-1]

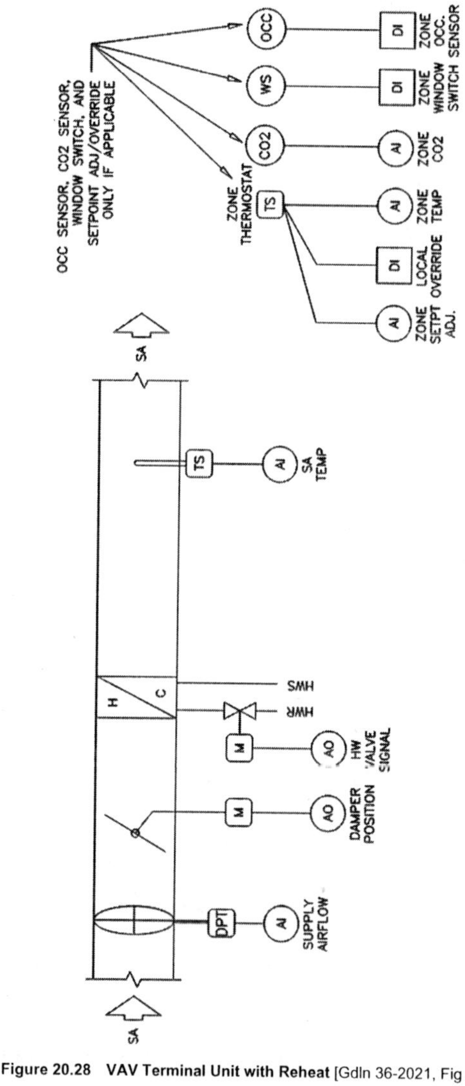

Figure 20.28 VAV Terminal Unit with Reheat [Gdln 36-2021, Fig A-2]

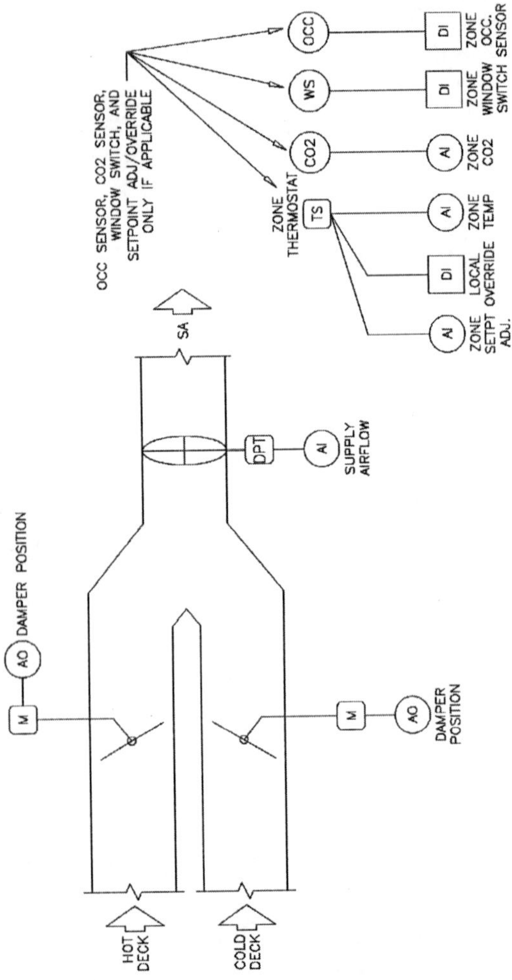

Figure 20.29 Fan-Coil Unit with Variable-Volume Fan, Heating and Cooling Coils [Gdln 36-2021, Fig A-28]

21. OCCUPANT COMFORT

The *operative temperature* (t_o) is the uniform temperature of an imaginary black enclosure, and the air within in, in which an occupant would exchange the same amount of heat by radiation plus convection as in the actual nonuniform environment; calculated in accordance with Normative Appendix A of ANSI/ASHRAE Standard 55-2020, *Thermal Environmental Conditions for Human Occupancy*. (See the complete standard for detailed guidance.)

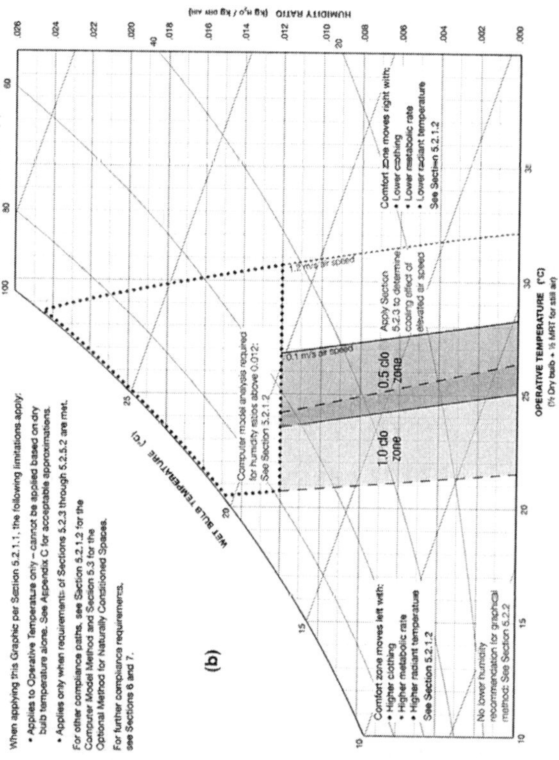

Acceptable ranges of operative temperature and humidity for people in 0.5 to 1.0 clo clothing, activity between 1.0 met and 1.3 met. The operative temperature ranges are based on a 80% satisfaction criterion; 10% general dissatisfaction and 10% partial (local) dissatisfaction.

Figure 21.1 Graphic Comfort Zone Method [Std 55-2013, Fig 5.3.1]

Table 21.1 Acceptable Thermal Environment for General Comfort [Std 55-2020, Tbl H-1]

PPD	PMV Range
< 10	−0.5 < PMV < + 0.5

Table 21.2 Limits on Temperature Drifts and Ramps [Std 55-2020, Tbl 5-12]

Time Period	0.25 h	0.5 h	1 h	2 h	4 h
Maximum Operative Temperature Change Allowed	1.1°C	1.7°C	2.2°C	2.8°C	3.3°C

Table 21.3 Typical Insulation and Permeation Efficiency Values for Western Clothing Ensembles [2021F, Ch 9, Tbl 7]

Ensemble Description[a]	I_{cl}, clo	I_t,[b] clo	f_{cl}	i_{cl}	i_m[b]
Walking shorts, short-sleeved shirt	0.36	1.02	1.10	0.34	0.42
Trousers, short-sleeved shirt	0.57	1.20	1.15	0.36	0.43
Trousers, long-sleeved shirt	0.61	1.21	1.20	0.41	0.45
Same as above, plus suit jacket	0.96	1.54	1.23		
Same as above, plus vest and T-shirt	1.14	1.69	1.32	0.32	0.37
Trousers, long-sleeved shirt, long-sleeved sweater, T-shirt	1.01	1.56	1.28		
Same as above, plus suit jacket and long underwear bottoms	1.30	1.83	1.33		
Sweat pants, sweat shirt	0.74	1.35	1.19	0.41	0.45
Long-sleeved pajama top, long pajama trousers, short 3/4 sleeved robe, slippers (no socks)	0.96	1.50	1.32	0.37	0.41
Knee-length skirt, short-sleeved shirt, panty hose, sandals	0.54	1.10	1.26		
Knee-length skirt, long-sleeved shirt, full slip, panty hose	0.67	1.22	1.29		
Knee-length skirt, long-sleeved shirt, half slip, panty hose, long-sleeved sweater	1.10	1.59	1.46		
Same as above, replace sweater with suit jacket	1.04	1.60	1.30	0.35	0.40
Ankle-length skirt, long-sleeved shirt, suit jacket, panty hose	1.10	1.59	1.46		
Long-sleeved coveralls, T-shirt	0.72	1.30	1.23		
Overalls, long-sleeved shirt, T-shirt	0.89	1.46	1.27	0.35	0.40
Insulated coveralls, long-sleeved thermal underwear, long underwear bottoms	1.37	1.94	1.26	0.35	0.39

Sources: McCullough and Jones (1984) and McCullough et al. (1989).

[a] All ensembles include shoes and briefs or panties. All ensembles except those with panty hose include socks unless otherwise noted.

[b] For $t_r = t_a$ and air velocity less than 0.2 m/s (I_a = 0.72 clo and i_m = 0.48 when nude).

Table 21.4 Insulation and Permeability Values for a Selection of Non-Western Clothing Ensembles [2021F, Ch 9, Tbl 8]

Ensemble Description[a]	Country	I_{cl} clo	I_t, clo	f_{cl}	i_m
Shalwar (pants), kameez (shirt), scarf, sandals (f)	Pakistan	0.69	1.1	1.41	0.32
Shalwar (pants), kameez (shirt), socks, athletic shoes (m)	Pakistan	0.86	1.3	1.36	0.35
Dishdasha (thowb or caftan), short-sleeved t-shirt, long serwal (pants), tagiya (hat), iqal (cord), ghutra (headdress), socks, athletic shoes (m)	Kuwait	1.36	1.7	1.66	0.30
Full slip, double-layer abaya (dress), anta (head cover), hijab (headscarf), sandals (f)	Kuwait	1.27	1.7	1.65	0.33
Underskirt, blouse, sari, sandals (f)	India	0.74	1.2	1.46	0.33
Churidhar pants, churidhar dress, shawl, sandals (f)	India	0.58	1.1	1.28	0.36
Short shirt with long sleeves, long pants, boubou (wide-sleeved robe), kufi (hat), sandals (m)	Nigeria/ Ghana	1.40	1.7	1.96	0.42
Short shirt with long sleeves, long pants, sandals (f)	Nigeria/ Ghana	0.78	1.3	1.35	0.40
Long-sleeved shirt, skirt, headscarf, socks, athletic shoes (f)	Indonesia	0.97	1.4	1.43	0.31
Camisole, short-sleeved qipao (dress), sandals (f)	China	0.42	0.9	1.31	0.40

(f) = clothing traditionally worn by women
(m) = clothing traditionally worn by men
Source: Havenith et al. (2015). Values are the means of manikin-based measurements conducted in three laboratories. All ensembles include bra and panties (female) and briefs (male). For all women's ensembles, I_a = 0.64 clo; for all men's ensembles, I_a = 0.63 clo.

Table 21.5 Expected Percent Dissatisfied Due to Sources of Local Discomfort [Std 55-2020, Tbl I-1]

Draft	Vertical Air Temperature Difference	Warm or Cool Floors	Radiant Asymmetry
<20%	<5%	<10%	<5%

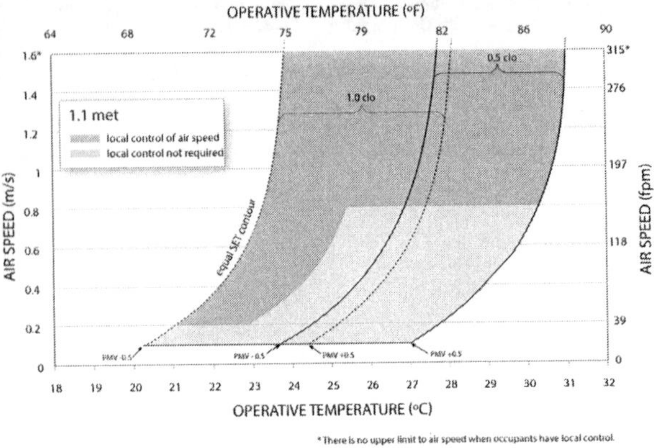

Figure 21.2 Acceptable Ranges of Operative Temperature t_o and Average Air Speed V_a for the 1.0 and 0.5 clo Comfort Zones at Humidity Ratio 0.010 [Std 55-2020, Fig 5-4]

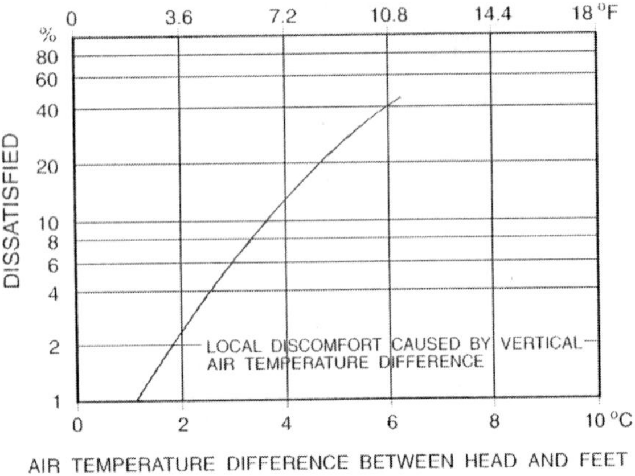

Figure 21.3 Local Thermal Discomfort caused by Vertical Temperature Differences [Std 55-2020, Fig I-3]

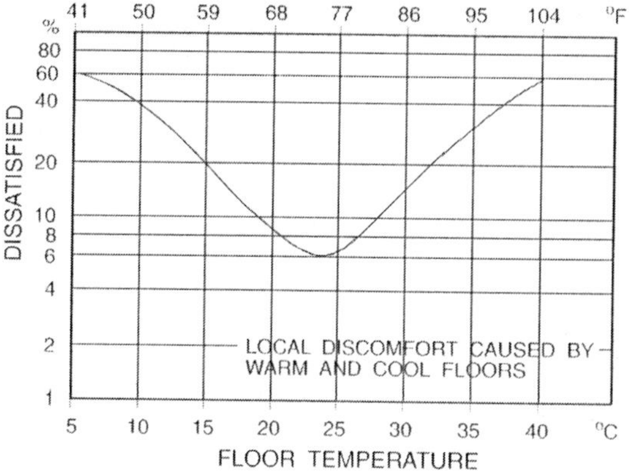

Figure 21.4 Local Discomfort caused by Warm and Cool Floors
[Std 55-2020, Fig I-4]

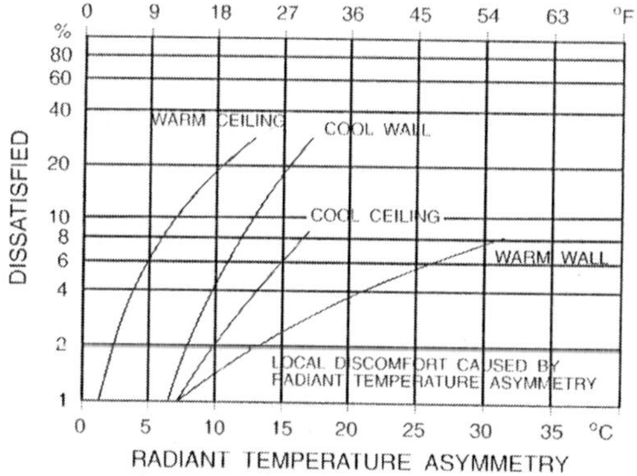

Figure 21.5 Local Thermal Discomfort caused by Radiant Asymmetry
[Std 55-2020, Fig I-1]

Table 21.6 Increases in Acceptable Operative Temperature Limits (Δt_0) in Occupant-Controlled, Naturally Conditioned Spaces (Figure 21.6) Resulting from Increasing Air Speed above 0.3 m/s [Std 55-2020, Tbl 5-13]

Average Air Speed (V_a) 0.6 m/s	Average Air Speed (V_a) 0.9 m/s	Average Air Speed (V_a) 1.2 m/s
1.2°C	1.8°C	2.2°C

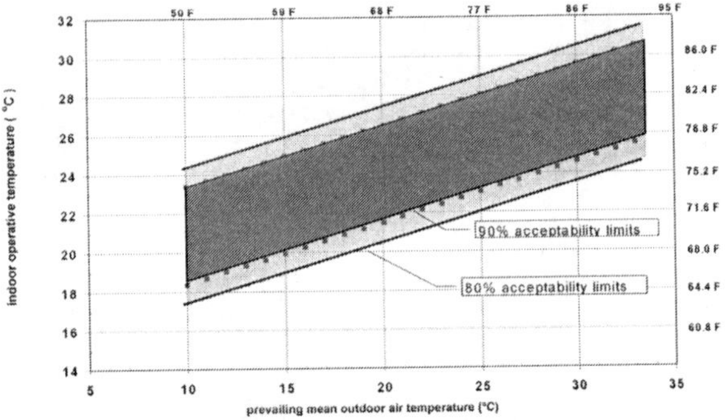

Figure 21.6 Acceptable Operative Temperature t_o Ranges for Naturally Conditioned Spaces [Std 55-2020, Fig 5-8]

Applicability of Figure 21.6. Figure 21.6 defines acceptable thermal environments only for occupant-controlled naturally conditioned spaces that meet all of the following criteria:

- There is no mechanical cooling system (e.g., refrigerated air conditioning, radiant cooling, or desiccant cooling) or heating system in operation.
- Representative occupants have metabolic rates ranging from 1.0 to 1.5 met.
- Representative occupants are free to adapt their clothing to the indoor and/or outdoor thermal conditions within a range at least as wide as 0.5 to 1.0 clo.
- The prevailing mean outdoor temperature is greater than 10°C and less than 33.5°C.

22. COMMISSIONING

Commissioning is a quality-oriented process for achieving, verifying, and documenting that the performance of facilities, systems, and assemblies meets defined objectives and criteria. The defined objectives and criteria are often referred to as the Owner's Project Requirements (OPR).

The commissioning (Cx) process uses the OPR as the reference to determine acceptance of the design and construction. Commissioning includes verifying and documenting that the project operational and maintenance documentation and training of operation and maintenance (O&M) personnel occur. The result should be fully functional systems that can be properly operated and maintained throughout the life of the building.

Recommissioning applies Cx to a project that has been previously delivered using the Cx process. This may be a scheduled recommissioning developed as part of ongoing Cx, or it may be triggered by use change, operational problems, or other needs. Existing building Cx (EBCx, often called *retrocommissioning*) applies Cx to an existing facility that may or may not have been previously commissioned.

Equipment, components, systems, and assemblies have become more complex. More specialization has occurred in the disciplines and trades, with increased interactions between all elements. This increased specialization and interaction requires increased integration between disciplines and specialized systems by the delivery team. Owners often use low-bid policies, and scopes of work for design professionals are often narrowed. The result has been buildings that do not meet owner expectations and often do not work as intended because of programming, design, and construction deficiencies. Commissioning is a value-added service that helps overcome these infrastructure inadequacies and fundamentally improve the performance of building systems and living conditions for occupants.

Key contributors to the commissioning process include the following:

- Owner/end user
- Architect
- Mechanical engineer
- Electrical engineer
- Commissioning provider (CxP)
- Mechanical contractor
- Electrical contractor
- Controls contractor
- Sheet metal contractor
- Testing, adjusting, and balancing (TAB) agency
- Owner's O&M staff
- General contractor/construction manager

New Building Commissioning Process

Commissioning begins during predesign, and formally continues through the first year of occupancy and operations. Although circumstances may require owners to begin Cx at the design or construction stage of a project, such a delayed implementation must capture the same information and verifications developed when Cx begins at project inception. Four phases of the new building and systems Cx process are defined below.

Predesign-Phase Commissioning

The primary activities and objectives of Cx during predesign are to:

- Develop the OPR
- Identify the scope and budget for the Cx process
- Develop the initial Cx plan
- Review and accept predesign-phase Cx-process activities
- Review and use lessons learned from previous projects

Design-Phase Commissioning

Design-phase Cx objectives include the following:

- Update the OPR
- Verify the Basis of Design (BOD) document against the OPR

- Update the design-phase Cx plan developed during predesign
- Develop and incorporate Cx requirements into project specifications
- Develop Cx plan for construction and occupancy/operations phases, including draft construction checklists
- Verify plans and specifications against BOD and OPR
- Begin codeveloping systems manual with relevant discipline design engineer
- Define training requirements for O&M personnel
- Perform Cx-focused design reviews
- Accept design-phase Cx

Construction-Phase Commissioning

Commissioning activities should take place throughout the construction phase and include verification and documentation that:

- All acceptance testing requirements are documented
- All systems and assemblies are provided and installed as specified
- All systems and assemblies are started and function properly
- All record documents are updated
- The systems manual is updated and provided to facility staff
- Facility staff and occupants receive specified training and orientation
- Acceptance testing occurs

Occupancy and Operations-Phase Commissioning

Occupancy and operations phase Cx typically begins with resolving the findings from performance monitoring over the first month or two into occupancy and ends with the completion of the first year of occupancy. The Cx process during this phase should ensure:

- Initial maintenance and operator training is complete
- Systems and assemblies received functional opposite-season verification
- Outstanding performance issues are identified and resolved before warranty expiration
- Commissioning process evaluation is conducted and satisfactorily resolved

Existing Building Commissioning Process (EBCx)

This is a quality-focused process for attaining the current facility requirements (CFR) of an existing facility and/or its systems and assemblies. The process focuses on planning, investigating, implementing, verifying, and documenting that the facility and/or its systems and assemblies are operated and maintained to meet the CFR, with a program to maintain the enhancements for the remaining life of the facility. System performance normally degrades with use and time, at a rate depending on the quality of maintenance and operations and the number of hours of operation. Quality of maintenance also affects equipment life expectancy. An EBCx effort should include updating or developing an owner's CFR, documenting existing systems, surveying the facility to identify operational inefficiencies, quantifying and prioritizing the inefficiencies found, determining how best to optimize equipment or operation, implementing changes, training operating personnel, documenting operations, and then reverifying with ongoing measurements that the EBCx process activities produced and continue to produce the desired effect.

EBCx is used by owners and facility decision makers to optimize the operations of their existing facilities to meet their CFR. The process has five basic steps, with an additional step for multifacility projects:

- Multifacility planning (if multiple buildings are involved)
- Assessment
- Investigation
- Implementation
- Handoff
- Ongoing Cx

ASHRAE Commissioning Documents

ASHRAE has developed numerous Cx process documents to address new and existing buildings and systems, to provide support for HVAC systems Cx efforts, and to assist in the development of Cx documentation. These include:

- ANSI/ASHRAE/IES Standard 202, *Commissioning Process for Buildings and Systems*
- ASHRAE Guideline 0, *The Commissioning Process*
- ASHRAE Guideline 0.2, *Commissioning Process for Existing Systems and Assemblies*
- ASHRAE Guideline 1.1, *HVAC&R Technical Requirements for the Commissioning Process*
- ASHRAE Guideline 1.2, *Technical Requirements for the Commissioning Process for Existing HVAC&R Systems and Assemblies*
- ASHRAE Guideline 1.3, *Building Operations and Maintenance Training for the HVAC&R Commissioning Process*
- ASHRAE Guideline 1.4, *Preparing Systems Manuals for Facilities*
- ASHRAE Guideline 1.5, *The Commissioning Process for Smoke Control Systems*
- *Commissioning Stakeholders' Guide*

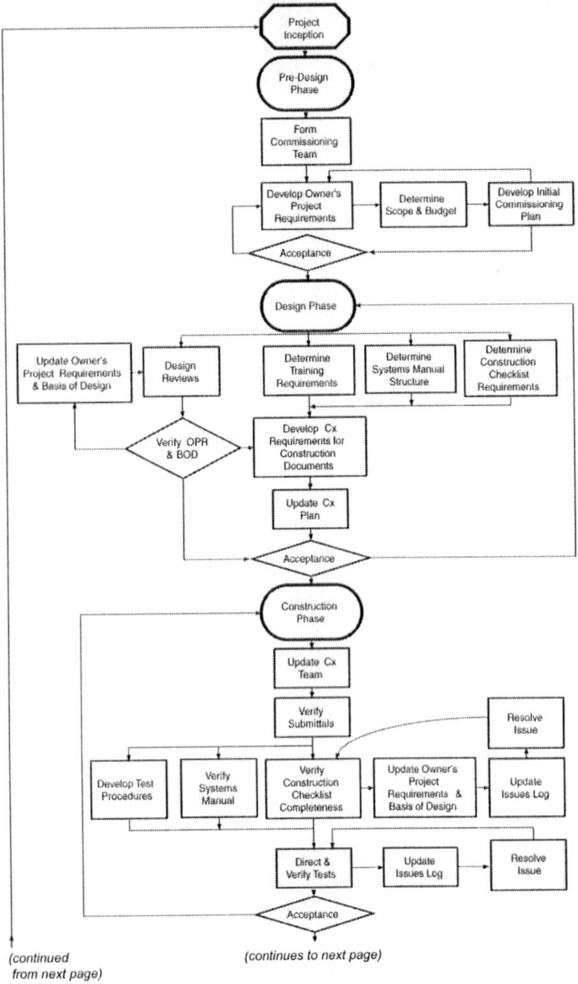

Figure 22.1 Commissioning Process Flowchart for New Buildings and Systems [Gdl 0-2019, Fig B-1]

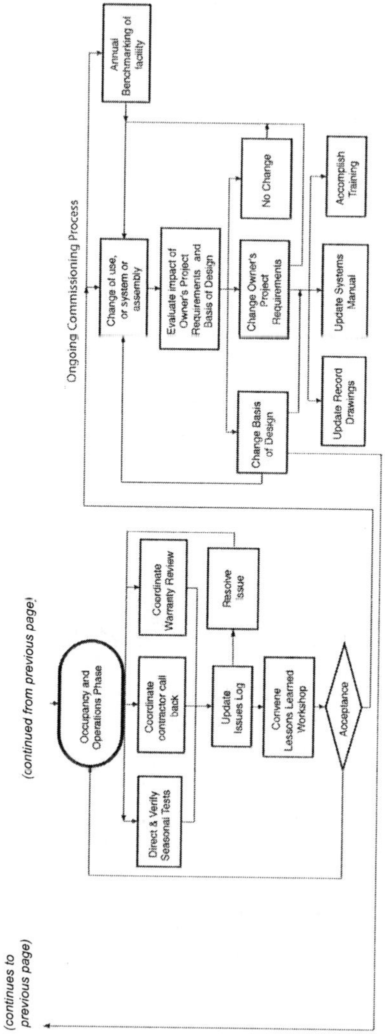

Figure 22.1 Commissioning Process Flow Chart for New Buildings and Systems [Gdl 0-2019, Fig B-1] ***(Continued)***

Table 22.1 Commissioning Process Documentation Matrix for New Building and Systems [Std 202-2018, Tbl A-1]

Phase	Document	Input By	Provided By	Reviewed/ Approved By	Used By	Notes
Project Initiation	OPR	Owner, CxP, O&M, Users, Design Team	Owner, CxP, or Design Team	Owner	CxP, Design Team	Design Team may not be hired yet.
Design	Initial Cx Plan	Owner, Design Team, CxP	Owner or CxP	Owner	CxP, Owner, Design Team, Construction Team	Design Team may not be hired yet.
	Systems Manual Outline	Owner, O&M, CxP	Owner or CxP	Owner	Design Team, Construction Team	Should follow ASHRAE Guideline 1.4.
	Training Requirements Outline	Owner, O&M, Users, CxP, Design Team	Owner or CxP	Owner, CxP	Design Team, Construction Team	Should follow ASHRAE Guideline 1.3 and as defined.
	Issues and Resolution Log Format	Owner, CxP	Owner or CxP	Owner	CxP, Design Team	May be only format at this time.
	Cx Report Format	Owner, CxP	CxP	Owner	Design Team, Owner	Follow requirement in this standard.
	OPR Update	Owner, CxP, O&M, Users, Design Team	CxP or Designer	Owner	CxP, Design Team	
	BoD	Design Team	Design Team	Owner, CxP	Design Team, CxP	
	Construction Specifications for Cx	Design Team, CxP, Owner, O&M	Design Team or CxP	Owner	Contractors, CxP, Design Team	May also be provided by project manager/Owner's representative.
	Systems Manual Outline—Expanded	Design Team, CxP, O&M, Contractor	Design Team or CxP	Owner, CxP	Design Team, Contractor	Contractor may not be hired yet.
	Training Requirements	Owner, CxP O&M, Users,	Owner, CxP, or O&M	Owner	Design Team	Contractor input
	Design Review Comments	CxP	CxP	Design Team, Owner	Design Team	
	Issues and Resolution Log	CxP	CxP	Design Team, Owner	CxP, Design Team	
	Design Cx Report	CxP	CxP	Owner	Owner	

Table 22.1 **Commissioning Process Documentation Matrix for New Building and Systems** [Std 202-2018, Tbl A-1] ***(Continued)***

Phase	Document	Input By	Provided By	Reviewed/ Approved By	Used By	Notes
Construction	OPR Update	Owner, O&M, Users, Design Team	CxP or Designer	Owner	CxP, Design Team, Contractors, Owner	
	BoD Update	Design Team	Design Team	CxP, Owner	Design Team, CxP, Owner	
	Cx Plan Update	Project Team	CxP	Owner	Project Team	
	Submittal Review Comments	CxP	Design Team	Design Team	Contractor	Copy to Owner.
	System Coordination Plans	Contractor, Design Team, CxP	Contractor	CxP, Design Team	Contractor, CxP	
	Evaluation Checklists	CxP, Contractor, Design Team	CxP or Contractor	CxP, Design Team	Contractor	
	Cx Progress Reports	CxP, Contractor	CxP	CxP, Owner	Contractor	
	Test Procedures	Project Team	CxP	CxP, Design Team	Contractor	
	Specified Test Data Submittals	Owner, Design Team, CxP	Contractors, CxP, or Owner	Design Team, CxP	Project Team	
	Cx Meeting Agendas and Minutes	CxP	CxP	All	All	
	Training Plans	Project Team	Contractor or CxP	Owner, CxP	O&M, Users, Contractor	
	Systems Manual	Project Team	Contractor or CxP	Owner, CxP	O&M, Users	See ASHRAE Guideline 1.4.
	Issues and Resolution Log	CxP	CxP	N/A	Project Team	
	Progress Report	CxP	CxP	Owner, Design Team	Project Team	
	Preliminary Construction Cx Report	CxP	CxP	Owner	Owner	Prior to occupancy
	Final Construction Cx Report	CxP	CxP	Owner	Owner	

Table 22.1 Commissioning Process Documentation Matrix for New Building and Systems [Std 202-2018, Tbl A-1] ***(Continued)***

Phase	Document	Input By	Provided By	Reviewed/ Approved By	Used By	Notes
Occupancy and Operations	OPR Update	Owner, O&M, Users, Design Team	CxP or Designer	Owner	CxP, Owner	See ASHRAE Guideline 0.2 for postoccupancy Cx.
	BoD Update	Design Team	Design Team	CxP, Owner	Owner, O&M, CxP	
	Maintenance Program	Owner, O&M, Contractor, CxP	Owner or CxP	Owner, CxP	O&M, Users	
	Test Procedures	CxP	CxP	Design Team, CxP	Contractor, O&M	See Systems Manual.
	Updated Issues and Resolution Log	CxP	CxP	N/A	Project Team	Must be included in final Cx Report.
	Prefinal	Project Team	CxP	N/A	Owner	Typically provided at substantial completion.
	Final Cx Report	Project Team	CxP	N/A	Owner	
	Final Issues and Resolution Log	Project Team	CxP	Owner	Owner	Included in final Cx Report.
	Recommissioning Plan	O&M, Users, CxP	CxP or Owner	Owner	Owner	

Note: See Section 3 of Standard 202-2018 for entities and deliverables contained in this table.

Table 22.2 Example Commissioning Process Design Checklist for Mechanical Engineer's First Submittal [Gdl 1.1-2021, Annex M, Example Checklist 2]

Instructions:	Step 1: Circle Yes or No and fill in with requested information. Step 2: Explain all "No" responses at the bottom of the checklist.		
Item	**Task Description**	**Location of Information in the Programming Document**	**Response**
1	**Owner's Project Requirements**		
A	**Key Owner's Project Requirements**		**Complete**
1	Commissioning Plan Updated, provide date and enclose with this submittal		Yes No
2	Basis of Design for controls completed		Yes No
3	Basis of Design for accessibility completed		Yes No
4	Sustainability and LEED issues coordination addressed		Yes No
5	Do the general HVAC&R requirements the current OPR requirements? Has justification been document and approved by owner's Project Manager?		Yes No
6	Control format, BACnet requirements complete and documented		Yes No
B	**Owner's Objectives**		
1	Preliminary mechanical room layout complete		Yes No
2	Energy analysis meeting goal of 30% less than ASHRAE 90?		Yes No
3	Single line diagrams developed for controls and systems		Yes No
4	Report on safety factors and tolerance for facility system operations		Yes No
5	Have chillers been sized and pre-order to meet occupancy goal		Yes No
6	Environmental and sustainability initial design complete		Yes No
C	**General Owner's Needs**		
1	Is current HVAC and control systems budget enclosed and within initial budget		Yes No
2	Mechanical rooms space and location coordinated with shops		Yes No
3	Has Electrical, Plumbing, Lighting and Communications coordination been completed?		Yes No
4	Does the initial design meet all Benchmark Established for HVAC&R? Document?		Yes No
5	Constructability and maintainability analysis completed		Yes No

"No" Responses

Item	**Date**	**Reason for "No" Response**

Table 22.3 Example Commissioning Process Construction Checklist for a Hot Water Unit Heater [Gdl 1.1-2021, Annex M, Example Checklist 22]

Instructions: Step 1: Circle Yes or No and fill in with requested information.
Step 2: Explain all "No" responses at the bottom of the checklist.

Item		Task Description	Response	
1		**Delivery Book**		
	A	**Model Verification**	**Submitted**	**Delivered**
	1	Manufacturer		
	2	Model		
	3	Serial number	N/A	
	4	Fan Motor Power (kW)		
	5	Fan Motor Voltage / Phase / Frequency (V / - / Hz)	/ /	/ /
	6	Total Heating Capacity (kW)		
	7	Heating Fluid Flow / Pressure Drop (L/s/Pa)	/	/
	B	**Physical Checks**		
	1	Unit is free from physical damage	Yes	No
	2	All components present	Yes	No
	3	The water openings are sealed with plastic plugs	Yes	No
	4	Manufacturer's data readable/accurate	Yes	No
	5	Unit identification attached and visible	Yes	No
2		**Construction Checklist**		
	A	**Installation of Unit Heater**		
	1	Unit supported using adequately sized mounting anchors	Yes	No
	2	Adequate clearance around unit for service	Yes	No
	3	All components accessible for maintenance	Yes	No
	4	Unit can be removed from building	Yes	No
	5	Unit identification attached and visible	Yes	No
	B	**Piping**		
	1	All piping components have been installed (in the correct order) as required by detail drawing	Yes	No
	2	Piping arranged for ease of unit removal	Yes	No
	3	Piping supported as required by specifications	Yes	No
	4	Piping is clean	Yes	No
	5	Piping insulation complete and installed as per specifications	Yes	No
	6	All valves and test ports are easily accessible	Yes	No
	C	**Electrical**		
	1	Local disconnect installed in an accessible location	Yes	No
	2	Motor rotation in the proper direction	Yes	No
	3	All electrical connections are tight	Yes	No
	4	All electrical components are grounded	Yes	No
	D	**Controls**		
	1	Room thermostat installed and calibration verified	Yes	No
	2	Hot water actuator calibration verified	Yes	No
	3	Heating sequence of control verified	Yes	No
	4	Valve tags are attached	Yes	No

Table 22.3 Example Commissioning Process Construction Checklist for a Hot Water Unit Heater [Gdl 1.1-2021, Annex M, Example Checklist 22] ***(Continued)***

Item	Task Description	Response	
E	TAB		
1	Motor rotation in the proper direction	Yes	No
2	Motor overloads verified	Yes	No
3	Motor voltage and amps verified - each phase	Yes	No
4	Entering and leaving air temperatures (°C)	/	/
5	Flow and air/water pressure drops verified	Yes	No

"No" Responses

Item	Date	Reason for "No" Response

Commissioning

Table 22.4 Example Commissioning Process Construction Checklist for a Split System A/C Unit [Gdl 1.1-2021, Annex M, Example Checklist 20]

Instructions:	Step 1: Circle Yes or No and fill in with requested information. Step 2: Explain all "No" responses at the bottom of the checklist.		

Item	Task Description	Response	
1	**Delivery Book**		
A	**Model Verification**	**Submitted**	**Delivered**
1	Manufacturer		
2	Model		
3	Serial Number	N/A	
4	Airflow (L/s)		
5	Fan Motor Power (kW)		
6	Fan Motor Voltage / Phase / Frequency (V / - / Hz)	/ /	/ /
7	Total Cooling Capacity (kW)		
B	**Physical Checks**		
1	Unit is free from physical damage	Yes	No
2	All components present	Yes	No
3	The refrigerant line openings are sealed	Yes	No
4	Installation and startup manual provided	Yes	No
5	Unit tags affixed	Yes	No
2	**Construction Checklist**		
A	**Installation of Split System Coil**		
1	Unit supported using adequately sized mounting anchors	Yes	No
2	Adequate clearance around unit for service	Yes	No
3	All components are accessible for maintenance	Yes	No
4	Unit can be removed from building	Yes	No
5	Condensate drain piping un-trapped and runs to open sight drain	Yes	No
6	Unit labeled and is easy to see	Yes	No
B	**Piping**		
1	All piping components have been installed (in the correct order) as required by detail drawing	Yes	No
2	Piping arranged for ease of unit removal	Yes	No
3	Piping supported as required by specifications	Yes	No
4	Refrigerant lines connected to indoor and outdoor units	Yes	No
5	Piping is clean and free from leaks	Yes	No
6	Piping insulation is complete and installed as per specifications	Yes	No
7	All valves and test ports are easily accessible	Yes	No
8	Valve tags attached	Yes	No
9	Unit filled with correct refrigerant	Yes	No
C	**Electrical**		
1	Local disconnect installed in an accessible location	Yes	No
2	Fan motor rotation in the proper direction	Yes	No
3	All electrical connections are tight	Yes	No
4	All electrical components are grounded	Yes	No

Table 22.4 Example Commissioning Process Construction Checklist for a Split System A/C Unit [Gdl 1.1-2021, Annex M, Example Checklist 20] ***(Continued)***

Item	Task Description	Response	
D	**Controls - Installation**		
1	Room thermostat installed and calibration verified	Yes	No
2	Control wiring provided to outdoor (compressor) unit	Yes	No
3	Communication with outdoor unit verified	Yes	No
E	**Controls - Startup**		
1	Cooling sequence of control verified	Yes	No
2	System starts and runs with no unusual noise or vibration	Yes	No
3	Manufacturer's startup checklist completed and attached	Yes	No
F	**TAB**		
1	Filters installed and are clean	Yes	No
2	Entering and leaving air temperatures (°C)	/	/
3	Airflow (L/s)	/	/

"No" Responses

Item	Date	Reason for "No" Response

Commissioning

Table 22.5 Example Commissioning Process Construction Checklist for an HVAC Pump [Gdl 1.1-2021, Annex M, Example Checklist 19]

Instructions: Step 1: Circle Yes or No and fill in with requested information.
Step 2: Explain all "No" responses at the bottom of the checklist.

Item		Task Description	Response	
1		**Delivery Book**		
	A	**Model Verification**	**Submitted**	**Delivered**
	1	Manufacturer		
	2	Model		
	3	Serial Number	N/A	
	4	Pump Type		
	5	Impeller diameter (mm)		
	6	Inlet / Outlet Sizes (mm)	/	/
	7	Capacity / Head (L/s/Pa)	/	/
	8	Motor Speed / Power (rpm/kW)	/	/
	9	Motor Voltage / Phase / Frequency (V / - / Hz)	/ /	/ /
	B	**Physical Checks**		
	1	Unit is free from physical damage	Yes	No
	2	All components present	Yes	No
	3	The water openings are sealed with plastic plugs	Yes	No
	4	Unit tags affixed	Yes	No
	5	Installation and startup manual provided	Yes	No
	6	Manufacturer's ratings readable/accurate	Yes	No
2		**Construction Checklist**		
	A	**Installation of Pump**		
	1	Unit secured as required by manufacturer and specifications	Yes	No
	2	Adequate clearance around unit for service	Yes	No
	3	All components accessible for maintenance	Yes	No
	4	Unit can be removed from building	Yes	No
	5	Unit labeled and is easy to see	Yes	No
	B	**Piping**		
	1	All piping components have been installed (in the correct order) as required by detail drawing	Yes	No
	2	Piping arranged for ease of unit removal	Yes	No
	3	Shut-off valves and unions installed on inlet and outlet of pump	Yes	No
	4	Pressure gauges installed on inlet and outlet of pump	Yes	No
	5	Piping supported as required by specifications	Yes	No
	6	Piping is clean	Yes	No
	7	Piping insulation complete and installed as per specifications	Yes	No
	8	All valves and test ports are easily accessible	Yes	No
	9	Valve tags attached	Yes	No
	C	**Electrical**		
	1	Safety disconnect installed in an accessible location	Yes	No
	2	Motor rotation in the proper direction	Yes	No
	3	All electrical connections are tight	Yes	No
	4	All electrical components are grounded	Yes	No

Table 22.5 Example Commissioning Process Construction Checklist for an HVAC Pump [Gdl 1.1-2021, Annex M, Example Checklist 19] ***(Continued)***

Item	Task Description	Response	
D	**Mechanical - Startup**		
1	Unit checked, aligned, and certified prior to startup and report submitted	Yes	No
2	Unit and motor lubricated before startup	Yes	No
3	Pump shaft rotates easily with power turned off	Yes	No
4	System starts and runs without any unusual noise or vibration	Yes	No
5	Manufacturer's startup checklist completed and attached	Yes	No
E	**TAB**		
1	Flow Rate, L/s		
2	Inlet pressure (Pa) / Outlet pressure (Pa)	/	/
3	Motor rotation in the proper direction	Yes	No
4	Motor overload verified	Yes	No
5	Motor voltage and amps verified - each phase	Yes	No
6	Start-up strainer removed (after 24 hours)	Yes	No

"No" Responses

Item	Date	Reason for "No" Response

Table 22.6 Example Commissioning Process Construction Checklist for a Fire Damper [Gdl 1.1-2021, Annex M, Example Checklist 14]

Instructions: Step 1: Circle Yes or No and fill in with requested information.
Step 2: Explain all "No" responses at the bottom of the checklist.

Item	Task Description	Response	
1	**Delivery Book**		
A	**Model Verification**	**Submitted**	**Delivered**
1	Manufacturer		
2	Model		
3	Style		
4	Width (mm)		
5	Height (mm)		
6	Orientation		
B	**Physical Checks**		
1	Unit is free from physical damage	Yes	No
2	All components/accessories present	Yes	No
3	Installation manual provided	Yes	No
2	**Construction Checklist**		
A	**Installation of Fire Damper**		
1	Unit secured as required by manufacturer and specifications	Yes	No
2	Adequate clearance around unit for maintenance	Yes	No
3	Unit mounted in correct orientation	Yes	No

"No" Responses

Item	Date	Reason for "No" Response

Table 22.7 Example Commissioning Process Test Protocol for Verification of Thermal Comfort OPR [Adapted from Gdl 1.1-2021, Tbl U-1]

Test	Step	Expected	Actual	Pass/Fail
Morning Warm-up – Cloudy Day				
	Set Perimeter Spaces to 17.8°C (5.6°C cooler than set point)			
	Set Interior Spaces to 20.6°C (2.8°C cooler than set point)			
	Record Results			
	Record Time to Stability	20 min.		
	Record Issues Identified			
Morning Warm-up – Clear Day				
	Set South, West, and North Perimeter Spaces to 17.8°C (5.6°C cooler than set point)			
	Set East Perimeter to 26.1°C (2.8°C warmer than set point)			
	Set Interior Spaces to 20.6°C (2.8°C cooler than set point)			
	Record Results			
	Record Time to Stability	20 min.		
	Record Issues Identified			
Morning Cool-down – Cloudy Day				
	Set Perimeter Spaces to 28.9°C (5.6°C warmer than set point)			
	Set Interior Spaces to 26.1°C (2.8°C warmer than set point)			
	Record Results			
	Record Time to Stability	20 min.		
	Record Issues Identified			
Morning Cool-down – Clear Day				
	Set South and North Perimeter and Interior Spaces to 26.1°C (2.8°C warmer than set point)			
	Set East Perimeter to 28.9°C (5.6°C warmer than set point)			
	Set West Perimeter to 24.4°C (1.1°C warmer than set point)			
	Record Results			
	Record Time to Stability	20 min.		
	Record Issues Identified			
Peak Cooling				
	Set All Spaces to 28.9°C (5.6°C warmer than set point)			
	Record Results			
	Record Time to Stability	20 min.		
	Record Issues Identified			
Peak Heating				
	Set All Spaces to 17.8°C (5.6°C cooler than set point)			
	Record Results			
	Record Time to Stability	20 min.		
	Record Issues Identified			
Afternoon Cooling				
	Set West Perimeter to 26.1°C (2.8°C warmer than set point)			
	Sell All Other Spaces to 23.3°C (at set point)			
	Record Results			
	Record Time to Stability	20 min.		
	Record Issues Identified			
Conference Room Load				
	Set Conference Room to 28.9°C (5.6°C warmer than set point)			
	Sell All Other Spaces to 25.6°C (1.7°C warmer than set point)			
	Record Results			
	Record Time to Stability	20 min.		
	Record Issues Identified			

Table 22.7 Example Commissioning Process Test Protocol for Verification of Thermal Comfort OPR [Adapted from Gdl 1.1-2021, Tbl U-1] ***(Continued)***

Other Observations: The following observations were noted during the implementation of this comfort OPR commissioning process test procedure:

#	Observation
1	
2	
3	
4	
5	
6	

Issues: The following issues were identified during the implementation of this comfort OPR commissioning process test procedure and require follow-up (the detailed issue should be entered into the issues database).

#	Issue Description	Person Responsible
1		
2		
3		
4		
5		

Signatures: The following shall be completed after implementation of the test procedure.

Signature	Signature
Commissioning Authority	BAMS Contractor
Date/Time	Date/Time
Signature	Signature
Mechanical Contractor	Electrical Contractor
Date/Time	Date/Time

Commissioning

Table 23.1 General Design Criteria[a, b] [Adapted from 2007A, Ch 3, Tbl 1]

General Category	Specific Category	Inside Design Conditions: Winter	Inside Design Conditions: Summer	Air Movement	Circulation, ach	Noise[c]	Filtering Efficiencies	Load Profile	Comments
Dining and Entertainment Centers	Cafeterias and Luncheonettes	21 to 23°C 20 to 30% rh	25°C[d] 50% rh	0.25 m/s at 1.8 m above floor	12 to 15	NC 40 to 50[e]	MERV 6–8	Peak at 1 to 2 PM 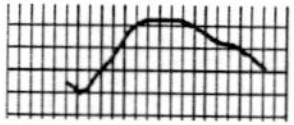	Prevent draft discomfort for patrons waiting in serving lines
	Restaurants	21 to 23°C 20 to 30% rh	23 to 26°C 55 to 60% rh	0.13 to 0.15 m/s	8 to 12	NC 35 to 40	MERV 6–8	Peak at 1 to 2 PM 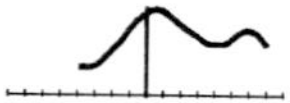	
	Bars	21 to 23°C 20 to 30% rh	23 to 26°C 50 to 60% rh	0.15 m/s at 1.8 m above floor	15 to 20	NC 35 to 50	Use charcoal for odor control with manual purge control for 100% outside air to exhaust MERV 6–8 prefilters	Peak at 5 to 7 PM 	
	Nightclubs and Casinos	21 to 23°C 20 to 30% rh	23 to 26°C 50 to 60% rh	below 0.13 m/s at 1.5 m above floor	20 to 30	NC 35 to 45[f]	Use charcoal for odor control with manual purge control for 100% outside air to exhaust MERV 6–8 prefilters	Nightclubs peak at 8 PM to 2 AM; Casinos peak at 4 PM to 2 AM; Equipment, 24 h/day	Provide good air movement but prevent cold draft discomfort for patrons
	Kitchens	21 to 23°C	29 to 31°C	0.15 to 0.25 m/s	12 to 15[g]	NC 40 to 50	N/A	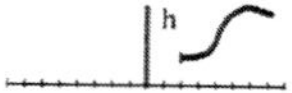	Negative air pressure required for odor control (also see 2019A, Ch 34)

Table 23.1 General Design Criteria[a, b] [Adapted from 2007A, Ch 3, Tbl 1] (*Continued*)

General Category	Specific Category	Inside Design Conditions: Winter	Inside Design Conditions: Summer	Air Movement	Circulation, ach	Noise[c]	Filtering Efficiencies	Load Profile	Comments
Office Buildings		21 to 23°C 20 to 30% rh	23 to 26°C 50 to 60% rh	0.13 to 0.23 m/s 4 to 10 L/(s·m^2)	4 to 10	NC 30 to 40	MERV 6–8	Peak at 4 PM 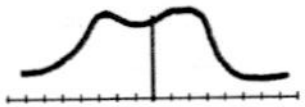	
Museums, Galleries, Libraries and Archives (also see 2019A, Ch 24)	Average	20 to 22°C 40 to 55% rh		below 0.13 m/s	8 to 12	NC 35 to 40	MERV 6–8	Peak at 3 PM 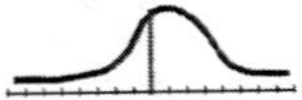	
	Archival	See 2019A, Ch 24		below 0.13 m/s	8 to 12	NC 35	MERV 8 prefilters plus charcoal filters MERV 11–13 final[i]	Peak at 3 PM 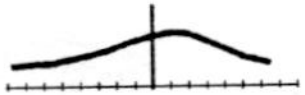	
Bowling Centers		21 to 23°C 20 to 30% rh	24 to 26°C 50 to 55% rh	0.25 m/s at 1.8 m above floor	10 to 15	NC 40 to 50	MERV 3–5	Peak at 6 to 8 PM 	
Communication Centers	Telephone Terminal Rooms	22 to 26°C 40 to 50% rh	22 to 26°C 40 to 50% rh	0.13 to 0.15 m/s	8 to 20	to NC 60	MERV 11–13	Varies with location and use	Constant temperature and humidity required
	Radio and Television Studios	21 to 23°C 40 to 50% rh	23 to 26°C 45 to 55% rh	0.13 to 0.15 m/s	15 to 40	NC 15 to 25	MERV 6–8	Varies widely because of changes in lighting and people	Constant temperature and humidity required

Table 23.1 General Design Criteria[a, b] [Adapted from 2007A, Ch 3, Tbl 1] (*Continued*)

General Category	Specific Category	Inside Design Conditions: Winter	Inside Design Conditions: Summer	Air Movement	Circulation, ach	Noise[c]	Filtering Efficiencies	Load Profile	Comments
Transportation Centers (also see 2019A, Ch 16)	Airport Terminals	23 to 26°C 30 to 40% rh	23 to 26°C 40 to 55% rh	below 0.13 m/s at 3.7 m above floor	8 to 12	NC 35 to 50	MERV 6–8 and charcoal filters	Peak at 10 AM to 9 PM 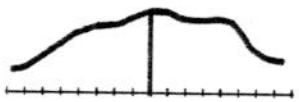	Positive air pressure required in terminal
	Ship Docks	21 to 23°C 20 to 30% rh	23 to 26°C 50 to 60% rh	0.13 to 0.15 m/s at 1.8 m above floor	8 to 12	NC 35 to 50	MERV 3–5	Peak at 10 AM to 5 PM 	Positive air pressure required in waiting area
	Bus Terminals	21 to 23°C 20 to 30% rh	23 to 26°C 50 to 50% rh	0.13 to 0.15 m/s at 1.8 m above floor	8 to 12	NC 35 to 50	MERV 6–8 with exfiltration	Peak at 10 AM to 5 PM 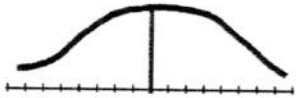	Positive air pressure required in terminal
	Garages[j]	4 to 13°C	27 to 38°C	0.15 to 0.38 m/s	4 to 6	NC 35 to 50	MERV 3–5	Peak at 10 AM to 5 PM 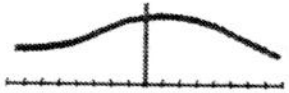	Negative air pressure required to remove fumes; positive air in pressure adjacent occupied spaces
Warehouses		Inside design temperatures for warehouses often depend on the materials stored.			1 to 4	to NC 75	MERV 3–5	Peak at 10 AM to 3 PM 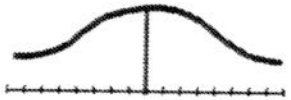	

Notes for Table 23.1:

[a]This table shows design criteria differences between various commercial and public buildings. It should not be used as sole source for design criteria. Each type of data contained here can be determined from ASHRAE Handbook and standards. MERV values are approximate equivalents for ASHRAE Standard 52.1 efficiencies listed in the original table.

[b]Consult governing codes to determine minimum allowable requirements. Outside air requirements may be reduced if high-efficiency adsorption equipment or other odor- or gas-removal equipment is used. See ANSI/ASHRAE Standard 62.1 for calculation procedures.

[c]Refer to Chapter 49 of the 2019 *ASHRAE Handbook—HVAC Applications*.

[d]Food in these areas is often eaten more quickly than in a restaurant; therefore, turnover of diners is much faster. Because diners seldom remain for long periods, they do not require the degree of comfort necessary in restaurants. Thus, it may be possible to lower design criteria standards and still provide reasonably comfortable conditions. Although space conditions of 27°C and 50% rh may be satisfactory for patrons when it is 35°C and 50% rh outside, inside conditions of 26°C and 40% rh are better.

[e]Cafeterias and luncheonettes usually have some or all food preparation equipment and trays in the same room with diners. These establishments are generally noisier than restaurants, so noise transmission from air-conditioning equipment is not as critical.

[f]In some nightclubs, air-conditioning system noise must be kept low so patrons can hear the entertainment.

[g]Usually determined by kitchen hood requirements.

[h]Peak kitchen heat load does not generally occur at peak dining load, although in luncheonettes and some cafeterias where cooking is done in dining areas, peaks may be simultaneous.

[i]Methods for removing chemical pollutants must also be considered.

[j]Also includes service stations.

Air-Conditioning Formulas

Sensible heat, SH = 1.2 Q Δt

Total heat, TH = 1.2 Q Δh

Latent heat, LH = 3.0 Q ΔW

where heat is in W:

Δt = temperature difference, K or °C

Δh = enthalpy difference, kJ/kg

ΔW = moisture concentration difference, g/kg dry air

Q = flow rate of air, L/s

Pump Power

$$P = Q \times h \times \rho / nC_w$$

where

P = kW

Q = L/s flow rate

h = kPa head

ρ = density ratio to water

n = efficiency, 50% to 85% usually

C_w = constant, 101

Fan Power

$$P = Q \times h \times \rho / nC_a$$

P = kW

Q = L/s flow rate

h = kPa head

ρ = density ratio to air

n = efficiency, 40% to 70% usually

C_a = constant, 40350

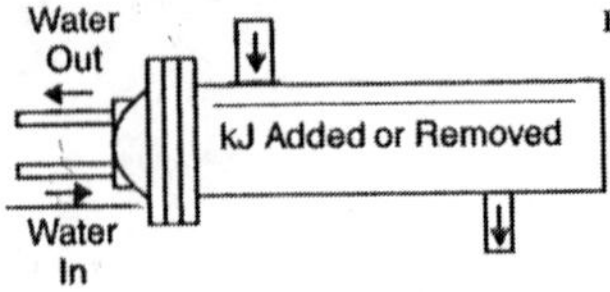

For heating or cooling water:

$$\text{L/s} = \frac{0.24\ \text{kW}}{(\text{water temp. rise or drop, °C})}$$

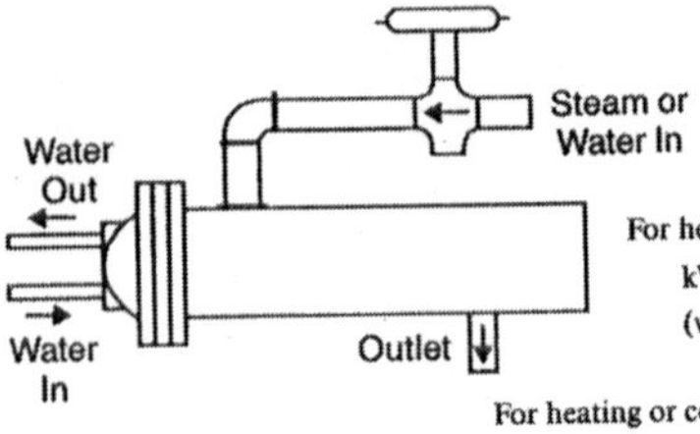

For heating water with steam:

$$\text{kW steam} = 4.187\ \text{L/s}\ (\text{water temp. rise, °C})$$

For heating or cooling water with water:

$$(\text{L/s})_1 = (\text{L/s})_2 \left(\frac{\text{water}_2\ \text{temp. rise or drop, °C}}{\text{water}_1\ \text{temp. rise or drop, °C}}\right)$$

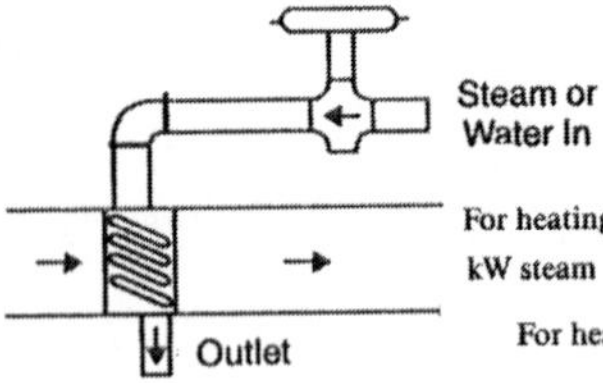

For heating air with steam coils:

$$\text{kW steam} = 0.012\ (\text{L/s})\ (\text{air temp. rise, °C})$$

For heating air with water coils:

$$(\text{L/s})_{\text{water}} = (1/3450)\ (\text{L/s})_{\text{air}} \left(\frac{\text{air temp. rise, °C}}{\text{water temp. drop, °C}}\right)$$

Figure 23.1 Sizing Formulas for Heating/Cooling

Table 23.2 Conversions to I-P and SI Units [2021F, Ch 40, Tbl 1]
(Multiply I-P values by conversion factors to obtain SI;
divide SI values by conversion factors to obtain I-P)

Multiply I-P	By	To Obtain SI
acre (43,560 ft^2)	0.4047	ha
	4046.873	m^2
atmosphere (standard)	*101.325	kPa
bar	*100	kPa
barrel (42 U.S. gal, petroleum)	159.0	L
	0.1580987	m^3
Btu (International Table)	1055.056	J
Btu (thermochemical)	1054.350	J
Btu/ft^2 (International Table)	11,356.53	J/m^2
Btu/ft^3 (International Table)	37,258.951	J/m^3
Btu/gal	278,717.1765	J/m^3
Btu·ft/h·ft^2·°F	1.730735	W/(m·K)
Btu·in/h·ft^2·°F (thermal conductivity k).	0.1442279	W/(m·K)
Btu/h	0.2930711	W
Btu/h·ft^2	3.154591	W/m^2
Btu/h·ft^2·°F (overall heat transfer coefficient U)	5.678263	W/(m^2·K)
Btu/lb	*2.326	kJ/kg
Btu/lb·°F (specific heat c_p)	*4.1868	kJ/(kg·K)
bushel (dry, U.S.)	0.0352394	m^3
calorie (thermochemical)	*4.184	J
centipoise (dynamic viscosity μ)	*1.00	mPa·s
centistokes (kinematic viscosity ν)	*1.00	mm^2/s
clo	0.155	(m^2·K)/W
dyne	1.0×10^{-5}	N
dyne/cm^2	*0.100	Pa
EDR hot water (150 Btu/h)	43.9606	W
EDR steam (240 Btu/h)	70.33706	W
EER	0.293	COP
ft	*0.3048	m
	*304.8	mm
ft/min, fpm	*0.00508	m/s
ft/s, fps	*0.3048	m/s
ft of water	2989	Pa
ft of water per 100 ft pipe	98.1	Pa/m
ft^2	0.092903	m^2
ft^2·h·°F/Btu (thermal resistance R)	0.176110	(m^2·K)/W
ft^2/s (kinematic viscosity ν)	92,900	mm^2/s
ft^3	28.316846	L
	0.02832	m^3
ft^3/min, cfm	0.471947	L/s
To Obtain I-P	**By**	**Divide SI**

Table 23.2 Conversions to I-P and SI Units [2021F, Ch 40, Tbl 1] ***(Continued)***
(Multiply I-P values by conversion factors to obtain SI; divide SI values by conversion factors to obtain I-P)

Multiply I-P	By	To Obtain SI
ft^3/s, cfs	28.316845	L/s
$ft \cdot lb_f$ (torque or moment)	1.355818	$N \cdot m$
$ft \cdot lb_f$ (work)	1.356	J
$ft \cdot lb_f/lb$ (specific energy)	2.99	J/kg
$ft \cdot lb_f/min$ (power)	0.0226	W
footcandle	10.76391	lx
gallon (U.S., *231 in^3)	3.785412	L
gph	1.05	mL/s
gpm	0.0631	L/s
gpm/ft^2	0.6791	$L/(s \cdot m^2)$
gpm/ton refrigeration	0.0179	mL/J
grain (1/7000 lb)	0.0648	g
gr/gal	17.1	g/m^3
gr/lb	0.143	g/kg
horsepower (boiler) (33 470 Btu/h)	9.81	kW
horsepower (550 $ft \cdot lb_f/s$)	0.7457	kW
inch	*25.4	mm
in. of mercury (60°F)	3.3864	kPa
in. of water (60°F)	248.84	Pa
in/100 ft, thermal expansion coefficient	0.833	mm/m
$in \cdot lb_f$ (torque or moment)	113	$mN \cdot m$
in^2	645.16	mm^2
in^3 (volume)	16.3874	mL
in^3/min (SCIM)	0.273117	mL/s
in^3 (section modulus)	16387	mm^3
in^4 (section moment)	416 231	mm^4
kWh	*3.60	MJ
kW/1000 cfm	2.118880	kJ/m^3
kilopond (kg force)	9.81	N
kip (1000 lb_f)	4.45	kN
kip/in^2 (ksi)	6.895	MPa
litre	*0.001	m^3
met	58.15	W/m^2
micron (μm) of mercury (60°F)	133	mPa
mile	1.609	km
mile, nautical	*1.852	km
mile per hour (mph)	1.609344	km/h
	0.447	m/s
millibar	*0.100	kPa
mm of mercury (60°F)	0.133	kPa
mm of water (60°F)	9.80	Pa
ounce (mass, avoirdupois)	28.35	g
To Obtain I-P	**By**	**Divide SI**

General

Table 23.2 Conversions to I-P and SI Units [2021F, Ch 40, Tbl 1] ***(Continued)***

(Multiply I-P values by conversion factors to obtain SI; divide SI values by conversion factors to obtain I-P)

Multiply I-P	By	To Obtain SI
ounce (force or thrust)	0.278	N
ounce (liquid, U.S.)	29.6	mL
ounce inch (torque, moment)	7.06	mN·m
ounce (avoirdupois) per gallon	7.489152	kg/m^3
perm (permeance at 32°F)	5.72135×10^{-11}	kg/(Pa·s·m^2)
perm inch (permeability at 32°F)	1.45362×10^{-12}	kg/(Pa·s·m)
pint (liquid, U.S.)	4.73176×10^{-4}	m^3
pound		
lb (avoirdupois, mass)	0.453592	kg
	453.592	g
lb$_f$ (force or thrust)	4.448222	N
lb$_f$/ft (uniform load)	14.59390	N/m
lb/ft·h (dynamic viscosity μ)	0.4134	mPa·s
lb/ft·s (dynamic viscosity μ)	1490	mPa·s
lb$_f$·s/ft^2 (dynamic viscosity μ)	47.88026	Pa·s
lb/h	0.000126	kg/s
lb/min	0.007559	kg/s
lb/h [steam at 212°F (100°C)]	0.2843	kW
lb$_f$/ft^2	47.9	Pa
lb/ft^2	4.88	kg/m^2
lb/ft^3 (density ρ)	16.0	kg/m^3
lb/gallon	120	kg/m^3
ppm (by mass)	*1.00	mg/kg
psi	6.895	kPa
quad (10^{15} Btu)	1.055	EJ
quart (liquid, U.S.)	0.9463	L
square (100 ft^2)	9.2903	m^2
tablespoon (approximately)	15	mL
teaspoon (approximately)	5	mL
therm (U.S.)	105.5	MJ
ton, long (2240 lb)	1.016046	Mg
ton, short (2000 lb)	0.907184	Mg; t (tonne)
ton, refrigeration (12 000 Btu/h)	3.517	kW
torr (1 mm Hg at 0°C)	133	Pa
watt per square foot	10.76	W/m^2
yd	*0.9144	m
yd^2	0.8361	m^2
yd^3	0.7646	m^3
To Obtain I-P	**By**	**Divide SI**

*Conversion factor is exact.

Notes: 1. Units are U.S. values unless noted otherwise.

2. Litre is a special name for the cubic decimetre. 1 L = 1 dm^3 and 1 mL = 1 cm^3.

General

24. APPENDIX

DB: Dry-bulb temperature, °C
MCWB: Mean coincident wet-bulb temperature, °C
WB: Wet-bulb temperature, °C
HDD and CDD 18.3: Annual heating and cooling degree-days, base 18.3°C, °C-day
DP: Dew-point temperature, °C

Climatic Design Conditions for Selected Locations

Station	Heating DB		Cooling DB/MCWB						Heat./Cool. Degree-Days	
			0.4%		1%		2%			
	99.6%	99%	DB / MCWB		DB / MCWB		DB / MCWB		HDD / CDD 18.3	
United States of America										
Alabama										
AUBURN UNIVERSITY	-4.9	-2.5	34.1	23.2	32.9	23.4	32.3	23.3	1290	1099
BIRMINGHAM SHUTTLESWORTH	-6.3	-3.9	35.3	23.6	34.0	23.6	32.9	23.5	1411	1191
CAIRNS AAF	-3.0	-1.3	35.4	24.4	34.1	24.5	33.0	24.3	977	1357
DOTHAN	-2.7	-0.8	35.9	24.2	34.4	24.0	33.4	23.9	944	1432
HUNTSVILLE INTL	-7.5	-5.2	35.2	23.8	34.0	23.7	32.9	23.4	1649	1083
MAXWELL AFB	-3.8	-1.8	36.2	24.3	35.1	24.6	34.0	24.5	1084	1418
MOBILE	-2.5	-0.6	34.6	25.1	33.5	24.9	32.6	24.7	889	1439
MONTGOMERY	-4.3	-2.3	35.9	24.5	34.8	24.4	33.8	24.3	1133	1365
NORTHEAST ALABAMA	-7.3	-5.2	34.5	23.9	33.1	23.8	32.4	23.7	1714	934
NORTHWEST ALABAMA	-6.9	-4.8	35.6	24.1	34.3	24.0	33.2	23.8	1629	1102
TUSCALOOSA	-5.5	-3.0	36.3	24.1	34.8	24.3	33.7	24.1	1331	1266
Alaska										
ANCHORAGE BRYANT AAF	-27.8	-24.5	24.3	16.0	22.5	15.2	20.7	14.3	5833	6
ANCHORAGE ELMENDORF AFB	-26.0	-22.8	23.6	15.0	22.1	14.5	20.2	13.7	5619	9
ANCHORAGE INTL	-21.9	-19.4	22.5	15.4	20.7	14.4	19.3	13.8	5477	6
ANCHORAGE LAKE HOOD	-22.0	-19.2	23.7	15.6	21.9	14.7	20.2	14.0	5342	12
ANCHORAGE MERRILL FIELD	-23.3	-21.1	23.4	15.5	21.9	14.9	20.3	14.0	5449	11
FAIRBANKS	-41.2	-38.6	27.1	16.1	25.4	15.5	23.6	14.7	7426	37
JUNEAU	-14.6	-12.3	23.5	15.4	21.4	14.7	19.4	13.7	4608	3
Arizona										
CASA GRANDE	-0.2	1.9	42.5	20.9	41.4	20.7	40.5	20.5	814	2013
DAVIS-MONTHAN AFB	0.1	1.7	40.7	18.5	39.5	18.4	38.2	18.2	769	1812

Climatic Design Conditions for Selected Locations *(Continued)*

Station	Heating DB		Cooling DB/MCWB						Heat./Cool. Degree-Days	
			0.4%		1%		2%			
	99.6%	99%	DB	MCWB	DB	MCWB	DB	MCWB	HDD	CDD 18.3
FLAGSTAFF PULLIAM	-15.4	-12.3	29.9	12.9	28.7	12.7	27.4	12.6	3747	75
LUKE AFB	1.7	3.1	43.9	20.9	42.6	20.9	41.4	20.8	643	2282
PHOENIX SKY HARBOR	4.0	5.5	43.6	20.7	42.5	20.6	41.5	20.5	486	2610
PRESCOTT	-7.6	-6.0	34.9	15.7	33.6	15.5	32.4	15.4	2240	596
TUCSON	0.0	1.5	40.9	18.9	39.8	18.7	38.6	18.6	738	1874
WINDOW ROCK	-17.5	-14.4	32.5	13.5	31.2	13.3	30.0	13.0	3468	191
YUMA	5.7	7.1	43.9	22.4	42.7	22.3	42.0	22.0	354	2644
Arkansas										
BENTONVILLE	-12.2	-8.9	35.2	23.9	33.2	24.0	32.4	23.6	2196	846
DRAKE FIELD	-12.0	-8.9	35.1	23.5	33.4	23.8	32.3	23.7	2208	786
FORT SMITH	-7.8	-5.4	37.8	24.4	36.1	24.6	34.6	24.5	1702	1218
GRIDER FIELD	-5.8	-3.8	36.2	25.3	34.9	25.2	33.7	24.9	1526	1237
JONESBORO	-8.4	-6.3	36.0	24.9	34.6	24.7	33.5	24.5	1923	1099
LITTLE ROCK AFB	-8.0	-5.7	37.5	25.1	36.0	25.4	34.5	25.2	1732	1181
LITTLE ROCK CLINTON	-6.6	-4.5	36.9	25.0	35.3	25.0	34.1	24.8	1601	1249
NORTH LITTLE ROCK	-7.5	-4.8	35.2	24.8	33.9	24.6	32.7	24.2	1754	1076
ROGERS	-12.2	-9.1	34.9	23.3	33.2	23.6	32.2	23.3	2183	837
SMITHFIELD	-12.0	-8.8	36.1	23.7	33.9	23.7	32.5	23.6	2149	857
TEXARKANA	-4.6	-2.8	37.3	24.3	35.9	24.3	34.5	24.2	1351	1330
California										
ALAMEDA	4.8	5.9	27.4	17.8	25.2	17.2	23.3	16.7	1380	94
BEALE AFB	-0.5	1.0	38.4	21.2	36.9	20.7	35.2	20.0	1341	846
BROWN FIELD	3.9	5.4	32.2	17.9	29.9	18.1	28.0	18.2	865	406
CAMARILLO	3.2	4.5	31.0	17.2	28.7	17.1	27.2	17.4	954	282
CAMP PENDLETON MCAS	0.0	1.4	32.9	18.3	30.8	18.3	28.9	18.4	1058	337
DESERT RESORTS	-0.2	1.6	44.8	22.2	43.2	21.9	42.1	21.7	577	2249
EL TORO MCAS	6.2	7.4	33.3	19.9	31.6	19.7	30.0	19.3	617	651

Climatic Design Conditions for Selected Locations (*Continued*)

	Heating DB		Cooling DB/MCWB						Heat./Cool. Degree-Days	
			0.4%		1%		2%			
Station	99.6%	99%	DB / MCWB		DB / MCWB		DB / MCWB		HDD / CDD 18.3	
FRESNO YOSEMITE	0.3	1.6	39.9	20.9	38.5	20.3	37.1	19.9	1188	1235
FULLERTON	4.5	6.0	35.2	19.3	33.2	19.3	31.5	19.0	579	802
HAYWARD	2.8	4.0	30.8	18.4	28.2	17.8	26.0	17.3	1341	177
HOLLYWOOD BURBANK	3.8	5.2	36.6	19.9	34.6	19.3	32.9	19.0	747	834
IMPERIAL COUNTY AP	2.2	3.6	44.4	22.6	43.0	22.4	42.1	22.2	496	2368
LANCASTER FOX	-5.9	-3.9	39.7	18.7	38.3	18.1	37.0	17.6	1602	1093
LEMORE NAS	-2.0	-0.5	39.8	21.1	38.5	20.6	37.2	20.2	1274	1034
LIVERMORE	-0.6	0.8	37.2	19.7	35.0	18.8	32.9	18.3	1437	491
LOMPOC	0.2	2.0	27.8	16.7	25.9	16.3	23.8	16.0	1534	41
LONG BEACH	5.4	6.6	33.2	18.9	31.2	18.9	29.5	18.6	633	644
LOS ANGELES HAWTHORNE	6.9	7.7	31.4	17.5	29.2	17.6	27.5	17.7	577	507
LOS ANGELES INTL	7.2	8.3	29.3	17.4	27.4	17.9	25.9	18.0	698	373
MARCH AFB	-0.3	1.3	38.5	19.8	37.1	19.3	35.5	19.0	1080	857
MCCLELLAN-PALOMAR	6.1	7.2	29.1	17.2	27.3	17.8	25.8	17.9	846	330
MEADOWS FIELD	0.7	2.2	39.6	21.0	38.3	20.5	37.0	20.0	1087	1337
MERCED CASTLE	-1.1	-0.1	39.2	20.9	37.8	20.2	36.8	19.7	1324	986
MIRAMAR MCAS	4.0	5.4	33.4	18.8	31.4	18.7	29.6	18.6	774	527
MODESTO CITY	-0.4	0.9	38.7	20.8	37.0	20.2	35.5	19.5	1256	939
MONTEREY	2.7	3.9	26.2	15.6	23.6	15.2	22.0	15.1	1729	36
MONTGOMERY-GIBBS	4.9	6.2	32.7	18.7	30.7	18.2	28.8	18.4	755	528
MOUNTAIN VIEW MOFFETT	2.5	3.8	31.1	18.7	28.8	18.2	26.9	17.8	1214	251
NAPA COUNTY AP	-1.3	0.1	32.8	18.9	30.2	18.3	28.0	17.8	1709	141
NORTH ISLAND NAS	7.1	8.0	29.5	17.7	27.5	18.4	26.2	18.8	607	445
OAKLAND INTL	2.7	4.0	28.8	18.0	26.4	17.3	24.1	16.7	1446	102
ONTARIO	3.7	5.0	38.0	20.7	36.5	20.2	34.9	19.7	735	1043
PALM SPRINGS	5.1	6.6	44.6	21.2	43.2	21.2	42.1	21.0	402	2506

Climatic Design Conditions for Selected Locations (*Continued*)

Station	Heating DB		Cooling DB/MCWB						Heat./Cool. Degree-Days	
			0.4%		1%		2%			
	99.6%	99%	DB	MCWB	DB	MCWB	DB	MCWB	HDD	CDD 18.3
POINT ARGUELLO	7.7	8.6	22.5	N/A	20.3	N/A	18.9	N/A	1895	16
POINT MUGU NAS	3.9	5.1	28.1	16.5	26.3	17.3	24.8	17.4	1129	147
PORTERVILLE	-1.0	0.6	38.7	21.6	37.5	20.8	36.3	20.2	1346	992
REDDING	-1.5	-0.1	40.8	20.0	39.1	19.5	37.4	18.9	1455	1089
RIVERSIDE	2.6	3.8	38.2	20.6	36.8	20.3	35.0	19.7	754	1013
SACRAMENTO EXECUTIVE	-0.3	1.1	37.9	20.9	36.1	20.2	34.3	19.6	1353	688
SACRAMENTO INTL	-0.7	0.9	38.0	21.2	36.4	20.6	34.8	19.9	1375	749
SACRAMENTO MATHER	-2.1	-0.6	38.4	20.4	36.8	19.7	34.9	19.2	1541	665
SACRAMENTO MCCLELLAN	-0.2	1.1	38.8	20.7	37.3	19.9	35.2	19.1	1295	848
SALINAS	1.3	2.6	28.7	16.6	26.1	16.1	24.0	15.9	1467	68
SAN BERNARDINO	1.4	2.8	39.7	20.4	38.0	19.9	36.8	19.7	790	1138
SAN DIEGO INTL	7.4	8.5	29.3	18.0	27.5	18.7	26.1	18.8	612	436
SAN FRANCISCO INTL	4.6	5.7	28.3	17.1	25.6	16.7	23.6	16.4	1448	96
SAN JOSE INTL	2.2	3.5	33.0	19.0	31.0	18.5	28.9	17.9	1168	344
SAN LUIS OBISPO	1.2	2.5	32.2	17.8	29.6	17.4	27.7	17.2	1197	189
SANTA BARBARA	1.9	3.0	28.6	17.5	26.7	17.5	25.1	17.2	1225	128
SANTA MARIA	0.9	2.1	29.5	16.8	27.2	16.6	25.2	16.3	1419	82
SONOMA COUNTY AP	-1.5	-0.3	34.9	19.2	32.8	18.7	30.7	18.1	1612	214
SOUTHERN CALIFORNIA	-3.6	-2.1	38.9	17.9	37.7	17.3	36.4	16.9	1442	1125
STOCKTON	-0.7	0.6	38.6	21.0	36.8	20.5	35.1	20.0	1321	788
TRAVIS AFB	-1.0	0.6	37.4	19.7	35.2	19.2	33.0	18.7	1379	552
VISALIA	-1.2	0.3	37.9	22.1	37.1	21.7	36.0	21.2	1341	952
Colorado										
BUCKLEY AFB	-16.9	-13.3	34.1	14.7	32.7	14.8	31.3	14.7	3149	416
CENTENNIAL	-17.3	-13.9	33.4	15.2	32.2	15.0	30.8	14.8	3286	374

Climatic Design Conditions for Selected Locations *(Continued)*

Station	Heating DB		Cooling DB/MCWB						Heat./Cool. Degree-Days	
			0.4%		1%		2%		HDD / CDD 18.3	
	99.6%	99%	DB / MCWB		DB / MCWB		DB / MCWB			
COLORADO SPRINGS	-16.5	-13.7	32.8	14.8	31.5	14.7	30.1	14.6	3341	306
DENVER INTL	-17.9	-14.6	34.9	15.5	33.5	15.4	32.0	15.3	3263	459
DENVER STAPLETON	-18.6	-14.9	34.4	15.9	32.9	15.6	31.4	15.3	3148	401
FORT COLLINS	-19.2	-15.1	32.3	16.1	30.7	15.8	29.1	15.6	3387	257
GRAND JUNCTION	-15.0	-11.7	36.6	16.2	35.3	15.7	34.0	15.4	3009	703
GREELEY-WELD COUNTY AP	-22.0	-17.6	35.9	17.1	33.8	16.9	32.3	16.8	3595	381
NORTHERN COLORADO	-17.8	-14.7	34.9	16.2	32.9	16.2	32.0	16.1	3387	379
PUEBLO	-17.2	-13.8	37.1	16.8	35.7	16.7	34.2	16.5	2988	569
Connecticut										
BRIDGEPORT SIKORSKY	-11.8	-9.3	31.3	22.9	29.7	22.2	28.3	21.6	2885	503
HARTFORD BRADLEY	-15.6	-12.7	33.1	23.0	31.5	22.2	29.9	21.3	3238	459
HARTFORD-BRAINARD	-13.7	-11.2	32.9	23.0	31.4	22.4	29.9	21.6	3051	507
WATERBURY-OXFORD	-16.1	-13.0	30.9	23.1	28.9	22.0	27.5	21.1	3523	284
WINDHAM	-16.1	-12.8	32.1	23.0	30.6	22.3	29.0	21.3	3279	379
Delaware										
DOVER AFB	-10.1	-7.8	33.5	24.5	32.2	24.0	30.8	23.6	2463	691
NEW CASTLE	-10.6	-8.4	33.4	24.0	31.9	23.4	30.6	22.9	2598	668
Florida										
CECIL FIELD	-1.0	1.1	35.9	24.9	34.6	24.7	33.5	24.5	638	1532
DAYTONA BEACH	2.1	4.2	33.7	25.1	32.7	25.1	31.9	25.0	394	1719
FORT MEYERS SW FLORIDA INTL	5.4	7.5	34.1	24.8	33.4	24.8	32.8	24.7	164	2131
FORT MYERS PAGE FIELD	6.0	8.1	34.1	24.8	33.4	24.9	32.8	24.8	148	2220
FT LAUDERDALE HOLLYWOOD	8.8	11.0	33.0	25.8	32.5	25.7	32.1	25.7	71	2578
GAINESVILLE	-1.4	0.7	34.4	24.5	33.6	24.4	32.7	24.2	616	1539
HOMESTEAD AFB	7.9	10.1	32.9	26.3	32.5	26.2	32.1	26.1	80	2372

Climatic Design Conditions for Selected Locations (*Continued*)

Station	Heating DB		Cooling DB/MCWB						Heat./Cool. Degree-Days	
			0.4%		1%		2%			
	99.6%	99%	DB	MCWB	DB	MCWB	DB	MCWB	HDD	CDD 18.3
JACKSONVILLE CRAIG	0.5	2.4	34.7	24.9	33.5	24.8	32.5	24.7	609	1582
JACKSONVILLE INTL	-1.4	0.5	34.7	25.0	33.7	24.9	32.7	24.7	704	1487
JACKSONVILLE NAS	1.3	3.3	35.6	24.8	34.4	24.6	33.4	24.4	503	1883
KENNEDY SPACE CENTER	4.0	6.2	33.3	25.6	32.5	25.6	32.0	25.5	287	1797
MACDILL AFB	3.9	6.2	34.1	26.1	33.5	26.0	32.8	25.8	272	2039
MAYPORT NAF	1.5	3.7	34.1	25.0	32.9	25.1	32.0	25.1	562	1659
MIAMI EXECUTIVE	7.6	9.8	33.8	25.6	33.0	25.5	32.5	25.4	90	2337
MIAMI NHC	9.4	11.5	33.3	25.4	32.7	25.4	32.2	25.3	62	2589
NAPLES	6.9	8.8	33.3	25.6	32.7	25.5	32.2	25.6	138	2205
OCALA	-1.3	1.1	34.0	24.2	33.0	24.3	32.5	24.2	562	1573
ORLANDO EXECUTIVE	3.9	6.3	34.3	24.6	33.6	24.5	32.8	24.4	271	2020
ORLANDO INTL	3.6	5.8	34.3	24.8	33.5	24.6	32.8	24.4	284	1933
ORLANDO MELBOURNE INTL	3.7	6.2	33.2	25.4	32.5	25.4	32.0	25.5	251	1975
ORLANDO SANFORD	2.9	5.2	34.7	24.4	33.8	24.4	33.0	24.3	321	1921
PALM BEACH INTL	7.0	9.2	33.2	25.4	32.5	25.4	31.9	25.4	115	2341
PANAMA CITY	-0.1	2.1	33.8	24.9	32.8	24.9	32.3	24.9	690	1582
PENSACOLA INTL	-1.2	1.0	34.4	25.3	33.3	25.2	32.4	25.1	759	1557
PENSACOLA NAS	-1.6	0.5	33.8	25.9	32.8	25.7	32.1	25.6	807	1464
SARASOTA BRADENTON INTL	4.5	6.9	33.5	25.6	32.8	25.6	32.4	25.5	241	1992
ST PETE-CLEARWATER	5.6	7.5	33.6	25.2	32.9	25.2	32.4	25.2	241	2083
TALLAHASSEE NWS	-3.1	-1.1	35.7	24.6	34.6	24.3	33.6	24.2	801	1538
TAMPA INTL	4.3	6.5	33.6	24.9	33.0	25.0	32.4	25.0	267	2074
TYNDALL AFB	-0.3	1.9	32.9	25.8	32.3	25.9	31.7	25.8	702	1504
VENICE PIER	5.4	7.6	31.3	24.7	30.6	25.2	30.2	25.2	260	1736
VERO BEACH	3.8	6.2	33.4	25.1	32.7	25.3	32.2	25.3	231	1969

Climatic Design Conditions for Selected Locations *(Continued)*

Station	Heating DB		Cooling DB/MCWB						Heat./Cool. Degree-Days	
			0.4%		1%		2%			
	99.6%	99%	DB / MCWB		DB / MCWB		DB / MCWB		HDD / CDD 18.3	
Georgia										
ATHENS	-5.3	-3.2	35.2	23.7	34.0	23.5	32.7	23.3	1499	1038
ATLANTA HARTSFIELD-JACKSON	-5.7	-3.1	34.3	23.2	33.1	23.1	32.0	22.9	1432	1094
AUGUSTA	-5.2	-3.3	36.2	24.4	34.9	24.2	33.8	24.0	1292	1190
COLUMBUS	-3.5	-1.4	35.6	23.6	34.5	23.5	33.5	23.4	1106	1345
DANIEL FIELD	-2.5	-1.0	36.1	23.7	34.8	23.5	33.7	23.2	1112	1368
DEKALB-PEACHTREE	-6.0	-3.5	34.5	23.1	33.3	23.0	32.4	22.7	1559	1038
DOBBINS AFB	-7.2	-4.4	33.9	23.4	32.9	23.3	31.9	23.1	1579	1033
FULTON COUNTY AP	-6.1	-3.6	34.3	23.2	33.2	23.1	32.3	23.0	1537	1012
HUNTER AAF	-2.3	-0.2	35.2	25.3	34.0	25.2	32.8	25.0	882	1448
LAWSON AAF	-5.4	-3.5	36.0	24.4	34.9	24.4	33.7	24.3	1252	1201
LEE GILMER	-5.9	-3.2	33.5	22.9	32.5	22.9	31.4	22.6	1626	945
MIDDLE GEORGIA	-4.6	-2.7	35.9	24.0	34.7	23.9	33.6	23.8	1235	1220
MOODY AFB	-2.0	0.1	35.4	24.7	34.4	24.6	33.6	24.4	802	1477
PEACHTREE CITY	-6.7	-4.4	34.3	23.4	33.2	23.3	32.3	23.2	1596	933
ROBINS AFB	-4.1	-2.2	36.0	24.3	34.9	24.3	33.7	24.1	1158	1263
ROME RUSSELL	-7.0	-4.8	35.4	23.6	34.1	23.3	32.9	23.3	1648	1026
SAVANNAH HILTON HEAD INTL	-2.5	-0.7	35.3	25.1	34.1	24.9	33.0	24.7	938	1404
SOUTHWEST GEORGIA	-3.0	-1.4	36.0	24.4	34.9	24.4	33.8	24.2	945	1461
VALDOSTA	-2.4	-0.8	35.8	24.8	34.7	24.6	33.7	24.4	809	1479
Hawaii										
HILO INTL	16.5	17.2	30.2	23.7	29.6	23.4	29.0	23.1	0	1857
HONOLULU INTL	17.5	18.4	32.0	23.4	31.5	23.2	31.0	22.9	0	2623
KALAELOA	16.0	17.0	32.1	23.4	31.3	23.1	31.0	23.0	0	2344
KANEOHE MCAS	17.7	18.8	29.9	24.0	29.2	23.7	28.8	23.6	0	2346

Climatic Design Conditions for Selected Locations *(Continued)*

Station	Heating DB		Cooling DB/MCWB						Heat./Cool. Degree-Days	
			0.4%		1%		2%			
	99.6%	99%	DB / MCWB		DB / MCWB		DB / MCWB		HDD / CDD 18.3	
Idaho										
BOISE	-11.5	-8.7	37.0	17.6	35.5	17.2	33.9	16.6	2951	590
CALDWELL	-12.5	-9.1	36.3	19.0	34.1	18.3	32.8	17.8	3123	422
COEUR D'ALENE	-14.1	-11.4	33.3	17.4	32.1	17.1	29.7	16.4	3741	199
IDAHO FALLS	-20.9	-17.5	33.4	16.1	32.1	15.9	30.5	15.4	4234	164
LEWISTON	-10.1	-7.0	37.1	18.5	35.2	18.0	33.2	17.4	2791	507
MAGIC VALLEY	-13.9	-11.3	35.0	17.0	33.4	16.7	32.1	16.3	3353	429
POCATELLO	-18.0	-15.1	34.9	16.4	33.3	16.0	31.7	15.5	3814	254
Illinois										
AURORA	-21.1	-17.7	32.5	23.6	31.2	23.0	29.8	22.4	3594	421
CHAMPAIGN WILLARD	-18.7	-15.7	33.0	23.9	32.0	23.7	30.6	23.0	3142	566
CHICAGO DUPAGE	-19.9	-17.1	32.4	23.7	31.1	23.0	29.5	22.1	3558	441
CHICAGO EXECUTIVE	-18.3	-15.8	33.0	23.3	31.6	22.6	30.1	21.9	3446	498
CHICAGO MIDWAY	-17.8	-15.0	33.2	23.7	32.0	22.9	30.5	22.2	3230	611
CHICAGO O'HARE	-18.7	-16.0	32.9	23.4	31.4	22.7	30.0	21.9	3421	511
CHICAGO ROCKFORD	-21.2	-18.0	32.6	23.5	31.1	22.8	29.7	22.1	3628	452
DECATUR	-17.1	-14.0	33.7	24.6	32.5	24.1	31.2	23.4	2986	624
PEORIA	-18.6	-15.7	33.4	24.7	32.1	24.0	30.7	23.2	3149	621
QUAD CITY	-20.2	-17.1	33.5	24.6	32.1	23.8	30.7	22.9	3356	571
QUINCY	-18.0	-15.0	33.9	24.7	32.5	24.1	31.1	23.4	3027	652
SCOTT AFB	-14.1	-11.3	35.0	25.4	33.7	25.0	32.4	24.5	2561	803
SPRINGFIELD LINCOLN	-17.5	-14.4	33.8	24.9	32.6	24.4	31.3	23.6	2936	671
ST LOUIS DOWNTOWN	-13.2	-10.8	35.1	24.8	33.7	24.5	32.5	24.0	2507	829
Indiana										
EVANSVILLE	-13.1	-10.1	34.3	24.3	33.0	24.1	31.9	23.7	2418	827

Climatic Design Conditions for Selected Locations (*Continued*)

Station	Heating DB		Cooling DB/MCWB						Heat./Cool. Degree-Days	
			0.4%		1%		2%			
	99.6%	99%	DB / MCWB		DB / MCWB		DB / MCWB		HDD / CDD 18.3	
FORT WAYNE INTL	-18.4	-15.3	32.6	23.5	31.2	22.8	29.7	22.1	3304	474
GRISSOM AFB	-19.8	-16.1	32.4	24.0	31.1	23.4	29.8	22.7	3264	519
INDIANAPOLIS INTL	-16.8	-13.6	32.9	23.8	31.6	23.3	30.3	22.6	2902	644
MONROE COUNTY AP	-16.1	-12.7	32.8	23.9	31.7	23.8	30.5	23.0	2791	592
PURDUE UNIVERSITY	-18.0	-15.1	33.0	24.1	31.8	23.5	30.3	22.6	3089	558
SOUTH BEND	-18.0	-15.0	32.3	23.3	30.8	22.4	29.4	21.8	3437	447
TERRE HAUTE	-17.0	-13.5	33.4	24.3	32.2	24.1	31.0	23.3	2869	619
Iowa										
AMES	-21.5	-18.6	32.7	24.5	31.3	23.8	29.9	22.9	3659	459
ANKENY	-20.2	-17.5	33.8	24.2	32.4	24.0	31.0	23.3	3451	532
BOONE	-21.0	-17.8	32.8	24.6	31.6	24.1	30.0	23.1	3597	494
DAVENPORT	-21.7	-18.2	32.7	23.8	31.3	23.4	30.0	22.6	3508	514
DES MOINES	-20.2	-17.5	33.9	24.5	32.3	23.9	30.8	23.1	3369	626
DUBUQUE	-22.6	-19.6	31.3	23.9	29.8	23.0	28.4	22.0	3892	363
EASTERN IOWA	-22.6	-19.6	32.3	24.6	30.9	23.7	29.3	22.8	3732	435
SIOUX GATEWAY	-21.7	-19.1	33.8	24.1	32.3	23.6	30.8	22.9	3716	520
WATERLOO	-23.3	-20.4	32.7	24.2	31.2	23.2	29.7	22.4	3876	434
Kansas										
JOHNSON COUNTY EXECUTIVE	-15.5	-12.7	35.2	24.3	33.3	24.2	32.1	23.9	2679	781
LAWRENCE	-16.0	-13.0	37.1	24.8	35.0	24.5	33.2	24.2	2757	826
MANHATTAN	-16.4	-13.3	37.7	24.2	36.0	24.2	33.9	23.8	2807	846
MARSHALL AAF	-15.3	-13.0	37.8	24.0	35.7	24.1	34.1	23.8	2703	914
MCCONNELL AFB	-13.0	-10.7	37.6	22.8	35.9	23.2	34.0	23.2	2359	981
SALINA	-15.2	-12.6	38.7	23.3	37.0	23.3	35.2	23.0	2637	973
TOPEKA BILLARD	-15.6	-12.9	36.8	24.6	35.0	24.5	33.3	24.0	2679	865

Climatic Design Conditions for Selected Locations (*Continued*)

Station	Heating DB		Cooling DB/MCWB						Heat./Cool. Degree-Days	
	99.6%	99%	0.4%		1%		2%		HDD / CDD 18.3	
			DB / MCWB		DB / MCWB		DB / MCWB			
TOPEKA FORBES	-15.8	-12.9	37.2	24.6	35.0	24.5	33.3	24.0	2712	844
WICHITA EISENHOWER	-13.4	-10.9	38.1	23.2	36.3	23.4	34.5	23.3	2428	993
WICHITA JABARA	-14.0	-11.4	37.5	23.3	35.9	23.4	33.8	23.3	2474	912
Kentucky										
BOWLING GREEN	-11.4	-8.4	34.5	24.0	33.2	24.0	32.2	23.7	2162	878
CAMPBELL AAF	-11.6	-8.6	34.5	24.4	33.0	24.3	32.2	24.0	2094	916
CINCINNATI NORTHERN KENTUCKY	-14.9	-11.7	33.0	23.4	31.7	23.0	30.4	22.5	2711	648
HENDERSON	-13.0	-10.1	34.1	24.7	32.9	24.4	32.2	24.1	2455	796
LAKE CUMBERLAND	-11.2	-8.0	34.0	23.5	32.7	23.1	31.7	22.7	2218	738
LEXINGTON BLUE GRASS	-13.5	-10.4	33.1	23.2	31.9	23.0	30.7	22.6	2491	701
LOUISVILLE BOWMAN	-12.3	-9.1	34.1	24.0	32.9	23.7	31.9	23.2	2287	864
LOUISVILLE INTL	-12.1	-9.1	34.4	24.0	33.2	23.8	32.0	23.3	2228	932
Louisiana										
ALEXANDRIA ESLER	-3.4	-1.9	36.6	24.9	35.4	25.[illegible]	34.2	25.0	1073	1435
ALEXANDRIA INTL	-2.8	-1.2	36.2	24.9	35.0	24.9	33.9	24.8	1021	1485
BARKSDALE AFB	-4.3	-2.5	37.3	24.2	35.9	24.3	34.7	24.4	1221	1391
BATON ROUGE	-2.1	-0.3	35.0	25.2	34.1	25.2	33.3	25.0	854	1550
LAFAYETTE	-1.1	0.9	35.0	25.4	34.0	25.3	33.1	25.2	771	1636
LAKE CHARLES NWS	-0.9	0.9	35.0	25.4	34.0	25.4	33.1	25.3	771	1634
MONROE	-3.9	-2.2	36.8	25.4	35.5	25.2	34.3	25.1	1192	1423
NEW ORLEANS INTL	0.6	2.6	34.7	25.5	33.8	25.4	32.9	25.3	663	1746
NEW ORLEANS LAKEFRONT	2.0	3.8	34.2	25.9	33.7	25.8	32.8	25.6	581	1932
NEW ORLEANS NAS	-0.8	1.2	34.0	25.4	33.2	25.3	32.5	25.2	734	1567
SHREVEPORT DOWNTOWN	-2.9	-1.3	37.4	24.6	36.0	24.6	34.7	24.6	1165	1513
SHREVEPORT REGIONAL	-3.4	-1.7	37.4	24.2	36.1	24.4	34.8	24.4	1135	1497

Climatic Design Conditions for Selected Locations (*Continued*)

Station	Heating DB		Cooling DB/MCWB						Heat./Cool. Degree-Days	
			0.4%		1%		2%			
	99.6%	99%	DB / MCWB		DB / MCWB		DB / MCWB		HDD / CDD 18.3	
Maine										
AUBURN-LEWISTON	-21.0	-17.7	31.0	21.9	28.8	21.1	27.3	19.8	4188	187
BANGOR	-21.6	-18.7	30.8	21.7	28.9	20.7	27.3	19.5	4223	207
BRUNSWICK	-19.1	-16.8	29.9	21.4	28.0	20.4	26.6	19.4	4023	197
PORTLAND INTL JETPORT	-17.7	-15.0	30.3	22.0	28.5	21.0	26.9	20.1	3828	223
SANFORD	-21.1	-17.6	32.1	22.4	29.9	21.6	28.0	20.3	4103	211
Maryland										
ANDREWS AFB	-10.0	-7.7	34.3	23.8	32.8	23.4	31.4	22.8	2416	722
BALTIMORE-WASHINGTON	-10.3	-8.0	34.5	23.9	32.9	23.4	31.5	22.7	2486	730
THOMAS POINT	-8.4	-6.1	30.5	24.7	29.4	24.4	28.5	24.0	2307	706
Massachusetts										
BARNSTABLE	-12.3	-9.7	29.2	22.9	27.7	22.1	26.5	21.5	3188	309
BOSTON LOGAN	-13.5	-10.6	32.7	22.8	30.9	22.0	29.1	21.1	3054	451
BUZZARDS BAY	-11.0	-8.5	24.7	N/A	23.8	N/A	23.1	N/A	2996	197
CHATHAM	-11.3	-8.7	28.6	23.0	27.2	22.2	25.9	21.5	3101	284
LAWRENCE	-15.2	-12.5	32.7	22.8	31.2	22.1	29.4	21.2	3319	407
MARTHA'S VINEYARD	-12.4	-9.8	29.0	22.7	27.6	22.0	26.3	21.4	3192	261
NEW BEDFORD	-13.6	-11.0	31.3	23.2	29.5	22.2	28.0	21.3	3193	339
NORWOOD	-16.2	-12.8	32.8	23.1	31.3	22.4	29.6	21.5	3359	359
PLYMOUTH	-14.7	-12.1	31.7	22.8	29.8	22.1	28.1	21.1	3327	335
SOUTH WEYMOUTH NAS	-14.5	-12.0	32.9	23.2	31.0	22.4	29.3	21.5	3240	359
WORCESTER	-16.8	-14.0	30.0	21.9	28.6	20.9	27.2	20.1	3645	291
Michigan										
DETROIT CITY	-15.5	-12.9	32.5	22.9	31.0	22.2	29.6	21.4	3327	501
DETROIT WAYNE COUNTY AP	-16.6	-13.7	32.4	23.3	30.8	22.5	29.4	21.6	3353	482

Climatic Design Conditions for Selected Locations *(Continued)*

Station	Heating DB		Cooling DB/MCWB						Heat./Cool. Degree-Days	
			0.4%		1%		2%			
	99.6%	99%	DB	MCWB	DB	MCWB	DB	MCWB	HDD	CDD 18.3
FLINT BISHOP	-18.4	-15.6	32.1	22.8	30.6	22.2	29.1	21.2	3687	359
GRAND RAPIDS FORD	-16.7	-14.0	31.9	22.7	30.4	22.0	29.0	21.2	3603	388
GROSSE ILE	-15.9	-12.5	31.6	23.5	29.8	23.0	28.0	22.1	3294	472
JACKSON COUNTY AP	-18.0	-15.2	31.6	22.9	30.0	22.1	28.6	21.3	3632	343
KALAMAZOO BATTLE CREEK	-17.0	-13.9	32.3	22.9	30.9	22.1	29.1	21.2	3465	421
LANSING	-18.3	-15.4	31.8	22.7	30.2	21.9	28.8	21.0	3723	346
MBS INTL	-17.9	-15.5	31.9	22.8	30.2	21.8	28.7	21.2	3763	347
MUSKEGON COUNTY AP	-15.2	-12.8	30.2	22.3	28.9	21.6	27.7	20.9	3608	319
OAKLAND COUNTY INTL	-17.6	-15.1	32.1	22.7	30.3	21.7	28.8	20.9	3667	381
SELFRIDGE AFB	-17.5	-14.8	32.2	23.4	30.3	22.4	28.9	21.7	3561	383
ST CLAIR COUNTY INTL	-18.6	-15.2	32.0	23.0	29.9	21.9	27.9	20.9	3777	257
WESTERN MICHIGAN	-14.7	-12.3	31.7	23.0	30.2	22.2	28.6	21.5	3445	373
WILLOW RUN	-17.5	-14.6	33.3	23.3	31.8	22.6	30.1	21.6	3441	455
Minnesota										
DULUTH INTL	-27.4	-24.5	29.1	21.0	27.5	19.7	25.9	18.7	5096	134
DULUTH SKY HARBOR	-23.8	-21.4	30.0	21.9	27.9	20.7	27.0	20.0	4751	172
MANKATO	-24.4	-22.3	32.1	23.2	30.2	22.3	28.6	21.5	4214	368
MINNEAPOLIS ANOKA COUNTY AP	-22.7	-20.3	32.3	23.7	30.9	23.1	28.8	22.0	4180	347
MINNEAPOLIS CRYSTAL	-23.1	-21.2	32.5	22.9	31.1	22.2	29.1	21.0	4197	406
MINNEAPOLIS FLYING CLOUD	-23.7	-21.3	32.5	23.4	31.1	22.6	29.2	21.5	4093	449
MINNEAPOLIS-ST PAUL	-23.6	-21.1	32.7	23.0	31.1	22.2	29.5	21.3	4109	463
ROCHESTER INTL	-24.9	-22.3	31.0	23.1	29.3	22.1	27.9	21.4	4322	303
SOUTH ST PAUL	-22.7	-20.3	32.5	22.7	31.0	21.9	29.1	20.9	4077	427
ST CLOUD	-27.1	-24.1	31.9	22.6	30.2	21.5	28.6	20.5	4639	279
ST PAUL DOWNTOWN	-23.2	-21.0	32.4	23.2	30.9	22.5	29.0	21.4	4101	427

Climatic Design Conditions for Selected Locations *(Continued)*

Station	Heating DB		Cooling DB/MCWB						Heat./Cool. Degree-Days	
			0.4%		1%		2%			
	99.6%	99%	DB / MCWB		DB / MCWB		DB / MCWB		HDD / CDD 18.3	
Mississippi										
HATTIESBURG-LAUREL	-3.9	-2.4	35.8	24.4	34.0	24.1	32.9	24.0	1121	1291
JACKSON INTL	-4.8	-2.9	35.8	24.4	34.6	24.4	33.5	24.3	1228	1323
KEESLER AFB	-1.0	1.2	34.6	26.7	33.6	26.4	32.6	26.1	779	1606
MERIDIAN	-5.4	-3.4	35.7	24.4	34.5	24.5	33.4	24.4	1282	1239
MERIDIAN NAS	-5.6	-3.3	35.6	24.3	34.5	24.3	33.5	24.3	1302	1259
TUPELO	-7.2	-4.8	35.5	24.3	34.3	24.3	33.2	24.1	1592	1151
Missouri										
CAPE GIRARDEAU	-12.4	-9.2	34.4	25.1	33.3	24.8	32.2	24.4	2331	851
COLUMBIA	-16.0	-12.8	34.9	24.3	33.3	24.4	31.8	23.9	2691	745
JEFFERSON CITY	-14.4	-11.3	35.2	24.5	33.7	24.1	32.4	23.8	2512	826
KANSAS CITY INTL	-16.5	-13.8	35.3	24.8	33.6	24.6	32.0	24.2	2765	783
KANSAS CITY WHEELER	-14.6	-12.1	36.2	24.7	34.5	24.4	33.0	24.1	2506	959
SPIRIT OF ST LOUIS	-14.7	-11.5	35.2	25.2	33.8	24.7	32.5	24.1	2584	803
SPRINGFIELD-BRANSON	-13.9	-10.8	35.2	23.4	33.5	23.5	32.1	23.4	2427	814
ST LOUIS LAMBERT	-14.0	-11.0	35.6	24.9	34.2	24.5	32.9	23.9	2433	964
WEBB CITY JOPLIN	-12.9	-10.2	36.3	23.9	34.6	24.2	33.1	24.0	2228	954
Montana										
BILLINGS LOGAN	-22.5	-19.3	34.9	17.0	33.1	16.6	31.3	16.3	3748	382
BOZEMAN YELLOWSTONE	-25.2	-21.4	33.4	16.2	31.5	15.7	29.6	15.2	4533	137
BUTTE MOONEY	-26.7	-22.6	31.2	14.1	29.5	13.6	27.8	13.3	5039	47
GREAT FALLS	-26.2	-22.7	33.8	16.1	31.9	15.6	30.0	15.2	4218	194
MALMSTROM AFB	-25.7	-22.7	34.6	16.9	32.6	16.2	30.7	15.6	3979	256
MISSOULA	-18.7	-15.4	34.0	16.5	32.2	16.2	30.3	15.7	4073	196

Climatic Design Conditions for Selected Locations *(Continued)*

Station	Heating DB		Cooling DB/MCWB						Heat./Cool. Degree-Days	
			0.4%		1%		2%			
	99.6%	99%	DB / MCWB		DB / MCWB		DB / MCWB		HDD / CDD 18.3	
Nebraska										
CENTRAL NEBRASKA	-19.4	-16.7	35.3	23.5	33.6	23.0	31.9	22.4	3362	604
EPPLEY FIELD	-19.3	-16.9	35.0	24.5	33.4	24.1	31.8	23.3	3304	685
LINCOLN	-19.1	-16.6	35.7	24.0	34.0	23.3	32.4	23.2	3285	683
NORTH OMAHA	-21.2	-17.8	34.5	23.9	32.7	23.7	31.1	22.8	3323	607
OFFUTT AFB	-19.1	-16.9	35.0	24.7	33.1	24.4	31.9	23.7	3297	660
Nevada										
LAS VEGAS MCCARRAN	0.4	1.9	42.8	19.4	41.5	19.[illegible]	40.4	18.7	1023	2045
NELLIS AFB	-1.1	0.3	42.9	19.3	41.7	19.[illegible]	40.5	18.6	1090	1949
RENO-TAHOE	-8.9	-7.0	36.1	16.3	34.7	15.[illegible]	33.3	15.1	2676	541
New Hampshire										
CONCORD	-19.5	-16.7	32.3	22.0	30.6	21.1	29.1	20.3	3894	279
JAFFREY	-19.2	-16.7	30.7	21.3	28.9	20.4	27.5	19.6	4019	223
MANCHESTER-BOSTON	-16.8	-13.9	32.8	22.0	31.3	21.3	29.8	20.6	3439	423
PORTSMOUTH PEASE	-16.7	-13.7	32.0	22.6	30.1	21.8	28.4	21.0	3550	322
New Jersey										
ATLANTIC CITY INTL	-11.3	-8.9	33.5	24.1	31.9	23.3	30.5	22.8	2630	615
MCGUIRE AFB	-12.0	-9.4	33.8	24.3	32.4	23.7	31.0	23.0	2700	611
MILLVILLE	-11.8	-9.2	33.3	24.0	31.8	23.5	30.5	22.9	2686	604
MONMOUTH JET CENTER	-11.8	-9.0	32.8	23.5	31.4	22.9	29.8	22.3	2768	542
NEWARK INTL	-11.0	-8.8	34.5	23.5	32.8	22.7	31.3	22.2	2581	714
TETERBORO	-11.4	-9.1	33.7	23.4	32.2	22.7	30.9	22.1	2704	644
TRENTON-MERCER	-11.3	-9.1	33.5	23.5	32.1	22.9	30.7	22.4	2707	613
New Mexico										
ALAMOGORDO WHITE SANDS	-6.2	-3.9	37.7	17.3	37.0	17.4	35.1	17.2	1603	1068

Climatic Design Conditions for Selected Locations (*Continued*)

Station	Heating DB		Cooling DB/MCWB						Heat./Cool. Degree-Days	
			0.4%		1%		2%		HDD / CDD 18.3	
	99.6%	99%	DB / MCWB		DB / MCWB		DB / MCWB			
ALBUQUERQUE INTL	-7.2	-5.3	35.3	15.4	34.1	15.3	32.9	15.2	2152	827
CANNON AFB	-10.0	-7.6	37.1	17.3	35.4	17.4	34.0	17.7	2032	829
CLOVIS	-11.0	-8.0	36.3	18.0	34.9	17.9	33.2	17.9	2213	712
FOUR CORNERS	-12.7	-10.4	35.6	15.2	34.2	15.0	33.0	14.8	2883	569
HOLLOMAN AFB	-7.3	-5.3	38.0	16.9	36.9	17.0	35.7	17.0	1709	1103
ROSWELL	-8.0	-5.8	38.8	17.9	37.3	18.1	36.0	18.2	1695	1134
WHITE SANDS	-7.5	-5.3	37.2	17.6	35.8	17.7	34.6	17.7	1638	1006
New York										
ALBANY INTL	-18.1	-15.4	31.7	22.7	30.1	21.7	28.7	21.0	3566	381
AMBROSE LIGHT	-10.6	-8.4	28.9	N/A	27.2	N/A	25.8	N/A	2710	405
BUFFALO NIAGARA	-16.4	-14.0	30.2	21.8	28.8	21.1	27.6	20.6	3583	337
CHAUTAUQUA COUNTY AP	-17.7	-15.3	28.0	20.9	27.3	20.3	26.1	19.5	3944	179
ELMIRA CORNING	-18.2	-15.5	32.0	22.0	30.3	21.3	28.8	20.6	3707	276
FARMINGDALE REPUBLIC	-11.0	-8.6	32.1	23.2	30.2	22.3	28.7	21.8	2782	526
GREATER BINGHAMTON	-18.1	-15.6	29.5	21.1	28.0	20.3	26.6	19.7	3908	231
GREATER ROCHESTER INTL	-16.6	-14.1	31.6	22.8	30.0	21.9	28.5	21.0	3558	342
GRIFFISS INTL	-21.8	-18.2	31.0	22.4	29.6	21.6	28.1	20.7	3903	289
HUDSON VALLEY	-16.4	-13.4	32.9	23.1	31.3	22.4	29.9	21.6	3298	421
LONG ISLAND MACARTHUR	-11.5	-9.1	31.5	23.1	29.9	22.3	28.5	21.8	2872	487
NEW YORK KENNEDY	-10.3	-8.0	32.1	22.8	30.4	22.3	29.0	21.8	2645	587
NEW YORK LA GUARDIA	-10.2	-7.8	33.7	23.3	32.1	22.5	30.6	22.0	2487	740
NEW YORK STEWART	-16.3	-13.4	32.4	22.3	30.9	22.0	29.0	20.9	3354	391
NIAGARA FALLS INTL	-16.7	-14.1	31.1	22.4	29.6	21.7	28.1	20.9	3651	338
ONEIDA COUNTY AP	-20.9	-17.3	30.9	22.7	29.1	21.6	27.8	20.8	3879	277
PLATTSBURGH	-22.7	-19.5	30.9	22.0	28.8	21.3	27.4	20.3	4194	226

Climatic Design Conditions for Selected Locations (*Continued*)

Station	Heating DB		Cooling DB/MCWB						Heat./Cool. Degree-Days	
			0.4%		1%		2%			
	99.6%	99%	DB / MCWB		DB / MCWB		DB / MCWB		HDD / CDD 18.3	
SYRACUSE HANCOCK	-18.6	-15.5	31.7	22.8	30.2	21.9	28.8	21.1	3613	358
WESTCHESTER COUNTY AP	-12.9	-10.6	31.9	23.0	30.2	22.2	28.8	21.5	3021	453
North Carolina										
ASHEVILLE	-9.5	-7.0	31.1	21.5	29.9	21.2	28.8	20.9	2222	507
CHARLOTTE DOUGLAS	-6.1	-4.0	34.6	23.7	33.3	23.4	32.1	23.1	1683	968
FAYETTEVILLE	-5.4	-3.1	35.8	24.5	34.3	24.0	33.1	23.7	1477	1157
HICKORY	-7.0	-4.8	33.3	22.6	32.2	22.4	31.0	22.1	1899	796
NEW RIVER MCAS	-5.1	-3.0	33.7	25.4	32.6	25.2	31.5	24.7	1399	1098
PIEDMONT TRIAD	-7.5	-5.4	33.6	23.3	32.4	23.1	31.2	22.7	1942	841
PITT-GREENVILLE	-6.3	-3.9	35.0	24.8	33.8	24.1	32.6	23.9	1644	1058
POPE AFB	-6.4	-4.2	36.0	24.5	34.6	24.2	33.1	23.9	1568	1147
RALEIGH-DURHAM	-6.9	-4.6	35.0	24.2	33.6	23.9	32.4	23.6	1771	969
RICHLANDS ELLIS	-7.0	-4.1	34.8	25.3	33.3	24.8	32.4	24.4	1598	1014
SEYMOUR JOHNSON AFB	-6.0	-3.7	35.8	24.6	34.2	24.5	33.0	24.1	1493	1142
SIMMONS AAF	-6.0	-3.8	35.6	24.1	34.2	24.0	33.0	23.7	1518	1158
WILMINGTON	-4.3	-2.5	34.1	25.4	32.9	25.0	31.7	24.6	1308	1153
WINSTON-SALEM REYNOLDS	-7.1	-4.9	33.4	23.0	32.4	22.8	31.3	22.4	1888	853
North Dakota										
BISMARCK	-27.5	-24.4	33.9	21.3	31.9	20.7	30.1	20.0	4674	308
FARGO HECTOR	-28.2	-25.5	32.2	22.4	30.6	21.3	29.1	20.5	4825	317
GRAND FORKS AFB	-29.0	-26.3	31.7	22.3	30.0	21.3	28.4	20.3	5156	230
GRAND FORKS INTL	-29.8	-27.1	31.6	22.2	30.0	21.0	28.5	20.1	5192	237
MINOT AFB	-30.1	-27.3	32.4	20.6	30.4	20.1	28.7	19.2	5198	198
MINOT INTL	-27.6	-25.0	32.8	20.7	30.9	20.3	29.0	19.2	4852	265

Climatic Design Conditions for Selected Locations *(Continued)*

Station	Heating DB		Cooling DB/MCWB						Heat./Cool. Degree-Days	
			0.4%		1%		2%			
	99.6%	99%	DB / MCWB		DB / MCWB		DB / MCWB		HDD / CDD 18.3	
Ohio										
AKRON-CANTON	-16.5	-13.7	31.5	22.7	30.2	22.0	28.9	21.2	3291	427
CINCINNATI LUNKEN	-13.8	-10.8	33.5	23.9	32.1	23.5	30.9	23.0	2653	631
CLEVELAND HOPKINS	-15.8	-12.8	32.1	23.1	30.7	22.4	29.2	21.7	3187	474
COLUMBUS GLENN	-15.3	-12.3	32.8	23.0	31.6	22.5	30.3	21.9	2867	610
COLUMBUS RICKENBACKER	-15.2	-12.2	33.4	23.3	32.3	22.9	31.1	22.4	2794	640
DAYTON INTL	-16.8	-13.6	32.3	23.2	31.0	22.7	29.7	21.9	3023	552
FAIRFIELD COUNTY AP	-17.1	-13.0	32.4	23.4	31.2	22.9	29.9	22.2	2988	473
FINDLAY	-17.5	-14.5	32.5	23.1	31.2	22.5	29.8	21.7	3236	494
MANSFIELD LAHM	-17.4	-14.5	31.2	22.8	30.0	22.1	28.7	21.4	3364	398
OHIO STATE UNIVERSITY	-15.8	-12.7	32.4	23.1	31.2	22.7	29.9	22.0	2999	537
TOLEDO EXPRESS	-17.5	-14.5	32.9	23.4	31.4	22.4	29.9	21.6	3349	466
WRIGHT-PATTERSON AFB	-16.4	-13.0	32.7	23.5	31.4	22.9	30.1	22.2	2959	544
YOUNGSTOWN-WARREN	-16.6	-13.8	31.2	22.6	29.8	21.8	28.5	20.9	3396	344
Oklahoma										
LAWTON FORT SILL NORTH	-9.0	-6.6	39.8	22.7	38.0	23.1	36.5	23.2	1719	1302
LAWTON FORT SILL SOUTH	-8.7	-6.5	40.0	22.8	38.5	22.9	37.1	23.1	1735	1322
OKLAHOMA CITY POST	-9.8	-7.4	37.9	23.1	36.6	23.2	35.0	23.4	1899	1193
OKLAHOMA CITY ROGERS	-9.7	-7.2	38.1	23.3	36.5	23.4	34.9	23.5	1888	1132
STILLWATER	-10.9	-7.9	38.9	23.7	37.3	23.9	35.3	24.1	1987	1155
TINKER AFB	-10.1	-7.5	37.8	23.0	36.1	23.2	34.4	23.4	1879	1114
TULSA INTL	-10.2	-7.6	37.8	24.2	36.2	24.5	34.8	24.4	1895	1196
TULSA JONES	-9.7	-7.4	38.1	24.4	36.8	24.8	35.2	24.7	1919	1171
VANCE AFB	-11.7	-9.1	38.7	23.0	37.3	23.1	35.7	23.4	2171	1121

Climatic Design Conditions for Selected Locations (*Continued*)

Station	Heating DB		Cooling DB/MCWB						Heat./Cool. Degree-Days	
			0.4%		1%		2%			
	99.6%	99%	DB / MCWB		DB / MCWB		DB / MCWB		HDD / CDD 18.3	
Oregon										
AURORA	-3.6	-2.1	33.6	19.5	31.4	19.2	29.1	18.3	2433	228
CORVALLIS	-4.0	-2.5	33.8	19.5	32.1	19.1	29.7	18.1	2383	221
EUGENE	-4.6	-2.6	33.4	19.3	31.3	18.7	29.2	18.0	2572	166
MCMINNVILLE	-3.4	-2.1	33.8	19.4	31.6	19.0	29.1	18.2	2544	179
MEDFORD	-4.5	-2.9	37.2	19.3	35.3	18.6	33.5	18.0	2331	512
PORTLAND HILLSBORO	-4.9	-3.0	33.6	19.8	31.3	19.2	28.9	18.3	2649	161
PORTLAND INTL	-3.4	-1.4	33.2	19.6	30.8	19.1	28.8	18.3	2322	269
REDMOND	-14.4	-10.9	34.2	16.3	32.6	15.9	31.0	15.4	3587	147
SALEM	-4.1	-2.3	33.7	19.3	31.4	18.8	29.2	18.0	2464	211
Pennsylvania										
ALLEGHENY COUNTY AP	-14.9	-12.3	31.4	22.3	30.1	21.7	28.9	21.0	2992	471
ALTOONA-BLAIR COUNTY AP	-14.7	-12.3	31.2	22.2	29.8	21.5	28.4	20.9	3240	362
CAPITAL CITY	-11.7	-9.2	33.5	23.4	32.1	22.7	30.7	22.1	2763	631
ERIE	-15.1	-12.4	30.5	22.8	29.1	22.1	27.9	21.4	3345	389
HARRISBURG	-12.0	-9.6	33.2	24.0	32.0	23.3	30.4	22.7	2792	644
LEHIGH VALLEY	-13.3	-10.9	32.9	23.4	31.4	22.6	30.0	21.9	3019	501
NORTHEAST PHILADELPHIA	-10.7	-8.2	34.0	23.9	32.6	23.2	31.3	22.6	2571	711
PHILADELPHIA INTL	-10.1	-7.9	34.1	23.9	32.7	23.3	31.3	22.5	2450	779
PITTSBURG-BUTLER	-16.3	-13.4	31.2	22.4	29.2	21.6	28.0	20.8	3345	331
PITTSBURGH INTL	-15.4	-12.7	31.6	22.2	30.2	21.6	29.0	20.9	3066	453
READING	-12.5	-10.2	33.5	23.7	32.0	23.1	30.6	22.4	2847	582
WASHINGTON COUNTY AP	-17.1	-13.4	31.3	22.1	29.8	21.6	28.5	20.9	3249	326
WILKES-BARRE SCRANTON	-15.4	-12.8	31.8	22.3	30.3	21.5	28.8	20.7	3314	382
WILLOW GROVE NAS	-10.8	-8.6	33.3	23.3	31.9	22.6	30.5	21.9	2742	574

Climatic Design Conditions for Selected Locations *(Continued)*

Station	Heating DB		Cooling DB/MCWB						Heat./Cool. Degree-Days	
			0.4%		1%		2%			
	99.6%	99%	DB / MCWB		DB / MCWB		DB / MCWB		HDD / CDD 18.3	
Rhode Island										
PROVIDENCE GREEN	-13.3	-10.7	32.2	23.0	30.5	22.3	28.9	21.4	3043	443
South Carolina										
CHARLESTON INTL	-2.7	-0.8	34.7	25.4	33.5	25.2	32.4	24.9	1012	1349
COLUMBIA METRO	-4.7	-2.8	36.2	24.1	34.9	23.8	33.8	23.7	1319	1276
FLORENCE	-4.7	-2.8	35.5	24.7	34.2	24.4	33.0	24.2	1329	1204
FOLLY ISLAND	-0.6	1.3	30.8	25.7	30.1	25.7	29.5	25.5	1043	1207
GREENVILLE-SPARTANBURG	-5.8	-3.7	34.5	23.0	33.2	22.9	32.0	22.6	1639	954
SHAW AFB	-4.8	-2.8	35.7	24.3	34.3	24.2	33.0	23.9	1328	1202
South Dakota										
ELLSWORTH AFB	-22.0	-19.0	35.2	19.2	33.0	18.9	31.2	18.7	3890	396
RAPID CITY	-22.5	-19.4	35.9	18.9	33.6	18.8	31.6	18.5	3952	367
SIOUX FALLS	-23.9	-21.3	32.9	23.5	31.3	22.8	29.8	22.0	4134	423
Tennessee										
CHATTANOOGA	-7.1	-4.8	34.9	23.3	33.7	23.2	32.5	23.0	1687	1040
KNOXVILLE TYSON	-8.7	-6.1	33.7	23.1	32.5	22.9	31.4	22.6	1937	881
MCKELLAR-SIPES	-9.4	-7.0	34.9	24.9	33.8	24.9	32.6	24.6	1913	979
MEMPHIS INTL	-7.2	-4.9	35.8	25.0	34.6	24.7	33.6	24.5	1587	1289
MILLINGTON-MEMPHIS	-7.9	-6.0	37.5	27.1	36.2	26.4	34.1	25.5	1713	1171
NASHVILLE INTL	-9.5	-7.0	34.7	23.7	33.5	23.6	32.4	23.3	1906	1000
TRI-CITIES	-10.7	-8.1	32.5	22.2	31.3	21.9	30.3	21.6	2295	608
Texas										
ABILENE	-6.6	-4.2	38.1	21.6	36.9	21.6	35.7	21.7	1341	1417
AMARILLO NWS	-11.8	-8.9	37.1	18.7	35.6	18.9	34.2	18.9	2196	848
ANGELINA COUNTY AP	-2.8	-1.2	37.2	24.4	35.8	24.6	34.6	24.6	996	1533

Climatic Design Conditions for Selected Locations (*Continued*)

Station	Heating DB		Cooling DB/MCWB						Heat./Cool. Degree-Days	
			0.4%		1%		2%			
	99.6%	99%	DB / MCWB		DB / MCWB		DB / MCWB		HDD / CDD 18.3	
AUSTIN-BERGSTROM	-3.0	-1.2	37.9	23.5	36.9	23.6	35.8	23.7	916	1683
BROWNSVILLE	3.7	5.8	35.6	25.8	34.9	25.7	34.3	25.7	277	2323
COLLEGE STATION EASTWOOD	-1.9	-0.1	37.7	24.2	36.7	24.2	35.6	24.3	848	1758
CORPUS CHRISTI INTL	1.4	3.4	36.1	25.5	35.2	25.5	34.4	25.5	456	2054
CORPUS CHRISTI NAS	2.8	5.0	33.8	26.7	33.1	26.6	32.7	26.6	397	2103
DALLAS EXECUTIVE	-4.0	-2.2	38.6	23.4	37.3	23.7	36.1	23.6	1157	1590
DALLAS FORT WORTH	-4.8	-2.5	38.6	23.4	37.3	23.6	36.1	23.7	1174	1642
DALLAS HENSLEY FIELD	-5.8	-2.7	37.6	24.2	36.4	24.1	35.1	23.9	1206	1513
DALLAS LOVE FIELD	-4.1	-2.0	38.7	23.7	37.5	24.0	36.3	23.9	1109	1724
DEL RIO	-0.3	1.4	39.1	22.4	38.1	22.5	37.1	22.6	705	1981
DRAUGHON-MILLER CENTRAL TEXAS	-3.9	-2.3	37.8	23.4	37.2	23.5	36.0	23.6	1085	1559
DYESS AFB	-7.2	-5.0	39.1	22.2	37.8	22.1	36.7	22.1	1379	1493
EAST TEXAS	-3.7	-2.1	37.7	24.0	36.4	24.2	35.0	24.3	1166	1481
EL PASO	-3.5	-1.8	38.5	17.6	37.3	17.4	36.1	17.4	1224	1462
ELLINGTON FIELD	0.9	2.6	36.1	25.5	35.0	25.8	33.9	25.8	644	1804
FORT WORTH ALLIANCE	-6.0	-3.7	39.0	23.3	37.7	23.6	36.4	23.5	1306	1541
FORT WORTH NAS	-4.1	-2.0	39.3	22.8	38.0	23.1	36.8	23.2	1122	1727
FT WORTH MEACHAM	-5.2	-2.9	38.7	23.4	37.5	23.7	36.3	23.6	1218	1584
GALVESTON	2.5	4.2	33.4	26.3	32.7	26.4	32.3	26.4	532	1910
GEORGETOWN	-3.2	-2.1	37.3	22.9	36.2	23.1	35.1	23.3	1066	1550
HOUSTON BUSH	-0.3	1.3	36.4	24.8	35.4	24.9	34.4	24.9	721	1778
HOUSTON HOBBY	0.8	2.6	35.7	25.1	34.7	25.2	33.8	25.1	619	1836
HOUSTON HOOKS	-1.2	0.8	36.7	24.4	35.4	24.7	34.2	24.7	780	1706
KILLEEN REGIONAL	-3.7	-1.5	38.0	22.8	37.2	22.9	36.0	22.9	1003	1627
KILLEEN SKYLARK	-3.6	-1.8	37.9	23.4	37.2	23.5	36.1	23.6	1026	1616

Climatic Design Conditions for Selected Locations *(Continued)*

Station	Heating DB		Cooling DB/MCWB						Heat./Cool. Degree-Days	
			0.4%		1%		2%			
	99.6%	99%	DB / MCWB		DB / MCWB		DB / MCWB		HDD / CDD 18.3	
LACKLAND AFB	-1.3	0.5	38.0	23.4	37.3	23.5	36.2	23.5	743	1857
LAREDO	2.2	3.9	40.6	23.7	39.4	23.7	38.4	23.7	427	2523
LAUGHLIN AFB	-0.8	1.1	40.2	22.2	38.9	22.5	37.8	22.6	671	2059
LUBBOCK	-8.9	-6.6	37.6	19.0	36.3	19.4	35.0	19.4	1787	1088
MCALLEN	3.6	5.5	38.5	24.7	37.6	24.9	36.7	24.8	287	2593
MCGREGOR	-4.0	-2.4	38.7	23.7	37.6	23.8	36.5	23.8	1129	1590
MCKINNEY	-6.2	-3.8	38.0	23.7	37.1	23.9	35.8	23.9	1373	1422
MIDLAND INTL	-6.3	-4.1	38.7	19.5	37.3	19.7	36.1	19.7	1375	1404
NACOGDOCHES	-3.9	-2.5	37.3	24.3	36.0	24.5	34.0	24.4	1163	1387
NEW BRAUNFELS	-1.9	-0.1	37.9	23.4	37.0	23.5	36.0	23.5	813	1736
PORT ARANSAS PIER	3.0	5.2	30.1	25.7	29.7	25.9	29.5	25.8	447	1742
PORT ARTHUR	-0.3	1.6	35.0	25.5	34.1	25.6	33.3	25.5	714	1680
RANDOLPH AFB	-1.9	0.0	37.9	23.2	37.0	23.2	35.9	23.3	796	1745
REESE AFB	-9.6	-7.0	38.3	19.4	36.6	19.6	35.1	19.6	1768	1017
SABINE PASS	0.1	2.2	31.8	25.2	30.9	25.4	30.5	25.4	782	1506
SAN ANGELO	-5.5	-3.5	39.0	21.3	37.8	21.2	36.6	21.2	1188	1517
SAN ANTONIO INTL	-1.1	0.6	37.4	23.1	36.5	23.1	35.5	23.2	751	1817
SAN ANTONIO STINSON	-0.6	1.1	38.5	23.3	37.5	23.4	36.4	23.3	688	1933
SAN MARCOS	-2.5	-1.0	37.8	23.5	37.2	23.5	36.1	23.6	889	1726
VALLEY	2.8	4.8	37.1	25.3	36.3	25.4	35.5	25.3	326	2312
VICTORIA	-0.4	1.3	36.7	24.8	35.6	24.8	34.6	24.8	642	1826
WACO	-4.2	-2.3	38.5	23.4	37.4	23.7	36.3	23.8	1113	1637
WICHITA FALLS SHEPPARD AFB	-7.5	-5.3	39.9	22.6	38.4	22.7	36.9	22.8	1541	1430

Climatic Design Conditions for Selected Locations *(Continued)*

Station	Heating DB		Cooling DB/MCWB						Heat./Cool. Degree-Days	
			0.4%		1%		2%			
	99.6%	99%	DB / MCWB		DB / MCWB		DB / MCWB		HDD / CDD 18.3	
Utah										
HILL AFB	-12.5	-10.6	34.6	16.0	33.2	15.6	32.1	15.4	3273	553
LOGAN-CACHE	-20.3	-17.2	34.9	16.4	33.6	16.1	32.2	15.7	3923	288
PROVO	-12.9	-10.4	35.1	17.0	33.6	16.8	32.3	16.7	3242	482
SALT LAKE CITY INTL	-11.5	-9.1	36.8	17.0	35.5	16.5	34.1	16.3	2961	750
ST GEORGE	-3.7	-2.3	41.2	18.9	39.8	18.5	38.3	18.0	1628	1513
Vermont										
BURLINGTON INTL	-21.7	-18.7	31.4	21.8	29.9	21.0	28.3	20.3	3969	318
Virginia										
DANVILLE	-7.8	-5.8	34.1	23.6	32.9	23.5	31.9	23.2	2010	814
DAVISON AAF	-10.1	-7.5	35.2	24.0	33.7	23.5	32.3	23.1	2337	765
DINWIDDIE	-9.0	-7.1	36.2	24.9	34.1	24.4	32.9	23.8	2051	883
LANGLEY AFB	-7.0	-4.6	33.6	24.6	32.4	24.3	31.3	23.9	1897	907
LESSBURG	-10.1	-7.7	35.0	24.8	33.0	24.0	32.3	23.6	2442	761
LYNCHBURG	-9.6	-7.3	33.2	23.1	31.9	22.7	30.7	22.2	2342	629
MANASSAS	-11.6	-8.9	33.9	23.8	32.7	23.6	31.4	23.0	2640	641
NEWPORT NEWS WILLIAMSBURG	-7.0	-4.8	34.6	25.0	33.2	24.5	32.1	24.1	1883	939
NORFOLK INTL	-5.6	-3.5	34.2	24.9	32.9	24.5	31.7	24.0	1744	991
NORFOLK NAS	-5.3	-2.9	34.4	25.0	33.1	24.5	32.1	24.2	1652	1070
OCEANA NAS	-6.2	-3.8	33.9	25.1	32.6	24.6	31.3	24.1	1811	925
QUANTICO MCAF	-8.9	-6.9	33.6	24.6	32.3	24.4	31.1	23.9	2283	798
RICHMOND	-8.1	-6.0	34.9	24.2	33.6	23.9	32.3	23.4	2019	894
ROANOKE-BLACKSBURG	-9.2	-6.8	33.4	22.6	32.1	22.1	30.8	21.8	2182	722
SHENANDOAH VALLEY	-11.9	-8.9	33.9	23.2	32.6	22.9	31.3	22.5	2489	618
VIRGINIA TECH MONTGOMERY	-12.0	-9.0	31.6	22.4	30.2	21.8	28.9	21.4	2628	454
WASHINGTON DULLES	-11.0	-8.5	34.0	23.6	32.6	23.1	31.3	22.6	2532	688
WASHINGTON RONALD REAGAN	-8.3	-6.3	34.7	24.2	33.3	23.7	32.0	23.1	2142	922

Climatic Design Conditions for Selected Locations *(Continued)*

Station	Heating DB		Cooling DB/MCWB						Heat./Cool. Degree-Days	
			0.4%		1%		2%			
	99.6%	99%	DB / MCWB		DB / MCWB		DB / MCWB		HDD / CDD 18.3	
Washington										
ARLINGTON	-7.0	-4.6	28.5	19.3	27.0	18.3	24.8	17.2	2993	41
BELLINGHAM	-6.4	-4.0	26.8	18.5	24.9	17.7	23.1	16.8	2925	36
BREMERTON	-5.1	-2.9	30.2	18.7	27.9	17.7	26.2	17.0	3079	61
FAIRCHILD AFB	-14.2	-11.3	33.7	16.7	32.0	16.2	30.0	15.8	3726	240
GRAY AFF	-6.2	-4.0	31.0	19.0	28.7	18.2	26.9	17.5	2843	91
KELSO-LONGVIEW	-4.8	-2.8	31.3	19.6	28.7	19.0	27.0	18.1	2629	118
MCCHORD AFB	-6.1	-4.0	30.5	18.7	28.3	18.0	26.4	17.2	2854	77
OLYMPIA	-6.1	-4.1	31.0	18.7	28.8	18.1	26.7	17.4	2955	68
PASCO TRI-CITIES	-11.6	-8.3	37.3	20.6	35.7	19.8	33.7	19.1	2751	461
SANDERSON FIELD	-5.5	-3.4	31.3	18.7	28.8	18.0	26.6	17.2	2976	69
SEATTLE KING COUNTY INTL	-2.9	-1.2	29.9	18.4	27.9	17.7	26.2	17.1	2388	163
SEATTLEPAINE	-3.7	-1.6	27.2	17.7	25.1	16.9	23.2	16.2	2827	56
SEATTLE TACOMA	-3.0	-1.1	30.0	18.4	27.9	17.7	26.0	17.0	2567	126
SEATTLE WEST POINT	-1.1	0.9	21.3	16.2	20.0	15.8	18.9	15.4	2733	4
SPOKANE FELTS FIELD	-12.5	-9.5	34.9	17.8	33.0	17.3	31.2	16.8	3357	282
SPOKANE INTL	-14.4	-11.2	34.0	17.0	32.3	16.4	30.3	15.8	3633	281
TACOMA NARROWS	-2.4	-0.6	28.8	17.9	27.0	17.2	25.0	16.5	2642	84
VANCOUVER PEARSON	-4.2	-2.6	33.0	19.3	31.0	18.9	28.7	18.2	2422	229
WALLAWALLA	-11.0	-7.7	37.0	19.0	34.9	18.2	32.8	17.6	2646	542
YAKIMA	-12.7	-9.8	36.3	18.9	34.5	18.5	32.6	17.7	3183	347
West Virginia										
HUNTINGTON TRI-STATE	-12.6	-9.6	33.0	23.3	31.7	23.0	30.5	22.4	2435	651
MID-OHIO VALLEY	-14.0	-11.1	32.5	23.2	31.2	22.6	30.0	22.0	2706	544
YEAGER	-12.4	-9.6	32.8	22.7	31.6	22.4	30.4	22.0	2436	616

Climatic Design Conditions for Selected Locations (Continued)

Station	Heating DB		Cooling DB/MCWB						Heat./Cool. Degree-Days	
			0.4%		1%		2%			
	99.6%	99%	DB / MCWB		DB / MCWB		DB / MCWB		HDD / CDD 18.3	
Wisconsin										
APPLETON	-21.1	-17.9	31.3	23.9	29.1	22.4	27.9	21.4	4006	334
CENTRAL WISCONSIN	-23.9	-21.8	30.2	22.4	28.6	21.4	27.3	20.1	4614	200
CHIPPEWA VALLEY	-25.4	-22.5	32.1	22.8	30.4	21.7	28.8	20.7	4339	329
DANE COUNTY REGIONAL	-21.5	-18.8	31.8	23.3	30.2	22.5	28.8	21.7	3918	363
FOND DU LAC	-21.3	-18.7	31.4	23.0	29.9	22.0	28.4	21.2	3969	337
GREEN BAY STRAUBEL	-22.3	-19.5	31.0	23.1	29.5	22.2	28.1	21.3	4177	278
KENOSHA AP	-19.6	-17.0	32.3	23.6	30.6	22.8	28.9	21.9	3718	355
LA CROSSE	-22.9	-20.3	33.0	23.9	31.4	22.8	30.0	21.9	3849	481
MANITOWOC COUNTY AP	-20.5	-17.8	29.5	22.1	27.8	21.4	26.6	20.5	4173	207
MILWAUKEE MITCHELL	-18.7	-16.2	32.0	23.5	30.2	22.4	28.6	21.5	3693	397
OSHKOSH WHITTMAN	-21.3	-18.8	31.2	23.1	29.5	22.1	28.0	21.2	4061	319
SHEBOYGAN	-19.2	-16.5	28.4	21.7	26.4	21.3	24.8	20.9	4033	191
SHEBOYGAN COUNTY AP	-20.6	-18.0	31.2	23.3	29.1	21.9	27.7	21.1	4133	246
WAUSAU	-24.5	-21.9	30.9	21.9	29.1	20.8	27.7	19.9	4431	261
Wyoming										
CASPER NATRONA COUNTY INTL	-22.2	-18.3	34.5	15.2	33.0	14.9	31.5	14.6	4076	262
CHEYENNE	-19.5	-16.1	32.1	14.5	30.7	14.2	29.2	14.1	3873	215
Canada										
Alberta										
BOW ISLAND	-29.3	-25.7	31.9	18.0	30.0	17.3	28.1	16.9	4732	120
CALGARY INTL	-27.7	-24.5	28.8	16.0	26.8	15.5	25.0	14.9	5054	43
CANADIAN OLYMPIC PARK UPPER	-27.2	-24.1	28.3	15.3	26.3	14.7	24.4	14.1	5003	46
EDMONTON BLATCHFORD	-28.4	-25.4	28.4	17.7	26.7	16.9	25.1	16.2	5209	77
EDMONTON INTL	-32.3	-29.0	27.8	18.0	26.1	17.3	24.5	16.4	5814	24

Climatic Design Conditions for Selected Locations *(Continued)*

Station	Heating DB		Cooling DB/MCWB 0.4%		Cooling DB/MCWB 1%		Cooling DB/MCWB 2%		Heat./Cool. Degree-Days	
	99.6%	99%	DB / MCWB		DB / MCWB		DB / MCWB		HDD / CDD 18.3	
EDMONTON NAMAO	-29.8	-26.7	27.8	17.7	26.1	16.9	24.4	16.0	5571	39
FORT MCMURRAY	-35.7	-32.9	28.9	17.4	26.9	16.5	25.1	15.7	6263	49
GRANDE PRAIRIE	-34.9	-30.7	27.4	16.5	25.6	15.8	24.0	15.1	5904	26
LACOMBE	-31.6	-28.0	28.1	18.0	26.2	17.2	24.5	16.3	5681	24
LETHBRIDGE CDA	-27.7	-24.4	31.8	17.0	29.8	16.4	27.8	16.0	4456	121
MEDICINE HAT	-29.8	-26.0	33.0	17.3	31.1	16.9	29.2	16.4	4676	182
RED DEER	-31.4	-27.7	27.9	17.2	26.1	16.4	24.4	15.6	5707	24
SPRINGBANK	-30.5	-26.9	27.5	15.6	25.6	14.9	23.8	14.4	5650	7
British Columbia										
ABBOTSFORD	-6.8	-4.4	30.1	19.6	28.0	18.9	26.1	18.1	2863	91
AGASSIZ	-6.8	-4.5	30.4	20.1	28.5	19.4	26.7	18.8	2816	129
BALLENAS ISLAND	-0.7	0.7	23.5	18.8	22.3	18.3	21.3	17.7	2622	61
COMOX	-4.3	-2.5	27.1	17.8	25.2	17.2	23.4	16.5	3036	67
DISCOVERY ISLAND	-1.1	1.0	22.7	15.8	20.7	14.9	19.1	14.3	2789	11
ENTRANCE ISLAND	-1.1	0.3	23.6	18.1	22.2	17.6	21.1	17.2	2651	62
ESQUIMALT HARBOUR	-2.2	-0.5	22.1	15.7	20.5	15.1	19.1	14.7	2983	7
HOWE SOUND PAM ROCKS	-2.5	-0.7	24.7	18.8	23.1	18.2	21.9	17.7	2652	82
KAMLOOPS	-18.1	-14.6	34.2	17.9	32.1	17.4	30.0	16.7	3497	293
KELOWNA	-16.9	-13.2	33.2	18.0	31.3	17.4	29.4	16.7	3843	152
MALAHAT	-5.3	-3.3	27.7	16.8	25.8	16.3	24.1	15.8	3233	104
PENTICTON	-12.8	-10.0	33.0	18.6	31.2	18.1	29.5	17.4	3386	241
PITT MEADOWS	-7.1	-4.7	30.4	19.6	28.4	19.0	26.5	18.3	2931	92
POINT ATKINSON	-1.8	-0.3	24.3	17.6	23.1	17.6	22.0	17.3	2466	97
PRINCE GEORGE	-28.6	-24.5	27.9	16.2	25.8	15.4	23.9	14.5	5123	20
SANDHEADS	-3.0	-1.0	22.4	N/A	21.3	N/A	20.4	N/A	2732	34

Climatic Design Conditions for Selected Locations (*Continued*)

Station	Heating DB		Cooling DB/MCWB						Heat./Cool. Degree-Days	
			0.4%		1%		2%			
	99.6%	99%	DB / MCWB		DB / MCWB		DB / MCWB		HDD / CDD 18.3	
SUMMERLAND	-13.9	-10.9	33.0	17.4	31.2	17.0	29.3	16.5	3480	272
UNIVERSITY OF VICTORIA	-2.4	-0.5	27.0	18.0	25.2	17.4	23.5	16.7	2762	38
VANCOUVER HARBOUR	-2.6	-0.8	26.0	18.1	24.5	17.6	23.2	17.0	2646	76
VANCOUVER INTL	-5.1	-3.0	25.1	18.5	23.7	17.9	22.5	17.3	2873	49
VERNON	-16.2	-13.2	33.0	18.3	31.0	17.8	29.0	17.0	3754	215
VICTORIA GONZALES	-2.6	-0.6	24.5	16.7	22.2	15.9	20.5	15.1	2860	22
VICTORIA HARTLAND	-3.7	-1.6	28.5	18.7	26.6	17.9	24.8	17.3	2847	96
VICTORIA INTL	-3.6	-2.0	26.8	17.7	24.8	16.9	23.1	16.2	2958	30
WEST VANCOUVER	-5.2	-3.0	27.3	18.0	25.4	17.7	23.8	17.1	2938	87
WHITE ROCK	-5.1	-2.9	25.0	18.6	23.4	17.9	22.1	17.3	2751	34
YOHO PARK	-29.8	-26.3	25.6	13.5	23.6	12.8	21.5	12.0	6419	1
Manitoba										
WINNIPEG INTL	-32.0	-29.6	30.4	21.2	28.7	20.3	27.1	19.3	5697	168
New Brunswick										
FREDERICTON INTL	-23.1	-20.3	29.9	21.0	28.1	20.0	26.5	19.1	4541	149
MONCTON ROMEO LEBLANC	-21.5	-19.2	28.8	21.0	27.2	19.9	25.7	19.1	4592	124
SAINT JOHN	-22.0	-19.3	26.3	18.9	24.7	18.0	23.2	17.2	4651	36
Newfoundland and Labrador										
ST JOHN'S INTL	-13.7	-12.0	25.0	19.3	23.5	18.5	22.1	17.8	4721	38
Northwest Territories										
YELLOWKNIFE	-40.0	-37.7	25.6	16.1	23.9	15.3	22.3	14.6	8051	38
Nova Scotia										
HALIFAX STANFIELD	-17.6	-15.2	27.9	20.5	26.2	19.5	24.8	18.8	4193	119
SHEARWATER	-15.7	-13.3	26.5	19.7	24.9	18.8	23.5	18.2	3996	89
SYDNEY	-16.6	-14.2	27.7	20.5	26.0	19.6	24.4	18.8	4432	93

Climatic Design Conditions for Selected Locations *(Continued)*

Station	Heating DB		Cooling DB/MCWB						Heat./Cool. Degree-Days	
			0.4%		1%		2%			
	99.6%	99%	DB / MCWB		DB / MCWB		DB / MCWB		HDD / CDD 18.3	
Nunavut										
IQALUIT	-37.0	-35.6	17.1	11.5	14.3	10.2	12.3	9.1	9638	0
Ontario										
BEAUSOLEIL	-24.5	-20.9	29.9	23.3	28.1	22.1	26.5	21.3	4364	215
BELLE RIVER	-14.6	-12.2	31.6	24.2	30.0	23.7	28.5	22.8	3307	454
CFB TRENTON	-22.0	-18.8	29.3	22.2	27.9	21.4	26.6	20.7	4078	233
ERIEAU	-15.6	-13.0	27.0	22.9	26.0	22.3	25.1	21.7	3587	296
GUELPH TURFGRASS INSTITUTE	-21.5	-18.5	29.7	21.5	28.2	20.8	26.7	20.0	4398	151
HAMILTON INTL	-18.7	-15.9	30.5	22.5	29.0	21.7	27.5	21.0	3902	261
LONDON INTL	-19.0	-16.1	30.3	22.3	28.8	21.6	27.3	20.7	3899	258
NORTH BAY	-27.7	-24.7	28.0	20.1	26.4	19.2	25.0	18.4	5126	131
OTTAWA INTL	-24.1	-21.4	30.9	22.0	29.2	20.9	27.6	20.2	4451	254
PETERBOROUGH TRENT UNIVERSITY	-23.2	-19.9	30.8	21.3	29.0	20.6	27.5	19.6	4324	193
PORT WELLER	-13.3	-11.1	29.4	22.6	27.8	22.0	26.5	21.4	3467	338
REGION OF WATERLOO INTL	-20.9	-17.6	30.5	21.8	28.9	21.1	27.4	20.4	4189	199
SAULT STE MARIE	-25.2	-22.0	28.5	21.1	26.8	20.0	25.2	19.1	4896	98
SUDBURY	-28.0	-24.8	29.0	20.1	27.3	19.0	25.7	18.1	5182	129
THUNDER BAY	-29.5	-26.7	29.0	20.4	27.2	19.2	25.5	18.3	5514	75
TIMMINS	-33.2	-29.9	29.5	19.8	27.6	18.5	25.9	17.8	5948	90
TORONTO BILLY BISHOP	-16.0	-13.3	28.3	21.6	26.8	21.1	25.4	20.6	3660	255
TORONTO BUTTONVILLE	-20.0	-16.9	31.6	22.3	29.8	21.2	28.1	20.5	3967	287
TORONTO PEARSON	-18.5	-15.7	31.5	22.4	29.7	21.5	28.1	20.8	3779	339
WINDSOR	-16.4	-13.6	32.0	23.2	30.5	22.4	29.1	21.6	3404	455
Prince Edward Island										
CHARLOTTETOWN	-19.3	-17.1	27.0	20.8	25.6	19.8	24.2	19.0	4511	119

Climatic Design Conditions for Selected Locations (*Continued*)

Station	Heating DB		Cooling DB/MCWB						Heat./Cool. Degree-Days	
			0.4%		1%		2%			
	99.6%	99%	DB / MCWB		DB / MCWB		DB / MCWB		HDD / CDD 18.3	
Québec										
BAGOTVILLE	-29.1	-26.5	29.4	19.6	27.4	18.6	25.6	17.9	5531	109
BIG TROUT LAKE	-36.6	-34.2	26.9	18.5	25.2	17.6	23.5	16.8	7303	51
JONQUIERE	-29.2	-26.5	29.0	19.7	27.1	19.0	25.3	18.3	5453	103
LA BAIE	-30.3	-27.6	29.0	19.6	27.0	19.1	25.2	18.3	5688	75
LAC SAINT-PIERRE	-24.0	-21.2	27.9	21.1	26.5	20.5	25.3	19.8	4525	214
L'ACADIE	-24.1	-21.5	30.1	21.7	28.6	21.0	27.2	20.3	4369	237
L'ASSOMPTION	-25.7	-22.6	30.5	22.0	28.8	20.9	27.3	20.2	4582	228
MONT JOLI	-23.0	-20.6	27.0	20.0	25.2	18.9	23.7	18.0	5223	77
MONT-ORFORD	-28.3	-25.1	25.1	18.5	23.5	17.7	22.0	17.1	5628	54
MONTREAL MCTAVISH	-21.6	-19.1	30.2	21.8	28.7	20.9	27.3	20.1	4111	324
MONTREAL MIRABEL INTL	-25.7	-22.8	29.7	22.0	28.1	20.9	26.7	20.0	4669	196
MONTREAL ST-HUBERT	-23.2	-20.6	30.2	22.2	28.7	21.2	27.3	20.4	4361	256
MONTREAL TRUDEAU	-22.9	-20.2	30.3	22.1	28.7	21.1	27.4	20.4	4257	301
NICOLET	-25.3	-22.5	28.8	22.4	27.3	21.4	25.9	20.5	4649	178
POINTE-AU-PERE	-21.8	-19.2	23.1	18.6	21.5	17.7	20.2	16.7	5312	12
QUEBEC CITY JEAN LESAGE	-25.8	-23.0	28.8	21.1	27.3	20.2	25.8	19.3	4963	138
QUEBEC CITY SAINTE-FOY	-24.0	-21.5	29.1	20.6	27.5	19.7	26.0	18.9	4810	162
SHERBROOKE AP	-27.1	-23.9	29.0	21.4	27.5	20.4	26.1	19.6	4849	117
SHERBROOKE LENNOXVILLE	-25.4	-22.3	29.5	21.5	27.9	20.6	26.6	19.9	4564	158
ST-ANICET	-24.6	-21.7	30.7	22.6	29.2	21.8	27.8	21.0	4386	234
STE-ANNE-DE-BELLEVUE	-23.6	-20.8	30.1	21.9	28.5	21.0	27.1	20.2	4373	246
TROIS-RIVIERES	-23.8	-21.2	27.4	21.3	26.2	20.8	25.1	20.2	4560	199
VARENNES	-23.5	-21.1	30.3	21.8	28.6	20.9	27.2	20.1	4451	217

Climatic Design Conditions for Selected Locations *(Continued)*

Station	Heating DB		Cooling DB/MCWB						Heat./Cool. Degree-Days	
			0.4%		1%		2%			
	99.6%	99%	DB / MCWB		DB / MCWB		DB / MCWB		HDD / CDD 18.3	
Saskatchewan										
MOOSE JAW	-31.0	-28.0	31.8	18.6	29.7	18.0	27.7	17.3	5354	123
PRINCE ALBERT	-35.0	-32.1	28.8	19.1	27.0	18.2	25.5	17.1	6158	67
REGINA	-32.7	-29.9	30.9	19.2	28.8	18.4	27.1	17.6	5720	112
SASKATOON INTL	-33.7	-30.7	30.2	18.9	28.3	18.0	26.5	17.4	5853	97
SASKATOON KERNEN FARM	-33.5	-30.6	30.6	17.7	28.6	16.9	26.8	16.1	5878	102
Yukon Territory										
WHITEHORSE	-38.1	-33.5	25.8	14.3	23.5	13.3	21.6	12.6	6657	9
Albania										
TIRANA RINAS	-3.3	-1.8	35.1	22.5	33.8	22.7	32.7	22.9	1513	739
Algeria										
CONSTANTINE BOUDIAF INTL	-0.2	0.8	39.1	20.1	37.2	20.0	35.3	19.8	1623	878
DAR EL BEIDA	1.9	3.0	35.5	22.4	33.8	22.5	32.1	22.6	964	919
ORAN ES SENIA	2.6	3.9	34.4	21.2	32.8	21.4	31.1	21.6	861	953
Argentina										
BUENOS AIRES EZEIZA	0.0	1.7	33.9	22.5	32.3	22.1	31.1	21.7	1155	684
BUENOS AIRES NEWBERY	4.8	5.9	31.6	23.2	30.1	23.0	29.0	22.5	862	784
CORDOBA	-0.7	1.1	35.0	21.5	33.2	21.1	31.9	20.9	973	741
CORRIENTES	4.5	6.0	36.9	24.5	35.7	24.6	34.2	24.3	377	1654
MAR DEL PLATA	-1.1	0.1	31.3	21.2	29.3	20.5	27.5	20.0	1833	254
MENDOZA	-0.3	1.0	35.9	19.8	34.2	19.7	33.0	19.3	1198	932
PARANA	2.7	4.0	34.4	23.2	33.0	22.8	31.8	22.4	811	924
POSADAS	5.2	7.0	36.1	24.0	35.1	24.0	34.1	23.9	293	1808
RESISTENCIA	2.0	3.9	37.2	24.2	36.0	24.5	34.8	24.4	446	1622
ROSARIO	-0.4	1.1	34.5	23.3	33.1	22.9	31.9	22.5	980	834

Climatic Design Conditions for Selected Locations (*Continued*)

Station	Heating DB		Cooling DB/MCWB						Heat./Cool. Degree-Days	
			0.4%		1%		2%			
	99.6%	99%	DB / MCWB		DB / MCWB		DB / MCWB		HDD / CDD 18.3	
SALTA	-1.0	0.8	33.1	18.6	31.8	18.3	30.1	18.8	901	597
SAN JUAN	-2.1	-0.5	38.1	19.8	36.7	19.7	35.2	19.3	1152	1163
SAN MIGUEL DE TUCUMAN	3.3	4.9	36.7	23.5	35.1	23.4	33.8	23.2	542	1286
SANTA FE	0.9	2.9	35.1	24.2	33.6	23.5	32.2	23.0	769	1066
SANTIAGO DEL ESTERO	-0.6	1.6	39.2	23.6	37.6	23.5	36.1	23.1	571	1520
Armenia										
YEREVAN ARABKIR	-13.1	-10.7	36.1	21.6	34.8	21.0	33.2	20.4	2740	771
Aruba										
QUEEN BEATRIX INTL	23.9	24.4	33.9	27.2	33.2	27.0	33.0	26.9	0	3891
Australia										
ADELAIDE AP	3.9	5.0	36.8	18.5	34.1	17.9	32.0	17.5	1129	513
ADELAIDE KENT TOWN	4.6	5.7	38.1	19.1	35.8	18.6	33.5	18.2	1050	644
ADELAIDE MOUNT LOFTY	2.5	3.1	31.8	16.2	29.7	15.6	27.8	14.9	2477	206
BRISBANE AP	6.0	7.6	30.8	23.1	29.7	23.1	28.9	22.6	321	1038
BRISBANE ARCHERFIELD	5.5	6.7	33.0	22.8	31.6	22.6	30.4	22.2	350	1106
CANBERRA AP	-3.6	-2.4	34.5	18.1	32.2	17.6	30.2	17.0	2026	304
CANBERRA TUGGERANONG	-3.7	-2.6	34.3	18.5	32.1	18.0	30.0	17.3	2037	306
COOLANGATTA	6.0	7.8	29.7	23.3	28.9	23.3	28.1	22.7	302	977
GOLD COAST SEAWAY	9.6	10.8	30.5	23.4	29.4	23.2	28.5	22.9	182	1151
MELBOURNE AP	2.8	3.9	35.4	18.1	32.9	17.8	30.2	17.4	1635	293
MELBOURNE LAVERTON	1.8	2.9	35.4	19.1	32.5	18.5	29.6	18.1	1618	263
MELBOURNE MOORABBIN	2.7	4.0	34.6	19.3	31.9	18.6	29.3	18.3	1541	259
MELBOURNE REGIONAL OFFICE	4.7	5.7	35.0	18.8	32.5	18.3	30.0	17.9	1241	361
MELBOURNE SCORESBY	2.3	3.4	34.5	19.5	32.1	19.0	29.9	18.6	1609	291
NEWCASTLE NOBBYS HEAD	7.6	8.6	31.0	19.6	28.1	19.5	26.1	20.4	573	602

Climatic Design Conditions for Selected Locations *(Continued)*

Station	Heating DB		Cooling DB/MCWB						Heat./Cool. Degree-Days	
			0.4%		1%		2%			
	99.6%	99%	DB / MCWB		DB / MCWB		DB / MCWB		HDD / CDD 18.3	
NEWCASTLE WILLIAMTOWN	3.9	5.0	35.2	20.8	32.6	20.7	30.5	20.6	800	639
PERTH AP	3.8	5.0	37.5	19.4	35.7	19.2	33.8	19.0	754	840
PERTH JANDAKOT	2.0	3.4	36.5	19.8	34.7	19.6	32.8	19.2	916	708
PERTH METRO	3.9	5.1	36.4	20.2	34.5	19.8	32.6	19.5	738	789
PERTH SWANBOURNE	6.5	7.6	35.0	19.9	32.7	19.7	30.7	19.4	623	710
SYDNEY AP	6.7	7.6	33.3	19.5	30.9	20.1	29.0	20.1	636	717
SYDNEY BANKSTOWN	3.3	4.4	34.4	20.4	32.0	20.5	30.0	20.1	880	608
SYDNEY CANTERBURY	3.7	4.7	33.0	20.1	30.7	20.3	28.9	20.1	882	549
SYDNEY OBSERVATORY	7.2	8.0	31.7	19.8	29.3	20.1	27.7	20.2	611	661
SYDNEY OLYMPIC PARK ARCHERY	5.0	6.1	34.1	20.6	31.8	20.3	29.9	20.2	736	678
Austria										
GUMPOLDSKIRCHEN	-9.4	-7.2	31.9	21.5	30.0	20.7	28.2	19.8	2901	301
TULLN LANGENLEBARN	-10.7	-7.9	32.2	21.7	30.2	20.9	28.5	20.0	3037	257
WIEN HOHE WARTE	-8.9	-6.8	31.8	21.8	29.9	21.0	28.2	20.1	2883	307
WIEN INNERE STADT	-7.7	-5.7	32.4	21.7	30.6	21.0	28.9	20.2	2614	429
WIEN SCHWECHAT	-10.0	-7.8	31.8	21.0	29.9	20.4	28.1	19.5	2995	273
Bahamas										
NASSAU INTL	14.9	16.5	33.9	26.5	33.0	26.3	32.2	26.0	6	2807
Bahrain										
BAHRAIN INTL	12.5	13.8	41.2	23.7	40.1	24.5	39.1	25.1	73	3477
Bangladesh										
DHAKA HAZRAT SHAHJALAL INTL	12.8	13.9	36.1	26.0	35.2	26.2	34.3	26.2	17	3174
Barbados										
GRANTLEY ADAMS INTL	22.7	23.1	31.2	26.4	31.1	26.3	31.0	26.3	0	3346

Climatic Design Conditions for Selected Locations (*Continued*)

Station	Heating DB		Cooling DB/MCWB						Heat./Cool. Degree-Days	
	99.6%	99%	0.4%		1%		2%		HDD / CDD 18.3	
			DB / MCWB		DB / MCWB		DB / MCWB			
Belarus										
BREST SHEBRIN	-17.1	-13.4	30.7	20.0	28.8	19.[illegible]	26.9	18.3	3667	167
GOMEL	-20.3	-17.1	30.9	20.3	29.1	19.6	27.4	18.8	4024	202
GRODNO	-19.1	-15.7	29.4	20.0	27.3	19.[illegible]	25.7	18.1	4009	111
MINSK	-19.2	-16.1	29.6	20.0	27.9	18.9	26.0	18.1	4181	131
MOGILEV	-21.8	-18.5	29.1	19.9	27.2	19.0	25.5	18.4	4409	107
VITEBSK	-21.6	-18.2	28.9	20.0	27.0	19.0	25.2	18.0	4387	118
Belgium										
ANTWERP INTL	-6.1	-4.1	29.8	20.6	27.6	19.5	25.6	18.6	2721	123
BRUSSELS AP	-6.1	-4.1	29.4	20.3	27.2	19.3	25.3	18.4	2820	107
UCCLE	-5.7	-3.9	29.3	19.6	27.3	18.8	25.3	17.8	2784	124
Belize										
LADYVILLE GOLDSON INTL	17.2	18.5	33.0	27.1	32.2	26.9	32.1	26.9	0	3169
Benin										
COTONOU	22.0	22.8	33.0	26.9	32.4	27.0	32.1	27.1	0	3458
Bermuda										
BERMUDA INTL	12.9	13.8	30.4	25.5	30.0	25.4	29.2	25.0	102	1555
Bolivia										
COCHABAMBA	2.1	3.8	30.1	14.6	29.2	14.4	28.2	14.2	463	310
LA PAZ EL ALTO	-5.0	-3.9	18.1	5.7	17.1	5.6	16.2	5.5	3841	0
SANTA CRUZ DE LA SIERRA	9.8	11.1	35.1	23.5	34.1	23.7	33.2	23.9	79	2252
Bosnia and Herzegovina										
BJELASNICA	-18.7	-16.3	19.5	12.7	18.1	12.3	16.8	11.8	5954	2
SARAJEVO	-12.9	-9.9	33.1	19.7	31.2	19.6	29.2	19.2	3067	248
SARAJEVO-BJELAVE	-11.2	-8.6	33.2	19.8	31.3	19.4	29.4	18.9	2909	308

Climatic Design Conditions for Selected Locations *(Continued)*

Station	Heating DB		Cooling DB/MCWB						Heat./Cool. Degree-Days	
			0.4%		1%		2%			
	99.6%	99%	DB / MCWB		DB / MCWB		DB / MCWB		HDD / CDD 18.3	
Botswana										
GABORONE	1.9	3.2	36.9	16.9	35.5	17.0	34.2	17.1	447	1387
Brazil										
ANAPOLIS	12.9	14.0	32.9	18.1	31.8	18.4	30.8	18.9	9	1584
ARACAJU	20.9	21.8	32.1	26.6	31.5	26.4	31.0	26.2	0	3065
BELEM	22.8	22.9	33.2	25.8	33.0	25.8	32.2	25.7	0	3389
BELO HORIZONTE CONFINS	10.8	11.8	32.1	19.6	31.0	19.9	30.1	19.9	61	1231
BELO HORIZONTE PAMPULHA	11.0	12.1	33.1	19.6	32.1	19.8	31.1	19.9	22	1601
BRASILIA	10.1	11.2	32.8	17.4	31.8	17.6	30.9	18.1	16	1437
CAMPINAS	9.0	10.8	33.8	20.2	32.8	20.5	31.8	20.6	92	1462
CAMPO GRANDE	8.0	10.0	35.0	20.2	34.0	20.6	33.0	20.9	87	2163
CUIABA	13.1	15.0	38.4	21.8	37.4	22.0	36.6	22.4	12	3362
CURITIBA	3.1	5.2	31.0	20.1	29.9	20.3	28.8	20.2	608	599
FLORIANOPOLIS	8.2	10.0	32.2	25.3	31.0	25.0	30.0	24.5	195	1327
FORTALEZA	22.8	23.0	32.1	24.8	31.8	24.7	31.2	24.5	0	3368
GOIANIA	12.9	14.0	35.9	19.1	34.8	19.3	33.9	19.8	3	2418
GUARULHOS	7.9	9.2	32.9	21.6	31.8	21.5	30.8	21.4	206	1077
LONDRINA	8.1	10.0	34.2	21.6	33.2	21.9	32.1	22.0	109	1686
MACAPA	22.8	22.9	35.2	26.4	35.1	26.4	34.8	26.5	0	3710
MACEIO	18.5	19.0	33.0	25.5	32.2	25.1	31.8	25.0	0	2667
MANAUS GOMES	21.8	21.9	35.8	26.0	35.0	25.7	34.1	25.7	0	3376
MANAUS PONTA PELADA	22.3	22.9	35.0	25.8	34.2	25.8	33.7	25.8	0	3461
NATAL	20.8	21.1	32.7	25.4	32.1	25.2	31.8	25.1	0	3093
PORTO ALEGRE	4.2	6.0	34.9	24.6	33.1	24.0	31.9	23.7	455	1172
PORTO VELHO	18.9	20.1	36.0	24.0	35.1	24.3	34.2	24.6	1	3341

Climatic Design Conditions for Selected Locations (*Continued*)

Station	Heating DB		Cooling DB/MCWB						Heat./Cool. Degree-Days	
			0.4%		1%		2%			
	99.6%	99%	DB	MCWB	DB	MCWB	DB	MCWB	HDD	CDD 18.3
RECIFE	21.5	22.0	34.0	27.1	33.2	26.5	32.8	26.4	0	3314
RIO DE JANEIRO GALEAO	14.9	15.9	36.1	24.6	34.9	24.7	33.8	24.7	6	2348
RIO DE JANEIRO SANTOS DUMONT	16.8	17.2	34.8	25.7	33.1	25.5	32.1	25.2	3	2353
SALVADOR	20.8	21.2	32.2	26.6	31.9	26.5	31.2	26.1	0	3008
SAO LUIS	22.8	23.1	33.3	25.5	33.0	25.4	32.2	25.0	0	3587
SAO PAULO CONGONHAS	9.1	10.8	32.2	20.1	31.2	20.1	30.1	20.1	201	1181
TERESINA	21.9	22.6	39.0	22.3	38.1	22.5	37.2	22.9	0	4027
VITORIA	16.6	17.5	34.1	25.5	33.2	25.2	32.5	25.0	0	2601
Brunei Darussalam										
BRUNEI INTL	22.9	23.2	33.8	25.8	33.1	26.0	32.8	26.0	0	3541
Bulgaria										
CHERNI VRAH	-19.1	-17.2	17.3	11.0	15.9	10.5	14.7	10.1	6386	0
PLOVDIV	-10.0	-7.2	34.9	20.9	33.2	20.5	31.9	20.1	2471	578
SOFIA	-12.2	-9.8	33.1	19.2	31.1	18.9	29.6	18.6	2958	323
VARNA	-9.1	-6.8	32.0	22.1	30.6	22.0	29.1	21.6	2467	467
Burkina Faso										
BOBO-DIOULASSO	18.2	19.3	38.2	20.4	37.7	20.4	36.9	20.6	0	3502
OUAGADOUGOU	16.2	17.2	41.0	20.5	40.1	20.5	39.1	20.6	0	3892
Cameroon										
YAOUNDE	16.9	18.1	32.2	22.7	31.8	22.8	31.0	23.0	0	2274
Central African Republic										
BANGUI M'POKO INTL	15.9	17.1	36.2	23.0	35.5	23.1	34.9	23.3	0	3068
Chad										
N'DJAMENA INTL	13.3	14.9	43.1	21.6	42.2	21.7	41.2	21.5	0	3937

Climatic Design Conditions for Selected Locations *(Continued)*

Station	Heating DB		Cooling DB/MCWB						Heat./Cool. Degree-Days	
			0.4%		1%		2%			
	99.6%	99%	DB / MCWB		DB / MCWB		DB / MCWB		HDD / CDD 18.3	
Chile										
ANTOFAGASTA	10.0	10.9	24.2	19.0	23.8	18.6	23.0	18.1	711	184
SANTIAGO PUDAHUEL	-0.8	0.4	32.1	17.0	31.0	16.9	29.9	16.8	1429	306
China										
ANQING	-2.1	-0.9	35.7	27.3	34.6	27.2	33.6	26.8	1553	1332
ANYANG	-8.5	-6.7	35.5	23.0	34.1	23.7	32.8	23.7	2306	1023
BAODING	-10.8	-8.9	35.4	22.9	33.9	23.1	32.6	23.2	2642	966
BAOJI	-5.7	-4.4	35.3	21.6	34.0	21.4	32.6	21.4	2282	862
BEIJING	-11.6	-9.8	35.2	22.0	33.9	22.2	32.2	22.3	2842	896
BENGBU	-4.7	-3.3	35.6	26.6	34.4	26.2	33.1	25.5	1874	1148
BENXI	-22.1	-19.9	31.9	22.2	30.5	21.9	29.3	21.6	4046	525
CANGZHOU	-9.4	-7.7	34.2	23.2	33.0	23.5	31.9	23.4	2649	927
CHANGCHUN	-25.8	-23.2	31.2	20.9	30.0	21.0	28.8	20.7	4832	441
CHANGDE	-1.2	-0.1	36.5	26.5	35.2	26.5	34.0	26.2	1459	1338
CHANGSHA	-1.2	-0.1	36.4	26.0	35.3	26.0	34.2	25.9	1435	1394
CHAOYANG	-18.9	-16.7	34.2	21.3	32.5	21.3	31.2	21.0	3654	673
CHENGDE	-18.3	-16.4	33.4	20.7	31.9	20.7	30.5	20.4	3786	539
CHENGDU SHUANGLIU	0.9	2.0	34.2	24.9	33.1	24.5	32.0	24.0	1284	1081
CHIFENG	-20.5	-18.7	33.1	19.5	31.5	19.2	30.0	18.7	4208	456
CHONGQING	2.8	3.8	37.4	25.0	36.1	25.1	34.8	25.0	1152	1321
DANDONG	-16.1	-14.2	30.2	24.0	28.9	23.3	27.7	22.7	3599	475
DATONG	-21.2	-19.0	32.0	17.7	30.5	17.2	29.1	17.1	4204	349
DEZHOU	-8.3	-6.8	34.2	24.3	32.9	24.3	31.9	24.0	2496	966
FUZHOU	4.8	5.9	35.5	26.6	34.2	26.4	33.1	26.2	684	1669
GANYU	-7.0	-5.4	33.5	26.5	32.1	26.0	30.8	25.7	2274	886

Climatic Design Conditions for Selected Locations (*Continued*)

Station	Heating DB		Cooling DB/MCWB						Heat./Cool. Degree-Days	
			0.4%		1%		2%			
	99.6%	99%	DB	MCWB	DB	MCWB	DB	MCWB	HDD	CDD 18.3
GAOYAO	6.4	7.6	35.1	26.4	34.3	26.3	33.5	26.2	372	2098
GUANGZHOU	5.9	7.0	36.0	26.2	35.0	26.2	34.0	26.1	371	2130
GUILIN	1.0	2.3	35.2	25.6	34.2	25.5	33.2	25.4	1002	1526
GUIYANG	-3.2	-1.9	30.6	21.7	29.5	21.4	28.4	21.0	1721	663
HAIKOU	10.7	12.3	35.1	26.8	34.2	26.7	33.4	26.6	102	2542
HANGZHOU	-2.0	-0.8	36.9	26.5	35.8	26.4	34.4	26.2	1507	1334
HARBIN	-27.4	-25.2	31.5	20.6	30.1	21.1	28.8	20.9	5192	424
HEFEI	-4.2	-2.9	35.9	27.8	34.7	27.4	33.3	26.8	1786	1210
HOHHOT	-23.2	-20.3	32.0	17.4	30.6	17.1	29.2	16.8	4416	350
HUAIYIN	-5.4	-4.1	34.1	27.0	33.0	26.6	31.9	25.8	2043	997
HUICHUAN	-0.5	0.6	32.6	22.7	31.6	22.5	30.5	22.3	1606	859
JIANGLING	-1.9	-0.7	35.2	27.8	34.2	27.3	33.2	26.7	1551	1264
JINAN	-8.4	-6.6	35.2	23.3	34.0	23.5	32.7	23.2	2234	1103
JINGDEZHEN	-1.1	0.1	36.4	26.6	35.4	26.3	34.4	26.1	1284	1466
JINZHOU	-15.9	-14.0	32.1	22.1	30.8	21.8	29.6	21.6	3454	652
JIXI	-24.6	-22.7	30.6	20.9	29.2	20.5	27.7	20.2	5206	298
KUNMING WUJIABA	0.1	1.9	28.1	16.4	27.1	16.4	26.1	16.5	1126	358
LANZHOU	-11.3	-9.8	32.6	18.0	31.1	17.4	29.7	16.8	3064	451
LINGXIAN	-10.6	-8.6	35.0	23.7	33.7	24.0	32.4	24.0	2567	930
LIUZHOU	3.2	4.6	35.2	25.8	34.4	25.7	33.6	25.6	703	1885
MENGJIN	-6.2	-5.0	35.1	22.1	33.6	22.3	32.2	22.4	2137	973
MUDANJIANG	-26.2	-24.1	31.5	21.3	30.0	20.7	28.6	20.3	5131	354
NANCHANG	-0.6	0.5	35.9	26.7	35.0	26.6	34.0	26.4	1317	1504
NANJING	-4.8	-3.1	36.0	26.8	34.8	26.5	33.3	26.2	1806	1163
NANNING	4.8	6.1	35.1	26.4	34.1	26.3	33.2	26.0	489	1964

Climatic Design Conditions for Selected Locations *(Continued)*

Station	Heating DB		Cooling DB/MCWB						Heat./Cool. Degree-Days	
			0.4%		1%		2%			
	99.6%	99%	DB / MCWB		DB / MCWB		DB / MCWB		HDD / CDD 18.3	
NEIJIANG	2.3	3.5	35.5	25.9	34.2	25.6	33.0	25.2	1174	1194
SANJIAZI	-27.9	-25.8	32.2	20.9	30.4	20.5	29.0	20.2	5382	424
SHANGHAI BAOSHAN	-1.9	-0.6	35.5	26.7	34.2	26.5	33.0	26.2	1564	1221
SHANGHAI HONGQIAO INTL	-2.5	-1.1	36.1	27.2	34.9	27.0	33.8	26.8	1533	1302
SHANTOU	7.3	8.8	34.9	27.0	33.9	27.0	32.9	26.9	314	1967
SHAOGUAN	2.3	3.5	35.3	25.9	34.4	25.8	33.5	25.6	759	1744
SHENGYANG TAOXIAN	-23.9	-21.2	32.1	23.3	30.9	23.1	29.9	22.5	4106	578
SHENYANG	-22.8	-20.4	31.9	23.3	30.7	22.7	29.7	22.2	4087	566
SHENZHEN	7.2	8.8	34.0	26.4	33.1	26.4	32.3	26.3	236	2254
SHIJIAZHUANG	-8.0	-6.5	36.2	22.1	34.6	22.6	33.2	22.8	2387	1080
SIPING	-23.9	-21.7	31.2	21.7	30.0	21.5	28.9	21.2	4496	494
TAI SHAN	-16.7	-14.6	22.9	17.5	21.9	17.6	21.1	17.9	4399	53
TAIYUAN	-14.3	-12.2	33.9	19.7	32.2	19.6	31.0	19.4	3120	591
TANGSHAN	-13.7	-11.6	33.8	23.2	32.5	23.2	31.3	22.9	2951	841
TIANHE	-2.5	-1.1	36.2	27.6	35.2	27.3	34.1	27.0	1570	1357
TIANJIN	-10.5	-8.8	34.8	23.2	33.4	23.3	32.2	23.1	2689	978
TIANJIN BINHAI INTL	-10.9	-9.0	35.0	22.8	33.8	23.1	32.1	22.8	2725	959
URUMQI	-22.1	-19.8	33.7	16.7	32.1	16.3	30.7	15.9	4291	573
URUMQI DIWOPU INTL	-24.1	-21.9	35.9	18.3	34.2	17.8	33.0	17.5	4246	819
WEIFANG	-10.4	-8.7	34.6	24.2	33.3	24.0	32.0	23.7	2611	869
WENZHOU	1.3	2.8	33.9	27.3	32.9	27.0	32.0	26.7	1076	1306
WUHUXIAN	-3.1	-1.8	36.2	27.4	35.1	27.2	33.8	26.8	1658	1242
XIAMEN	6.9	8.0	34.9	26.3	34.0	26.4	33.0	26.2	417	1857
XIANYANG	-8.9	-7.0	36.8	23.3	35.1	23.0	33.8	22.8	2348	920
XIHUA	-5.2	-3.9	35.3	25.6	34.1	25.5	32.9	25.0	1999	1081

Station	Heating DB		Cooling DB/MCWB						Heat./Cool. Degree-Days	
			0.4%		1%		2%			
	99.6%	99%	DB / MCWB		DB / MCWB		DB / MCWB		HDD / CDD 18.3	
XINGTAI	-7.3	-5.9	35.9	22.3	34.3	22.7	33.0	23.0	2305	1073
XINING	-17.3	-15.6	28.0	15.4	26.2	14.5	24.6	13.7	4208	54
XINYANG	-4.4	-3.1	35.0	26.6	33.8	25.9	32.6	25.3	1837	1108
XINZHENG	-6.1	-4.9	35.7	23.5	34.2	23.8	33.0	23.7	2099	1049
XUZHOU	-6.1	-4.6	35.0	25.9	33.8	25.5	32.6	24.9	2075	1078
YANGJIANG	7.2	8.5	33.0	26.7	32.2	26.5	31.5	26.3	265	2065
YANJI	-21.8	-19.8	31.4	21.8	29.7	21.0	28.1	20.4	4702	308
YICHANG	-1.2	0.0	35.7	26.6	34.5	26.1	33.2	25.5	1471	1217
YINCHUAN	-15.8	-13.9	32.9	18.7	31.7	18.5	30.4	18.0	3373	546
YINGKOU	-17.5	-15.4	30.6	24.4	29.6	23.9	28.6	23.4	3602	633
YUEYANG	-1.1	0.0	34.5	27.3	33.7	27.0	33.0	26.7	1436	1362
YUNCHENG	-8.7	-6.8	36.5	22.3	35.2	22.3	33.8	22.0	2278	1058
ZHANGJIAKOU	-17.2	-15.3	33.1	18.9	31.5	18.6	30.1	18.5	3708	548
ZHANJIANG	7.6	9.1	34.0	26.8	33.2	26.8	32.5	26.7	213	2251
ZHOUSHUIZI	-12.2	-10.4	31.9	23.6	30.2	23.2	29.1	22.8	3060	671
Colombia										
CALI	17.7	18.0	32.8	22.2	32.0	22.1	31.1	22.0	0	2238
BARRANQUILLA	22.8	23.1	34.5	27.4	33.9	27.3	33.1	27.1	0	3706
BOGOTA	3.9	5.2	21.5	13.1	20.9	13.1	20.1	13.0	1614	0
CARTAGENA	23.2	23.9	33.2	27.6	32.9	27.5	32.2	27.1	0	3676
RIONEGRO	10.2	11.1	24.0	15.8	23.2	15.6	23.0	15.6	358	29
Congo										
BRAZZAVILLE MAYA MAYA INTL	18.1	19.0	34.2	24.6	33.7	24.6	33.0	24.5	0	2876
Congo, the Democratic Republic of the										
KINSHASA N'DJILI INTL	18.9	19.8	34.1	25.1	33.5	24.9	32.9	24.8	0	2947

Climatic Design Conditions for Selected Locations *(Continued)*

Station	Heating DB		Cooling DB/MCWB						Heat./Cool. Degree-Days	
			0.4%		1%		2%		HDD / CDD 18.3	
	99.6%	99%	DB / MCWB		DB / MCWB		DB / MCWB			
Costa Rica										
SAN JOSE SANTAMARIA INTL	16.9	17.7	30.9	20.6	30.0	20.4	29.2	20.4	0	1871
Côte d'Ivoire										
ABIDJAN	21.7	22.2	32.9	27.3	32.2	27.0	31.9	26.9	0	3284
Croatia										
ZAGREB MAKSIMIR	-9.3	-6.7	32.9	21.7	31.2	21.2	29.6	20.5	2657	379
ZAGREB PLESO	-10.8	-8.0	33.0	22.0	31.2	21.4	29.9	21.0	2769	347
Cuba										
CAMAGUEY INTL	15.0	16.6	33.8	23.6	33.1	23.8	32.6	23.8	3	2700
HAVANA JOSE MARTI	10.2	12.1	33.1	24.9	32.8	24.9	32.1	24.8	27	2313
SANTIAGO DE CUBA	18.8	19.8	32.1	24.9	31.8	25.0	31.2	25.0	0	2881
Cyprus										
NICOSIA ERCAN	2.1	3.8	39.1	21.2	37.9	21.1	36.9	21.1	862	1503
Czech Republic										
BRNO-TURANY	-11.8	-9.4	31.3	20.4	29.5	19.8	27.7	19.0	3291	224
OSTRAVA MOSNOV	-14.5	-11.5	30.8	20.3	28.8	19.7	27.0	18.8	3474	148
PRAHA-KBELY	-12.1	-9.5	30.3	19.7	28.4	18.9	26.7	18.3	3347	164
PRAHA-LIBUS	-11.7	-9.2	31.4	19.5	29.2	18.8	27.4	18.0	3306	182
PRAHA-RUZYNE	-12.7	-10.0	30.1	19.3	28.1	18.6	26.2	17.9	3554	125
Denmark										
DROGDEN FYR	-5.2	-3.9	22.8	18.7	21.7	18.1	20.6	17.4	3323	41
KOEBENHAVNS AP	-6.9	-5.2	25.8	18.5	24.2	18.0	22.8	17.3	3453	60
ROSKILDE AP	-9.1	-7.0	26.0	18.4	24.2	18.0	22.8	17.4	3655	38
VAERLOSE	-10.0	-7.4	26.8	18.5	25.1	18.0	23.2	17.4	3773	46

Climatic Design Conditions for Selected Locations (*Continued*)

Station	Heating DB		Cooling DB/MCWB						Heat./Cool. Degree-Days	
	99.6%	99%	0.4%		1%		2%		HDD / CDD 18.3	
			DB / MCWB		DB / MCWB		DB / MCWB			
Dominican Republic										
LAS AMERICAS INTL	18.2	19.0	33.1	26.3	32.2	26.1	32.0	26.1	0	2923
SANTO DOMINGO	19.9	20.7	32.7	27.2	32.2	27.1	31.7	27.0	0	3104
Ecuador										
GUAYAQUIL	18.9	19.2	32.9	24.0	32.1	24.	31.8	24.1	0	2818
QUITO	6.3	7.2	21.9	11.8	21.1	11.9	20.8	11.9	1417	0
Egypt										
ALEXANDRIA INTL	7.1	8.1	33.9	22.0	32.1	23.2	31.1	23.6	423	1420
ASSIUT INTL	4.8	5.9	41.7	21.1	40.1	21.0	38.8	20.8	457	2186
CAIRO INTL	8.2	9.2	38.8	21.0	37.1	21.4	36.0	21.6	311	1986
LUXOR INTL	6.0	7.2	43.8	23.4	42.5	23.0	41.4	22.7	242	2925
PORT SAID EL GAMIL	9.6	10.8	32.2	25.3	31.2	25.4	30.8	25.3	292	1607
El Salvador										
EL SALVADOR INTL	19.8	20.2	35.2	23.0	34.8	23.3	34.0	23.7	0	3439
Equatorial Guinea										
MALABO	21.8	21.9	32.2	26.6	32.0	26.6	31.6	26.5	0	3024
Estonia										
TALLINN	-18.5	-15.1	26.9	19.4	25.0	18.2	23.1	17.2	4506	45
Ethiopia										
ADDIS ABABA BOLE INTL	7.5	8.9	26.2	13.8	25.5	13.9	24.9	13.9	464	109
Faroe Islands										
TORSHAVN	-3.6	-2.3	14.3	12.4	13.5	11.9	13.0	11.5	4143	0
Fiji										
SUVA	19.4	20.1	30.8	N/A	30.3	N/A	29.8	N/A	0	2559

Climatic Design Conditions for Selected Locations *(Continued)*

Station	Heating DB		Cooling DB/MCWB						Heat./Cool. Degree-Days	
			0.4%		1%		2%			
	99.6%	99%	DB / MCWB		DB / MCWB		DB / MCWB		HDD / CDD 18.3	
Finland										
HELSINKI VANTAA	-21.5	-18.2	27.2	18.4	25.6	17.4	23.9	16.6	4637	59
ISOSAARI	-18.2	-14.9	23.0	19.0	21.8	18.5	20.7	17.7	4491	36
France										
CAP COURONNE	-2.6	0.4	30.7	22.6	29.5	22.3	28.3	21.8	1568	567
CAP FERRAT	4.0	5.2	29.1	22.4	28.1	22.4	27.1	22.1	1270	560
CAP POMEGUES	-1.4	1.8	28.6	22.0	27.3	21.9	26.3	21.5	1523	457
LYON ST EXUPERY AP	-5.8	-4.0	33.0	20.0	31.1	19.6	29.3	19.2	2401	364
LYON-BRON AP	-5.3	-3.6	33.8	20.2	31.8	19.8	30.0	19.4	2323	400
MARSEILLE PROVENCE AP	-2.2	-0.8	33.1	21.0	31.8	20.7	30.6	20.4	1588	675
NICE COTE D'AZUR AP	2.1	3.4	29.7	22.7	28.8	22.4	27.9	22.2	1342	588
PARIS CHARLES DE GAULLE AP	-4.9	-3.2	31.1	19.9	29.0	19.2	27.0	18.4	2510	191
PARIS LE BOURGET AP	-4.3	-2.7	31.3	20.1	29.1	19.3	27.1	18.5	2469	181
PARIS MONTSOURIS	-2.9	-1.6	31.6	20.4	29.4	19.5	27.5	18.7	2256	254
PARIS ORLY AP	-4.7	-3.0	31.4	20.2	29.3	19.5	27.4	18.7	2512	202
TOULOUSE BLAGNAC AP	-3.7	-2.0	33.3	20.8	31.5	20.3	29.8	19.8	1989	417
TRAPPES	-4.6	-3.1	30.5	19.9	28.2	19.0	26.3	18.3	2638	151
VELIZY-VILLACOUBLAY AB	-4.6	-3.0	30.4	19.8	28.3	19.0	26.5	18.3	2668	168
French Guiana										
CAYENNE MATOURY	21.6	21.9	32.9	25.2	32.2	25.2	32.0	25.2	0	3093
Gabon										
LIBREVILLE INTL	22.0	22.8	31.8	27.3	31.2	27.1	30.9	27.0	0	3058
Gambia										
BANJUL YUNDUM	16.5	17.2	37.9	20.3	36.1	20.4	35.0	20.9	0	3158

Climatic Design Conditions for Selected Locations (Continued)

Station	Heating DB		Cooling DB/MCWB						Heat./Cool. Degree-Days	
	99.6%	99%	0.4% DB / MCWB		1% DB / MCWB		2% DB / MCWB		HDD / CDD 18.3	
Georgia										
TBILISI INTL	-6.8	-4.6	35.1	21.9	33.9	21.5	32.3	21.5	2240	753
Germany										
BERLIN DAHLEM	-12.0	-9.1	29.3	19.0	27.3	18.1	25.6	17.4	3390	118
BERLIN SCHONEFELD	-11.8	-8.2	31.2	19.4	29.2	18.7	27.2	18.3	3163	184
BERLIN TEGEL	-10.2	-7.2	31.2	19.2	29.2	18.6	27.2	18.1	3041	214
BERLIN TEMPELHOF	-11.3	-8.3	30.2	19.1	28.2	18.4	26.4	17.7	3182	163
BREMEN	-8.8	-6.8	29.8	19.6	27.3	19.0	25.4	18.1	3192	107
CELLE	-10.1	-7.6	30.8	19.4	28.7	18.6	26.6	17.8	3169	138
DRESDEN	-11.2	-8.2	31.0	19.3	28.9	18.5	26.9	18.2	3202	180
DUSSELDORF	-6.2	-4.1	31.1	19.7	28.9	18.9	26.9	18.2	2721	172
ESSEN MULHEIM	-9.9	-6.9	28.2	19.3	26.6	18.4	24.8	17.6	3178	103
FRANKFURT AM MAIN	-8.2	-5.9	32.1	20.0	30.0	19.2	28.0	18.4	2854	233
FURSTENFELDBRUCK	-15.1	-12.1	29.1	18.9	27.1	18.1	25.2	17.2	3706	82
GUTERSLOH	-8.8	-6.2	30.2	19.3	28.0	18.7	26.0	18.0	3077	122
HAMBURG FUHLSBUTTEL	-8.2	-6.1	29.1	19.2	27.1	19.0	25.1	17.9	3149	101
HANNOVER	-9.0	-6.8	30.2	19.7	28.1	19.0	26.1	18.1	3064	128
HEIDELBERG	-8.2	-5.8	32.1	20.4	30.2	19.7	28.2	18.7	2721	276
ITZEHOE	-9.6	-7.3	28.4	18.7	26.1	18.3	24.1	17.7	3486	62
KOELN BONN	-8.0	-5.8	31.1	19.8	28.9	18.9	26.9	18.2	2882	151
LEIPZIG HALLE	-11.1	-8.0	31.2	19.6	29.1	19.0	27.1	18.4	3109	188
LEIPZIG HOLZHAUSEN	-11.0	-8.2	30.3	19.3	28.3	18.4	26.5	17.9	3169	153
MUNICH INTL	-12.9	-9.8	30.1	19.3	28.1	18.7	26.2	18.1	3436	119
NORVENICH	-7.2	-5.1	31.1	19.7	28.8	19.1	26.8	18.2	2834	145
NURNBERG	-11.0	-7.9	31.9	19.6	29.8	19.0	27.8	18.3	3223	171

Climatic Design Conditions for Selected Locations (*Continued*)

Station	Heating DB		Cooling DB/MCWB						Heat./Cool. Degree-Days	
			0.4%		1%		2%			
	99.6%	99%	DB / MCWB		DB / MCWB		DB / MCWB		HDD / CDD 18.3	
POTSDAM	-12.6	-10.0	29.7	18.9	27.7	18.3	25.9	17.7	3437	119
QUICKBORN	-9.7	-7.4	28.3	18.9	26.2	18.5	24.3	17.7	3461	56
ROTH	-13.1	-10.0	31.2	19.3	29.0	18.4	27.1	17.8	3520	120
STUTTGART ECHTERDINGEN	-10.0	-7.2	31.1	19.6	29.1	19.0	27.2	18.5	3062	177
STUTTGART SCHNARREN	-11.5	-9.0	29.6	19.6	27.8	18.6	26.1	17.9	3152	160
WUNSTORF	-9.2	-6.9	30.9	19.6	28.5	18.8	26.4	18.0	3058	144
Ghana										
ACCRA	22.4	22.9	33.7	25.7	33.1	25.8	32.8	25.8	0	3493
Greece										
ATHINAI HELLINIKON	1.9	3.6	35.8	21.2	34.3	21.1	33.2	21.2	1046	1194
ELEFSIS	0.9	2.2	37.0	20.7	35.3	20.5	34.1	20.2	1171	1213
THESSALONIKI MAKEDONIA	-3.0	-1.2	34.9	21.7	33.2	21.5	32.1	21.2	1711	887
Guatemala										
GUATEMALA LA AURORA	10.9	12.0	28.1	17.9	27.1	18.0	26.2	17.9	58	730
Guyana										
TIMEHRI	20.8	21.4	33.6	25.8	33.0	25.7	32.2	25.5	0	3099
Honduras										
SAN PEDRO SULA MORALES	17.8	18.8	36.8	25.9	35.6	26.1	34.8	26.2	0	3227
TEGUCIGALPA TONCONTIN	11.9	13.0	32.0	19.4	31.0	19.8	30.1	19.7	11	1528
Hong Kong										
HONG KONG INTL	8.9	10.2	34.1	26.5	33.2	26.4	32.9	26.3	173	2404
HONG KONG OBSERVATORY	9.6	10.9	32.2	26.5	31.7	26.4	31.2	26.3	237	1976
Hungary										
BUDAORS	-11.2	-9.0	31.0	20.1	29.3	19.7	27.8	19.2	3072	246
BUDAPEST FERIHEGY	-11.8	-9.0	33.1	22.0	31.2	21.1	29.6	20.5	3052	311

Climatic Design Conditions for Selected Locations (*Continued*)

Station	Heating DB		Cooling DB/MCWB						Heat./Cool. Degree-Days	
			0.4%		1%		2%			
	99.6%	99%	DB / MCWB		DB / MCWB		DB / MCWB		HDD / CDD 18.3	
BUDAPEST PESTSZENTLORINC	-9.8	-7.7	33.4	20.8	31.6	20.2	29.8	19.5	2840	399
India										
AHMEDABAD	10.9	12.1	43.1	23.0	41.9	22.9	40.5	22.8	10	3548
AKOLA	11.9	13.3	43.5	22.0	42.5	21.7	41.3	21.6	1	3627
AURANGABAD	11.1	12.5	40.3	22.8	39.4	22.7	38.4	22.5	4	2876
BELGAUM	13.1	14.3	36.5	19.4	35.5	19.5	34.6	19.6	0	2269
BENGALURU	15.5	16.2	34.3	20.0	33.6	19.9	32.7	19.9	0	2202
BHOPAL	9.2	10.5	42.0	21.9	40.8	21.5	39.6	21.4	66	2753
BHUBANESHWAR	13.8	14.9	39.3	27.3	38.0	27.2	36.8	26.9	1	3417
BIKANER	6.2	7.6	44.1	21.6	42.8	22.0	41.6	22.4	176	3482
CHENNAI INTL	20.0	20.9	39.0	26.0	37.9	26.2	36.8	26.0	0	3908
COIMBATORE INTL	18.5	19.4	36.7	22.2	35.8	22.2	34.9	22.6	0	3208
GUWAHATI INTL	11.0	11.9	35.2	27.8	34.4	27.6	33.6	27.4	43	2524
GWALIOR	5.9	7.0	43.8	23.4	42.6	23.4	41.3	23.3	191	3056
HYDERABAD BEGUMPET	13.4	14.9	41.0	22.0	39.9	22.0	38.7	21.9	1	3148
INDORE INTL	9.7	11.0	40.8	20.9	39.7	20.4	38.6	20.3	45	2666
JABALPUR	8.3	9.6	42.4	20.8	41.3	20.7	40.1	20.8	82	2860
JAIPUR	7.2	8.7	42.7	20.9	41.2	21.1	40.0	21.1	158	3041
JAMSHEDPUR	10.1	11.2	42.1	22.6	40.6	22.9	39.0	23.4	25	3118
JODHPUR	8.7	9.9	42.8	21.7	41.5	21.9	40.2	22.0	72	3392
KOLKATA BOSE INTL	11.2	12.2	37.9	27.3	36.9	27.5	35.9	27.2	21	3164
KOZHIKODE	22.6	23.1	35.2	28.2	34.5	27.8	33.8	27.5	0	3661
LUCKNOW	6.5	7.9	42.8	23.4	41.2	23.5	39.9	23.7	200	2885
MANGALORE INTL	21.0	21.8	34.4	24.9	34.0	25.0	33.4	24.9	0	3381
MUMBAI SHIVAJI INTL	17.2	18.8	36.0	22.7	35.0	23.0	34.1	23.4	0	3571

Climatic Design Conditions for Selected Locations *(Continued)*

Station	Heating DB		Cooling DB/MCWB						Heat./Cool. Degree-Days	
			0.4%		1%		2%			
	99.6%	99%	DB / MCWB		DB / MCWB		DB / MCWB		HDD / CDD 18.3	
NAGPUR AMBEDKAR INTL	11.2	12.8	44.2	22.6	43.1	22.4	42.0	22.3	5	3367
NELLORE	20.8	21.4	40.9	27.1	39.5	27.2	38.2	27.0	0	4164
NEW DELHI INDIRA GANDHI INTL	5.9	7.0	43.8	22.2	42.1	22.3	40.8	22.4	284	2988
NEW DELHI SAFDARJUNG	6.0	7.1	42.3	23.2	41.0	23.6	39.6	23.7	264	2872
PATIALA	5.2	6.3	41.8	24.6	40.3	24.4	38.7	24.5	386	2495
PATNA	7.7	8.9	41.3	23.2	39.9	23.4	38.1	24.2	144	2943
PUNE	10.0	11.2	38.5	19.9	37.4	19.8	36.4	19.9	6	2428
RAJKOT	12.1	13.3	41.4	22.1	40.2	22.0	39.1	22.5	4	3492
SOLAPUR	15.3	16.7	41.3	22.3	40.3	22.3	39.3	22.2	0	3574
SURAT	14.1	15.4	38.2	22.6	36.7	22.9	35.5	23.3	1	3458
THIRUVANANTHAPURAM	22.3	22.8	34.2	26.2	33.6	26.0	33.1	25.8	0	3477
TIRUCHIRAPPALLI	20.0	20.9	39.2	26.0	38.8	25.9	37.9	25.7	0	4111
VISHAKHAPATNAM CWC	20.1	20.8	34.0	27.1	33.2	27.5	32.7	27.5	0	3480
Indonesia										
DENPASAR NGURAH RAI	21.9	22.9	32.5	26.6	32.1	26.5	31.6	26.4	0	3388
JAKARTA SOEKARNO-HATTA	22.3	22.9	34.0	25.4	33.2	25.6	33.0	25.6	0	3518
JUANDA SURABAYA	21.2	22.1	34.1	24.7	33.5	24.8	33.1	25.0	0	3570
KUALANAMU MEDAN	22.6	22.9	34.2	26.1	33.7	26.1	33.1	26.1	0	3460
MENADO SAM RATULANGI	20.8	21.5	33.9	24.7	33.1	24.7	32.7	24.7	0	3166
MIA PADANG	21.7	22.3	32.2	26.1	31.9	26.1	31.6	26.0	0	3177
PEKAN BARU SIMPANGTIGA	22.1	22.5	34.6	26.2	34.1	26.2	33.6	26.2	0	3537
UJUNG PANDANG HASANUDDIN	20.1	21.0	34.6	23.9	34.0	24.3	33.2	24.8	0	3338
Iran, Islamic Republic of										
ABADAN	4.1	5.8	48.2	22.6	47.2	22.6	46.1	22.2	389	3386
AHWAZ	4.9	6.2	48.1	23.5	47.1	23.3	46.1	22.9	401	3387

Climatic Design Conditions for Selected Locations (*Continued*)

Station	Heating DB		Cooling DB/MCWB						Heat./Cool. Degree-Days	
	99.6%	99%	0.4% DB / MCWB		1% DB / MCWB		2% DB / MCWB		HDD / CDD 18.3	
ANZALI	1.3	2.8	31.6	25.6	30.7	25.3	29.9	25.0	1461	943
ARAK	-13.7	-9.4	36.7	15.9	35.5	15.5	34.4	15.2	2353	918
BANDAR ABBASS INTL	9.1	10.8	42.0	24.4	40.2	25.6	39.0	26.1	65	3288
HAMEDAN	-17.1	-13.0	36.2	16.1	35.1	15.8	34.0	15.5	2778	570
ISFAHAN SHAHID BEHESHTI INTL	-8.2	-6.8	39.1	16.8	38.1	16.3	37.0	16.0	2023	1031
KASHAN	-4.2	-2.1	42.2	19.1	41.0	18.9	39.8	18.6	1439	1889
KERMAN	-7.0	-5.1	38.1	15.7	37.0	15.2	36.0	14.9	1594	1034
KERMANSHAH	-7.2	-5.1	40.1	16.9	39.0	16.7	37.9	16.3	2014	1043
MASHHAD INTL	-8.9	-5.9	37.8	17.5	36.2	17.3	35.1	16.9	1983	1084
MEHRABAD INTL	-3.2	-1.2	39.0	17.6	37.9	17.4	36.8	17.3	1554	1582
SHIRAZ SHAHID DASTGHAIB INTL	-2.1	-0.9	39.2	17.2	38.2	16.7	37.3	16.4	1343	1438
TABRIZ INTL	-11.3	-9.1	36.2	16.8	35.0	16.6	33.7	16.2	2610	862
URMIA INTL	-11.9	-9.2	34.1	17.6	32.8	17.5	31.2	17.3	2825	493
ZAHEDAN INTL	-5.0	-3.1	39.2	16.5	38.1	16.2	37.1	15.6	1151	1487
ZANJAN	-13.7	-10.8	35.1	16.1	33.7	15.9	32.1	15.8	2859	505
Iraq										
BAGHDAD INTL	1.8	3.1	47.1	21.9	46.0	21.6	44.8	21.1	651	2865
Ireland										
CASEMENT	-3.0	-1.5	22.9	17.5	21.2	16.6	20.0	16.0	3104	8
DUBLIN AP	-2.7	-1.2	22.1	17.0	20.6	16.3	19.4	15.7	3164	4
Israel										
TEL AVIV BEN GURION	6.2	7.7	35.2	20.5	33.5	21.9	32.2	22.8	474	1513
TEL AVIV SDE DOV	8.5	9.5	31.2	24.4	30.5	24.5	30.0	24.4	449	1368
Italy										
BARI PALESE	0.9	2.1	34.2	22.2	32.3	21.9	31.0	21.6	1451	710

Climatic Design Conditions for Selected Locations *(Continued)*

Station	Heating DB		Cooling DB/MCWB						Heat./Cool. Degree-Days	
			0.4%		1%		2%			
	99.6%	99%	DB / MCWB		DB / MCWB		DB / MCWB		HDD / CDD 18.3	
BOLOGNA	-3.9	-2.2	34.9	22.6	33.2	22.2	31.9	21.9	2067	711
CATANIA FONTANAROSSA	1.4	2.8	34.7	22.8	32.9	22.8	31.3	22.5	1102	834
CATANIA SIGONELLA	1.9	3.1	36.9	22.2	35.1	22.2	33.8	22.1	1091	986
FIRENZE PERETOLA	-3.1	-1.4	35.8	21.8	34.1	21.5	32.8	21.2	1658	739
GENOVA SESTRI	1.8	3.0	30.2	22.8	29.1	23.2	28.1	23.0	1316	679
GRAZZANISE	-0.7	0.8	33.3	24.5	32.1	24.2	31.0	24.0	1480	691
MILANO LINATE	-4.0	-2.7	33.7	23.8	32.2	22.9	31.0	22.3	2068	677
NAPOLI CAPODICHINO	1.1	2.7	33.1	22.9	32.0	22.8	30.9	22.8	1243	819
PALERMO PUNTA RAISI	6.8	7.8	33.1	22.3	31.1	23.0	29.9	23.6	792	962
PRATICA DI MARE	0.8	2.1	31.1	23.2	30.1	23.3	29.1	23.4	1324	637
ROMA CIAMPINO	-0.9	0.8	34.0	21.5	32.9	21.5	31.2	21.2	1517	708
ROMA FIUMICINO	-0.1	1.0	31.6	21.9	30.3	22.1	29.5	22.1	1450	589
TORINO BRIC DELLA CROCE	-4.8	-3.1	28.2	20.8	27.1	20.4	26.0	19.8	2576	297
TORINO CASELLE	-5.0	-3.8	31.2	22.4	30.0	21.7	28.9	21.1	2360	434
TRIESTE	-1.6	0.1	32.0	23.5	30.8	23.4	29.7	22.9	1714	688
Jamaica										
KINGSTON MANLEY	22.7	23.0	33.2	26.1	33.0	26.2	32.3	25.9	0	3627
Japan										
AKITA	-4.8	-3.9	31.9	24.5	30.4	23.9	28.9	23.3	2779	549
ASAHIKAWA	-17.5	-15.2	29.9	22.7	28.3	21.4	26.9	20.6	4276	259
ASHIYA AB	-0.9	0.2	32.8	25.8	31.2	25.8	30.2	25.5	1703	854
ATSUGI AB	-0.9	0.1	33.1	25.6	32.1	25.1	31.0	24.8	1641	932
CHIBA	0.4	1.3	32.6	25.8	31.6	25.5	30.7	25.2	1592	926
FUJISAN	-28.7	-26.8	12.1	5.7	10.7	5.2	9.3	4.7	8834	0
FUKUOKA AP	0.0	1.0	34.0	25.7	33.0	25.5	32.0	25.1	1509	1089
FUKUYAMA	-2.6	-1.5	34.1	25.5	33.1	25.4	32.1	25.1	1834	962

Climatic Design Conditions for Selected Locations (*Continued*)

Station	Heating DB		Cooling DB/MCWB						Heat./Cool. Degree-Days	
			0.4%		1%		2%			
	99.6%	99%	DB	MCWB	DB	MCWB	DB	MCWB	HDD	CDD 18.3
FUSHIKI	-2.4	-1.5	33.6	25.1	32.1	25.1	30.6	24.8	2177	768
FUTENMA MCAS	11.0	11.9	32.4	26.6	32.0	26.6	31.2	26.4	187	1890
GIFU AB	-3.2	-2.1	34.9	25.7	33.2	25.1	32.1	24.7	1944	952
HAMAMATSU AB	-0.9	0.1	33.1	25.8	31.9	25.4	30.8	25.3	1574	939
HANEDA AP	1.0	1.9	33.0	26.1	31.9	25.6	30.9	25.4	1552	933
HIMEJI	-2.1	-1.2	33.6	25.7	32.6	25.4	31.5	25.0	1854	930
HIROSHIMA	-0.8	0.1	34.0	25.1	33.0	25.1	32.0	24.7	1628	1068
IIZUKA	-1.8	-0.7	33.7	25.8	32.7	25.6	31.6	25.2	1707	966
IRUMA AB	-3.1	-2.0	34.8	25.4	33.1	25.0	31.9	24.6	1966	834
KADENA AB	8.9	10.2	33.1	26.9	32.2	26.9	31.8	27.0	225	1868
KAGOSHIMA	1.6	2.7	33.6	26.0	32.7	25.8	31.9	25.6	1062	1329
KANAZAWA	-1.2	-0.5	33.2	24.8	32.2	24.7	31.2	24.4	1992	855
KANSAI INTL	1.8	2.8	33.0	25.5	32.1	25.3	31.1	25.2	1499	1072
KOBE	0.2	1.2	33.4	25.3	32.3	25.0	31.3	24.8	1548	1121
KOCHI	-0.9	0.2	33.1	25.5	32.2	25.3	31.3	25.1	1367	1089
KOMATSU AB	-1.9	-1.0	33.2	24.5	32.1	24.5	30.9	24.3	2083	776
KUMAGAYA	-2.1	-1.2	35.7	25.5	34.2	25.2	32.7	24.6	1827	943
KUMAMOTO	-1.6	-0.5	34.7	25.5	33.7	25.2	32.7	24.9	1483	1188
KURE	0.0	1.0	32.6	25.3	31.7	25.1	30.8	24.8	1586	996
KYOTO	-0.8	0.0	35.3	24.8	34.1	24.5	32.9	24.2	1723	1099
MATSUYAMA	0.0	0.9	33.6	25.0	32.7	24.8	31.9	24.6	1546	1054
MINAMITORISHIMA	18.1	18.8	31.8	26.4	31.4	26.3	31.0	26.2	0	2789
MIYAZAKI	-0.1	1.1	33.7	25.7	32.5	25.7	31.4	25.6	1234	1116
NAGANO	-6.7	-5.5	33.2	23.8	31.8	23.4	30.3	22.8	2690	677
NAGASAKI	0.8	1.8	33.0	25.8	32.0	25.7	31.1	25.4	1344	1096
NAGOYA AP	-2.0	-1.0	35.2	25.3	34.1	25.0	32.9	24.6	1742	1100
NAHA AP	12.2	13.1	32.2	26.5	32.0	26.5	31.2	26.5	112	2086

Climatic Design Conditions for Selected Locations *(Continued)*

Station	Heating DB		Cooling DB/MCWB						Heat./Cool. Degree-Days	
			0.4%		1%		2%			
	99.6%	99%	DB / MCWB		DB / MCWB		DB / MCWB		HDD / CDD 18.3	
NARA	-2.1	-1.3	34.2	25.1	33.2	24.8	32.1	24.5	1908	914
NAZE	9.5	10.3	32.9	26.1	32.2	26.1	31.6	26.0	343	1646
NIIGATA	-1.7	-1.0	33.0	25.2	31.7	24.9	30.4	24.5	2229	757
NYUTABARU AB	-1.1	0.1	32.9	25.7	31.2	25.9	30.2	25.7	1335	988
OITA	-0.3	0.7	33.6	25.4	32.6	25.3	31.6	25.0	1506	1009
OKAYAMA	-1.3	-0.3	34.6	25.4	33.6	25.1	32.6	24.8	1720	1079
OMAEZAKI	0.2	1.2	30.5	26.2	29.8	25.9	29.1	25.5	1438	875
ONAHAMA	-2.3	-1.3	29.2	24.4	28.1	24.1	27.2	23.7	2118	528
OSAKA	0.7	1.5	34.5	25.0	33.6	24.7	32.5	24.5	1528	1183
OSAKA INTL	-1.2	-0.2	34.9	25.6	33.9	25.2	32.9	25.0	1734	1108
OTARU	-9.5	-8.3	28.4	22.4	26.8	21.6	25.4	20.8	3678	236
OZUKI AB	-0.2	0.8	32.9	26.1	31.9	25.9	30.9	25.9	1683	917
SAPPORO	-9.9	-8.6	29.5	22.8	27.9	21.8	26.5	20.9	3569	314
SENDAI	-3.6	-2.6	31.7	24.5	30.2	24.1	28.8	23.4	2454	536
SHIMOFUSA AB	-2.0	-1.0	33.8	26.1	32.2	25.4	31.1	25.2	1812	852
SHIMONOSEKI	1.6	2.7	32.4	25.8	31.5	25.6	30.6	25.3	1407	1028
SHIZUHAMA AB	-0.2	0.8	32.2	26.0	31.1	25.9	30.1	25.4	1501	898
SHIZUOKA	-0.2	0.9	33.0	25.3	31.7	25.2	30.7	24.9	1421	969
SUMOTO	0.0	0.9	32.0	25.6	31.0	25.3	29.9	25.0	1723	870
TADOTSU	0.0	1.0	33.7	25.3	32.7	25.0	31.7	24.8	1619	1036
TAKAMATSU	-0.3	0.6	34.6	25.3	33.6	25.1	32.5	25.0	1618	1100
TOKUSHIMA	0.4	1.4	33.5	25.6	32.4	25.4	31.5	25.2	1522	1058
TOKYO	0.7	1.5	33.7	25.6	32.6	25.2	31.5	24.7	1559	986
TOYAMA	-2.4	-1.5	34.0	25.5	32.7	25.3	31.4	24.9	2114	822
TSUIKI AB	-2.0	-1.0	32.9	25.7	31.8	25.6	30.8	25.6	1819	849
UTSUNOMIYA	-4.1	-3.1	33.8	25.5	32.4	25.1	31.0	24.5	2104	781

Climatic Design Conditions for Selected Locations (*Continued*)

Station	Heating DB		Cooling DB/MCWB						Heat./Cool. Degree-Days	
			0.4%		1%		2%			
	99.6%	99%	DB / MCWB		DB / MCWB		DB / MCWB		HDD / CDD 18.3	
WAKAYAMA	0.4	1.3	33.6	24.9	32.5	24.9	31.5	24.8	1536	1094
YOKOHAMA	0.7	1.5	32.8	25.7	31.7	25.2	30.7	24.9	1566	921
YOKOSUKA	1.9	2.9	34.2	25.9	32.1	25.1	30.2	24.8	1415	959
YOKOTA AB	-3.8	-2.5	34.0	25.9	32.8	25.4	31.2	24.7	1963	803
Jordan										
AMMAN	1.8	3.0	36.1	18.3	34.8	18.2	33.2	18.0	1148	1189
IRBID	1.8	3.4	34.8	19.4	33.1	19.4	31.9	19.3	1073	1098
QUEEN ALIA INTL	-0.2	0.9	37.1	19.5	35.8	19.0	34.2	18.7	1313	874
Kazakhstan										
ALMATY	-19.8	-16.8	34.4	18.5	32.9	18.1	31.2	17.8	3472	518
KARAGANDY	-33.1	-29.9	32.1	16.8	30.1	16.3	28.2	15.7	5654	153
NUR-SULTAN	-33.0	-30.1	32.3	17.8	30.5	17.3	28.7	16.9	5702	193
PAVLODAR	-35.1	-31.8	33.0	18.3	31.1	18.0	29.1	17.5	5704	244
SHYMKENT	-15.4	-12.0	38.2	18.8	36.9	18.5	35.2	18.1	2531	898
TARAZ	-20.5	-17.0	36.1	17.7	34.8	17.5	33.2	17.1	3127	629
Kenya										
MOMBASA INTL	20.4	20.9	33.2	25.3	32.8	25.2	32.2	25.1	0	3082
NAIROBI JOMO KENYATTA INTL	10.2	11.7	29.2	16.0	28.5	16.0	27.8	16.0	70	639
Korea, Democratic People's Republic of										
CHONGJIN	-12.3	-10.8	27.7	22.9	26.4	22.3	25.3	21.9	3727	251
HAMHUNG	-12.8	-11.0	32.2	24.4	30.4	23.8	28.9	22.9	3123	470
KAESONG	-12.9	-11.0	31.6	25.8	30.2	24.7	28.9	23.9	3011	622
NAMPO	-12.9	-11.0	30.9	26.2	29.6	25.2	28.5	24.4	3120	624
PYONGYANG SUNAN INTL	-14.6	-12.7	31.7	25.0	30.5	24.2	29.4	23.7	3199	661
SINUIJU	-15.5	-13.6	31.3	24.3	29.9	23.6	28.6	23.1	3446	559

Climatic Design Conditions for Selected Locations (*Continued*)

Station	Heating DB		Cooling DB/MCWB						Heat./Cool. Degree-Days	
			0.4%		1%		2%			
	99.6%	99%	DB / MCWB		DB / MCWB		DB / MCWB		HDD / CDD 18.3	
WONSAN	-10.3	-8.4	32.2	23.8	30.5	23.3	28.9	22.7	2854	504
Korea, Republic of										
BUSAN	-5.1	-3.4	31.5	25.6	30.4	25.2	29.3	24.7	1843	748
BUSAN GIMHAE INTL	-6.1	-4.9	33.0	25.9	31.9	25.5	30.2	24.8	2071	808
CHANGWON	-5.4	-3.6	32.6	25.9	31.4	25.4	30.1	24.8	1979	809
CHEONGJU	-10.5	-8.6	33.1	24.5	31.9	23.8	30.6	23.1	2607	796
CHEONGJU INTL	-13.2	-11.1	33.2	26.0	32.1	25.2	30.9	24.6	2842	733
DAEGU AFB	-7.0	-5.4	34.5	24.1	33.2	23.7	31.8	23.2	2178	884
DAEGU INTL	-8.1	-6.8	35.0	25.3	33.2	24.7	32.0	24.1	2313	843
GIMPO INTL	-13.2	-11.2	32.9	25.2	31.2	24.4	30.0	23.5	2966	701
GWANGJU	-6.6	-5.0	33.0	25.5	31.8	24.9	30.6	24.2	2227	844
GWANGJU AP	-7.1	-5.8	34.1	26.3	32.9	25.6	31.2	24.8	2382	838
INCHEON	-10.4	-8.4	31.6	25.3	30.2	24.7	29.0	24.0	2678	678
JEJU	0.3	1.3	32.2	25.5	31.1	25.5	30.2	25.3	1612	862
JEJU INTL	-0.1	0.9	32.1	25.9	31.0	26.1	30.0	25.7	1728	788
JEONJU	-8.7	-6.8	33.4	25.3	32.2	24.9	31.0	24.0	2405	840
JINJU	-8.5	-7.0	33.2	24.8	31.9	24.4	30.6	23.8	2382	737
OSAN AB	-13.1	-11.1	33.4	26.0	32.0	25.4	30.8	24.7	2849	738
POHANG AP	-7.1	-5.8	34.0	25.9	32.7	25.4	31.0	24.8	2233	686
PYEONGTAEK AB	-12.1	-10.2	33.1	26.1	31.9	25.3	30.6	24.7	2803	749
SEOGWIPO	0.4	1.7	31.5	26.6	30.6	26.4	29.7	26.1	1374	918
SEOUL AB	-12.5	-10.2	33.8	25.8	32.1	24.9	30.9	24.2	2796	747
SEOUL OBSERVATORY	-11.6	-9.5	32.8	24.4	31.4	23.6	30.2	22.7	2666	786
SEOUL SINYONGSAN	-11.8	-10.1	33.2	25.2	32.0	24.5	30.8	24.1	2620	796
SUWON	-11.3	-9.3	32.7	25.2	31.4	24.6	30.1	23.7	2738	756

Climatic Design Conditions for Selected Locations *(Continued)*

Station	Heating DB		Cooling DB/MCWB						Heat./Cool. Degree-Days	
			0.4%		1%		2%			
	99.6%	99%	DB / MCWB		DB / MCWB		DB / MCWB		HDD / CDD 18.3	
TAEJON	-10.7	-8.8	33.1	25.8	31.8	24.9	30.4	24.0	2648	736
ULSAN	-5.9	-4.3	33.2	25.0	31.9	24.7	30.6	24.2	2068	743
WANDO	-4.0	-2.7	31.7	26.4	30.5	25.9	29.4	25.3	2085	736
YEOSU	-4.9	-3.4	30.9	25.3	29.8	25.0	28.7	24.6	1991	728
Kosovo										
PRISTINA INTL	-13.2	-10.0	33.8	18.6	31.9	18.4	30.0	17.9	2938	307
Kuwait										
KUWAIT INTL	4.1	5.9	48.1	21.2	47.1	20.9	46.1	20.7	371	3532
Kyrgyzstan										
BISHKEK	-17.1	-13.8	35.5	18.7	34.0	18.1	32.6	17.7	2923	672
Lao People's Democratic Republic										
VIENTIANE WATTAY INTL	14.3	16.0	37.4	26.2	36.2	26.0	35.1	25.8	4	3273
Latvia										
RIGA	-18.3	-14.8	28.8	20.0	26.9	19.5	25.0	18.2	4105	89
Lebanon										
BEIRUT RAFIC HARIRI INTL	8.4	9.8	32.8	23.4	31.5	24.3	30.9	24.5	381	1536
Libyan Arab Jamahiriya										
BENGHAZI	6.9	7.9	37.1	20.5	35.2	20.5	33.8	20.3	577	1391
MISRATA	8.2	9.2	37.1	21.8	34.8	21.6	32.8	21.8	421	1451
TRIPOLI INTL	4.5	5.8	42.1	22.6	40.0	22.3	38.0	21.9	614	1711
Liechtenstein										
VADUZ	-9.3	-7.1	30.2	20.0	28.4	19.3	26.8	18.6	2915	219
Lithuania										
KAUNAS	-18.7	-15.5	28.6	19.9	26.8	18.9	25.1	18.0	4053	93
VILNIUS	-19.3	-16.0	28.8	19.5	26.9	18.6	25.1	17.7	4217	98

Climatic Design Conditions for Selected Locations *(Continued)*

Station	Heating DB		Cooling DB/MCWB						Heat./Cool. Degree-Days	
			0.4%		1%		2%			
	99.6%	99%	DB / MCWB		DB / MCWB		DB / MCWB		HDD / CDD 18.3	
Luxembourg										
LUXEMBOURG AP	-7.7	-5.9	29.8	19.0	27.8	18.2	26.0	17.5	3183	147
Macao										
MACAU INTL	7.9	9.1	32.8	27.0	32.0	27.0	31.2	26.9	264	2034
Madagascar										
ANTANANARIVO IVATO	8.0	9.0	30.0	19.5	29.1	19.7	28.2	19.7	279	747
Malaysia										
KOTA KINABALU INTL	22.8	23.0	34.1	28.3	33.7	28.1	33.1	27.7	0	3553
KUALA LUMPUR SUBANG	22.9	23.4	35.0	26.3	34.2	26.2	34.0	26.2	0	3766
KUCHING INTL	22.1	22.6	34.1	26.2	33.5	26.2	33.0	26.2	0	3342
PAHANG	21.8	22.2	34.1	27.1	33.7	27.1	33.1	27.0	0	3408
SANDAKAN	22.9	23.3	33.9	26.7	33.2	26.6	32.8	26.6	0	3497
TAWAU	22.3	22.8	33.0	26.1	32.2	26.2	32.0	26.3	0	3288
Mali										
BAMAKO	15.0	16.2	40.5	19.7	39.8	19.7	38.9	19.9	0	3536
Malta										
MALTA LUQA	6.9	7.9	34.2	21.4	32.8	21.8	31.2	22.0	722	1071
Mauritania										
NOUAKCHOTT	12.9	14.1	41.4	20.1	39.6	20.0	37.8	19.9	2	3008
Mexico										
ACAPULCO INTL	19.2	20.8	33.2	26.6	33.0	26.5	32.2	26.0	0	3258
CANCUN INTL	13.2	14.9	34.0	27.0	33.2	26.7	32.9	26.6	3	2847
CHETUMAL INTL	15.4	17.3	34.3	26.9	33.7	26.9	33.0	26.7	0	3341
GUADALAJARA INTL	1.9	3.1	33.1	15.2	32.2	14.9	31.2	14.6	348	733
GUANAJUATO INTL	4.0	5.8	34.1	14.7	32.9	14.6	31.8	14.6	279	772

Climatic Design Conditions for Selected Locations (*Continued*)

Station	Heating DB		Cooling DB/MCWB						Heat./Cool. Degree-Days	
	99.6%	99%	0.4%		1%		2%		HDD / CDD 18.3	
			DB / MCWB		DB / MCWB		DB / MCWB			
HERMOSILLO	4.9	6.8	42.8	22.7	41.7	23.0	40.2	22.8	188	2751
MAZATLAN INTL	8.7	9.9	33.9	25.3	33.1	25.2	32.8	25.2	31	2113
MERIDA INTL	13.8	15.7	38.8	24.4	37.4	24.5	36.4	24.6	1	3317
MEXICO CITY INTL	3.1	4.9	29.1	12.3	28.0	12.1	27.0	12.0	568	216
MONTERREY INTL	3.0	4.9	38.8	22.8	37.8	22.7	36.8	22.6	377	2136
PUERTO VALLARTA INTL	14.1	15.2	33.2	26.8	33.0	26.7	32.3	26.4	1	2561
SAN LUIS POTOSI	0.8	2.2	32.2	14.0	31.0	14.2	29.8	14.2	639	443
TAMPICO INTL	10.0	11.9	34.6	27.0	33.8	26.7	33.0	26.5	76	2641
TAPACHULA	20.1	20.9	35.6	26.1	34.9	26.1	34.3	26.0	0	3519
TIJUANA	5.8	6.8	32.8	17.5	30.8	17.3	29.0	17.3	699	519
TOLUCA INTL	-2.0	-0.8	26.1	12.1	25.0	11.9	24.0	11.6	1741	2
TORREON	4.0	5.9	38.0	20.4	37.0	20.2	36.0	20.1	301	2107
VERACRUZ INTL	14.8	16.0	35.2	26.8	34.2	26.8	33.2	26.6	3	2762
Moldova, Republic of										
CHISINAU	-14.2	-11.5	32.8	19.7	31.0	19.2	29.5	18.7	3138	442
Mongolia										
ULAANBAATAR	-36.2	-34.0	31.0	15.5	28.9	14.8	26.9	14.2	7066	102
Montenegro										
PODGORICA GRAD	-2.9	-1.4	37.0	21.9	35.3	21.5	34.0	21.2	1683	932
Morocco										
AGADIR AL MASSIRA INTL	5.0	6.1	38.9	20.0	35.1	19.2	32.2	19.0	394	962
AGADIR INEZGANE	5.0	6.4	35.2	19.4	31.8	18.9	29.0	18.4	520	651
CASABLANCA ANFA	6.8	7.9	29.5	21.5	27.6	21.9	26.4	21.8	599	680
CASABLANCA NOUASSEUR	3.0	4.2	36.0	21.6	33.4	21.4	31.3	21.0	793	836
FES SAIS	0.9	2.1	40.0	19.7	38.1	19.7	36.1	19.6	1123	954

Climatic Design Conditions for Selected Locations (*Continued*)

Station	Heating DB		Cooling DB/MCWB						Heat./Cool. Degree-Days	
			0.4%		1%		2%			
	99.6%	99%	DB / MCWB		DB / MCWB		DB / MCWB		HDD / CDD 18.3	
MARRAKECH MENARA	4.0	5.2	42.1	19.9	40.0	20.0	38.0	19.9	600	1486
MEKNES	2.6	3.8	39.4	20.4	37.1	20.4	35.1	20.2	1056	927
OUJDA ANGADS	1.0	2.3	38.1	20.6	36.1	20.7	34.2	20.5	1052	954
SALE	4.8	5.9	32.8	21.0	30.0	21.0	28.1	21.1	795	554
TANGIER IBN BATTUTA	4.2	5.8	33.2	21.3	31.9	21.2	30.2	20.9	763	768
TETOUAN	6.0	7.2	33.4	20.1	31.6	20.2	30.0	20.2	591	880
Mozambique										
MAPUTO	11.9	13.0	36.0	24.0	34.1	23.8	32.8	23.7	13	2049
Myanmar										
YANGON	17.6	18.8	38.2	25.9	37.2	25.6	36.2	25.3	0	3552
Namibia										
WINDHOEK	4.0	5.9	34.2	14.9	33.2	14.6	32.2	14.3	308	1177
Nepal										
TRIBHUVAN INTL	2.2	3.1	30.8	20.6	29.9	20.8	29.0	21.3	699	963
Netherlands										
AMSTERDAM AP SCHIPHOL	-5.8	-3.9	28.1	20.0	25.9	19.2	23.9	18.1	2867	81
HOEK VAN HOLLAND	-4.8	-3.1	27.5	19.4	25.0	18.5	23.0	18.0	2704	80
IJMUIDEN	-6.4	-4.1	25.6	18.7	23.6	17.8	21.8	17.6	2921	51
ROTTERDAM THE HAGUE AP	-5.9	-3.9	28.4	20.0	26.1	19.2	24.2	18.2	2861	81
VALKENBURG	-6.0	-4.1	27.2	19.8	24.9	18.8	22.9	17.9	2915	58
WOENSDRECHT	-7.2	-4.8	29.3	19.9	26.9	19.0	25.0	18.1	2928	83
New Caledonia										
NOUMEA	16.2	16.8	31.7	25.0	30.8	24.8	30.0	24.5	2	1876
New Zealand										
AUCKLAND	4.1	5.3	25.8	20.1	24.8	19.6	23.9	19.0	1194	182

Climatic Design Conditions for Selected Locations (Continued)

Station	Heating DB		Cooling DB/MCWB						Heat./Cool. Degree-Days	
			0.4%		1%		2%			
	99.6%	99%	DB	MCWB	DB	MCWB	DB	MCWB	HDD	CDD 18.3
CHRISTCHURCH	-2.8	-1.8	28.2	16.9	26.1	16.1	24.1	15.5	2566	61
Nicaragua										
MANAGUA INTL	20.0	20.9	36.0	24.4	35.2	24.2	34.8	24.2	0	3515
Niger										
NIAMEY	16.0	17.0	42.8	20.3	41.9	20.2	41.0	20.3	0	4299
Nigeria										
LAGOS IKEJA	21.1	21.9	34.8	25.9	34.1	26.1	33.7	26.2	0	3444
North Macedonia										
SKOPJE INTL	-11.2	-8.0	36.1	20.2	34.3	20.0	32.9	19.7	2528	559
Norway										
HAKADAL	-19.3	-16.7	27.0	17.4	25.1	16.8	23.1	15.8	4707	39
OSLO BLINDERN	-13.9	-11.7	26.9	17.3	25.0	16.7	23.2	15.7	4152	56
Oman										
AL BURAIMI	10.6	11.9	44.9	20.8	44.0	20.9	43.2	21.0	53	3843
Pakistan										
ISLAMABAD INTL	2.3	3.6	41.1	22.8	39.2	23.0	38.0	22.8	613	2095
JINNAH INTL	10.2	11.9	39.0	22.7	37.2	23.1	36.0	23.5	21	3276
LAHORE ALLAMA IQBAL INTL	3.2	4.9	43.2	23.3	41.9	23.3	40.1	23.3	433	2634
Palestinian Territory, Occupied										
JERUSALEM ATAROT	2.0	3.5	33.3	18.3	32.0	18.3	30.8	18.2	1219	841
Panama										
PANAMA PACIFICO	22.2	22.9	35.0	25.4	34.1	25.3	33.8	25.2	0	3566
TOCUMEN INTL	20.9	21.8	34.2	25.4	33.8	25.3	33.1	25.0	0	3393
Paraguay										
ASUNCION	5.2	7.2	37.2	24.1	36.2	24.2	35.2	24.2	240	2150

Climatic Design Conditions for Selected Locations *(Continued)*

Station	Heating DB		Cooling DB/MCWB						Heat./Cool. Degree-Days	
			0.4%		1%		2%			
	99.6%	99%	DB / MCWB		DB / MCWB		DB / MCWB		HDD / CDD 18.3	
Peru										
AREQUIPA	6.1	6.9	24.0	9.1	23.2	8.8	22.9	8.7	1132	3
CHICLAYO	14.9	15.2	32.2	24.2	31.2	23.8	30.7	23.5	1	1611
CUSCO	0.2	1.2	23.1	8.6	22.2	8.5	21.8	8.5	1953	0
IQUITOS	19.2	20.8	34.2	26.3	33.8	26.3	33.1	26.2	0	3097
LIMA	14.0	14.6	29.0	23.0	27.9	22.5	27.0	22.1	177	804
PIURA	15.9	16.4	34.1	25.3	33.3	24.9	32.7	24.7	0	2421
PUCALLPA	18.1	19.8	34.8	26.2	34.1	26.1	33.4	26.0	1	3167
TRUJILLO	14.4	14.8	28.1	23.6	27.1	23.2	26.2	22.6	116	709
Philippines										
CAGAYAN DE ORO	22.1	22.8	34.6	27.5	34.1	27.4	33.6	27.3	0	3601
DAVAO FRANCISCO BANGOY INTL	22.8	23.2	34.1	26.8	33.5	26.7	33.1	26.7	0	3599
GENERAL SANTOS	22.8	23.0	35.0	27.3	34.3	27.1	33.8	27.0	0	3548
ILOILO	22.8	23.2	34.7	27.9	33.9	27.7	33.2	27.5	0	3522
MACTAN CEBU INTL	23.4	23.9	33.2	27.0	32.8	26.9	32.2	26.9	0	3561
MANILA	23.2	23.8	34.5	26.4	33.8	26.4	33.2	26.3	0	3734
MANILA NINOY AQUINO INTL	21.8	22.8	35.1	25.9	34.3	25.8	33.9	25.8	0	3617
QUEZON CITY SCIENCE GARDEN	20.3	21.2	35.1	25.9	34.3	25.9	33.7	25.9	0	3423
SANGLEY POINT	23.3	23.9	35.3	28.6	34.7	28.4	34.1	28.2	0	3848
ZAMBOANGA	22.7	23.2	34.2	27.4	33.8	27.3	33.2	27.1	0	3642
Poland										
BALICE	-15.6	-12.5	30.4	20.5	28.5	19.8	26.8	19.0	3569	147
GDANSK LECHA WALESY	-15.0	-11.9	27.2	19.1	25.8	18.4	24.0	17.7	3877	61
GDANSK-SWIBNO	-16.3	-11.8	26.6	20.0	24.5	18.8	22.6	18.0	3823	52
HEL	-9.1	-7.0	25.6	20.4	24.1	19.5	22.6	18.7	3536	66

Climatic Design Conditions for Selected Locations (*Continued*)

Station	Heating DB		Cooling DB/MCWB						Heat./Cool. Degree-Days	
			0.4%		1%		2%			
	99.6%	99%	DB / MCWB		DB / MCWB		DB / MCWB		HDD / CDD 18.3	
KATOWICE MUCHOWEC	-14.8	-11.7	30.1	20.0	28.1	19.1	26.4	18.3	3606	125
LODZ	-14.7	-11.8	30.3	19.9	28.3	19.0	26.7	18.3	3632	141
LUBLIN RADAWIEC	-16.3	-13.2	29.5	20.4	27.7	19.8	25.9	18.8	3795	126
POZNAN LAWICA	-13.5	-10.3	30.8	19.5	28.8	18.8	27.0	18.1	3444	160
RACIBORZ	-14.8	-11.7	30.5	20.2	28.5	19.5	26.8	18.8	3424	147
SZCZECIN	-12.0	-8.9	29.4	20.2	27.4	19.3	25.7	18.6	3411	108
TERESPOL	-17.8	-14.2	30.2	20.6	28.4	19.8	26.7	18.9	3770	138
WARSZAWA OKECIE	-15.5	-12.2	30.2	20.3	28.4	19.4	26.8	18.7	3586	165
WROCLAW STRACHOWICE	-13.5	-10.3	30.9	20.5	29.0	19.5	27.2	18.8	3331	162
Portugal										
LISBOA GAGO COUTINHO	4.7	5.9	33.7	20.1	31.7	19.6	29.7	19.2	999	581
Puerto Rico										
ROOSEVELT ROADS	20.2	21.2	32.3	26.4	31.9	26.1	31.3	25.8	0	3176
SAN JUAN MARIN INTL	21.1	21.7	33.1	25.3	32.1	25.5	31.6	25.5	0	3197
Qatar										
DOHA INTL	12.3	13.8	44.3	22.3	43.1	22.5	42.0	23.0	41	3857
Romania										
BUCURESTI BANEASA	-13.0	-10.1	34.4	21.0	32.8	20.6	31.2	20.0	2932	432
BUCURESTI AFUMATI	-13.2	-10.2	34.1	21.3	32.6	21.2	31.0	20.5	2857	485
CLUJ NAPOCA	-14.1	-11.1	31.4	20.5	29.8	20.0	28.1	19.2	3350	218
CONSTANTA	-9.2	-6.8	30.4	23.3	29.2	22.8	28.1	22.3	2502	517
CRAIOVA	-12.2	-9.7	34.2	20.8	32.6	20.5	31.0	20.2	2799	500
IASI	-15.6	-12.4	33.5	20.9	31.7	20.4	30.1	20.0	3151	396
MIHAIL KOGALNICEANU	-11.1	-8.8	33.1	21.1	31.8	20.9	30.1	20.6	2749	484
TIMISOARA	-11.2	-8.8	34.2	20.7	32.8	20.9	31.0	20.2	2776	416

Climatic Design Conditions for Selected Locations *(Continued)*

Station	Heating DB		Cooling DB/MCWB						Heat./Cool. Degree-Days	
			0.4%		1%		2%			
	99.6%	99%	DB / MCWB		DB / MCWB		DB / MCWB		HDD / CDD 18.3	
Russian Federation										
ARKHANGELSK TALAGI	-32.2	-28.7	27.8	19.4	25.2	18.1	23.1	16.9	6133	47
ASTRAKHAN	-18.6	-15.1	36.9	20.8	35.1	20.4	33.7	20.1	3302	752
BARNAUL	-33.9	-30.9	30.1	18.8	28.4	18.2	26.9	17.6	5811	163
BOSTOVO CHEREPOVETS	-29.4	-26.1	28.2	20.0	26.3	19.1	24.4	18.1	5426	62
BRYANSK	-21.9	-18.4	29.6	19.8	27.8	18.9	26.1	18.0	4394	146
CHELYABINSK MEZHDUNARODNYY	-29.2	-26.2	30.8	19.2	28.9	18.7	27.0	18.0	5576	151
CHITA KADALA	-37.2	-35.1	31.1	18.5	29.1	17.7	27.1	16.8	6998	107
IM E K FEDOROVA	-31.1	-28.6	15.8	13.2	13.2	11.5	11.2	9.9	8266	0
IRKUTSK	-35.8	-32.1	29.1	17.9	27.2	17.5	25.7	16.9	6529	65
IZHEVSK	-29.0	-25.6	29.8	20.0	28.0	19.1	26.2	18.3	5599	132
KALUGA	-24.9	-21.4	28.5	19.5	27.0	18.7	25.4	18.3	4744	89
KAZAN	-27.6	-24.2	31.1	19.8	29.1	19.3	27.2	18.5	5169	196
KEMEROVO	-34.0	-31.0	29.0	19.1	27.1	18.3	25.7	17.6	6168	119
KHABAROVSK	-30.0	-28.1	30.2	22.6	28.8	21.8	27.1	21.0	6044	228
KHRABROVO	-16.4	-13.1	28.3	20.2	26.3	19.1	24.7	18.3	3784	82
KIROV	-29.2	-25.8	29.3	20.3	27.4	19.2	25.7	18.2	5584	119
KRASNODAR	-14.1	-10.3	34.9	22.5	33.1	22.0	31.3	21.4	2743	590
KRASNOYARSK	-33.7	-31.1	28.4	18.3	26.6	17.6	24.8	16.8	6254	68
KRASNOYARSK MININO	-36.6	-33.8	29.3	19.2	27.5	18.5	25.7	17.7	6133	107
KURGAN	-32.6	-29.5	31.5	19.3	29.6	18.9	27.8	18.3	5817	176
KURSK	-21.9	-18.8	30.9	19.4	28.9	18.6	27.2	18.0	4276	231
MAGNITOGORSK	-29.7	-27.1	30.8	18.4	29.0	17.8	27.3	17.1	5738	154
MAKHACHKALA	-12.1	-8.6	31.7	23.0	30.5	23.2	29.3	22.9	2684	608
MOSKVA SHEREMETYEVO	-23.9	-20.2	30.0	19.2	28.0	18.5	26.0	18.1	4772	128

Climatic Design Conditions for Selected Locations (*Continued*)

Station	Heating DB		Cooling DB/MCWB						Heat./Cool. Degree-Days	
			0.4%		1%		2%			
	99.6%	99%	DB / MCWB		DB / MCWB		DB / MCWB		HDD / CDD 18.3	
MOSKVA VDNH	-21.9	-18.6	29.9	21.1	28.0	20.2	26.2	19.4	4580	151
MOSKVA VNUKOVO	-23.0	-19.9	29.2	19.1	27.8	18.7	26.0	18.1	4743	136
MURMANSK	-33.0	-29.6	24.8	16.3	22.1	15.0	20.0	13.9	6681	12
NIZHNY NOVGOROD STRIGINO	-26.2	-23.0	30.2	20.1	28.2	19.4	26.8	18.8	4973	142
NIZHNY TAGIL	-31.3	-28.6	28.5	19.6	26.7	18.7	25.0	17.7	6081	66
NOVOKUZNETSK	-33.2	-30.6	29.5	19.4	27.7	18.6	26.0	18.0	5902	109
NOVOSIBIRSK TOLMACHEVO	-36.2	-33.1	29.8	18.9	28.0	18.2	26.3	17.5	6122	121
OMSK	-33.1	-30.5	31.0	18.7	29.0	18.3	27.2	17.6	6032	164
ORENBURG	-29.5	-26.2	34.9	19.4	33.0	19.0	31.0	18.5	5074	338
ORYOL	-23.4	-20.0	30.5	19.9	28.6	19.3	26.9	18.5	4409	181
PENZA	-27.1	-23.7	32.0	19.9	30.0	19.4	28.1	18.6	4849	208
PERM BOLSHOYE SAVINO	-30.8	-27.2	30.0	20.3	28.1	19.3	26.1	18.3	5701	116
PSKOV	-23.1	-19.2	28.9	20.4	27.0	19.2	25.1	18.3	4484	91
ROSTOV-ON-DON	-17.8	-14.9	35.1	21.1	33.1	20.8	31.2	20.1	3347	552
RYAZAN	-24.0	-20.9	30.6	19.8	28.4	19.0	26.7	18.3	4713	166
SAMARA OGMS	-26.2	-23.1	32.9	19.5	30.9	19.0	29.0	18.3	4884	275
SARATOV TSENTRALNY	-23.3	-20.7	33.5	19.8	31.6	19.5	29.8	19.0	4428	404
SMOLENSK	-22.0	-18.8	28.1	20.1	26.4	19.1	24.7	18.3	4628	90
SOCHI	-1.8	-0.2	30.9	23.8	29.8	23.7	28.8	23.2	1861	567
ST PETERSBURG PULKOVO	-22.9	-19.8	28.4	19.6	26.3	18.7	24.7	17.8	4697	75
STAVROPOL	-17.0	-13.1	33.8	19.4	32.0	19.0	30.1	18.6	3267	434
SURGUT	-40.2	-37.3	28.8	18.6	26.9	17.9	24.9	17.4	7314	86
TOMSK	-36.4	-33.3	28.7	19.6	27.1	18.7	25.5	17.9	6336	105
TRUBCHEVSK	-22.7	-19.1	30.1	20.4	28.3	19.6	26.6	18.9	4300	152
TULA	-24.3	-21.0	30.2	20.0	28.2	19.2	26.5	18.5	4663	147

Climatic Design Conditions for Selected Locations *(Continued)*

Station	Heating DB		Cooling DB/MCWB						Heat./Cool. Degree-Days	
			0.4%		1%		2%			
	99.6%	99%	DB / MCWB		DB / MCWB		DB / MCWB		HDD / CDD 18.3	
TVER	-24.9	-21.6	29.8	19.5	27.7	19.0	25.8	18.2	4816	119
TYUMEN	-32.2	-29.4	29.7	19.7	28.0	19.0	26.3	18.2	5967	123
UFA	-30.9	-27.8	31.8	20.5	29.9	20.0	28.0	19.0	5401	172
ULAN-UDE BAIKAL	-36.1	-33.8	32.1	18.0	29.9	17.6	27.9	17.0	6811	153
VELIKIYE LUKI	-22.8	-19.2	28.2	19.5	26.6	18.9	25.1	18.1	4499	86
VLADIKAVKAZ	-14.0	-10.9	31.1	20.3	29.3	19.9	27.7	19.3	3297	287
VLADIMIR	-25.2	-22.0	29.7	20.9	27.7	20.1	25.9	19.3	4930	138
VLADIVOSTOK	-24.1	-21.3	28.1	21.3	26.2	20.5	24.6	19.9	4926	162
VOLGOGRAD GUMRAK	-22.4	-19.5	35.8	18.8	33.8	18.7	31.9	18.3	3994	518
VORONEZH	-22.9	-19.7	32.2	20.0	30.2	19.4	28.4	18.6	4191	283
VORONEZH CHERTOVITSKY	-24.2	-21.0	33.0	19.1	30.9	18.6	28.9	18.1	4274	241
YEKATERINBURG KOLTSOVO	-31.0	-27.8	29.9	19.7	27.9	18.8	26.1	17.9	5851	101
YELABUGA TATARSTAN	-28.4	-25.1	31.3	20.4	29.2	20.0	27.4	18.8	5246	194
Rwanda										
KIGALI INTL	14.9	15.2	30.2	18.6	29.4	18.6	28.9	18.5	0	1266
Saudi Arabia										
ABHA	6.8	7.9	32.0	13.7	31.1	13.4	30.2	13.5	425	898
DHAHARAN KING ABDULAZIZ AB	8.1	9.5	45.9	22.9	44.8	22.9	43.2	23.0	151	3539
GASSIM	4.0	5.9	45.2	N/A	44.2	N/A	43.8	N/A	364	3120
JEDDAH KING ABDULAZIZ INTL	16.2	17.2	41.0	23.7	39.9	24.2	38.9	24.6	0	3939
KHAMIS MUSHAIT	7.5	8.8	32.6	14.7	31.9	15.0	31.1	15.0	277	1110
MAKKAH	17.0	18.2	45.2	24.5	44.2	24.5	43.2	24.4	0	4951
MEDINA PRINCE ABDULAZIZ INTL	9.9	11.3	45.2	18.8	44.3	18.5	43.8	18.3	59	3896
RIYADH KING SALMAN AB	6.2	8.0	44.9	19.5	44.0	19.1	43.1	18.7	254	3429
TABUK	2.1	3.8	41.8	18.9	40.2	18.4	39.1	18.1	615	2223

Climatic Design Conditions for Selected Locations (Continued)

Station	Heating DB		Cooling DB/MCWB						Heat./Cool. Degree-Days	
			0.4%		1%		2%			
	99.6%	99%	DB / MCWB		DB / MCWB		DB / MCWB		HDD / CDD 18.3	
Senegal										
DAKAR	16.9	17.2	32.8	22.2	31.6	24.7	31.0	25.4	0	2456
Serbia										
BEOGRAD	-8.6	-6.4	34.6	21.4	32.9	21.2	31.2	20.6	2386	596
BEOGRAD SURCIN	-10.1	-7.8	34.8	21.4	32.9	21.4	31.1	20.7	2571	488
Singapore										
SINGAPORE CHANGI INTL	23.3	23.8	33.2	26.3	33.0	26.2	32.2	26.2	0	3647
Slovakia										
BRATISLAVA-LETISKO	-10.1	-7.6	32.7	20.8	30.9	20.1	29.1	19.4	2935	324
Slovenia										
LJUBLJANA BEZIGRAD	-8.4	-6.5	32.0	21.2	30.4	20.5	28.8	19.8	2789	328
South Africa										
BLOEMFONTEIN INTL	-4.8	-3.4	34.1	15.4	32.9	15.4	31.8	15.4	1358	537
CAPE TOWN INTL	4.1	5.3	31.9	19.8	29.9	19.3	28.1	18.7	848	437
DE AAR	-0.6	0.8	35.0	15.6	33.8	15.6	32.7	15.4	1103	790
DURBAN	9.2	10.4	30.2	23.9	29.2	23.5	28.6	23.2	143	1094
EAST LONDON	8.0	9.0	30.8	20.0	28.9	20.5	27.3	20.7	403	592
JOHANNESBURG INTL	0.4	2.2	29.1	14.8	28.1	14.9	27.0	15.1	1038	301
PORT ELIZABETH INTL	5.2	6.8	29.5	18.8	27.6	19.4	26.1	19.7	646	412
PRETORIA EENDRACHT	2.8	3.9	32.2	17.3	31.1	17.3	30.1	17.4	598	854
Spain										
LA CORUNA	4.6	5.8	25.9	19.2	24.2	18.6	22.9	18.1	1360	129
ALICANTE AP	3.4	4.8	32.8	21.2	31.4	21.5	30.5	21.7	872	917
BARCELONA AP	1.9	3.1	30.9	23.5	29.8	23.4	28.9	23.1	1234	699
BILBAO AP	-0.2	1.1	32.2	20.8	29.7	20.3	27.7	19.7	1516	353

Climatic Design Conditions for Selected Locations *(Continued)*

	Heating DB		Cooling DB/MCWB						Heat./Cool. Degree-Days	
			0.4%		1%		2%			
Station	99.6%	99%	DB / MCWB		DB / MCWB		DB / MCWB		HDD / CDD 18.3	
GRAN CANARIA	13.8	14.5	30.5	19.8	28.8	20.1	27.7	20.6	54	1142
MADRID TORREJON AB	-4.8	-3.1	36.9	19.3	35.2	18.6	34.0	18.1	2031	636
MADRID-BARAJAS AP	-3.5	-2.1	36.8	18.4	35.5	18.1	34.1	17.6	1909	692
MALAGA AP	4.2	5.8	35.6	20.1	33.2	20.1	31.2	20.1	763	940
MURCIA	2.7	4.1	36.6	22.1	35.1	21.8	33.8	21.7	852	1174
PALMA DE MALLORCA AP	0.7	1.9	33.7	22.3	32.1	22.5	30.9	22.5	1208	753
SEVILLA AP	2.2	3.8	39.2	21.4	37.9	21.1	36.2	20.6	834	1276
VALENCIA AP	0.8	2.1	33.7	20.7	32.0	21.3	31.0	21.6	1058	864
VALLADOLID	-3.6	-2.4	34.6	18.2	32.9	17.8	31.2	17.3	2343	382
ZARAGOZA AP	-2.0	-0.3	36.6	20.8	34.9	20.4	33.1	19.9	1660	738
Sri Lanka										
KATUNAYAKE	21.0	22.0	33.1	24.7	32.5	25.1	32.1	25.3	0	3421
Sudan										
KHARTOUM INTL	14.8	16.1	43.2	19.5	42.7	19.5	41.8	19.4	1	4538
Suriname										
ZANDERIJ	20.8	21.4	34.2	24.6	33.9	24.7	33.2	24.8	0	3269
Sweden										
GOTEBORG	-11.5	-8.9	27.1	18.1	25.4	17.6	23.7	16.8	3547	68
MALMO	-9.1	-6.7	26.6	18.8	25.0	18.4	23.3	17.5	3444	49
STOCKHOLMN BROMMA	-15.0	-12.0	27.3	18.1	25.5	17.3	23.9	16.6	4110	59
UPPSALA	-19.0	-15.6	27.2	18.8	25.2	17.7	23.4	16.8	4457	35
Switzerland										
BERN ZOLLIKOFEN	-9.4	-7.3	30.1	19.5	28.3	19.1	26.5	18.4	3293	144
LAEGEREN	-10.5	-8.4	26.9	17.8	25.1	17.0	23.5	16.5	3823	84
ZUERICH-FLUNTERN	-8.1	-6.2	29.8	19.5	27.8	18.8	26.0	18.1	3144	168

Climatic Design Conditions for Selected Locations *(Continued)*

Station	Heating DB		Cooling DB/MCWB						Heat./Cool. Degree-Days	
	99.6%	99%	0.4%		1%		2%		HDD / CDD 18.3	
			DB / MCWB		DB / MCWB		DB / MCWB			
Syrian Arab Republic										
ALEPPO INTL	-2.0	-0.5	39.8	19.9	38.1	19.3	36.7	19.7	1471	1414
DAMASCUS INTL	-3.1	-1.2	40.0	18.9	38.5	18.5	37.1	18.3	1382	1237
DARAA	1.1	2.6	36.5	19.3	35.0	19.4	33.6	19.6	1117	1109
HAMA	-0.2	1.1	39.5	21.2	37.9	20.8	36.5	20.4	1208	1464
LATAKIA	4.3	5.9	32.6	23.6	31.6	24.5	30.9	24.6	657	1292
Taiwan										
GANGSHAN	9.9	11.2	33.2	27.2	32.9	27.1	32.1	26.9	79	2262
CHIANG KAI SHEK INTL	9.2	10.8	34.9	26.9	34.0	27.0	33.1	26.8	251	2000
HENGCHUN	16.1	17.0	32.7	27.0	32.2	26.8	31.8	26.6	6	2686
HSINCHU	9.1	10.2	33.1	27.9	32.6	27.7	32.1	27.5	276	1846
HSINCHU CITY	9.1	10.4	33.2	26.7	32.5	26.5	31.9	26.4	267	1812
KAOHSIUNG	12.9	14.1	32.7	27.2	32.2	27.1	31.7	26.9	31	2542
KAOHSIUNG INTL	12.1	13.8	33.5	26.8	33.1	26.8	32.4	26.5	28	2674
KEELUNG	10.5	11.5	33.9	25.9	33.0	25.8	32.2	25.8	237	1842
KINMEN	7.1	8.1	32.9	28.5	32.1	28.1	31.6	27.9	497	1592
MATSU NANGAN	4.5	5.6	31.6	27.7	30.9	27.5	30.0	27.1	997	1197
PINGTUNG NORTH	11.1	12.7	34.4	27.3	34.0	27.2	33.2	26.9	40	2511
PINGTUNG SOUTH	11.8	13.0	35.0	27.3	34.2	27.1	33.7	26.9	32	2623
TAICHUNG INTL	8.1	9.2	32.8	26.7	32.1	26.4	31.8	26.4	267	1803
TAICHUNG SHUINAN AP	8.0	9.2	34.2	27.8	33.9	27.8	33.1	27.5	181	2080
TAICHUNG WUQI	10.3	11.4	32.9	27.3	32.3	27.2	31.8	27.0	193	1986
TAINAN AP	10.8	12.0	33.2	27.6	32.9	27.4	32.2	26.9	71	2334
TAIPEI	9.8	10.9	34.8	26.3	34.0	26.2	33.2	26.0	233	2001
TAIPEI SONGSHAN	9.8	11.0	35.8	26.8	34.8	26.5	34.0	26.5	208	2182

Climatic Design Conditions for Selected Locations (*Continued*)

Station	Heating DB		Cooling DB/MCWB						Heat./Cool. Degree-Days	
			0.4%		1%		2%		HDD / CDD 18.3	
	99.6%	99%	DB / MCWB		DB / MCWB		DB / MCWB			
TAOYUAN	8.7	9.8	34.0	28.2	33.2	27.9	32.7	27.6	329	1807
Tajikistan										
DUSHANBE	-9.1	-5.9	38.2	19.4	37.1	18.8	35.9	18.5	1881	964
Tanzania, United Republic of										
DAR ES SALAAM INTL	18.3	19.0	33.5	26.4	33.0	26.1	32.3	25.7	0	3032
Thailand										
BANGKOK DON MUEANG INTL	19.5	21.0	37.2	26.5	36.4	26.6	35.8	26.5	0	4013
PHUKET	23.9	24.3	34.9	26.2	34.2	26.1	33.6	26.0	0	3859
Togo										
LOME	21.8	22.4	33.2	26.1	33.0	26.3	32.5	26.4	0	3478
Trinidad and Tobago										
PIARCO INTL	20.9	21.8	34.0	25.8	33.2	25.5	33.0	25.4	0	3397
Tunisia										
TUNIS CARTHAGE	5.3	6.8	37.9	22.3	35.9	22.4	34.1	22.2	713	1297
Turkey										
ADANA INCIRLIK	0.9	2.2	36.8	22.4	35.2	22.9	34.1	23.2	975	1388
ADANA SAKIRPASA	1.2	3.0	36.4	22.4	35.1	23.2	34.1	23.5	884	1494
ANKARA ESENBOGA	-13.2	-10.1	33.9	17.0	32.1	16.7	30.6	16.5	3037	326
ANKARA ETIMESGUT	-9.9	-7.7	35.0	17.6	33.2	17.5	31.8	17.2	2679	481
ANTALYA HAVALIMANI	2.4	3.8	38.6	20.1	36.9	20.1	35.1	20.3	933	1344
BURSA	-3.1	-1.8	34.6	22.2	33.2	22.0	31.9	21.6	1831	721
DIYARBAKIR	-8.2	-5.2	40.7	18.9	39.2	18.8	38.2	18.6	2053	1257
ERZURUM	-28.1	-25.1	31.0	14.6	29.3	14.4	27.9	14.4	4774	97
ESKISEHIR HAVALIMANI	-10.0	-7.8	33.5	18.9	31.9	18.5	30.2	18.0	2766	364
ISTANBUL ATATURK	-1.4	0.0	32.1	21.5	30.9	21.3	29.9	21.1	1716	776

Station	Heating DB		Cooling DB/MCWB						Heat./Cool. Degree-Days	
			0.4%		1%		2%			
	99.6%	99%	DB / MCWB		DB / MCWB		DB / MCWB		HDD / CDD 18.3	
IZMIR ADNAN MENDERES	-2.4	-0.9	37.2	20.8	35.9	20.6	34.8	20.5	1481	1071
IZMIR CIGLI	-1.2	-0.1	36.9	21.4	35.2	21.3	34.1	21.0	1291	1073
KAYSERI ERKILET	-14.8	-11.1	34.8	17.1	33.0	16.7	31.2	16.2	2937	331
KONYA	-11.1	-8.5	34.2	16.5	32.9	16.4	31.2	16.1	2701	531
MALATYA ERHAC	-10.6	-8.1	38.1	19.0	36.9	18.4	35.5	18.0	2507	851
OGUZELI	-4.3	-2.9	39.1	20.7	38.0	20.2	36.8	19.8	1851	1189
SAMSUN	-0.3	0.9	29.2	22.5	28.4	22.3	27.7	22.0	1684	561
VAN FERITMELEN	-12.8	-10.8	29.1	17.7	28.0	17.9	27.1	18.0	3400	230
Turkmenistan										
ASHGABAT	-7.8	-4.8	40.9	19.9	39.4	19.5	38.1	19.3	1794	1570
Uganda										
ENTEBBE INTL	17.1	17.8	29.1	21.2	28.7	21.5	28.0	21.7	2	1619
Ukraine										
CHERNIHIV	-19.3	-16.5	31.0	20.4	29.2	19.8	27.5	19.0	3951	215
DNIPROPETROVSK	-17.9	-14.8	34.1	20.9	32.1	20.4	30.4	19.9	3540	437
DONETSK	-19.1	-16.0	33.6	19.2	31.6	19.0	29.8	18.7	3748	361
KARHIV	-19.5	-16.5	33.0	19.4	31.0	18.9	29.2	18.5	3822	342
KHERSON	-15.5	-12.5	34.7	21.2	32.8	20.7	31.1	20.0	3149	486
KRYVYI RIH	-17.9	-14.9	33.9	20.3	32.0	19.8	30.1	19.4	3519	387
KYIV	-16.9	-14.1	31.2	20.3	29.6	19.8	28.0	19.0	3696	266
LUHANSK	-21.0	-17.5	35.0	20.6	33.0	20.2	30.9	19.5	3659	398
LVIV	-17.0	-13.7	29.9	20.4	28.0	19.4	26.2	18.5	3734	139
MARIUPOL'	-15.8	-12.9	32.2	21.4	30.7	21.2	29.2	20.8	3383	460
ODESA	-13.2	-10.3	32.9	20.6	31.0	20.2	29.3	19.9	3050	457
POLTAVA	-18.7	-15.8	32.0	20.1	30.2	19.6	28.6	19.0	3773	316

Climatic Design Conditions for Selected Locations (*Continued*)

Station	Heating DB		Cooling DB/MCWB						Heat./Cool. Degree-Days	
			0.4%		1%		2%		HDD / CDD 18.3	
	99.6%	99%	DB / MCWB		DB / MCWB		DB / MCWB			
SIMFEROPOL	-12.2	-9.3	33.8	20.5	32.0	20.1	30.2	19.5	2854	442
VINNYTSIA	-18.6	-15.2	30.2	19.7	28.4	19.1	26.9	18.4	3846	185
ZAPORIZHZHIA	-17.7	-14.6	34.4	20.3	32.8	20.0	30.9	19.4	3449	447
United Arab Emirates										
ABU DHABI INTL	12.0	13.2	45.1	23.0	43.9	23.3	42.7	23.4	19	3809
ABU DHABI BATEEN	14.0	15.0	44.7	23.4	43.1	23.7	41.9	23.9	6	3925
AL AIN INTL	11.0	12.1	46.0	22.5	45.1	22.6	44.1	22.6	38	4004
DUBAI INTL	13.8	14.8	43.3	23.6	42.1	23.8	41.0	24.1	9	3870
SHARJAH INTL	10.8	12.0	44.3	23.5	43.1	23.7	42.0	24.1	28	3609
United Kingdom										
AUGHTON	-2.9	-1.5	24.5	17.4	22.4	16.7	20.5	15.9	3193	18
BINGLEY	-3.8	-2.5	24.0	17.4	21.9	16.4	20.1	15.5	3525	11
BIRMINGHAM	-4.2	-2.9	26.8	18.1	24.3	17.2	22.8	16.6	3048	34
BRISTOL	-3.2	-2.0	24.8	17.8	22.8	17.0	20.9	16.3	3047	18
BRISTOL WEATHER CENTRE	-1.9	-0.5	26.7	18.3	24.8	17.2	23.0	16.6	2594	59
CARDIFF WEATHER CENTRE	-0.8	0.3	26.5	18.5	24.5	17.6	22.8	16.8	2500	63
CHURCH LAWFORD	-4.0	-2.6	26.8	18.7	24.6	17.6	22.8	16.7	3054	34
CILFYNYDD	-4.1	-2.6	25.5	18.0	23.4	16.8	21.5	16.1	3277	22
CROSBY	-3.3	-1.5	24.6	18.4	22.4	17.5	20.7	16.7	2857	22
EDINBURGH AP	-5.1	-3.2	22.2	16.8	20.8	16.2	19.2	15.2	3410	4
EMLEY MOOR	-3.2	-2.2	23.9	17.5	21.8	16.6	20.2	15.7	3453	14
GLASGOW AP	-6.0	-3.9	23.2	17.2	21.2	16.2	19.9	15.6	3352	7
GRAVESEND-BROADNESS	-2.5	-1.2	27.9	19.6	25.8	18.6	24.0	17.7	2593	78
HAWARDEN	-4.2	-2.5	25.3	18.2	23.2	17.4	21.6	16.6	2979	20
KENLEY AF	-3.2	-2.0	26.6	18.3	24.5	17.3	22.7	16.6	2952	44

Climatic Design Conditions for Selected Locations *(Continued)*

Station	Heating DB		Cooling DB/MCWB						Heat./Cool. Degree-Days	
	99.6%	99%	0.4%		1%		2%		HDD / CDD 18.3	
			DB / MCWB		DB / MCWB		DB / MCWB			
LECONFIELD	-3.3	-1.9	25.1	18.5	23.3	17.6	21.7	16.6	3107	19
LEEDS BRADFORD	-3.2	-2.1	24.0	17.8	22.0	16.7	20.1	15.8	3418	13
LEEDS WEATHER CENTRE	-2.1	-1.1	26.5	18.2	24.3	17.1	22.6	16.2	2916	47
LIVERPOOL JOHN LENNON	-2.1	-1.0	25.1	17.9	23.1	17.0	21.8	16.6	2807	31
LONDON HEATHROW	-2.3	-1.1	28.7	18.7	26.5	17.3	24.7	17.1	2534	102
LONDON WC CLERKENWELL	-0.6	0.4	28.4	18.5	26.5	17.7	24.7	17.0	2297	131
MANCHESTER AP	-3.9	-2.1	25.8	18.1	23.8	17.2	21.9	16.4	3049	29
NORTHOLT	-3.8	-2.4	28.4	18.6	26.2	17.7	24.3	17.0	2719	74
NOTTINGHAM EAST MIDLANDS	-3.1	-2.0	26.8	18.3	24.2	17.2	22.8	16.6	2985	41
VALLEY ANGLESEY	-1.4	-0.2	22.8	17.1	20.7	16.0	19.1	15.4	2840	9
Uruguay										
MONTEVIDEO CARRASCO	1.5	3.0	31.9	21.8	30.0	21.5	28.5	21.2	1175	489
MONTEVIDEO PRADO	3.0	4.3	31.7	22.7	30.2	22.1	28.9	21.8	1084	575
Uzbekistan										
NAMANGAN	-8.5	-6.1	37.2	20.8	36.0	20.5	34.8	20.2	2191	1156
SAMARKAND	-10.2	-7.1	36.9	19.0	35.4	18.5	34.2	18.1	2156	906
TASHKENT INTL	-9.9	-7.2	39.0	19.7	37.8	19.3	36.2	18.9	2056	1117
Venezuela										
SAN ANTONIO DEL TACHIRA	19.9	20.8	35.2	23.5	34.8	23.4	34.0	23.2	0	3312
MAIQUETIA	20.8	21.8	33.9	27.8	33.1	27.4	32.8	27.3	0	3408
Viet Nam										
DA NANG INTL	16.8	17.7	36.7	26.4	35.7	26.3	34.8	26.3	3	2997
HA NOI	10.0	11.0	36.2	27.4	35.1	27.4	34.2	27.3	163	2430
HO CHI MINH TAN SON NHAT INTL	20.1	21.2	35.8	25.7	34.9	25.7	34.1	25.7	0	3656
PHU LIEN	9.8	10.9	34.3	28.9	33.3	28.7	32.5	28.4	169	2224

Climatic Design Conditions for Selected Locations *(Continued)*

Station	Heating DB		Cooling DB/MCWB						Heat./Cool. Degree-Days	
			0.4%		1%		2%		HDD / CDD 18.3	
	99.6%	99%	DB / MCWB		DB / MCWB		DB / MCWB			
Western Sahara										
LAAYOUNE	10.2	11.7	36.1	20.7	33.8	20.6	31.2	20.3	115	1270
Yemen										
ADEN INTL	21.2	22.0	37.0	24.8	36.2	25.0	35.9	25.1	0	3982
Zambia										
LUSAKA	7.2	8.1	34.1	18.3	33.0	18.6	31.9	18.5	181	1172
Zimbabwe										
HARARE	5.9	7.0	31.2	16.3	30.2	16.2	29.2	16.3	361	764

INDEX